Methods in Enzymology

Volume 342

RIBONUCLEASES

Part B

METHODS IN ENZYMOLOGY

Methods in Enzymology

Volume 342

Ribonucleases

Part B

EDITED BY

Allen W. Nicholson

DEPARTMENT OF BIOLOGICAL SCIENCES
WAYNE STATE UNIVERSITY
DETROIT, MICHIGAN

ACADEMIC PRESS
San Diego London Boston New York Sydney Tokyo Toronto

This book is printed on acid-free paper. ∞

Academic Press
A Harcourt Science and Technology Company
525 B Street, Suite 1900, San Diego, California 92101-4495, USA
http://www.academicpress.com

Academic Press
Harcourt Place, 32 Jamestown Road, London NW1 7BY, UK
http://www.academicpress.com

International Standard Book Number: 0-12-182243-5

PRINTED IN THE UNITED STATES OF AMERICA
01 02 03 04 05 06 07 SB 9 8 7 6 5 4 3 2 1

Table of Contents

Section II. Processing and Degradative Exoribonucleases

A. 5′ → 3′-Exoribonucleases

B. 3′ → 5′-Exoribonucleases

Section III. Ribonuclease Complexes

Section IV. Organellar Ribonucleases

Section V. Viral Ribonucleases

Contributors to Volume 342

Article numbers are in parentheses following the names of contributors.
Affiliations listed are current.

SHERIF ABOU ELELA (14), *Département de Microbiologie et d'Infectiologie, Faculté de Médecine, Université de Sherbrooke, Sherbrooke, Québec J1H 5N4, Canada*

SIDNEY ALTMAN (9), *Department of Molecular, Cellular and Developmental Biology, Yale University, New Haven, Connecticut 06520*

ASOKA K. AMARASINGHE (13), *Department of Molecular and Cell Biology, University of California, Berkeley, California 94720*

SACHA BAGINSKY (33), *Institute of Plant Sciences, Swiss Federal Institute of Technology, CH-8092 Zurich, Switzerland*

JEF D. BOEKE (19), *Department of Molecular Biology and Genetics, Johns Hopkins University School of Medicine, Baltimore, Maryland 21205*

SERGEI BORUKHOV (7), *Department of Microbiology and Immunology, State University of New York Health Science Center, Brooklyn, New York 11203*

DOUGLAS J. BRIANT (28), *Department of Biochemistry and Molecular Biology, University of British Columbia, Vancouver, British Columbia V6T IZ3, Canada*

TI CAI (12), *Department of Biochemistry and Molecular Biology, State University of New York Upstate Medical University, Syracuse, New York 13210*

IRINA CALIN-JAGEMAN (13), *Department of Biological Sciences, Wayne State University, Detroit, Michigan 48202*

VINCENT J. CANNISTRARO (26), *Department of Molecular Microbiology, Washington University School of Medicine, St. Louis, Missouri 63110*

AGAMEMNON J. CARPOUSIS (27), *Laboratoire de Microbiologie et Génétique Moléculaire, UMR 5100, Centre National de la Recherche Scientifique, 31062 Toulouse cédex 04, France*

IGOR V. CHERNUKHIN (24), *Department of Biological Sciences, University of Essex, Colchester, Essex CO4 3SQ, United Kingdom*

GLEN A. COBURN (28), *Department of Biochemistry and Molecular Biology, University of British Columbia, Vancouver, British Columbia V6T IZ3, Canada*

JORGE CRUZ-REYES (31), *Department of Biological Chemistry, Johns Hopkins University School of Medicine, Baltimore, Maryland 21205*[*]

KRISTOPHER S. CUNNINGHAM (4), *Department of Molecular and Cellular Biochemistry, Ohio State Biochemistry Program, Ohio State University, Columbus, Ohio 43210*

MASAD J. DAMHA (19), *Department of Chemistry, McGill University, Montreal, Québec H3A-2K6, Canada*

CHARLES DANN III (19), *Department of Biophysics, Johns Hopkins University School of Medicine, Baltimore, Maryland 21205*

[*]Current address: Department of Biochemistry and Biophysics, Texas A&M University, College Station, Texas 77843.

SONJA DIEGEL (16), *Department of Genetics, University of Kassel, D-34132 Kassel, Germany*

BEIHUA DONG (2), *Department of Cancer Biology, Lerner Research Institute, Cleveland Clinic Foundation, Cleveland, Ohio 44195*

TRAVIS DUNCKLEY (18), *Department of Neurobiology, Barrow Neurological Institute, Phoenix, Arizona 85013*

ANDRZEJ DZIEMBOWSKI (30), *Department of Genetics, University of Warsaw, and Institute of Biochemistry and Biophysics, Polish Academy of Sciences, 02-106 Warsaw, Poland*

MABROUK M. ELGADI (36), *Biology Department, McMaster University, Hamilton, Ontario L8N 3Z5, Canada*

YAETA ENDO (5), *Department of Applied Chemistry, Faculty of Engineering, Ehime University, Matsuyama 790-8577, Japan*

DAVID R. ENGELKE (10), *Department of Biological Chemistry, University of Michigan Medical School, Ann Arbor, Michigan 48109*

DAVID FRENDEWEY (15), *Regeneron Pharmaceuticals, Inc., Tarrytown, New York 10591*

TANIA N. GONZALEZ (1), *Department of Biochemistry and Biophysics, Howard Hughes Medical Institute, University of California, San Francisco, California 94143*

PAMELA J. GREEN (22), *DOE-Plant Research Laboratory, Department of Biochemistry and Molecular Biology, Michigan State University, East Lansing, Michigan 48824*

WILHELM GRUISSEM (33), *Institute of Plant Sciences, Swiss Federal Institute of Technology, CH-8092 Zurich, Switzerland*

BALRAJ HANDA (37), *Roche Products Ltd., Welwyn Garden City, Hertfordshire AL7 3AY, United Kingdom*

MARK N. HANSON (4), *Department of Molecular and Cellular Biochemistry, Molecular, Cellular and Developmental Biology Program, Ohio State University, Columbus, Ohio 43210*

AHMED HARMOUCH (13), *Department of Biological Sciences, Wayne State University, Detroit, Michigan 48202*

CHRISTIAN HEUBECK (11), *Institut für Biochemie, Universität Würzburg, Biozentrum D-97074 Würzburg, Germany*

MARTIN HILDEBRANDT (16), *Roche Diagnostics GmbH, D-83272 Penzberg, Germany*

LISA HOOKER (37), *Roche Products Ltd., Welwyn Garden City, Hertfordshire AL7 3AY, United Kingdom*

FELICIA HOUSER-SCOTT (10), *Department of Biological Chemistry, University of Michigan Medical School, Ann Arbor, Michigan 48109*

MARCEL M. HULST (35), *Research Branch Houtribweg, Institute for Animal Science and Health, NL-8200 AB Lelystad, The Netherlands*

NAYEF JARROUS (9), *Department of Molecular Biology, The Hebrew University-Hadassah Medical School, Jerusalem 91120, Israel*

ARLEN W. JOHNSON (21), *Section of Molecular Genetics and Microbiology, and Institute for Cellular and Molecular Biology, University of Texas, Austin, Texas 78712*

MARK A. JOHNSON (22), *DOE-Plant Research Laboratory, Department of Biochemistry and Molecular Biology, Michigan State University, Lansing, Michigan 48823**

*Current address: University of Chicago, Department of Molecular Genetics and Cell Biology, Chicago, Illinois 360637.

JAMES P. KASTENMAYER (22), *DOE-Plant Research Laboratory, Department of Biochemistry and Molecular Biology, Michigan State University, East Lansing, Michigan 48823*

DAVID KENNELL (26), *Department of Molecular Microbiology, Washington University School of Medicine, St. Louis, Missouri 63110*

VANESSA KHEMICI (27), *Laboratoire de Microbiologie et Génétique Moléculaire, UMR 5100, Centre National de la Recherche Scientifique, 31062 Toulouse cédex 04, France*

LEIF A. KIRSEBOM (8), *Department of Cell and Molecular Biology, Uppsala University, SE-751 24 Uppsala, Sweden*

KLAUS KLUMPP (37), *Roche Products Ltd., Welwyn Garden City, Hertfordshire AL7 3AY, United Kingdom*

BRUNO LAMONTAGNE (14), *Département de Microbiologie et d'Infectiologie, Faculté de Médecine, Université de Sherbrooke, Sherbrooke, Québec J1H 5N4, Canada*

OLEG LAPTENKO (7), *Department of Microbiology and Immunology, State University of New York Health Science Center, Brooklyn, New York 11203*

DANIEL J. LEAHY (19), *Department of Biophysics, Johns Hopkins University School of Medicine, Baltimore, Maryland 21205*

JOOKYUNG LEE (7), *Department of Microbiology and Immunology, State University of New York Health Science Center, Brooklyn, New York 11203*

ANNE LEROY (27), *Laboratoire de Microbiologie et Génétique Moléculaire, UMR 5100, Centre National de la Recherche Scientifique, 31062 Toulouse cédex 04, France*

KARSTEN LIERE (34), *Genetics Institute, Humboldt-University Berlin, D-10115 Berlin, Germany*

GERHARD LINK (34), *Department of Plant Cell Physiology, University of Bochum, D-44780 Bochum, Germany*

SIEW LOON OOI (19), *Department of Molecular Biology and Genetics, Johns Hopkins University School of Medicine, Baltimore, Maryland 21205*

GEORGE A. MACKIE (28), *Department of Biochemistry and Molecular Biology, University of British Columbia, Vancouver, British Columbia V6T 1Z3, Canada*

JAVIER MARTÎNEZ (25), *Department of Cell and Molecular Biology, Uppsala University, SE-751 24 Uppsala, Sweden*

XIN MIAO (28), *Department of Biochemistry and Molecular Biology, University of British Columbia, Vancouver, British Columbia V6T 1Z3, Canada*

PHILIP MITCHELL (29), *Institute of Cell and Molecular Biology, University of Edinburgh, Edinburgh EH9 3JR, United Kingdom*

AXEL MÖHRLE (16), *MWG-Biotech AG, D-85560 Ebersberg, Germany*

ROB J. M. MOORMANN (35), *Research Branch Houtribweg, Institute for Animal Science and Health, NL-8200 AB Lelystad, The Netherlands*

KAZUO NAGAI (6), *Department of Biological Chemistry, Chubu University, Kasugai Aichi 487-8501, Japan*

KIEBANG NAM (19), *Laboratory of Eukaryotic Gene Regulation, National Institute of Child Health and Human Development, National Institutes of Health, Baltimore, Maryland 20892*

WOLFGANG NELLEN (16), *Department of Genetics, University of Kassel, D-34132 Kassel, Germany*

SARAH F. NEWBURY (24), *Genetics Unit, Department of Biochemistry, University of Oxford, Oxford OX1 3QU, United Kingdom*

ALLEN W. NICHOLSON (13), *Department of Biological Sciences, Wayne State University, Detroit, Michigan 48202*

JÖRG NICKELSEN (34), *Botany Department, University of Bochum, D-44780 Bochum, Germany*

MAHO NIWA (1), *Department of Biochemistry and Biophysics, Howard Hughes Medical Institute, University of California, San Francisco, California 94143*

SILKE NOCK (1), *Department of Biochemistry and Biophysics, Howard Hughes Medical Institute, University of California, San Francisco, California 94143*

JINDRICH NOVOTNY (16), *Department of Genetics, University of Kassel, D-34132 Kassel, Germany*

JÜRGEN OBERSTRASS (16), *Department of Genetics, University of Kassel, D-34132 Kassel, Germany*

TATSUZO OKA (5), *Department of Veterinary Physiology, Faculty of Agriculture, Kagoshima University, Kagoshima 890-0065, Japan*

ROY PARKER (18), *Department of Molecular and Cellular Biology, Howard Hughes Medical Institute, University of Arizona, Tucson, Arizona 85721*

ANNIE PRUD'HOMME-GENEREUX (28), *Department of Biochemistry and Molecular Biology, University of British Columbia, Vancouver, British Columbia V6T 1Z3, Canada*

YAN-GUO REN (25), *Department of Cell and Molecular Biology, Uppsala University, SE-751 24 Uppsala, Sweden*

GUISEPPE ROTONDO (15), *Department of Molecular Biology and Genetics, Johns Hopkins University School of Medicine, Baltimore, Maryland 21218*

LORRAINE RUSCH (2), *Department of Cancer Biology, Lerner Research Institute, Cleveland Clinic Foundation, Cleveland, Ohio 44195*

LAURA N. RUSCHÉ (31), *Department of Molecular and Cell Biology, University of California, Berkeley, California 94720*

HOLLY A. SAFFRAN (36), *Department of Medical Microbiology and Immunology, University of Alberta, Edmonton, Alberta T6G 2H7, Canada*

TATSUYA SAWASAKI (5), *Department of Applied Chemistry, Faculty of Engineering, Ehime University, Matsuyama 790-8577, Japan*

HEIKE SCHIRMACHER (16), *Amgen GmbH, D-80992 Munich, Germany*

MARK E. SCHMITT (12), *Department of Biochemistry and Molecular Biology, State University of New York Upstate Medical University, Syracuse, New York 13210*

ASTRID SCHÖEN (11), *Institut für Biochemie, Universität Würzburg, Biozentrum, D-97074 Würzburg, Germany*

DANIEL R. SCHOENBERG (4), *Department of Molecular and Cellular Biochemistry, Ohio State University, Columbus, Ohio 43210*

JULIAN E. SEAGO (24), *Genetics Unit, Department of Biochemistry, University of Oxford, Oxford OX1 3QU, United Kingdom*

CARMELA SIDRAUSKI (1), *Department of Biochemistry and Biophysics, Howard Hughes Medical Institute, University of California, San Francisco, California 94143*

ROBERT H. SILVERMAN (2), *Department of Cancer Biology, Lerner Research Institute, Cleveland Clinic Foundation, Cleveland, Ohio 44195*

CLARE SIMPSON (32), *Boyce Thompson Institute for Plant Research, Cornell University, Ithaca, New York 14853*

MARIA A. SIWECKA (17), *Institute of Biochemistry and Biophysics, Polish Academy of Sciences, 02-106 Warsaw, Poland*

LAWRENCE I. SLOBIN (23), *Department of Biochemistry, University of Mississippi School of Medicine, Jackson, Mississippi 39216*

JAMES R. SMILEY (36), *Department of Medical Microbiology and Immunology, University of Alberta, Edmonton, Alberta T6G 2H7, Canada*

BARBARA SOLLNER-WEBB (31), *Department of Biological Chemistry, Johns Hopkins University School of Medicine, Baltimore, Maryland 21205*

PIOTR P. STEPIEN (30), *Department of Genetics, University of Warsaw, and Institute of Biochemistry and Biophysics, Polish Academy of Sciences, 02-106 Warsaw, Poland*

DAVID STERN (32), *Boyce Thompson Institute for Plant Research, Cornell University, Ithaca, New York 14853*

AUDREY STEVENS (20), *Life Sciences Division, Oak Ridge National Laboratory, Oak Ridge, Tennessee 37831*

WEIMEI SUN (13), *Department of Biological Sciences, Wayne State University, Detroit, Michigan 48202*

PAUL F. TORRENCE (3), *Department of Chemistry, Northern Arizona University, Flagstaff, Arizona 86011*

MARC UZAN (38), *Institut Jacques Monod, UMR 7592-CNRS-Universités Paris 6 and 7, 75251 Paris cédex 05, France*

NATHALIE VANZO (27), *Developmental Biology Programme, European Molecular Biology Laboratory, 69117 Heidelberg, Germany*

ANDERS VIRTANEN (25), *Department of Cell and Molecular Biology, Uppsala University, SE-751 24 Uppsala, Sweden*

MASAAKI WACHI (6), *Department of Bioengineering, Tokyo Institute of Technology, Midori-ku, Yokohama, 226-8501, Japan*

PETER WALTER (1), *Department of Biochemistry and Biophysics, Howard Hughes Medical Institute, University of California, San Francisco, California 94143*

ZHENGFU WANG (3), *Department of Cancer Biology, Lerner Research Institute, Cleveland Clinic Foundation, Cleveland, Ohio 44195*

WILLIAM A. ZIEHLER (10), *Department of Biological Chemistry, University of Michigan Medical School, Ann Arbor, Michigan 48109*

Preface

The cleavage of RNA has remarkably diverse biological consequences. The agents that catalyze this reaction—ribonucleases—have been the objects of intensive study for many years and from many angles. As noted by one of the contributing authors, research on ribonucleases has produced no less than four Nobel prizes. The archetypal member, pancreatic ribonuclease, was the first protein whose covalent structure was determined, as well as the first protein to be chemically synthesized. There has been a major expansion in our knowledge of the structures, mechanisms, and evolutionary relatedness of ribonucleases, as well as an understanding of ribonuclease function in RNA maturation, RNA degradation, gene regulation, and cellular physiology. Engineered ribonucleases, ribozymes, and synthetic ribonucleases are firmly established, and their applications in medicine and biotechnology are underway. In view of the distinguished history and new research developments it was surprising to find that a *Methods in Enzymology* volume on ribonucleases was not on the shelves. The enthusiastic response to the invitation to contribute to such a volume led to over 80 chapters, which now appear in Volumes 341 and 342 of *Methods in Enzymology*. These volumes should be of lasting value in providing techniques and tools for identifying, characterizing, and applying ribonucleases, and will impart some of the current excitement in research on these fascinating enzymes.

ALLEN W. NICHOLSON

METHODS IN ENZYMOLOGY

VOLUME I. Preparation and Assay of Enzymes
Edited by SIDNEY P. COLOWICK AND NATHAN O. KAPLAN

VOLUME II. Preparation and Assay of Enzymes
Edited by SIDNEY P. COLOWICK AND NATHAN O. KAPLAN

VOLUME III. Preparation and Assay of Substrates
Edited by SIDNEY P. COLOWICK AND NATHAN O. KAPLAN

VOLUME IV. Special Techniques for the Enzymologist
Edited by SIDNEY P. COLOWICK AND NATHAN O. KAPLAN

VOLUME V. Preparation and Assay of Enzymes
Edited by SIDNEY P. COLOWICK AND NATHAN O. KAPLAN

VOLUME VI. Preparation and Assay of Enzymes (*Continued*)
Preparation and Assay of Substrates
Special Techniques
Edited by SIDNEY P. COLOWICK AND NATHAN O. KAPLAN

VOLUME VII. Cumulative Subject Index
Edited by SIDNEY P. COLOWICK AND NATHAN O. KAPLAN

VOLUME VIII. Complex Carbohydrates
Edited by ELIZABETH F. NEUFELD AND VICTOR GINSBURG

VOLUME IX. Carbohydrate Metabolism
Edited by WILLIS A. WOOD

VOLUME X. Oxidation and Phosphorylation
Edited by RONALD W. ESTABROOK AND MAYNARD E. PULLMAN

VOLUME XI. Enzyme Structure
Edited by C. H. W. HIRS

VOLUME XII. Nucleic Acids (Parts A and B)
Edited by LAWRENCE GROSSMAN AND KIVIE MOLDAVE

VOLUME XIII. Citric Acid Cycle
Edited by J. M. LOWENSTEIN

VOLUME XIV. Lipids
Edited by J. M. LOWENSTEIN

VOLUME XV. Steroids and Terpenoids
Edited by RAYMOND B. CLAYTON

VOLUME XVI. Fast Reactions
Edited by KENNETH KUSTIN

VOLUME XVII. Metabolism of Amino Acids and Amines (Parts A and B)
Edited by HERBERT TABOR AND CELIA WHITE TABOR

VOLUME XVIII. Vitamins and Coenzymes (Parts A, B, and C)
Edited by DONALD B. MCCORMICK AND LEMUEL D. WRIGHT

VOLUME XIX. Proteolytic Enzymes
Edited by GERTRUDE E. PERLMANN AND LASZLO LORAND

VOLUME XX. Nucleic Acids and Protein Synthesis (Part C)
Edited by KIVIE MOLDAVE AND LAWRENCE GROSSMAN

VOLUME XXI. Nucleic Acids (Part D)
Edited by LAWRENCE GROSSMAN AND KIVIE MOLDAVE

VOLUME XXII. Enzyme Purification and Related Techniques
Edited by WILLIAM B. JAKOBY

VOLUME XXIII. Photosynthesis (Part A)
Edited by ANTHONY SAN PIETRO

VOLUME XXIV. Photosynthesis and Nitrogen Fixation (Part B)
Edited by ANTHONY SAN PIETRO

VOLUME XXV. Enzyme Structure (Part B)
Edited by C. H. W. HIRS AND SERGE N. TIMASHEFF

VOLUME XXVI. Enzyme Structure (Part C)
Edited by C. H. W. HIRS AND SERGE N. TIMASHEFF

VOLUME XXVII. Enzyme Structure (Part D)
Edited by C. H. W. HIRS AND SERGE N. TIMASHEFF

VOLUME XXVIII. Complex Carbohydrates (Part B)
Edited by VICTOR GINSBURG

VOLUME XXIX. Nucleic Acids and Protein Synthesis (Part E)
Edited by LAWRENCE GROSSMAN AND KIVIE MOLDAVE

VOLUME XXX. Nucleic Acids and Protein Synthesis (Part F)
Edited by KIVIE MOLDAVE AND LAWRENCE GROSSMAN

VOLUME XXXI. Biomembranes (Part A)
Edited by SIDNEY FLEISCHER AND LESTER PACKER

VOLUME XXXII. Biomembranes (Part B)
Edited by SIDNEY FLEISCHER AND LESTER PACKER

VOLUME XXXIII. Cumulative Subject Index Volumes I-XXX
Edited by MARTHA G. DENNIS AND EDWARD A. DENNIS

VOLUME XXXIV. Affinity Techniques (Enzyme Purification: Part B)
Edited by WILLIAM B. JAKOBY AND MEIR WILCHEK

VOLUME XXXV. Lipids (Part B)
Edited by JOHN M. LOWENSTEIN

VOLUME XXXVI. Hormone Action (Part A: Steroid Hormones)
Edited by BERT W. O'MALLEY AND JOEL G. HARDMAN

VOLUME XXXVII. Hormone Action (Part B: Peptide Hormones)
Edited by BERT W. O'MALLEY AND JOEL G. HARDMAN

VOLUME XXXVIII. Hormone Action (Part C: Cyclic Nucleotides)
Edited by JOEL G. HARDMAN AND BERT W. O'MALLEY

VOLUME XXXIX. Hormone Action (Part D: Isolated Cells, Tissues, and Organ Systems)
Edited by JOEL G. HARDMAN AND BERT W. O'MALLEY

VOLUME XL. Hormone Action (Part E: Nuclear Structure and Function)
Edited by BERT W. O'MALLEY AND JOEL G. HARDMAN

VOLUME XLI. Carbohydrate Metabolism (Part B)
Edited by W. A. WOOD

VOLUME XLII. Carbohydrate Metabolism (Part C)
Edited by W. A. WOOD

VOLUME XLIII. Antibiotics
Edited by JOHN H. HASH

VOLUME XLIV. Immobilized Enzymes
Edited by KLAUS MOSBACH

VOLUME XLV. Proteolytic Enzymes (Part B)
Edited by LASZLO LORAND

VOLUME XLVI. Affinity Labeling
Edited by WILLIAM B. JAKOBY AND MEIR WILCHEK

VOLUME XLVII. Enzyme Structure (Part E)
Edited by C. H. W. HIRS AND SERGE N. TIMASHEFF

VOLUME XLVIII. Enzyme Structure (Part F)
Edited by C. H. W. HIRS AND SERGE N. TIMASHEFF

VOLUME XLIX. Enzyme Structure (Part G)
Edited by C. H. W. HIRS AND SERGE N. TIMASHEFF

VOLUME L. Complex Carbohydrates (Part C)
Edited by VICTOR GINSBURG

VOLUME LI. Purine and Pyrimidine Nucleotide Metabolism
Edited by PATRICIA A. HOFFEE AND MARY ELLEN JONES

VOLUME LII. Biomembranes (Part C: Biological Oxidations)
Edited by SIDNEY FLEISCHER AND LESTER PACKER

VOLUME LIII. Biomembranes (Part D: Biological Oxidations)
Edited by SIDNEY FLEISCHER AND LESTER PACKER

VOLUME LIV. Biomembranes (Part E: Biological Oxidations)
Edited by SIDNEY FLEISCHER AND LESTER PACKER

VOLUME LV. Biomembranes (Part F: Bioenergetics)
Edited by SIDNEY FLEISCHER AND LESTER PACKER

VOLUME LVI. Biomembranes (Part G: Bioenergetics)
Edited by SIDNEY FLEISCHER AND LESTER PACKER

VOLUME LVII. Bioluminescence and Chemiluminescence
Edited by MARLENE A. DELUCA

VOLUME LVIII. Cell Culture
Edited by WILLIAM B. JAKOBY AND IRA PASTAN

VOLUME LIX. Nucleic Acids and Protein Synthesis (Part G)
Edited by KIVIE MOLDAVE AND LAWRENCE GROSSMAN

VOLUME LX. Nucleic Acids and Protein Synthesis (Part H)
Edited by KIVIE MOLDAVE AND LAWRENCE GROSSMAN

VOLUME 61. Enzyme Structure (Part H)
Edited by C. H. W. HIRS AND SERGE N. TIMASHEFF

VOLUME 62. Vitamins and Coenzymes (Part D)
Edited by DONALD B. MCCORMICK AND LEMUEL D. WRIGHT

VOLUME 63. Enzyme Kinetics and Mechanism (Part A: Initial Rate and Inhibitor Methods)
Edited by DANIEL L. PURICH

VOLUME 64. Enzyme Kinetics and Mechanism (Part B: Isotopic Probes and Complex Enzyme Systems)
Edited by DANIEL L. PURICH

VOLUME 65. Nucleic Acids (Part I)
Edited by LAWRENCE GROSSMAN AND KIVIE MOLDAVE

VOLUME 66. Vitamins and Coenzymes (Part E)
Edited by DONALD B. MCCORMICK AND LEMUEL D. WRIGHT

VOLUME 67. Vitamins and Coenzymes (Part F)
Edited by DONALD B. MCCORMICK AND LEMUEL D. WRIGHT

VOLUME 68. Recombinant DNA
Edited by RAY WU

VOLUME 69. Photosynthesis and Nitrogen Fixation (Part C)
Edited by ANTHONY SAN PIETRO

VOLUME 70. Immunochemical Techniques (Part A)
Edited by HELEN VAN VUNAKIS AND JOHN J. LANGONE

VOLUME 71. Lipids (Part C)
Edited by JOHN M. LOWENSTEIN

VOLUME 72. Lipids (Part D)
Edited by JOHN M. LOWENSTEIN

VOLUME 73. Immunochemical Techniques (Part B)
Edited by JOHN J. LANGONE AND HELEN VAN VUNAKIS

VOLUME 74. Immunochemical Techniques (Part C)
Edited by JOHN J. LANGONE AND HELEN VAN VUNAKIS

VOLUME 75. Cumulative Subject Index Volumes XXXI, XXXII, XXXIV–LX
Edited by EDWARD A. DENNIS AND MARTHA G. DENNIS

VOLUME 76. Hemoglobins
Edited by ERALDO ANTONINI, LUIGI ROSSI-BERNARDI, AND EMILIA CHIANCONE

VOLUME 77. Detoxication and Drug Metabolism
Edited by WILLIAM B. JAKOBY

VOLUME 78. Interferons (Part A)
Edited by SIDNEY PESTKA

VOLUME 79. Interferons (Part B)
Edited by SIDNEY PESTKA

VOLUME 80. Proteolytic Enzymes (Part C)
Edited by LASZLO LORAND

VOLUME 81. Biomembranes (Part H: Visual Pigments and Purple Membranes, I)
Edited by LESTER PACKER

VOLUME 82. Structural and Contractile Proteins (Part A: Extracellular Matrix)
Edited by LEON W. CUNNINGHAM AND DIXIE W. FREDERIKSEN

VOLUME 83. Complex Carbohydrates (Part D)
Edited by VICTOR GINSBURG

VOLUME 84. Immunochemical Techniques (Part D: Selected Immunoassays)
Edited by JOHN J. LANGONE AND HELEN VAN VUNAKIS

VOLUME 85. Structural and Contractile Proteins (Part B: The Contractile Apparatus and the Cytoskeleton)
Edited by DIXIE W. FREDERIKSEN AND LEON W. CUNNINGHAM

VOLUME 86. Prostaglandins and Arachidonate Metabolites
Edited by WILLIAM E. M. LANDS AND WILLIAM L. SMITH

VOLUME 87. Enzyme Kinetics and Mechanism (Part C: Intermediates, Stereochemistry, and Rate Studies)
Edited by DANIEL L. PURICH

VOLUME 88. Biomembranes (Part I: Visual Pigments and Purple Membranes, II)
Edited by LESTER PACKER

VOLUME 89. Carbohydrate Metabolism (Part D)
Edited by WILLIS A. WOOD

VOLUME 90. Carbohydrate Metabolism (Part E)
Edited by WILLIS A. WOOD

VOLUME 91. Enzyme Structure (Part I)
Edited by C. H. W. HIRS AND SERGE N. TIMASHEFF

VOLUME 92. Immunochemical Techniques (Part E: Monoclonal Antibodies and General Immunoassay Methods)
Edited by JOHN J. LANGONE AND HELEN VAN VUNAKIS

VOLUME 93. Immunochemical Techniques (Part F: Conventional Antibodies, Fc Receptors, and Cytotoxicity)
Edited by JOHN J. LANGONE AND HELEN VAN VUNAKIS

VOLUME 94. Polyamines
Edited by HERBERT TABOR AND CELIA WHITE TABOR

VOLUME 95. Cumulative Subject Index Volumes 61–74, 76–80
Edited by EDWARD A. DENNIS AND MARTHA G. DENNIS

VOLUME 96. Biomembranes [Part J: Membrane Biogenesis: Assembly and Targeting (General Methods; Eukaryotes)]
Edited by SIDNEY FLEISCHER AND BECCA FLEISCHER

VOLUME 97. Biomembranes [Part K: Membrane Biogenesis: Assembly and Targeting (Prokaryotes, Mitochondria, and Chloroplasts)]
Edited by SIDNEY FLEISCHER AND BECCA FLEISCHER

VOLUME 98. Biomembranes (Part L: Membrane Biogenesis: Processing and Recycling)
Edited by SIDNEY FLEISCHER AND BECCA FLEISCHER

VOLUME 99. Hormone Action (Part F: Protein Kinases)
Edited by JACKIE D. CORBIN AND JOEL G. HARDMAN

VOLUME 100. Recombinant DNA (Part B)
Edited by RAY WU, LAWRENCE GROSSMAN, AND KIVIE MOLDAVE

VOLUME 101. Recombinant DNA (Part C)
Edited by RAY WU, LAWRENCE GROSSMAN, AND KIVIE MOLDAVE

VOLUME 102. Hormone Action (Part G: Calmodulin and Calcium-Binding Proteins)
Edited by ANTHONY R. MEANS AND BERT W. O'MALLEY

VOLUME 103. Hormone Action (Part H: Neuroendocrine Peptides)
Edited by P. MICHAEL CONN

VOLUME 104. Enzyme Purification and Related Techniques (Part C)
Edited by WILLIAM B. JAKOBY

VOLUME 105. Oxygen Radicals in Biological Systems
Edited by LESTER PACKER

VOLUME 106. Posttranslational Modifications (Part A)
Edited by FINN WOLD AND KIVIE MOLDAVE

VOLUME 107. Posttranslational Modifications (Part B)
Edited by FINN WOLD AND KIVIE MOLDAVE

VOLUME 108. Immunochemical Techniques (Part G: Separation and Characterization of Lymphoid Cells)
Edited by GIOVANNI DI SABATO, JOHN J. LANGONE, AND HELEN VAN VUNAKIS

VOLUME 109. Hormone Action (Part I: Peptide Hormones)
Edited by LUTZ BIRNBAUMER AND BERT W. O'MALLEY

VOLUME 110. Steroids and Isoprenoids (Part A)
Edited by JOHN H. LAW AND HANS C. RILLING

VOLUME 111. Steroids and Isoprenoids (Part B)
Edited by JOHN H. LAW AND HANS C. RILLING

VOLUME 112. Drug and Enzyme Targeting (Part A)
Edited by KENNETH J. WIDDER AND RALPH GREEN

VOLUME 113. Glutamate, Glutamine, Glutathione, and Related Compounds
Edited by ALTON MEISTER

VOLUME 114. Diffraction Methods for Biological Macromolecules (Part A)
Edited by HAROLD W. WYCKOFF, C. H. W. HIRS, AND SERGE N. TIMASHEFF

VOLUME 115. Diffraction Methods for Biological Macromolecules (Part B)
Edited by HAROLD W. WYCKOFF, C. H. W. HIRS, AND SERGE N. TIMASHEFF

VOLUME 116. Immunochemical Techniques (Part H: Effectors and Mediators of Lymphoid Cell Functions)
Edited by GIOVANNI DI SABATO, JOHN J. LANGONE, AND HELEN VAN VUNAKIS

VOLUME 117. Enzyme Structure (Part J)
Edited by C. H. W. HIRS AND SERGE N. TIMASHEFF

VOLUME 118. Plant Molecular Biology
Edited by ARTHUR WEISSBACH AND HERBERT WEISSBACH

VOLUME 119. Interferons (Part C)
Edited by SIDNEY PESTKA

VOLUME 120. Cumulative Subject Index Volumes 81–94, 96–101

VOLUME 121. Immunochemical Techniques (Part I: Hybridoma Technology and Monoclonal Antibodies)
Edited by JOHN J. LANGONE AND HELEN VAN VUNAKIS

VOLUME 122. Vitamins and Coenzymes (Part G)
Edited by FRANK CHYTIL AND DONALD B. MCCORMICK

VOLUME 123. Vitamins and Coenzymes (Part H)
Edited by FRANK CHYTIL AND DONALD B. MCCORMICK

VOLUME 124. Hormone Action (Part J: Neuroendocrine Peptides)
Edited by P. MICHAEL CONN

VOLUME 125. Biomembranes (Part M: Transport in Bacteria, Mitochondria, and Chloroplasts: General Approaches and Transport Systems)
Edited by SIDNEY FLEISCHER AND BECCA FLEISCHER

VOLUME 126. Biomembranes (Part N: Transport in Bacteria, Mitochondria, and Chloroplasts: Protonmotive Force)
Edited by SIDNEY FLEISCHER AND BECCA FLEISCHER

VOLUME 127. Biomembranes (Part O: Protons and Water: Structure and Translocation)
Edited by LESTER PACKER

VOLUME 128. Plasma Lipoproteins (Part A: Preparation, Structure, and Molecular Biology)
Edited by JERE P. SEGREST AND JOHN J. ALBERS

VOLUME 129. Plasma Lipoproteins (Part B: Characterization, Cell Biology, and Metabolism)
Edited by JOHN J. ALBERS AND JERE P. SEGREST

VOLUME 130. Enzyme Structure (Part K)
Edited by C. H. W. HIRS AND SERGE N. TIMASHEFF

VOLUME 131. Enzyme Structure (Part L)
Edited by C. H. W. HIRS AND SERGE N. TIMASHEFF

VOLUME 132. Immunochemical Techniques (Part J: Phagocytosis and Cell-Mediated Cytotoxicity)
Edited by GIOVANNI DI SABATO AND JOHANNES EVERSE

VOLUME 133. Bioluminescence and Chemiluminescence (Part B)
Edited by MARLENE DELUCA AND WILLIAM D. MCELROY

VOLUME 134. Structural and Contractile Proteins (Part C: The Contractile Apparatus and the Cytoskeleton)
Edited by RICHARD B. VALLEE

VOLUME 135. Immobilized Enzymes and Cells (Part B)
Edited by KLAUS MOSBACH

VOLUME 136. Immobilized Enzymes and Cells (Part C)
Edited by KLAUS MOSBACH

VOLUME 137. Immobilized Enzymes and Cells (Part D)
Edited by KLAUS MOSBACH

VOLUME 138. Complex Carbohydrates (Part E)
Edited by VICTOR GINSBURG

VOLUME 139. Cellular Regulators (Part A: Calcium- and Calmodulin-Binding Proteins)
Edited by ANTHONY R. MEANS AND P. MICHAEL CONN

VOLUME 140. Cumulative Subject Index Volumes 102–119, 121–134

VOLUME 141. Cellular Regulators (Part B: Calcium and Lipids)
Edited by P. MICHAEL CONN AND ANTHONY R. MEANS

VOLUME 142. Metabolism of Aromatic Amino Acids and Amines
Edited by SEYMOUR KAUFMAN

VOLUME 161. Biomass (Part B: Lignin, Pectin, and Chitin)
Edited by WILLIS A. WOOD AND SCOTT T. KELLOGG

VOLUME 162. Immunochemical Techniques (Part L: Chemotaxis and Inflammation)
Edited by GIOVANNI DI SABATO

VOLUME 163. Immunochemical Techniques (Part M: Chemotaxis and Inflammation)
Edited by GIOVANNI DI SABATO

VOLUME 164. Ribosomes
Edited by HARRY F. NOLLER, JR., AND KIVIE MOLDAVE

VOLUME 165. Microbial Toxins: Tools for Enzymology
Edited by SIDNEY HARSHMAN

VOLUME 166. Branched-Chain Amino Acids
Edited by ROBERT HARRIS AND JOHN R. SOKATCH

VOLUME 167. Cyanobacteria
Edited by LESTER PACKER AND ALEXANDER N. GLAZER

VOLUME 168. Hormone Action (Part K: Neuroendocrine Peptides)
Edited by P. MICHAEL CONN

VOLUME 169. Platelets: Receptors, Adhesion, Secretion (Part A)
Edited by JACEK HAWIGER

VOLUME 170. Nucleosomes
Edited by PAUL M. WASSARMAN AND ROGER D. KORNBERG

VOLUME 171. Biomembranes (Part R: Transport Theory: Cells and Model Membranes)
Edited by SIDNEY FLEISCHER AND BECCA FLEISCHER

VOLUME 172. Biomembranes (Part S: Transport: Membrane Isolation and Characterization)
Edited by SIDNEY FLEISCHER AND BECCA FLEISCHER

VOLUME 173. Biomembranes [Part T: Cellular and Subcellular Transport: Eukaryotic (Nonepithelial) Cells]
Edited by SIDNEY FLEISCHER AND BECCA FLEISCHER

VOLUME 174. Biomembranes [Part U: Cellular and Subcellular Transport: Eukaryotic (Nonepithelial) Cells]
Edited by SIDNEY FLEISCHER AND BECCA FLEISCHER

VOLUME 175. Cumulative Subject Index Volumes 135–139, 141–167

VOLUME 176. Nuclear Magnetic Resonance (Part A: Spectral Techniques and Dynamics)
Edited by NORMAN J. OPPENHEIMER AND THOMAS L. JAMES

VOLUME 177. Nuclear Magnetic Resonance (Part B: Structure and Mechanism)
Edited by NORMAN J. OPPENHEIMER AND THOMAS L. JAMES

VOLUME 196. Molecular Motors and the Cytoskeleton
Edited by RICHARD B. VALLEE

VOLUME 197. Phospholipases
Edited by EDWARD A. DENNIS

VOLUME 198. Peptide Growth Factors (Part C)
Edited by DAVID BARNES, J. P. MATHER, AND GORDON H. SATO

VOLUME 199. Cumulative Subject Index Volumes 168–174, 176–194

VOLUME 200. Protein Phosphorylation (Part A: Protein Kinases: Assays, Purification, Antibodies, Functional Analysis, Cloning, and Expression)
Edited by TONY HUNTER AND BARTHOLOMEW M. SEFTON

VOLUME 201. Protein Phosphorylation (Part B: Analysis of Protein Phosphorylation, Protein Kinase Inhibitors, and Protein Phosphatases)
Edited by TONY HUNTER AND BARTHOLOMEW M. SEFTON

VOLUME 202. Molecular Design and Modeling: Concepts and Applications (Part A: Proteins, Peptides, and Enzymes)
Edited by JOHN J. LANGONE

VOLUME 203. Molecular Design and Modeling: Concepts and Applications (Part B: Antibodies and Antigens, Nucleic Acids, Polysaccharides, and Drugs)
Edited by JOHN J. LANGONE

VOLUME 204. Bacterial Genetic Systems
Edited by JEFFREY H. MILLER

VOLUME 205. Metallobiochemistry (Part B: Metallothionein and Related Molecules)
Edited by JAMES F. RIORDAN AND BERT L. VALLEE

VOLUME 206. Cytochrome P450
Edited by MICHAEL R. WATERMAN AND ERIC F. JOHNSON

VOLUME 207. Ion Channels
Edited by BERNARDO RUDY AND LINDA E. IVERSON

VOLUME 208. Protein–DNA Interactions
Edited by ROBERT T. SAUER

VOLUME 209. Phospholipid Biosynthesis
Edited by EDWARD A. DENNIS AND DENNIS E. VANCE

VOLUME 210. Numerical Computer Methods
Edited by LUDWIG BRAND AND MICHAEL L. JOHNSON

VOLUME 211. DNA Structures (Part A: Synthesis and Physical Analysis of DNA)
Edited by DAVID M. J. LILLEY AND JAMES E. DAHLBERG

VOLUME 212. DNA Structures (Part B: Chemical and Electrophoretic Analysis of DNA)
Edited by DAVID M. J. LILLEY AND JAMES E. DAHLBERG

VOLUME 213. Carotenoids (Part A: Chemistry, Separation, Quantitation, and Antioxidation)
Edited by LESTER PACKER

VOLUME 214. Carotenoids (Part B: Metabolism, Genetics, and Biosynthesis)
Edited by LESTER PACKER

VOLUME 215. Platelets: Receptors, Adhesion, Secretion (Part B)
Edited by JACEK J. HAWIGER

VOLUME 216. Recombinant DNA (Part G)
Edited by RAY WU

VOLUME 217. Recombinant DNA (Part H)
Edited by RAY WU

VOLUME 218. Recombinant DNA (Part I)
Edited by RAY WU

VOLUME 219. Reconstitution of Intracellular Transport
Edited by JAMES E. ROTHMAN

VOLUME 220. Membrane Fusion Techniques (Part A)
Edited by NEJAT DÜZGUÜNES

VOLUME 221. Membrane Fusion Techniques (Part B)
Edited by NEJAT DÜZGÜNES

VOLUME 222. Proteolytic Enzymes in Coagulation, Fibrinolysis, and Complement Activation (Part A: Mammalian Blood Coagulation Factors and Inhibitors)
Edited by LASZLO LORAND AND KENNETH G. MANN

VOLUME 223. Proteolytic Enzymes in Coagulation, Fibrinolysis, and Complement Activation (Part B: Complement Activation, Fibrinolysis, and Nonmammalian Blood Coagulation Factors)
Edited by LASZLO LORAND AND KENNETH G. MANN

VOLUME 224. Molecular Evolution: Producing the Biochemical Data
Edited by ELIZABETH ANNE ZIMMER, THOMAS J. WHITE, REBECCA L. CANN, AND ALLAN C. WILSON

VOLUME 225. Guide to Techniques in Mouse Development
Edited by PAUL M. WASSARMAN AND MELVIN L. DEPAMPHILIS

VOLUME 226. Metallobiochemistry (Part C: Spectroscopic and Physical Methods for Probing Metal Ion Environments in Metalloenzymes and Metalloproteins)
Edited by JAMES F. RIORDAN AND BERT L. VALLEE

VOLUME 227. Metallobiochemistry (Part D: Physical and Spectroscopic Methods for Probing Metal Ion Environments in Metalloproteins)
Edited by JAMES F. RIORDAN AND BERT L. VALLEE

VOLUME 228. Aqueous Two-Phase Systems
Edited by HARRY WALTER AND GÖTE JOHANSSON

VOLUME 229. Cumulative Subject Index Volumes 195–198, 200–227

VOLUME 230. Guide to Techniques in Glycobiology
Edited by WILLIAM J. LENNARZ AND GERALD W. HART

VOLUME 231. Hemoglobins (Part B: Biochemical and Analytical Methods)
Edited by JOHANNES EVERSE, KIM D. VANDEGRIFF, AND ROBERT M. WINSLOW

VOLUME 232. Hemoglobins (Part C: Biophysical Methods)
Edited by JOHANNES EVERSE, KIM D. VANDEGRIFF, AND ROBERT M. WINSLOW

VOLUME 233. Oxygen Radicals in Biological Systems (Part C)
Edited by LESTER PACKER

VOLUME 234. Oxygen Radicals in Biological Systems (Part D)
Edited by LESTER PACKER

VOLUME 235. Bacterial Pathogenesis (Part A: Identification and Regulation of Virulence Factors)
Edited by VIRGINIA L. CLARK AND PATRIK M. BAVOIL

VOLUME 236. Bacterial Pathogenesis (Part B: Integration of Pathogenic Bacteria with Host Cells)
Edited by VIRGINIA L. CLARK AND PATRIK M. BAVOIL

VOLUME 237. Heterotrimeric G Proteins
Edited by RAVI IYENGAR

VOLUME 238. Heterotrimeric G-Protein Effectors
Edited by RAVI IYENGAR

VOLUME 239. Nuclear Magnetic Resonance (Part C)
Edited by THOMAS L. JAMES AND NORMAN J. OPPENHEIMER

VOLUME 240. Numerical Computer Methods (Part B)
Edited by MICHAEL L. JOHNSON AND LUDWIG BRAND

VOLUME 241. Retroviral Proteases
Edited by LAWRENCE C. KUO AND JULES A. SHAFER

VOLUME 242. Neoglycoconjugates (Part A)
Edited by Y. C. LEE AND REIKO T. LEE

VOLUME 243. Inorganic Microbial Sulfur Metabolism
Edited by HARRY D. PECK, JR., AND JEAN LEGALL

VOLUME 244. Proteolytic Enzymes: Serine and Cysteine Peptidases
Edited by ALAN J. BARRETT

VOLUME 245. Extracellular Matrix Components
Edited by E. RUOSLAHTI AND E. ENGVALL

VOLUME 246. Biochemical Spectroscopy
Edited by KENNETH SAUER

VOLUME 247. Neoglycoconjugates (Part B: Biomedical Applications)
Edited by Y. C. LEE AND REIKO T. LEE

VOLUME 248. Proteolytic Enzymes: Aspartic and Metallo Peptidases
Edited by ALAN J. BARRETT

VOLUME 249. Enzyme Kinetics and Mechanism (Part D: Developments in Enzyme Dynamics)
Edited by DANIEL L. PURICH

VOLUME 250. Lipid Modifications of Proteins
Edited by PATRICK J. CASEY AND JANICE E. BUSS

VOLUME 251. Biothiols (Part A: Monothiols and Dithiols, Protein Thiols, and Thiyl Radicals)
Edited by LESTER PACKER

VOLUME 252. Biothiols (Part B: Glutathione and Thioredoxin; Thiols in Signal Transduction and Gene Regulation)
Edited by LESTER PACKER

VOLUME 253. Adhesion of Microbial Pathogens
Edited by RON J. DOYLE AND ITZHAK OFEK

VOLUME 254. Oncogene Techniques
Edited by PETER K. VOGT AND INDER M. VERMA

VOLUME 255. Small GTPases and Their Regulators (Part A: Ras Family)
Edited by W. E. BALCH, CHANNING J. DER, AND ALAN HALL

VOLUME 256. Small GTPases and Their Regulators (Part B: Rho Family)
Edited by W. E. BALCH, CHANNING J. DER, AND ALAN HALL

VOLUME 257. Small GTPases and Their Regulators (Part C: Proteins Involved in Transport)
Edited by W. E. BALCH, CHANNING J. DER, AND ALAN HALL

VOLUME 258. Redox-Active Amino Acids in Biology
Edited by JUDITH P. KLINMAN

VOLUME 259. Energetics of Biological Macromolecules
Edited by MICHAEL L. JOHNSON AND GARY K. ACKERS

VOLUME 260. Mitochondrial Biogenesis and Genetics (Part A)
Edited by GIUSEPPE M. ATTARDI AND ANNE CHOMYN

VOLUME 261. Nuclear Magnetic Resonance and Nucleic Acids
Edited by THOMAS L. JAMES

VOLUME 262. DNA Replication
Edited by JUDITH L. CAMPBELL

VOLUME 263. Plasma Lipoproteins (Part C: Quantitation)
Edited by WILLIAM A. BRADLEY, SANDRA H. GIANTURCO, AND JERE P. SEGREST

VOLUME 264. Mitochondrial Biogenesis and Genetics (Part B)
Edited by GIUSEPPE M. ATTARDI AND ANNE CHOMYN

VOLUME 265. Cumulative Subject Index Volumes 228, 230–262

VOLUME 266. Computer Methods for Macromolecular Sequence Analysis
Edited by RUSSELL F. DOOLITTLE

VOLUME 267. Combinatorial Chemistry
Edited by JOHN N. ABELSON

VOLUME 268. Nitric Oxide (Part A: Sources and Detection of NO; NO Synthase)
Edited by LESTER PACKER

VOLUME 269. Nitric Oxide (Part B: Physiological and Pathological Processes)
Edited by LESTER PACKER

VOLUME 270. High Resolution Separation and Analysis of Biological Macromolecules (Part A: Fundamentals)
Edited by BARRY L. KARGER AND WILLIAM S. HANCOCK

VOLUME 271. High Resolution Separation and Analysis of Biological Macromolecules (Part B: Applications)
Edited by BARRY L. KARGER AND WILLIAM S. HANCOCK

VOLUME 272. Cytochrome P450 (Part B)
Edited by ERIC F. JOHNSON AND MICHAEL R. WATERMAN

VOLUME 273. RNA Polymerase and Associated Factors (Part A)
Edited by SANKAR ADHYA

VOLUME 274. RNA Polymerase and Associated Factors (Part B)
Edited by SANKAR ADHYA

VOLUME 275. Viral Polymerases and Related Proteins
Edited by LAWRENCE C. KUO, DAVID B. OLSEN, AND STEVEN S. CARROLL

VOLUME 276. Macromolecular Crystallography (Part A)
Edited by CHARLES W. CARTER, JR., AND ROBERT M. SWEET

VOLUME 277. Macromolecular Crystallography (Part B)
Edited by CHARLES W. CARTER, JR., AND ROBERT M. SWEET

VOLUME 278. Fluorescence Spectroscopy
Edited by LUDWIG BRAND AND MICHAEL L. JOHNSON

VOLUME 279. Vitamins and Coenzymes (Part I)
Edited by DONALD B. MCCORMICK, JOHN W. SUTTIE, AND CONRAD WAGNER

VOLUME 280. Vitamins and Coenzymes (Part J)
Edited by DONALD B. MCCORMICK, JOHN W. SUTTIE, AND CONRAD WAGNER

VOLUME 281. Vitamins and Coenzymes (Part K)
Edited by DONALD B. MCCORMICK, JOHN W. SUTTIE, AND CONRAD WAGNER

VOLUME 282. Vitamins and Coenzymes (Part L)
Edited by DONALD B. MCCORMICK, JOHN W. SUTTIE, AND CONRAD WAGNER

VOLUME 283. Cell Cycle Control
Edited by WILLIAM G. DUNPHY

VOLUME 303. cDNA Preparation and Display
Edited by SHERMAN M. WEISSMAN

VOLUME 304. Chromatin
Edited by PAUL M. WASSARMAN AND ALAN P. WOLFFE

VOLUME 305. Bioluminescence and Chemiluminescence (Part C)
Edited by THOMAS O. BALDWIN AND MIRIAM M. ZIEGLER

VOLUME 306. Expression of Recombinant Genes in Eukaryotic Systems
Edited by JOSEPH C. GLORIOSO AND MARTIN C. SCHMIDT

VOLUME 307. Confocal Microscopy
Edited by P. MICHAEL CONN

VOLUME 308. Enzyme Kinetics and Mechanism (Part E: Energetics of Enzyme Catalysis)
Edited by DANIEL L. PURICH AND VERN L. SCHRAMM

VOLUME 309. Amyloid, Prions, and Other Protein Aggregates
Edited by RONALD WETZEL

VOLUME 310. Biofilms
Edited by RON J. DOYLE

VOLUME 311. Sphingolipid Metabolism and Cell Signaling (Part A)
Edited by ALFRED H. MERRILL, JR., AND YUSUF A. HANNUN

VOLUME 312. Sphingolipid Metabolism and Cell Signaling (Part B)
Edited by ALFRED H. MERRILL, JR., AND YUSUF A. HANNUN

VOLUME 313. Antisense Technology (Part A: General Methods, Methods of Delivery, and RNA Studies)
Edited by M. IAN PHILLIPS

VOLUME 314. Antisense Technology (Part B: Applications)
Edited by M. IAN PHILLIPS

VOLUME 315. Vertebrate Phototransduction and the Visual Cycle (Part A)
Edited by KRZYSZTOF PALCZEWSKI

VOLUME 316. Vertebrate Phototransduction and the Visual Cycle (Part B)
Edited by KRZYSZTOF PALCZEWSKI

VOLUME 317. RNA–Ligand Interactions (Part A: Structural Biology Methods)
Edited by DANIEL W. CELANDER AND JOHN N. ABELSON

VOLUME 318. RNA–Ligand Interactions (Part B: Molecular Biology Methods)
Edited by DANIEL W. CELANDER AND JOHN N. ABELSON

VOLUME 319. Singlet Oxygen, UV-A, and Ozone
Edited by LESTER PACKER AND HELMUT SIES

VOLUME 320. Cumulative Subject Index Volumes 290–319

VOLUME 321. Numerical Computer Methods (Part C)
Edited by MICHAEL L. JOHNSON AND LUDWIG BRAND

VOLUME 322. Apoptosis
Edited by JOHN C. REED

VOLUME 323. Energetics of Biological Macromolecules (Part C)
Edited by MICHAEL L. JOHNSON AND GARY K. ACKERS

VOLUME 324. Branched-Chain Amino Acids (Part B)
Edited by ROBERT A. HARRIS AND JOHN R. SOKATCH

VOLUME 325. Regulators and Effectors of Small GTPases (Part D: Rho Family)
Edited by W. E. BALCH, CHANNING J. DER, AND ALAN HALL

VOLUME 326. Applications of Chimeric Genes and Hybrid Proteins (Part A: Gene Expression and Protein Purification)
Edited by JEREMY THORNER, SCOTT D. EMR, AND JOHN N. ABELSON

VOLUME 327. Applications of Chimeric Genes and Hybrid Proteins (Part B: Cell Biology and Physiology)
Edited by JEREMY THORNER, SCOTT D. EMR, AND JOHN N. ABELSON

VOLUME 328. Applications of Chimeric Genes and Hybrid Proteins (Part C: Protein-Protein Interactions and Genomics)
Edited by JEREMY THORNER, SCOTT D. EMR, AND JOHN N. ABELSON

VOLUME 329. Regulators and Effectors of Small GTPases (Part E: GTPases Involved in Vesicular Traffic)
Edited by W. E. BALCH, CHANNING J. DER, AND ALAN HALL

VOLUME 330. Hyperthermophilic Enzymes (Part A)
Edited by MICHAEL W. W. ADAMS AND ROBERT M. KELLY

VOLUME 331. Hyperthermophilic Enzymes (Part B)
Edited by MICHAEL W. W. ADAMS AND ROBERT M. KELLY

VOLUME 332. Regulators and Effectors of Small GTPases (Part F: Ras Family I)
Edited by W. E. BALCH, CHANNING J. DER, AND ALAN HALL

VOLUME 333. Regulators and Effectors of Small GTPases (Part G: Ras Family II)
Edited by W. E. BALCH, CHANNING J. DER, AND ALAN HALL

VOLUME 334. Hyperthermophilic Enzymes (Part C)
Edited by MICHAEL W. W. ADAMS AND ROBERT M. KELLY

VOLUME 335. Flavonoids and Other Polyphenols (in preparation)
Edited by LESTER PACKER

VOLUME 336. Microbial Growth in Biofilms (Part A: Developmental and Molecular Biological Aspects)
Edited by RON J. DOYLE

VOLUME 337. Microbial Growth in Biofilms (Part B: Special Environments and Physicochemical Aspects)
Edited by RON J. DOYLE

VOLUME 338. Nuclear Magnetic Resonance of Biological Macromolecules (Part A)
Edited by THOMAS L. JAMES, VOLKER DÖTSCH, AND ULI SCHMITZ

VOLUME 339. Nuclear Magnetic Resonance of Biological Macromolecules (Part B)
Edited by THOMAS L. JAMES, VOLKER DÖTSCH, AND ULI SCHMITZ

VOLUME 340. Drug–Nucleic Acid Interactions
Edited by JONATHAN B. CHAIRES AND MICHAEL J. WARING

VOLUME 341. Ribonucleases (Part A)
Edited by ALLEN W. NICHOLSON

VOLUME 342. Ribonucleases (Part B)
Edited by ALLEN W. NICHOLSON

VOLUME 343. G Protein Pathways (Part A: Receptors) (in preparation)
Edited by RAVI IYENGAR AND JOHN D. HILDEBRANDT

VOLUME 344. G Protein Pathways (Part B: G Proteins and Their Regulators) (in preparation)
Edited by RAVI IYENGAR AND JOHN D. HILDEBRANDT

VOLUME 345. G Protein Pathways (Part C: Effector Mechanisms) (in preparation)
Edited by RAVI IYENGAR AND JOHN D. HILDEBRANDT

VOLUME 346. Gene Therapy Methods (in preparation)
Edited by M. IAN PHILLIPS

VOLUME 347. Protein Sensors and Reactive Oxygen Species (Part A: Selenoproteins and Thioredoxin) (in preparation)
Edited by HELMUT SIES AND LESTER PACKER

VOLUME 348. Protein Sensors and Reactive Oxygen Species (Part B: Thiol Enzymes and Proteins) (in preparation)
Edited by HELMUT SIES AND LESTER PACKER

VOLUME 349. Superoxide Dismutase (in preparation)
Edited by LESTER PACKER

Section I

Processing and Degradative Endoribonucleases

Articles 1 through 7

A. Ribonucleoprotein Ribonucleases
Articles 8 through 12

B. Double-Strand-Specific Ribonucleases
Articles 13 through 17

C. Ribonucleases That Cleave Atypical Phosphodiesters
Articles 18 and 19

[1] Purification and Activity Assays of the Catalytic Domains of the Kinase/Endoribonuclease Ire1p from *Saccharomyces cerevisiae*

By SILKE NOCK, TANIA N. GONZALEZ, CARMELA SIDRAUSKI, MAHO NIWA, and PETER WALTER

Introduction

Ire1p is a single-spanning transmembrane protein of the endoplasmic reticulum (ER) of all eukaryotic cells. It is a bifunctional enzyme that exhibits both kinase- and site-specific endoribonuclease activities. Work with the yeast *Saccharomyces cerevisiae* showed that Ire1p provides a key regulatory switch during an intracellular signaling pathway that originates in the lumen of the ER.[1,2] In brief, when unfolded proteins accumulate in the ER, a signal is sent to induce a transcriptional program, termed the unfolded protein response or UPR, that causes an increase in the protein-folding capacity of the ER. An accumulation of unfolded proteins is initially sensed by the ER–lumenal domain of Ire1p by an unknown mechanism.[3–5] Ire1p molecules are then thought to laterally oligomerize in the plane of the membrane, which leads to *trans*-autophosphorylation of their kinase domains and concomitant activation of an endoribonuclease activity.[6] Activated Ire1p cleaves the mRNA encoding Hac1p, a UPR-specific transcription factor, at two positions, thereby excising an intron from the RNA.[7] A second enzyme, tRNA ligase, joins the two exons liberated by Ire1p cleavage to produce spliced *HAC1* mRNA that is now efficiently translated to produce Hac1p, which in turn drives the transcriptional programs that comprise the UPR.[8] The presence of the intron in the unspliced *HAC1* mRNA blocks its translation.[9] Removal of the intron by the spliceosome-independent Ire1p/tRNA ligase-mediated splicing reaction is necessary to induce the UPR.

Ire1 in *S. cerevisiae* is a 1115-amino acid protein. It is initially synthesized with an N-terminal signal sequence followed by an ER–lumenal "unfolded protein-sensing" domain [amino acids (aa) 1–526], a single-transmembrane α helix

[1] J. S. Cox, C. E. Shamu, and P. Walter, *Cell* **73,** 1197 (1993).
[2] K. Mori, W. Ma, M. J. Gething, and J. Sambrook, *Cell* **74,** 743 (1993).
[3] C. E. Shamu, J. S. Cox, and P. Walter, *Trends Cell Biol.* **4,** 56 (1994).
[4] R. Chapman, C. Sidrauski, and P. Walter, *Annu. Rev. Cell Dev. Biol.* **14,** 459 (1998).
[5] C. Sidrauski, R. Chapman, and P. Walter, *Trends Cell Biol.* **8,** 245 (1998).
[6] C. E. Shamu and P. Walter, *EMBO J.* **15,** 3028 (1996).
[7] C. Sidrauski and P. Walter, *Cell* **90,** 1 (1997).
[8] C. Sidrauski, J. S. Cox, and P. Walter, *Cell* **87,** 405 (1996).
[9] R. E. Chapman and P. Walter, *Curr. Biol.* **7,** 850 (1997).

0076-6879/01 $35.00

(aa 527–556), a linker domain (aa 557–678), a kinase domain (aa 679–973), and a C-terminal domain (aa 974–1115) that is presumed to harbor the endonuclease activity. Homologs of Ire1p have been identified in *Schizosaccharomyces pombe, Caenorhabditis elegans, Drosophila melanogaster, Arabidopsis thaliana,* and mammalian cells. By sequence analysis, the enzymes are most conserved in the kinase and presumed nuclease domains; and site-specific endoribonuclease activity has been demonstrated for two mammalian isoforms[10] and shown to be indistinguishable from that of *S. cerevisiae* Ire1p. In *HAC1* mRNA, Ire1p recognizes conserved RNA stem–loop structures, which it cleaves after invariant G residues found in the third position of a seven-nucleotide loop. RNA oligonucleotides that form such stem–loop structures are cleaved by Ire1p and hence provide convenient "minisubstrates" to characterize the reaction. In this way, it was determined that Ire1p cleaves RNA to leave a 2′,3′-phosphate on the 5′ fragment and a free 5′-hydroxyl group on the 3′ fragment.[11]

We here describe the overexpression in two different systems, purification, and activity assays for fusion proteins containing the cytosolic domains of Ire1p from *S. cerevisiae*. All expressed proteins truncate Ire1p just after the transmembrane region and hence contain the linker, kinase, and presumed nuclease domains. Because C-terminal tags reduce Ire1p activity *in vivo* (C. Shamu and P. Walter, unpublished data, 1995), all fusion proteins were constructed containing N-terminal glutathione *S*-transferase (GST) or hexahistidine (His_6) tags to facilitate purification. Detailed methods for analysis of Ire1p endonuclease cleavage products can be found elsewhere.[11]

Solutions

Buffer A: 20 m*M* HEPES (pH 7.5), 150 m*M* KCl, 1 m*M* dithiothreitol (DTT), 5 m*M* $MgCl_2$, 10% (v/v) glycerol
Buffer B: 20 m*M* HEPES (pH 7.5), 1 m*M* DTT
Buffer C: 20 m*M* HEPES (pH 7.5), 1 m*M* DTT, 500 m*M* KCl
Buffer D: 50 m*M* HEPES (pH 7.5), 150 m*M* KCl, 1 m*M* EDTA, 1 m*M* DTT
Buffer E: 20 m*M* HEPES (pH 7.5), 10 m*M* magnesium acetate, 50 m*M* potassium acetate, 1 m*M* DTT
Buffer F: 0.3 *M* sodium acetate (pH 5.2), 10 m*M* magnesium acetate
Urea buffer (2×): Combine 28 ml of 5 *M* NaCl, 4 ml of 1 *M* Tris (pH 7.5), 8 ml of 0.5 *M* EDTA, 40 ml of 10% (w/v) sodium dodecyl sulfate (SDS); add water to 200 ml

[10] M. Niwa, C. Sidrauski, R. J. Kaufman, and P. Walter, *Cell* **99,** 691 (1999).
[11] T. N. Gonzalez, C. Sidrauski, S. Dörfler, and P. Walter, *EMBO J.* **18,** 3119 (1999).

Buffer G: Combine 21 g of urea and 25 ml of 2× urea buffer; add water up to 50 ml; heat (~50°) to dissolve

Buffer H: 50 m*M* Tris (pH 8.5), 1 m*M* phenylmethylsulfonyl fluoride (PMSF), 10 m*M* 2-mercaptoethanol, 1% (v/v) Triton X-100

Buffer I: 50 m*M* sodium phosphate (pH 8), 300 m*M* NaCl, 10% (v/v) glycerol

Cell Growth and Protein Expression in *Escherichia coli*

For expression of GST-Ire1(l+k+t) (linker plus kinase plus tail, amino acids 556–1115) in *Escherichia coli,* the vector pCF210[7] is constructed by subcloning the Ire1p fragment into pGEX-6P-2 (Amersham Pharmacia, Piscataway, NJ). This construct contains a PreScission protease cleavage site that allows removal of the GST tag linked to the amino-terminal end of Ire1(l+k+t).

We use BL21(DE3)pLysS cells (Stratagene, La Jolla, CA), which consistently give better expression yields than DH5α or BL21(DE3) cells. Irep1 expression vectors can be unstable in *E. coli* cells; we therefore always transform cells freshly with the expression vector. Five colonies are used to start a 50-ml preculture in Luria–Bertani (LB) medium, containing carbenicillin (100 μg/ml) (GIBCO-BRL Gaithersburg, MD). The preculture is grown for 14 hr to late log phase and used to inoculate larger batches at a dilution of 1 : 200 (5 to 1000 ml) in LB medium containing carbenicillin at 100 μg/ml. Typically, cells are grown in six 1-liter batches of LB medium in shaking flasks (225 rpm) at 37° to an OD_{600} of 0.6 to 0.8. Cells are then induced with 0.7 m*M* isopropyl-β-D-thiogalactopyranoside (IPTG) (Denville Scientific, Metuchen, NJ) and shifted to 30°. About 4 hr after induction, cells are harvested by centrifugation with a GSA rotor (Sorvall, Newtown, CT) at 16,000*g* at 4° for 15 min, yielding about 20 g of *E. coli* cell paste (wet weight). The cell pellet is resuspended in buffer A (150 ml), supplemented with protease inhibitors (one tablet/50 ml, Protease Inhibitor Complete; Boehringer Mannheim, Indianapolis, IN) and 1% (v/v) Triton X-100 (Calbiochem, La Jolla, CA). The suspension is quick-frozen in liquid nitrogen and stored at −80°.

Purification of GST-Ire1(l+k+t)

The cell suspension is thawed and kept on ice. Fresh DTT (to a final concentration of 1 m*M*) and additional protease inhibitor mix (three inhibitor cocktail tablets) are added, and the cells are lysed in an ice-cold Microfluidizer (Microfluidics, Newton, MA) for three cycles. The BL21(DE3)pLysS cells used do not require the addition of lysozyme. The crude extract is centrifuged in an SS-34 rotor at 31,000*g* at 4° for 30 min to remove the cell debris. To the resulting supernatant is added a 2-ml aliquot of a slurry of 50% glutathione–Sepharose 4B (Amersham Pharmacia; equilibrated in 20 m*M* HEPES, pH 7.0) and the mixture is incubated with gentle overhead mixing on a rotating wheel for 1 hr in three 50-ml conical tubes at 4°. The Sepharose beads are collected by centrifugation at 1000*g* for 5 min at 4°; the supernatant is discarded. The resin is washed four times with 40 ml of ice-cold buffer A, and recollected by centrifugation at 1000*g* as described above, washed a second time (four times with 10 ml of buffer A) and recollected, and resuspended in a small amount of buffer A and transferred to a 15-ml conical tube. The resin is then washed with 10 ml of buffer B (low salt), followed by a final wash with 10 ml of buffer C (high salt). All washes are done for 15 min at 4° by gently mixing.

We have found that Ire1p(l+k+t) prepared in this way, that is, without any further washes, copurifies with an almost equimolar amount of the *E. coli* heat shock protein 70 (HSP70)-like chaperone DnaK. To remove DnaK, we add another wash in 10 ml of buffer A containing 5 m*M* ATP.

Next, the resin is equilibrated with buffer D. Excess buffer is removed to produce a 50% slurry (2 ml). A 60-μl aliquot of PreScission protease (120 U; Amersham Pharmacia) is added to sever Ire1p(l+k+t) from the GST tag. The mixture is incubated at 4° for 12 hr on a rotating wheel. The completeness of the cleavage is monitored by removing a sample and subjecting it to SDS–polyacrylamide gel electrophoresis.

The resin is collected by centrifugation and washed with 500 μl of buffer A. Both supernatants are combined. To remove residual traces of uncleaved protein and PreScission protease (which also has a GST tag), 300 μl of a 50% glutathione–Sepharose 4B slurry in buffer A is added to the supernatant and gently mixed at 4° for 30 min. The resin is removed by centrifugation, and glycerol is added to a final concentration of 20% (v/v). The protein is aliquoted, quick-frozen in liquid nitrogen, and stored at −80°. The final concentration of Ire1p(l+k+t) is estimated by comparing Coomassie blue-stained bands of Ire1p (l+k+t) with bovine serum albumin (BSA) standards of known amounts after SDS–polyacrylamide gel electrophoresis (PAGE). The yield from a 6-liter culture is about 1–2 mg.

Baculovirus Expression and Purification of Yeast Ire1p(l+k+t)

Expression of Ire1p(l+k+t) in *Spodoptera frugiperda* (Sf9) insect cells, using a baculovirus vector, is more cumbersome than expression in *E. coli,* but has the advantage that Ire1p(l+k+t) does not copurify with chaperones. Similar observations have also been made with other kinases, indicating that eukaryotic cells may better promote the correct folding of kinase domains (D. Morgan, personal communication, 2000).

Baculovirus expression vector pPW463 contains the same portion of *S. cerevisiae* Ire1 as pCF210 described above. It is made by subcloning the Ire1p(l+k+t)-encoding DNA fragment after *Eco*RI–*Hin*dIII cleavage into pFastBacHTb (GIBCO-BRL). Ire1p(l+k+t) is preceded by a His_6 tag.

The expression of His_6-Ire1p(l+k+t) from the pPW463 plasmid is performed as described in the *Bac-to-Bac Baculovirus Expression Systems Manual* (GIBCO-BRL). In brief, Sf9 cells are transfected with pPW463 DNA according to the protocol. Serum-free SF-900 II SFM medium (GIBCO-BRL) is used to grow the cells at 28°, while stirring at 100 rpm. The virus is harvested after 72 hr, the titer is determined, and the virus is amplified two more times at a low multiplicity of infection (MOI of 0.01–0.1).

For the final protein expression culture, infection is done with a high MOI of 5–10 in 250 ml of SF-900 II SFM medium. The cell density is about 2×10^7 cells, and cells are harvested after 48 hr by centrifugation at room temperature, 1200*g* for 10 min. To the harvested cells, 50 μl of buffer H is added per milliliter of culture. Cells are resuspended and quick-frozen in liquid nitrogen.

For purification of His_6-Ire1p(1+k+t), the cell suspension is thawed. Cells are lysed by sonication (two 20-sec pulses, small tip—avoid heating of sample), and the resulting extract is centrifuged at 100,000*g* for 1 hr at 4°. The supernatant is diluted into 60 ml of buffer I and passed over a Hi-Trap chelating column (Amersham Pharmacia) loaded with Co^{2+}. Co^{2+}-loaded resin results in lower background binding of cellular proteins than Ni^{2+}-loaded resin. Thus, prior to use, Hi-Trap chelating resin is prepared by washing it first with 50 m*M* EDTA to strip bound cations, after which it is loaded with a solution of 0.2 *M* $CoCl_2$. The resin is then washed with water and equilibrated with buffer I.

The extract (60 ml) is chromatographed on a 1-ml column at 0.5 ml/min, washed extensively with buffer I (10 ml), followed by buffer I at pH 6 (10 ml), buffer I–1 *M* NaCl (10 ml), buffer I (10 ml), and finally buffer I containing 20 m*M* imidazole. His_6-Ire1p(l+k+t) is eluted with buffer I containing 200 m*M* imidazole. Eluted protein is analyzed and roughly quantified by Coomassie blue staining after SDS–PAGE. An 800-ml culture yields approximately 0.4 mg of His_6-Ire1p(l+k+t).

The α and β isoforms of human His_6-Ire1p(l+k+t) are expressed and purified in the same way.[10]

Kinase Assay

The kinase activity of Ire1p(l+k+t) is determined by autophosphorylation. Kinase reactions (30 μl) contain 0.1 m*M* ATP, 5 μCi of [γ-^{32}P]ATP, buffer E, and 0.5 μg of Ire1p(l+k+t). A cocktail containing buffer E, ATP, and [γ-^{32}P]ATP (3000 Ci/mmol; Amersham Pharmacia) is dispensed into 1.5-ml microcentrifuge tubes and the tubes are placed in a water bath at 30°. Reactions are initiated by adding the enzyme and incubated at 30° for 30 min. Aprotinin (3 μg) is added as carrier, and reactions are stopped with 120 μl of 20% (w/v) trichloroacetic acid (TCA) for 30 min on ice. The protein is displayed on a 7% (w/v) SDS–polyacrylamide gel and visualized by autoradiography and Coomassie blue staining.

Ire1p(l+k+t) preparations purified from both *E. coli* and Sf9 cells are already phosphorylated. This is apparent as a marked mobility shift on SDS–polyacrylamide gels when the enzymes are treated with Ptc2p (a phosphatase shown to interact with Ire1p *in vivo*[12]). Thus the kinase activity measured in our assays reflects phosphate exchange and/or phosphorylation at sites that are still available. Probably as a consequence of its already activated phosphorylation state, Ire1p(l+k+t) preparations are constitutively active and do not require the addition of hydrolyzable ATP to exhibit nuclease activity (although the addition of some form of adenosyl nucleotide, such as ADP, is required). In contrast, preparations of mutant forms of Ire1(l+k+t) designed to inactivate the kinase activity are unphosphorylated and inactive for nuclease function (M. Niwa and P. Walter, unpublished data, 2000).

In Vitro Transcription of *HAC1* 508 RNA

HAC1 508 RNA is a 508-nucleotide-long fragment of unspliced *HAC1* RNA that contains both splice junctions (181 nucleotides of the 5′ exon, 252 nucleotides of the intron, and 75 nucleotides of the 3′ exon). The template for *in vitro* transcription is prepared by linearizing 5 μg of pCF187 by digestion with *Sac*I for 2 hr at 37°. The reaction is then treated for 5 min at 37° with 10 U of T4 DNA polymerase (New England BioLabs, Beverly, MA) to produce blunt ends (to give a more homogeneous transcript), phenol extracted, and ethanol precipitated. The transcription reaction (20 μl) contains T7 RNA polymerase buffer, 1 m*M* ATP, 1 m*M* CTP, 1 m*M* GTP, 0.1 m*M* UTP (nucleotides from Boehringer Mannheim), 25 μCi of [α-^{32}P]UTP (3000 Ci/mmol; Amersham Pharmacia), 40 U of RNasin (Promega, Madison, WI), and 1 μg of linearized pCF187, and is initiated with 20 U of T7 RNA polymerase. The reaction is incubated at 37° for 2 hr. At that time, 180 μl of water and sodium acetate, pH 5.2, to a final concentration of 0.3 *M*, is added. The reaction is phenol–chloroform extracted and ethanol precipitated, using 40 μg of glycogen (Boehringer Mannheim) as carrier. The transcript is dissolved in 25 μl of gel loading buffer followed by heating to 95° for 3 min and purified on a 5% (w/v) polyacrylamide gel containing 7 *M* urea. Bands are visualized by exposing the gel to X-ray film for 5 min. The film is aligned with the gel, and the most dominant band is excised. The RNA is eluted from the gel slice by adding 400 μl of buffer F and 400 μl of phenol–chloroform followed by shaking overnight at 4°. The RNA in the extract is ethanol precipitated and dissolved in 15 μl of water. A 20-μl reaction yields 4×10^6 cpm of purified transcript, which is enough for 200 standard cleavage reactions.

[12] A. A. Welihinda, W. Tirasophon, S. R. Green, and R. J. Kaufman, *Mol. Cell. Biol.* **18,** 1967 (1998).

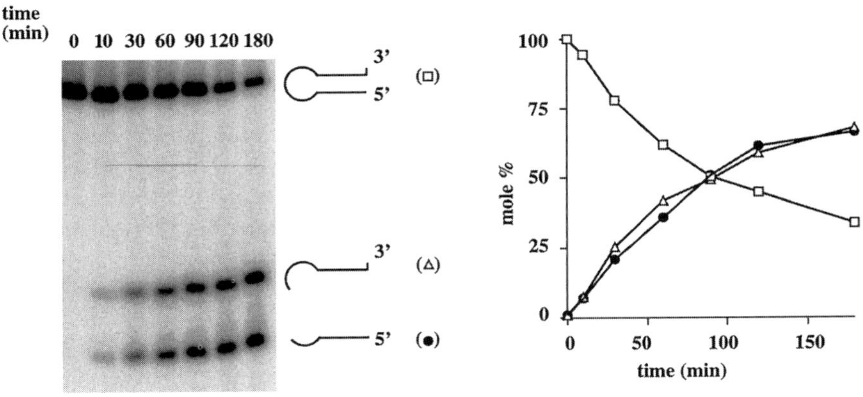

FIG. 1. Time course for hactng-10 minisubstrate cleavage by Ire1p(l+k+t). The cleavage reaction (20 μl) was performed as described in text and contained 400 n*M* Ire1(l+k+t) (expressed in *E. coli*, and with the GST tag removed) and 255 p*M* hactng-10 substrate. The symbols represent: hactng-10 minisubstrate □, 3′ cleavage product △, 5′ cleavage product ●.

In Vitro Transcription of Stem–Loop hactng-10 Minisubstrate

For preparation of the *HAC1* stem–loop minisubstrates,[11] a single-stranded DNA oligonucleotide (hactng-10; 42 nucleotides long, comprising the 3′ splice site stem–loop) is synthesized to contain a T7 promoter sequence followed by the sequences specifying the desired RNA molecule. A short oligonucleotide (T7 promoter oligonucleotide, 15 p*M*) is annealed to template (hactng-10 oligonucleotide, 0.25 p*M*) to create a double-stranded T7 RNA polymerase promoter.[13] Both oligonucleotides are gel purified prior to use. The mixture is heated to 95° for 3 min and immediately cooled on ice. The transcription reaction is performed as described above. The resulting stem–loop RNA is purified on 15% (w/v) denaturing polyacrylamide gels.

hactng-10 oligonucleotide: 5′-TGAGGTCAAA CCTGACTGCG CTTCG-GACAG TACAAGCTTG ACCTATAGTG AGTCGTATTA-3′
T7 promoter oligonucleotide: 5′-TAATACGACTCACTATAG-3′

RNA Cleavage Assay

Cleavage reactions (20 μl) contain buffer E, 2 m*M* ADP, and 215 p*M* (20000 cpm) *HAC1* 508 RNA or 255 p*M* (2000 cpm) *HAC1* stem–loop hactng-10

[13] J. F. Milligan, D. R. Groebe, G. W. Witherell, and O. C. Uhlenbeck, *Nucleic Acids Res.* **15,** 8783 (1987).

minisubstrate. Reactions are initiated by adding 400 n*M* (0.5 μg/20 μl) Ire1p (l+k+t), incubated at 30° for varying lengths of time, and stopped with 200 μl of buffer G. RNA is extracted with 200 μl of phenol–chloroform, ethanol precipitated with glycogen as carrier, dissolved in 6 μl of RNA gel loading buffer, heated to 95° for 3 min, and loaded onto a denaturing urea–polyacrylamide gel to separate the cleavage products. Gels containing polyacrylamide at 5 and 15% (w/v) are used for reactions containing *HAC1* 508 RNA and the hactng-10 minisubstrate, respectively. The reaction products are visualized by autoradiography and quantitated by phosphorimaging.

On a mole-by-mole basis, the Ire1p(l+k+t)-catalyzed cleavage of the *HAC1* 508 RNA is about three times more efficient than cleavage of the stem–loop minisubstrate. Under standard reaction conditions, Ire1p(l+k+t) (400 n*M*) cleaves 50% of *HAC1* 508 RNA (215 p*M*) in 30 min and 50% of stem–loop hactng-10 minisubstrate (255 p*M*) in 90 min (see Fig. 1).

Acknowledgments

This work was supported by a UCSF Biomedical Science Research Career Enhancement Fellowship from the National Institute of General Medical Science to T.N.G.; by NIH training grant NIH-NCI T32 CA09270, Molecular Biology of Eukaryotic Cells and Viruses, to M.N.; and by grants from the National Institutes of Health to P.W. P.W. is an Investigator of the Howard Hughes Medical Institute.

[2] Monitoring Activation of Ribonuclease L by 2′,5′-Oligoadenylates Using Purified Recombinant Enzyme and Intact Malignant Glioma Cells

By LORRAINE RUSCH, BEIHUA DONG, and ROBERT H. SILVERMAN

Introduction

RNase L is a regulated endoribonuclease present in a wide range of mammalian cell types, where it mediates antiviral, antiproliferative, and apoptotic effects of the interferons (IFNs) (reviewed in Stark *et al.*[1]). The most unusual property of RNase L is its requirement for allosteric activators consisting of 5′-phosphorylated, 2′,5′-linked oligoadenylates of the general formula: $p_x(A2'p)_nA$, where $x = 1$ to 3 and $n \geq 2$.[2] RNase L has a relatively broad specificity for cleaving single-stranded

[1] G. R. Stark, I. M. Kerr, B. R. G. Williams, R. H. Silverman, and R. D. Schreiber, *Annu. Rev. Biochem.* **67,** 227 (1998).

[2] I. M. Kerr and R. E. Brown, *Proc. Natl. Acad. Sci. U.S.A.* **75,** 256 (1978).

0076-6879/01 $35.00

RNA but prefers UU and UA dinucleotides, leaving a 3′-phosphoryl group and a 5′-OH.[3,4] 2′,5′-oligoadenylate (2-5A) is produced in cells in response to both IFN and double-stranded RNA (dsRNA), a type of nucleic acid often associated with virus infections. Exposure of mammalian cells to IFN-α or IFN-β results in the induction of a family of 2-5A synthetases. Double-stranded RNA produced by virus infections activates the 2-5A synthetases, resulting in the production of 2-5A from ATP. 2-5A then binds with high affinity to RNase L, converting it from its inactive monomeric state to its activated dimeric state.[5,6] The human and murine forms of RNase L contain 741 and 735 amino acid residues, respectively[7] (A. Zhou and R. H. Silverman, unpublished data, 2000). In the active, dimeric form of the enzyme, each subunit of RNase L is bound to one molecule of 2-5A.[6] The regulatory domains of RNase L, including the 2-5A-binding domain, are present in the N-terminal half. A series of nine ankyrin repeats present in the N-terminal part of RNase L is responsible for enzyme repression.[8] Two conserved lysine residues in P-loop motifs present in ankyrin repeats 7 and 8 are implicated in mediating 2-5A-binding activity.[7] The C-terminal half of RNase L contains protein kinase-like domains and the ribonuclease domain. The ribonuclease domains of RNase L and IRE1 kinase/endoribonucleases are homologous.[9] RNase L can be harnessed for the purpose of degrading targeted RNA molecules by chemically coupling 2-5A to antisense oligonucleotides.[10] Here we describe methods for measuring some of the biochemical and biological activities of RNase L. One of these methods describes a functional assay for 2-5A that utilizes an isolated, recombinant RNase L. Other techniques that are described directly or indirectly monitor effects of RNase L activation within living cells. These cellular methods use a human malignant glioma cell line (U373) that contains relatively high levels of endogenous RNase L. However, the same methods can also be applied to many other mammalian cell lines.

Preparation of 2′,5′-Oligoadenylates

2-5A is prepared with 2-5A synthetase immobilized on an activating affinity matrix of poly(I) : poly(C)–agarose [AGPoly(I) : poly(C), type 6; Pharmacia

[3] D. H. Wreschner, J. W. McCauley, J. J. Skehel, and I. M. Kerr, *Nature (London)* **289,** 414 (1981).

[4] G. Floyd-Smith, E. Slattery, and P. Lengyel, *Science* **212,** 1030 (1981).

[5] B. Dong and R. H. Silverman, *J. Biol. Chem.* **270,** 4133 (1995).

[6] J. L. Cole, S. S. Carroll, and L. C. Kuo, *J. Biol. Chem.* **271,** 3979 (1996).

[7] A. Zhou, B. A. Hassel, and R. H. Silverman, *Cell* **72,** 753 (1993).

[8] B. Dong and R. H. Silverman, *J. Biol. Chem.* **272,** 22236 (1997).

[9] C. Sidrauski and P. Walter, *Cell* **90,** 1031 (1997).

[10] P. F. Torrence, R. K. Maitra, K. Lesiak, S. Khamnei, A. Zhou, and R. H. Silverman, *Proc. Natl. Acad. Sci. U.S.A.* **90,** 1300 (1993).

Biotech, Piscataway, NJ] (see also Silverman and Krause[11]). HeLa cells are grown in Dulbecco's modified Eagle's medium (DMEM) with 10% (v/v) fetal bovine serum in ten 150-cm^2 plates at 37° to about 90% confluency and treated with IFN-α2a (Roferon, 1000 U/ml; Roche, Nutley, NJ) for 16 hr. Cells are washed in phosphate-buffered saline (PBS) and harvested by scraping. A postmitochondrial fraction is prepared by lysing the cells in 1.5-packed cell volumes of buffer A [0.5% nonidet P-40 (NP-40), 90 m*M* KCl, 1 m*M* magnesium acetate, 10 m*M* HEPES (pH 7.6); leupeptin (10 μg/ml), and 2 m*M* 2-mercaptoethanol], vortexing briefly, and centrifuging at 10,000*g* for 10 min at 4°. The supernatant containing the 2-5A synthetase in crude extract is removed. Poly(I) : poly(C)–agarose [1.2-ml suspension containing 6 mg of poly(I) : poly(C)] is washed twice by centrifuging at 1500*g* for 5 min at 4°, discarding the supernatant, and resuspending in 50 ml of buffer B [10 m*M* HEPES (pH.7.5), 50 m*M* KCl, 1.5 m*M* magnesium acetate, 7 m*M* 2-mercaptoethanol, and 20% (v/v) glycerol]. The final pellet of poly(I) : poly(C)–Sepharose is resuspended in the postmitochondrial fraction of HeLa cells and incubated for 1 hr at 4° on a rotating shaker at a low speed. The complex of 2-5A synthetase bound to the resin is centrifuged at 1500*g* for 5 min at 4° and the supernatant is discarded. The pellet is washed three times in 15 ml of buffer B by centrifuging at 1500*g* for 5 min, discarding the supernatant, and resuspending. The final pellet is resuspended in 6 ml of buffer C [buffer B supplemented with 4 m*M* ATP (pH 7) (Sigma, St. Louis, MO) and 10 m*M* magnesium acetate]. The first round of 2-5A synthesis is performed by incubation for 20 hr at 37° with gentle, continuous shaking. The resin complex is separated from the crude 2-5A preparation by centrifugation and removal of the supernatant containing the 2-5A. A second round of 2-5A synthesis is performed by resuspension in 6 ml of fresh buffer C and incubation for an additional 72 hr at 37°. The 2-5A is stored at −70°.

The yield and oligomer distribution of 2-5A are determined by high-performance liquid chromatography (HPLC). The 2-5A mixture (20 μM in 50 μl of water) is analyzed by reversed-phase HPLC analysis (C_{18}, 60 Å, 5 μm, Princeton Sphere, Princeton Chromatography, Inc., Cranbury, NJ).[12] Elution is with a linear gradient from 0 to 20% of methanol–water (1 : 1, v/v) in 50 m*M* ammonium phosphate (pH 7.0) at a flow rate of 1 ml/min in 25 min, and then increasing from 20 to 50% of methanol–water (1 : 1, v/v) in 5 min. Results from a typical preparation of 2-5A are shown in Table I. About 90 and 95% of the ATP was converted to 2-5A in the first and second rounds of synthesis, respectively. The majority (57–58%) of the product was trimer and tetramer 2-5A. The pooled, crude 2-5A can be used

[11] R. H. Silverman and D. Krause, *in* "Lymphokines and Interferons: A Practical Approach" (M. J. Clemens, A. G. Morris, and A. J. H. Gearing, eds.), p. 149. IRL Press, Oxford, 1987.

[12] M. Knight, P. J. Cayley, R. H. Silverman, D. H. Wreschner, C. S. Gilbert, R. E. Brown, and I. M. Kerr, *Nature (London)* **288,** 189 (1980).

TABLE I
DISTRIBUTION OF VARIOUS OLIGOMERS OF 2′,5′-OLIGOADENYLATE PREPARED ENZYMATICALLY

Oligomer	Retention time (min)	Round 1 % Total	Retention time (min)	Round 2 % Total
ATP (residual)	4.0	9.7	3.9	4.7
pppA2′p5′A	5.6–6.3	8.7	4.9–5.7	6.6
$ppp(A2'p)_2A$	11.2–12.2	25.5	11.0–13.2	31.1
$ppp(A2'p)_3A$	17.6–18.2	32.3	19.0–19.6	26.6
$ppp(A2'p)_4A$	21.1–21.7	12.8	23.0–23.9	10.1
$ppp(A2'p)_5A$	23.8–24.2	4.3	25.6–26.0	5.8
$ppp(A2'p)_6A$	25.9	1.2	27.8	0.4

directly for activating RNase L, or, depending on the application, the individual 2-5A oligomers can be isolated by preparative-scale HPLC.

Assays with Purified, Recombinant RNase L Fusion Protein

One of the most convenient sources of RNase L for functional analysis of 2-5A involves the *Escherichia coli* expression and purification of a glutathione *S*-transferase (GST) fusion protein.[8] In a prior study, we described the cloning of a full-length coding sequence DNA for human RNase L downstream (3′) of the coding sequence for GST in expression vector pGEX-4T-3 (Pharmacia Biotech). The cDNA for RNase L in plasmid pGEX4-T-3 is transformed into *E. coli* DH5α and the bacteria are grown at 30° to an A_{595} of 0.5 before being induced with 0.1 m*M* isopropyl-β-D-thiogalacto pyranoside (IPTG) for 3 hr. The harvested cell pellets are washed with PBS, resuspended in 5 volumes of PBS-C [PBS with 10% (v/v) glycerol, 1 m*M* EDTA, 0.1 m*M* ATP, 5 m*M* $MgCl_2$, 14 m*M* 2-mercaptoethanol, leupeptin (1 μg/ml), and 1 m*M* phenylmethylsulfonyl fluoride (PMSF)] supplemented with lysozyme (1 μg/ml), and incubated at room temperature for 20 min. The suspended cells are lysed by sonicating on ice for 20 sec four times; TritonX-100 is added to a final concentration of 1% (v/v), and the cell lysates are incubated at room temperature for 20 min. The supernatants are collected after centrifugation at 16,700*g* for 20 min at 4°. Purification of fusion proteins is performed as described by the manufacturer of the glutathione–Sepharose 4B (Pharmacia Biotech), with modifications. Briefly, glutathione–Sepharose 4B [200 μl of a 50% (v/v) slurry in PBS-C] is added to extract from 200-ml cultures of bacteria at room temperature for 20 min with shaking. After washing the protein–bead complexes three times with 10 ml of PBS-C, the fusion proteins are eluted with 20 m*M* glutathione in 50 m*M* Tris-HCl, pH 8.0, containing leupeptin (1 μg/ml), with shaking at room temperature for 20 min.

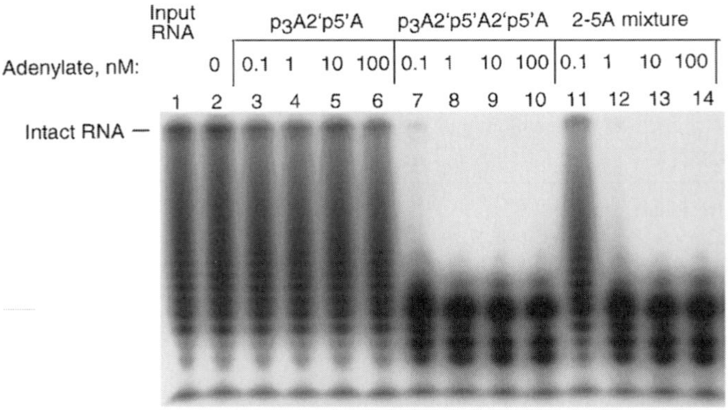

FIG. 1. Cleavage of U_{25}–[^{32}P]pCp by recombinant GST–RNase L activated with 2-5A. GST–RNase L (100 n*M*) was incubated in the absence or presence of various concentrations (in AMP equivalents) of $p_3A2'p5'A$, $p_3A2'p5'A2'p5'A$, or a 2-5A mixture (round 2, Table I) as indicated. Incubations with U_{25}–[^{32}P]pCp were for 30 min at 30°. An autoradiogram of the gel is shown.

Expression and purity of the protein preparations are determined by sodium dodecyl sulfate–polyacrylamide gel electrophoresis (SDS–PAGE) and Coomassie blue staining and by Western blots with monoclonal antibody to RNase L.[5]

As substrate, oligouridylic acid (U_{25}; Midland Certified Reagent Company, Midland, TX) is labeled at its 3′ terminus with [5′-^{32}P]pCp (3000 Ci/mmol; Du Pont NEN, Boston, MA) with T4 RNA ligase (GIBCO-BRL, Gaithersburg, MD). The U_{25}–[^{32}P]pCp, 80–160 n*M*, is incubated with 100 n*M* purified GST-RNase L (220 ng) in the presence and absence of different forms of 2-5A in a final volume of 20 μl at 30° for 30 min. Reaction mixtures are heated to 100° for 5 min in loading buffer and intact RNA is separated from the RNA degradation products in 20% (w/v) polyacrylamide–8% (w/v) urea sequencing gels. The amount of intact U_{25}–[^{32}P]pCp remaining after the incubations can be determined from autoradiograms of the gels with a Sierra Scientific (Sunnyvale, CA) high-resolution charge-coupled device (CCD) camera and the computer program NIH Image 1.6 (National Institute of Health, Bethesda, MD) or by PhosphorImager (Molecular Dynamics, Sunnyvale, CA).

In a representative assay, there was no RNA cleavage in the absence of 2-5A, or with $p_3A2'p5'A$ at concentrations as high as 100 n*M* (Fig. 1, lanes 1–6). In contrast, 0.1 n*M* $p_3A2'p5'A2'p5'A$ caused nearly complete cleavage of the U25 substrate (Fig. 1, lane 7). The crude 2-5A mixture, round 2 (Table I), was less active (by a factor of <10) than the purified trimer $p_3A2'p5'A2'p5'A$ (Fig. 1, lanes 11–14). However, the crude 2-5A mixture more closely mimicks what is present in cells exposed to IFN and infected with virus.[12]

2′,5′-Oligoadenylate Transfections

Transfection is necessary to cause 2-5A to transit the cell membrane. Methods that utilize the human malignant glioma U373 cell line (American Type Culture Collection, Manassas, VA) are described because these cells are easy to culture and have relatively high levels of RNase L. U373 cells are cultured in DMEM supplemented with streptomycin–penicillin and 10% (v/v) heat-inactivated fetal bovine serum. Transfections of 2-5A are performed with LipofectAMINE Plus (or other cationic lipid carriers) according to the manufacturer protocol (GIBCO-BRL). Briefly, the 2-5A is incubated with the Plus reagent for 15 min at room temperature in serum-free Opti-MEM prior to addition of LipofectAMINE. The combined mixture is incubated for an additional 15 min at room temperature and diluted to the appropriate volume in Opti-MEM. Medium is removed from cells and replaced with the transfection mixture. Cells are incubated at 37° for 3.5 hr or, to maintain cells for extend periods of time, an equal volume of medium–FBS is added at this time and the incubations are continued.

Monitoring RNase L-Mediated rRNA Cleavage in Intact Cells

One of the hallmarks of RNase L activation in cells is the specific cleavage of rRNA in intact ribosomes.[13,14] To monitor 18S rRNA breakdown, U373 cells in 100-mm^2 plates are incubated in the presence or absence of human IFN-β (1000 U/ml; Serono, Norwell, MA) for 16 hr prior to transfection with 0, 1, 3, or 6 μM 2-5A for 4 hr. While the IFN pretreatment is not required for RNase L activation, it does lead to slightly enhanced levels of RNase L protein and activity. The inducing effect of IFN on RNase L levels is much greater in most murine cell types than in human cells (data not shown). Cells are washed twice in PBS and total RNA is harvested in Trizol reagent (2 ml/sample) (GIBCO-BRL) prior to the addition of chloroform (20%, v/v) and vigorous vortexing for 30 sec. The samples are incubated at room temperature for 5 min and centrifuged at 12,000g for 25 min at 4°. The aqueous upper phase is removed to a fresh tube and an equal volume of 2-propanol is added to precipitate RNA. The samples are incubated for 15 min and centrifuged for 25 min at 12,000g at room temperature. The supernatants are discarded and the pellets are washed with ice-cold 70% (v/v) ethanol. RNA is dissolved in diethyl pyrocarbonate (DEPC)-treated water, absorbance at 260 nm is measured, and 20 μg per lane is separated by electrophoresis in 1.2% (w/v) agarose, 2.2 M formaldehyde gels. RNA is transferred to Nylon membranes (Amersham, Arlington Heights, IL) for 16 hr, cross-linked under UV for 90 sec, incubated in prehybridization solution at 42° for

[13] D. H. Wreschner, T. C. James, R. H. Silverman, and I. M. Kerr, *Nucleic Acids Res.* **9,** 1571 (1981).

[14] M. S. Iordanov, J. M. Paranjape, A. Zhou, J. Wong, B. R. Williams, E. F. Meurs, R. H. Silverman, and B. E. Magun, *Mol. Cell. Biol.* **20,** 617 (2000).

16 hr, and probed with cDNA to 18S rRNA (American Type Culture Collection) labeled with [α-^{32}P]dCTP by random priming with the Prime-a-Gene labeling system (Promega, Madison, WI). Membranes are washed in 1X salt–sodium citrate (SSC) containing 0.1% (w/v) SDS at room temperature and exposed to X-ray film.

In a representative experiment, the dimer form of 2-5A ($p_3A2'p5'A$), included as a control, was unable to activate RNase L to cleave 18S rRNA in either the presence or absence of IFN in U373 cells (Fig. 2, lanes 3 to 8). In contrast, about 19 and 23% of the 18S rRNA was cleaved after transfection with 3 or 6 μM 2-5A

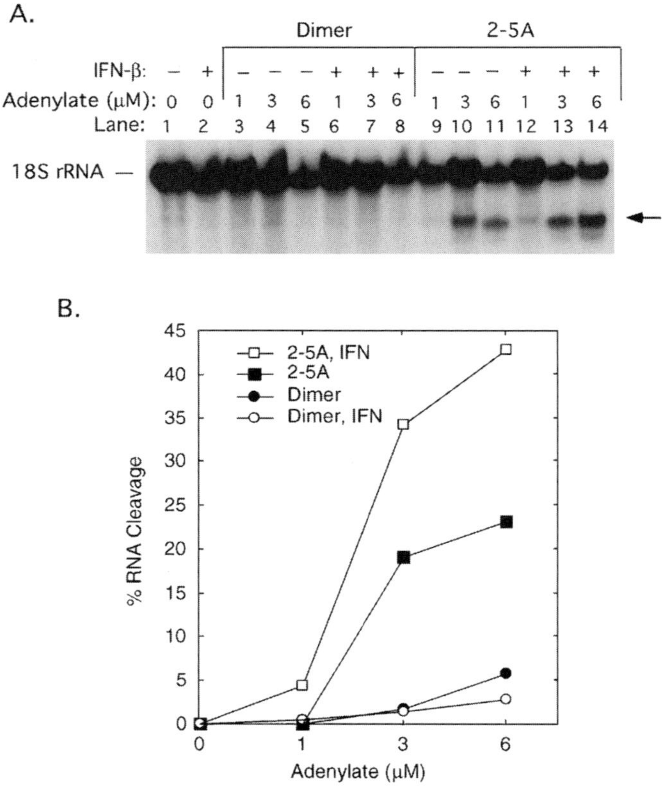

FIG. 2. RNase L-mediated cleavage of 18S rRNA in U373 cells after transfection with 2-5A. U373 cells were incubated in the absence or presence of IFN-β (1000 U/ml) for 16 hr prior to transfection with dimer ($p_3A2'p5'A$) or 2-5A mixture (round 2, Table I) (6 μM, 4 hr). RNA was separated on a 1.2% (w/v) agarose–2.2 *M* formaldehyde gel. (A) An autoradiogram of a Northern blot for 18S rRNA is shown. The arrow indicates the position of the 18S rRNA cleavage product induced by activation of RNase L. (B) A graph of the ratio of cleaved to total 18S rRNA determined by densitometry.

mixture, respectively. IFN pretreatment caused greater levels of 2-5A-mediated breakdown of 18S rRNA, with 34 and 43% cleavage at 3 and 6 μM 2-5A, respectively. It is apparent from these results that much higher levels of 2-5A are required than when using purified enzyme (compare Figs. 1 and 2). The apparent reasons include the relatively inefficiency of the transfection method and the metabolic lability of 2-5A due to the presence of phosphatases and phosphodiesterases in serum and in cells.

Measuring Protein Synthesis Inhibition by RNase L in Intact Cells

The effects of RNase L activation on rates of cellular protein synthesis are determined by transfecting cells with 2-5A and pulse labeling newly synthesized proteins with [^{35}S]methionine.[15] U373 cells (10^6 cells/well) are cultured in six-well plates and 2-5A transfections are performed as described (above). As a control, the dimer form of 2-5A, $p_3A2'p5'A$, which fails to bind or activate RNase L, is used. Cells are incubated at 37° for 3 hr after the transfections and prior to pulse labeling. The medium is removed and the cells are washed once with 1 ml of prewarmed PBS. RPMI 1640 medium lacking methionine (GIBCO-BRL) and supplemented with a 0.3-μCi/ml concentration of [^{35}S]methionine (1175 Ci/mmol; NEN) is added and cells are incubated for 2 hr. After removing the labeling medium ice-cold 5% (w/v) trichloroacetic acid (TCA) (1 ml) is added to precipitate the proteins. To remove unincorporated label, the TCA is removed and discarded, and fresh TCA (1 ml) is added and also discarded. The precipitated protein is solubilized with 0.25 *M* NaOH (1 ml) incubated at 30° for 10 min. The solution is neutralized with 0.25 *M* HCl (100 μl). Radioactivity in an aliquot (200 μl) of each well is measured by liquid scintillation counting.

In a typical assay, there was 57 and 83% inhibition of protein synthesis by 3 and 6 μM 2-5A mixture, respectively (Fig. 3). Dimer $p_3A2'p5'A$ did not inhibit protein synthesis at concentrations ≤ 3 μM, but it did cause a 21% nonspecific effect at 6 μM (Fig. 3).

Annexin V-Binding Assays to Measure Apoptosis in Response to 2′,5′-Oligoadenylate Activation of RNase L

Apoptosis is one of the biological consequences of excessive RNase L activation in living cells.[1] Annexin V-binding assays are performed to measure levels of apoptosis in cells transfected with 2-5A. The egress of phosphatidylserine (PS) from the cytoplasmic to the outer leaflet of the cell membrane is an early event in apoptosis. Annexin V is a calcium-dependent phospholipid-binding protein that is used to monitor translocation of PS to the cell surface. In this method, PS present on the cell surface is monitored by fluroscein isothiocyanate (FITC)-labeled

[15] P. L. Fox and P. E. DiCorleto, *J. Cell Physiol.* **121,** 298 (1984).

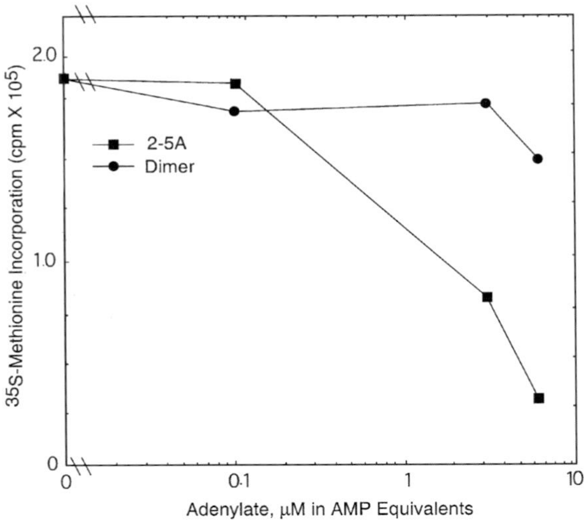

FIG. 3. Inhibition of protein synthesis in U373 cells after 2-5A transfections. Transfections were performed with 2-5A mixture (round 2, Table I) or dimer ($p_3A2'p5'A$) (6 μM, 4 hr) at concentrations ranging from 0.1 to 6 μM, and then pulse labeled with [^{35}S]methionine. Assay results are the average of duplicate determinations.

annexin V and fluroscence-activated cell sorting (FACS) analysis.[16] Annexin V analysis is performed in combination with propidium iodide (PI) to monitor the integrity of the cell membrane. Transfections are performed on U373 cells as described above, and samples are harvested at various times. Medium containing floating cells and debris is removed to labeled tubes. Cells that remain adherent to the wells are washed with PBS and removed by the addition of trypsin–EDTA (1 ml; GIBCO-BRL) at 37° for 2 min. The trypsinized cell suspension is combined with the medium containing the floating cells and centrifuged at 1000*g* for 5 min at room temperature. The cells are washed once and resuspended in PBS (1 ml). The cells are stained simultaneously with FITC–annexin V (green fluorescence) and PI (red fluorescence) for 15 min at room temperature in the dark as described (PharMingen-Becton Dickinson, San Diego, CA). Cells are gently resuspended, bivariate flow cytometry using a FACScan is performed, and the data are analyzed with CellQuest software (Becton Dickinson, San Jose, CA).

In a representative assay, apoptosis increased as a function of time (Fig. 4). Cells were incubated in the presence or absence of IFN-β (1000 U/ml; Serono)

[16] G. Koopman, C. P. Reutelingsperger, G. A. Kuijten, R. M. Keehnen, S. T. Pals, and M. H. van Oers, *Blood* **84,** 1415 (1994).

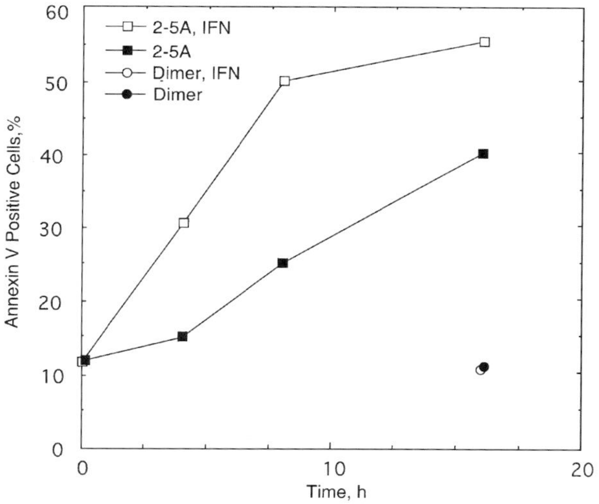

FIG. 4. Annexin V-binding analysis reveals increased levels of apoptosis in U373 cells transfected with 2-5A. U373 cells were incubated in the absence or presence of IFN-β (1000 U/ml) for 16 hr prior to transfection with either $p_3A2'p5'A$ or 2-5A (round 2, Table I) (6 μM). Cells were harvested at various times for annexin V-binding assays by FACS analysis.

for 16 hr prior to transfections. There was 41 and 56% apoptosis after 16 hr in the presence of 2-5A alone or 2-5A plus IFN, respectively. Dimer 2-5A failed to increase apoptosis over background levels in the presence or absence of IFN. Similarly, IFN by itself did not cause apoptosis during the initial 16 hr of treatment.

Summary

A wide range of methods have been developed to study RNase L activation in cell-free systems and in intact cells. Many of the original methods were developed in the laboratory of I. Kerr in the early 1980s (e.g., see Knight *et al.*[12]). Additional methods described in this article were developed or adapted from research in other fields after the cloning of RNase L and the appreciation of its role in apoptosis.[1,7,8,17–19] These methods provide the basic techniques needed to induce

[17] A. Zhou, J. Paranjape, T. L. Brown, H. Nie, S. Naik, B. Dong, A. Chang, B. Trapp, R. Fairchild, C. Colmenares, and R. H. Silverman, *EMBO J.* **16,** 6355 (1997).

[18] J. C. Castelli, B. A. Hassel, K. A. Wood, X. L. Li, K. Amemiya, M. C. Dalakas, P. F. Torrence, and R. J. Youle, *J. Exp. Med.* **186,** 967 (1997).

[19] J. C. Castelli, B. A. Hassel, A. Maran, J. Paranjape, J. A. Hewitt, X. L. Li, Y. T. Hsu, R. H. Silverman, and R. J. Youle, *Cell Death Differ.* **5,** 313 (1998).

RNase L activation *in vitro* and to measure some of its biological effects in living mammalian cells.

Acknowledgments

We thank Aimin Zhou and Fulvia Terezi for valuable discussions and Amy Raber for performing flow cytometry. This investigation was supported by United States Public Health Service Grant CA44059 from the Department of Health and Human Services, National Cancer Institute.

[3] Accelerating RNA Decay through Intervention of RNase L: Alternative Synthesis of Composite 2′,5′-Oligoadenylate–Antisense

By PAUL F. TORRENCE and ZHENGFU WANG

Introduction

The completion[1] of the 34 million-letter code for the smallest human chromosome (22) containing 545 genes and 134 pseudogenes presages the problems proteomics faces in ascribing function to sequence. Unraveling the role of various human genes in disease and choosing appropriate targets for drug discovery will require robust prediction and validation paradigms. One such approach is the use of antisense to "knock out" or downregulate gene expression through interference with the translation of mRNA.[2] This latter methodology relies on the use of an oligonucleotide, or analog thereof, with a sequence complementary (antisense) to that of a particular mRNA. Hybridization of the antisense reagent to the specific mRNA may produce a "steric blocking" of access to the translational mechanisms of the cell, or may lead to the degradation of the mRNA through the intervention of the cellular enzyme RNase H.

Optimization of the antisense paradigm is key to expression of its full potential for determining gene function as well as for the discovery of novel therapeutics. Oftentimes, the chemical structural modifications that can maximize target affinity, antisense metabolic stability, and cellular uptake may not be compatible with the catalytic mRNA degradation mediated by RNase H, which is specific for DNA–RNA hybrids.[2,3]

[1] I. Dunham, N. Shimizu, B. A. Roe, S. Chissoe, A. R. Hunt, U. E. Collins, R. Bruskiewich, D. M. Beare, M. Clamp, L. J. Smink, *et al., Nature (London)* **402,** 489 (1999).

[2] S. T. Crooke, *Methods Enzymol.* **313,** 3 (2000).

[3] C. F. Bennett, *Biochem. Pharmacol.* **55,** 9 (1998).

0076-6879/01 $35.00

RNase L is a unique latent RNase that is specifically activated by the equally unique 2′,5′-phosphodiester-linked oligoriboadenylates referred to as 2-5A or (pp) p5′A2′(p5′A2′)$_n$.[4-10] As a component of the antiviral action of interferons (IFNs), 2-5A is generated from ATP through the action of a multienzyme family of 2-5A synthetases, which are activated by double-stranded RNA (dsRNA). Chemical conjugation of 2-5A with antisense oligonucleotides provides a chimeric molecule that recruits RNase L for the specific degradation of a selected RNA. 2-5A–antisense represents an addition of a new approach to the targeted destruction of RNA.[4-10] One advantage of 2-5A–antisense is that it recruits the enzyme RNase L to cleave the targeted mRNA. RNase L is not restricted by the same structural requirements for RNA substrate cleavage as is RNase H. Therefore, a greater variety of chemically modified antisense oligonucleotide analogs can be used in conjunction with the 2-5A–antisense approach.

We have shown that peptide–nucleic acids (PNAs)[11-13] are effective partners for the 2-5A–antisense strategy.[14] This finding provided a means to endow PNAs with an enzyme-based catalytic turnover of RNA substrate. This was not possible with PNAs alone because RNase H does not cleave PNA–RNA complexes.

Previously, 2-5A–PNA antisense constructs were synthesized[14] through a solid-phase stepwise approach similar to that employed for standard DNA machine synthesis. We have reported a different approach that employs a convergent synthetic scheme wherein the PNA and 2-5A are prepared separately and then linked in a final step.[15] The procedures detailed herein outline this alternative preparation. In this example, 2-5A pentamer is joined to a tetrameric PNA; however, in principle, the length and base of the PNA can be varied as required.

There are three separate parts to this preparation: synthesis of a 2-5A oligomer, generation of a PNA antisense component containing an amino linker

[4] P. F. Torrence, R. K. Maitra, K. Lesiak, S. Khamnei, A. Zhou, and R. H. Silverman, *Proc. Natl. Acad. Sci. U.S.A.* **90,** 1300 (1993).

[5] K. Lesiak, S. Khamnei, and P. F. Torrence, *Bioconjug. Chem.* **4,** 467 (1993).

[6] A. Maran, R. K. Maitra, A. Kumar, B. Dong, W. Xiao, G. Li, B. R. Williams, P. F. Torrence, and R. H. Silverman, *Science* **265,** 789 (1994).

[7] N. M. Cirino, G. Li, W. Xiao, P. F. Torrence, and R. H. Silverman, *Proc. Natl. Acad. Sci. U.S.A.* **94,** 1937 (1997).

[8] M. R. Player, D. L. Barnard, and P. F. Torrence, *Proc. Natl. Acad. Sci. U.S.A.* **95,** 8874 (1998).

[9] A. Maran, C. F. Waller, J. M. Paranjape, G. Li, W. Xiao, K. Zhang, M. E. Kalaycio, R. K. Maitra, A. E. Lichtin, W. Brugger, P. F. Torrence, and R. H. Silverman, *Blood* **92,** 4336 (1998).

[10] S. Kondo, Y. Kondo, G. Li, R. H. Silverman, and J. K. Cowell, *Oncogene* **16,** 3323 (1998).

[11] P. E. Nielsen, M. Egholm, R. H. Berg, and O. Buchardt, *Science* **254,** 1497 (1991).

[12] E. Uhlmann, A. Peyman, G. Breipohl, and D. W. Will, *Angew. Chem. Int. Ed.* **37,** 2796 (1998).

[13] P. E. Nielsen, *in* "Biomedical Chemistry: Applying Chemical Principles to the Understanding and Treatment of Disease" (P. F. Torrence, ed.). John Wiley & Sons, New York, 2000.

[14] J. C. Verheijen, G. A. van der Marel, J. H. van Boom, S. F. Bayly, M. R. Player, and P. F. Torrence, *Bioorg. Med. Chem.* **7,** 449 (1999).

[15] Z. Wang, L. Chen, S. F. Bayly, and P. F. Torrence, *Bioorg. Med. Chem. Lett.* **10,** 1357 (2000).

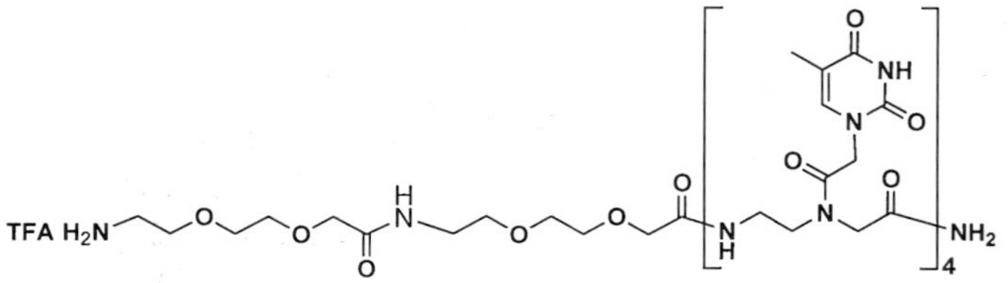

Compound 1

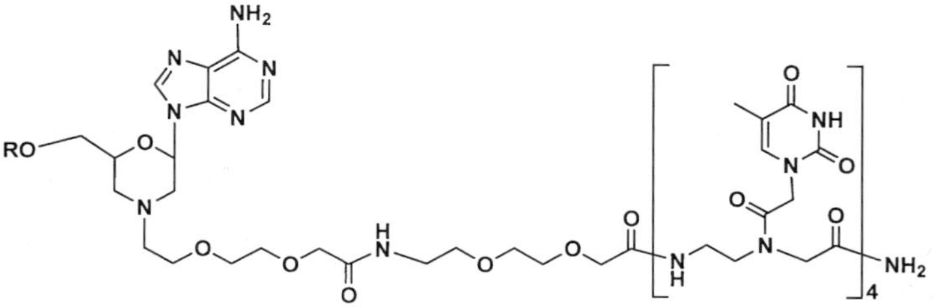

R= p5'A2'p5'A2'p5'A2'p5'A2'p-
Compound 2

FIG. 1. Structures of PNA tetramer with attached amino linker (compound **1**) and final product, 2-5A–PNA, compound **2.**

(compound **1,** Fig. 1) that is later used in reaction with the 2-5A component (pentameric 2-5A, p5′A2′p5′A2′p5′A2′p5′A2′p5′A), and conjugation of the two molecules to form a chimera (compound **2,** Fig. 1). The latter conjugation depends on cleavage of the 2′,3′-C—C bond of the ribonucleoside at the 3′ end of nucleic acids by sodium periodate to give a dialdehyde. Reaction of this with an amino group forms a Schiff base that can be reduced to a substituted morpholine. Sodium cyanoborohydride is used for the reduction because it reduces only the Schiff base; the aldehyde is not affected under these conditions.

Methods and Methodology

Preparation of 2′,5′-Oligoadenylates

The synthetic strategy is based on the phosphite–triester approach to DNA/RNA synthesis. The appropriately protected 2-cyanoethylphosphoramidite derivative of riboadenosine, 5′-*O*-(4,4′-dimethoxytrityl)-3′-*O*-(*tert*-butyldimethylsilyl)-N^6-benzoyladenosine), is used for chain elongation, and the solid-phase methodology is employed with linkage of riboadenosine joined to controlled pore glass (CPG) through a long-chain alkylamine (lcaa) and a succinyl moiety. 5′-Phosphorylation of the oligoadenylate completes the basic synthesis, after deprotection, and the oligonucleotide is purified.

The following reagents should be available.

Intermediate for stepwise elongation of 2′,5′-oligoadenylate chain: N^6-Benzoyl-5′-*O*-dimethoxytrityl-3′-*O*-*tert*-butyldimethylsilyladenosine 2′-(*N,N*-diisopropyl-2-cyanoethyl)phosphoramidite (ChemGenes, Waltham, MA), three 25-mg quantities, each in a separate vial that can be sealed with a rubber septum

Catalyst for coupling: Tetrazole (two 35-mg quantities, each in a separate vial as described above)

Solid support with attached adenosine: N^6-Benzoyl-5′-*O*-dimethoxytrityl-2′(3′)-*O*-acetyladenosine-3′(2′)-lcaa–CPG, approximately 35 mg (1 μmol) packed in a DNA synthesis column

Reagent for 5′-phosphorylation: 2-[2-(4,4′-Dimethoxytrityl)ethylsulfonyl] ethyl-2-cyanoethyl)-*N,N*-diisopropyl) phosphoramidite (Glen Research, Sterling, VA)

The vials described above are sealed with a rubber septum and a metal cap, and a hypodermic needle (e.g., 20-gauge) is inserted through the septum. The vials and contents are transferred to a vacuum desiccator containing fresh phosphorus(V) oxide (P_2O_5) and then dried *in vacuo* overnight. The desiccator is opened and the hypodermic needles are rapidly removed

The following solutions should then be prepared.

Solution A (for capping): 30% (v/v) acetic anhydride in tetrahydrofuran (THF), equivalent to 1.5 ml of acetic anhydride and 3.3 ml of dry THF. This should be prepared in an oven-dried vial previously septum sealed. Use will amount to about 0.4 ml per cycle (see below)

Solution B (for capping): 0.6 *M* dimethylaminopyridine (DMAP) in pyridine–THF (3 : 2, v/v) equivalent to 350 mg of DMAP, 3 ml of pyridine, 2 ml of THF. Use 0.4 ml per cycle

Solution C (for detritylation): 3% (w/v) dichloroacetic acid (DCA) in dichloromethane (CH_2Cl_2). Use 1 ml/cycle

Solution D (for coupling): 0.5 *M* tetrazole in dry acetonitrile prepared by injecting the acetonitrile (1 ml) with a dry syringe into the dried vial, described above, containing the 35 mg of tetrazole under a dry nitrogen atmosphere. Use will be about 0.15 ml/cycle

Solution E (for oxidation): prepare a 0.1 *M* iodine solution from 1.3 g of I_2, 10 ml of lutidine, 40 ml of THF, and 0.5 ml of water. Use will be 1 ml/cycle

TABLE I
SOLID-PHASE SYNTHESIS OF OLIGONUCLEOTIDES: COUPLING CYCLE

Step	Solvents/reagents[a]	Time	Volume (ml)
1. Detritylation	3% (w/v) DCA in CH_2Cl_2	90 sec	1
2. Washing	2% (v/v) Py in acetonitrile		1
3. Washing	Acetonitrile		3
4. Drying	Nitrogen or Argon	3 min	
5. Coupling	0.2 *M* monomer in 0.5 *M* tetrazole–acetonitrile	8 min	0.15
6. Washing	Acetonitrile		3
7. Drying	Nitrogen or Argon	2 min	
8. Capping	A + B, 1 : 1 (v/v) A: 30% (v/v) Ac_2O in THF B: 0.6 *M* DMAP in Py–THF (3 : 2, v/v)	2 min	1
9. Washing	Acetonitrile		3
10. Drying	Nitrogen or Argon	2 min	
11. Oxidation	0.1 *M* I_2 in lutidine–THF–water, 20 : 80 : 1 (v/v/v)	45 sec	1
12. Washing	Acetonitrile		3
13. Drying	Nitrogen or Argon	3 min	

[a] Abbreviations: DCA, Dichloroacetic acid; Py, Pyridine; Ac_2O, acetic anhydride; THF, tetrahydrofuran; DMAP, dimethylaminopyridine.

Solution F (wash): 1% (v/v) pyridine in acetonitrile prepared by mixing 200 μl of pyridine in 20 ml of THF. About 1 ml/cycle will be used

Solid-Phase Support Manual Synthesis of 2′,5′-Oligoadenylates: General Procedure

For simplicity, when a DNA synthesizer is not readily available, and when conservation of reagents may be important, 2′,5′-oligoadenylate synthesis can be carried out manually on DNA synthesis columns (1.5-cm length; American Bionetics, Hayward, CA) containing approximately 1 μmol of CPG-bound N^6-benzoyl-5′-*O*-dimethoxytrityl-2′-*O*-acetyladenosine, using commercially available adaptors and gas-tight syringes. The applied coupling cycle is presented in Table I. Syntheses are controlled by quantitating spectrophotometrically the release of trityl cation, by visible spectrophometry. The average coupling efficiency determined in this way should be 90%.

The typical experimental setup for manual synthesis is shown in Fig. 2.

Synthesis Procedure

See also Table I.

1. (Steps 8–10 in Table I): To cap the unreacted amino groups on the CPG solid support in the column, withdraw 400 μl of solution A and 400 μl of solution

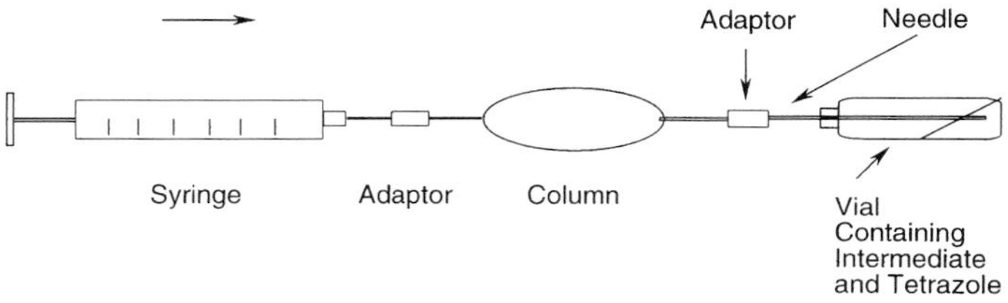

FIG. 2. Schematic showing assembly of manual oligonucleotide synthesis apparatus used in preparation of 2-5A.

B into the syringe and ensure thorough mixing by rotating the syringe a few times. Over a period of 2 min, inject the contents of the syringe into a column containing the CPG-bound protected adenosine. Then wash the column with three separate 1-ml injections of dry acetonitrile. Dry the column with argon gas, using a slight vacuum at the other end of the column. This should take about 2 min and the CPG powder in the column should be able to flow freely.

2. (Steps 1–4 in Table I): To remove the 5′-dimethoxytrityl moiety, use the syringe to inject solution C (1 ml) over a period of 1.5 min. Collect the solution that elutes for later determination of the coupling yield. Next, wash the column with 1 ml of solution F followed by a total of 3 ml of acetonitrile. Flush dry with argon as described above.

3. (Steps 5–7 in Table I): This step involves coupling to form the internucleotide bonds. Add 150 μl of solution D to the phosphoramidite such that the final concentration is about 0.2 *M*. Using Fig. 2 as a guide, draw the solution of the phosphorus(III) intermediate and tetrazole (described above) into the column by pulling back the piston of the empty syringe. Hold the remaining apparatus by the needle and roll the syringe and column back and forth gently on a horizontal surface to ensure thorough mixing. Allow the reaction to proceed for about 8 min, and then push in the syringe plunger to expel the remaining solution back into the vial. Wash the column by withdrawing three 1-ml volumes of acetonitrile and then dry the column with argon gas as above.

4. (Steps 8–10 in Table I): To "cap" unreacted hydroxyl groups, proceed as in step 1. After 2 min of reaction time, expel the reactants from the column and wash with three 1-ml volumes of dry acetonitrile and dry as usual with argon.

5. (Steps 11–13 in Table I): To carry out the oxidation of the phosphorus(III) intermediate, inject 1 ml of solution E over a period of 45 sec, but no longer than 60 sec. Wash the column with three 1-ml volumes of acetonitrile and dry with argon.

6. Repeat Steps 2–5 for each internucleotide bond desired. For example, for 2-5A tetramer, the process should be repeated three more times.

7. To phosphorylate the oligoadenylate at the 5′ terminus, carry out the 5′-detritylation as specified by step 2. After drying the column with argon, add 150 μl of solution D to the 5′-phosphorylating reagent such that the final concentration is 0.2 *M*. Otherwise, follow the procedure in step 3 to perform the phosphorylation. Dry the column as usual.

8. The final oxidation step is performed as in step 5. After the oxidation, wash the column with three 1-ml volumes of acetonitile and dry.

9. To remove the oligoadenylate from the CPG support, inject 1 ml of 30% (v/v) NH_4OH-CH_3OH (3 : 1, v/v) into the column, allowing two drops to be eluted from the end into a vial such as that used for nucleotide monomers. Allow the column to incubate at room temperature for 30 min and then push through (with the aid of the attached syringe) two more drops of ammonia solution. Allow the column to incubate an additional 3–4 hr (longer is acceptable). Push the remaining ammonia–methanol through the column and wash the column with an additional 1 ml of methanolic ammonia. Collect and save all the eluted solutions, including the initial drops eluted, in the capped vial.

10. Heat the contents (methanolic ammonia solution of eluted protected oligoadenylate) of the sealed vial at 55° for 8 hr in order to effect removal of the base protection. Carefully evaporate the solution to dryness in a rotary evaporator *in vacuo*. Use extreme care, as the high ammonia content can cause vigorous foaming and loss of product.

11. Remove the ribose silyl protection by adding 1 ml of 1 *M* tetrabutylammonium fluoride in THF and vortexing the solution until dissolution is complete. Leave the solution overnight at room temperature. Finally, remove the THF by *in vacuo* evaporation.

12. The oligonucleotides may be purified by high-performance liquid chromatography (HPLC) on a semipreparative Ultrasphere ODS column (Beckman, Fullerton, CA). A semipreparative Ultrasphere ODS column [reversed-phase C_{18}, 10 × 250 mm, flow rate of 2 ml/min, linear gradients of mobile phase B in buffer A, where buffer A is 50 m*M* ammonium acetate, pH 7.0, and buffer B consists of 50% (v/v) methanol in water] can be used for purification of synthetic oligonucleotides. The collected samples are then desalted on a DEAE-Sephadex A-25 anion-exchange column (HCO_3^- form), and eluted with required linear gradients of triethylammonium bicarbonate. After evaporation of the buffer, oligonucleotides can be converted into sodium salts by precipitation with a 2% (w/v) solution of sodium iodide in acetone.

For another purification method, a PRP-1 HPLC column (300 × 7 mm) can be used with the following elution protocol: solvent A is 10 m*M* tetrabutylammonium dihydrogen phosphate (TBAP) in water, pH 7.5; solvent B is 10 m*M* TBAP in acetonitrile–water (8 : 2, v/v), pH 7.5; elution is performed with a convex gradient of 5–85% solvent B in solvent A over 60 min at a flow rate of 1.5 ml/min.

Fractions containing the desired oligoadenylate may be pooled and evaporated to about 1 ml and desalted with a C_{18} SepPak cartidge (Waters, Milford, MA). The tetrabutylammonium salt can be converted into its sodium salt by Dowex 50W ion-exchange resin (Na^+ form). The purity should be checked by HPLC on a Beckman Ultrasphere ODS column.

Preparation of Peptide–Nucleic Acid T4 Bearing Terminal Amino Linker

PNA tetramer T4 may be prepared by 9-fluorenylmethoxycarbonyl (Fmoc) chemistry on a 2-μmol Fmoc-xanthenyl amide linker (XAL)-polyethylene glycol (PEG)-phosphatidylserine (PS) column with Fmoc-*N*-(2-aminoethyl)glycyl PNA T monomer on an Expedite 8909 PNA synthesizer according to the chemistry and protocols developed by the manufacturer[16] (PerSeptive Biosystems, Framingham, MA; now PE Biosystems, Foster City, CA). *O*-(7-Azabenzotriazol-1-yl)-*N,N,N′*,*N′*-tetramethyluronium hexafluorophosphate (HATU) acts as coupling reagent. PNA monomers, linker Fmoc-AEEA-OH (2-[2-(2-aminoethoxyl)-ethoxyl]acetic acid), HATU, base solution (diisopropylethylamine, DIEA), deblocking solution [20% (v/v) piperidine in dimethylformamide (DMF)], wash solution (DMF), and an Fmoc-XAL-PEG-PS column can be obtained from PerSeptive Biosystems. TFA and *m*-cresol are from Aldrich (Milwaukee, WI). Two linkers are introduced on the N terminus of PNA T4 to minimize the influence of the PNA portion on 2-5A activity. After synthesis and removal of the terminal Fmoc group, the column should be disassembled from the synthesizer and treated according to the protocol defined by the manufacturer.

The crude product is analyzed by reversed phase HPLC on a Beckman Ultrasphere ODS column (4.6 × 250 mm) with buffer A (0.1 *M* ammonium acetate) and buffer B [22.5% (v/v) CH_3CN–0.1 *M* ammonium acetate]. Linear gradient elution is performed at 45° with a flow rate of 1 ml/min. The purity of PNA T4 is more than 89%. It can be used for the 2-5A–PNA chimera synthesis without further purification according to the defined protocol in the manufacturer manual[16] (http://www.pebio.com/ds/pna/).

It is also possible to purchase PNAs on a custom basis from PE Biosystems (http://www.pebio.com/ab/).

Preparation of 2′,5′-Oligoadenylate–Peptide–Nucleic Acid Chimeric Adducts

Preparation of Compound **2**. To a cooled solution of pentamer p5′A2′p′5′A2′p5′A2′p5′A2′p5′A (7.34 OD units, 0.1218 μmol, in 50 μl of water), add 2.4 μl of 0.1 *M* $NaIO_4$. The oxidation should be carried out in the dark for 1.5 hr.

[16] PerSeptive Biosystems, "PNA Chemistry for the Expedite™ 8900 Nucleic Acid Synthesis System: User's Guide." PerSeptive Biosystems, Framingham, Massachusetts (1998).

Then add 4.8 μl of 0.1 *M* Na_2SO_3 solution to destroy the excess oxidant. After 20 min, add 9.36 OD units of compound **1** (dissolved in 60 μl of water), and then adjust the pH to about 8.6. After 1 hr, add 4.8 μl of 0.5 *M* $NaBH_3CN$, and once more adjust the pH to about 6.6. After overnight reaction, the compound can be purified by reversed phase HPLC on an ODS column (250 $\times$ 10 mm). Beckman Ultrasphere ODS analytic (4.6 $\times$ 250 mm) and semipreparative (10 $\times$ 250 mm) columns, respectively, can be used for the analysis and purification with the following buffers : buffer A, 0.1 *M* ammonium acetate; buffer B, 22.5% (v/v) CH_3CN–0.1 *M* ammonium acetate. Linear gradient elution may be performed at 45° with a flow rate of 1 or 3 ml/min. In the purification of all compounds by HPLC, the peak containing chimera (compound **2**) can be pooled and the solution may be dried in a Speed-Vac. The residue may be further desalted with a C_{18} Sep Pak cartridge (Waters) or by dialysis (Spectra/Por, molecular weight cutoff 500; spectrum Laboratories, Rancho Dominguez, CA). Its identity can be corroborated by determination of its molecular mass by electrospray ionization-mass spectroscopy (ESI-MS): calculated for $C_{106}H_{141}N_{44}O_{51}P_5$, 3002.4 Da; found, 3002.1 Da.

[4] Polysomal Ribonuclease 1

By KRISTOPHER S. CUNNINGHAM, MARK N. HANSON, and DANIEL R. SCHOENBERG

Introduction

Enzymes involved in mRNA decay can be classified by the pathways in which they participate. The general pathway for mRNA decay, in which deadenylation precedes decapping and 5′–3′ degradation, has been well characterized in yeast.[1] In vertebrate cells there is now clear evidence of a 3′ → 5′-deadenylase [termed deadenylating nuclease (DAN) or poly(A)-specific exoribonuclease (PARN)[2,3]], and for constituents of the exosome complex of 3′ → 5′-exoribonucleases.[4] While a vertebrate decapping pyrophosphatase has yet to be described, a vertebrate homolog

[1] S. Tharun and R. Parker, *in* "mRNA Metabolism and Post-transcriptional Gene Regulation" (J. Harford and D. R. Morris, eds.), p. 181. John Wiley & Sons, 1997.

[2] E. Dehlin, M. Wormington, C. G. Körner, and E. Wahle, *EMBO J.* **19,** 1079 (1998).

[3] G. C. Körner and E. Wahle, *J. Biol. Chem.* **272,** 10448 (1997).

[4] C. Allmang, E. Petfalski, A. Podtelejnikov, M. Mann, D. Tollervey, and P. Mitchell, *Genes Dev.* **13,** 2148 (1999).

0076-6879/01 $35.00

of the yeast 5′ → 3′-exonuclease Xrn1p has been identified,[5] and there is indirect evidence of deadenylation and decapping followed by 5′–3′ degradation of some vertebrate mRNAs.[6] Unlike yeast, vertebrate cells also make extensive use of endonuclease-mediated pathways in the process of mRNA decay.

In *Xenopus* liver estrogen stimulates the transcriptional induction and stabilization of mRNA encoding the yolk protein precursor vitellogenin, and activates a process that causes the pre-existing serum protein mRNAs to disappear from the cytoplasm.[7] In the early 1990s we determined that this process was mediated by the appearance of a sequence-selective endoribonuclease activity on polysomes,[8] and proceeded to purify the protein to obtain mass spectrometric sequence data[9] that were used for the subsequent cloning of its cDNA.[10] We named this new endoribonuclease PMR-1, or polysomal ribonuclease 1, both for the subcellular site at which it was identified, and for the fact that it is structurally unrelated to any known endoribonucleases. Rather, PMR-1 is a member of the peroxidase gene family. Like its closest homolog human myeloperoxidase, PMR-1 is made as a larger precursor and proteolytically processed to the enzymatically active 60-kDa form purified from polysomes. Although the 60-kDa protein cleaves preferentially within the sequence APyrUGA in single-stranded RNA, the 80-kDa precursor is catalytically inactive.

Although PMR-1 is the first vertebrate mRNA endonuclease to be identified, purified, and cloned, it is likely that there are numerous mRNA endonucleases yet to be found. The methods and approaches to purify and characterize PMR-1 that are described in this article should be of general use for others seeking to perform similar studies of other mRNA endonucleases.

In Vitro Assays for Polysomal Ribonuclease 1

Because PMR-1 was found in association with mRNA-containing complexes, the first steps in its analysis required facile approaches for the isolation of messenger ribonucleoprotein (mRNP) and polysome complexes. Furthermore, PMR-1 selectively associates with membrane-bound polysomes. A particular advantage of using a tissue as the source for purifying PMR-1 is that relatively large quantities

[5] V. I. Bashkirov, H. Scherthan, J. A. Solinger, J. M. Buerstedde, and W. D. Heyer, *J. Cell. Biol.* **136,** 761 (1997).

[6] P. Couttet, M. Fromont-Racine, D. M. Steel, R. Pictet, and T. Grange, *Proc. Natl. Acad. Sci. U.S.A.* **94,** 5628 (1997).

[7] R. L. Pastori, J. E. Moskaitis, S. W. Buzek, and D. R. Schoenberg, *Mol. Endocrinol.* **5,** 461 (1991).

[8] R. L. Pastori, J. E. Moskaitis, and D. R. Schoenberg, *Biochemistry* **30,** 10490 (1991).

[9] R. E. Dompenciel, V. R. Garnepudi, and D. R. Schoenberg, *J. Biol. Chem.* **270,** 6108 (1995).

[10] E. Chernokalskaya, A. N. DuBell, K. S. Cunningham, M. N. Hanson, R. E. Dompenciel, and D. R. Schoenberg, *RNA* **4,** 1537 (1998).

of material can be processed at a reasonable cost. The approaches for preparing polysomes and mRNP complexes described below were optimized for *Xenopus* liver, but can be readily adapted to other tissues or cell pellets.

Preparation of Postmitochondrial Extract

The protocol described below was created for use with approximately 5–10 g of tissue. It can be adjusted for the amount of sample.

1. Diced tissue is placed into 2.5 volumes (v/w) of ice-cold homogenization buffer containing 40 m*M* Tris-HCl (pH 7.5), 10 m*M* $MgCl_2$, 7% sucrose (w/v), 2 m*M* dithiothreitol, and protease inhibitors [0.2 m*M* phenylmethylsulfonyl fluoride (PMSF), leupeptin (0.5 μg/ml), pepstatin A (0.7 μg/ml), and aprotinin (2 μg/ml)] and homogenized with 10 strokes of a chilled Teflon–glass homogenizer.
2. The homogenate is filtered through a polyamide nylon mesh (Nitex 27621, 0.5 mm; Tetko, Elmsford, NY) although this can be substituted with a double layer of cheesecloth with consequent loss of some sample. The filtrate is then centrifuged at 4°, $1000g_{max}$ for 15 min in a swinging bucket rotor to remove nuclei and the resulting supernatant is centrifuged at 4°, $25,000g_{max}$ for 15 min in a swinging bucket rotor to pellet mitochondria.
3. The postmitochondrial supernatant is removed by piercing the side of the tube above the pellet with an 18-gauge needle and withdrawing the solution with a syringe, taking care to avoid contamination with the floating lipid layer. This material can be stored at −80° for months without apparent loss of activity.

Preparation of Polysomes

1. Polysomes are harvested from the postmitochondrial extract by centrifuging at $125,000g_{max}$ in a swinging bucket rotor [such as a Beckman (Fullerton, CA) SW41 or Sorvall (Newtown, CT) T-641] for 60 min at 4°. The "cytosolic" supernatant obtained in this step should also be saved and assayed for endonuclease activity. Gently resuspend the polysomes in 30 m*M* Tris-HCl (pH 7.4), 2 m*M* dithiothreitol, 5 m*M* $MgCl_2$. Try to keep the polysomes as concentrated as possible, because there is a 2-ml sample volume limit on a single gradient.
2. Prepare a 7-ml linear gradient of 15–40% (w/v) sucrose in 30 m*M* Tris-HCl (pH 7.4), 2 m*M* dithiothreitol, 5 m*M* EGTA, 5 m*M* $MgCl_2$, in a clear (not polyallomer) 14 × 89 mm ultracentrifuge tube. This is underlayered with a 3-ml pad of 60% (w/v) sucrose in 30 m*M* Tris-HCl (pH 7.4), 2 m*M* dithiothreitol, 10 m*M* EDTA. The postmitochondrial extract is gently loaded onto the gradient and centrifuged at $225,000g_{max}$ at 4° for 3.5 hr in a Sorvall T-641 or Beckman SW41 rotor. The EDTA present in the sucrose pad causes free polysomes to disaggregate at the boundary, thus preventing their further sedimentation to the

bottom of the tube and reducing the number of fractions that need to be analyzed. Membrane-bound polysomes sediment more slowly under these conditions and appear as a flocculent layer just above the interface with the EDTA-containing sucrose pad.

3. Alternatively, if membrane-bound and free polysomes are not to be separated, include 0.1% (v/v) Triton X-100 in the homogenization buffer. It should be noted that Triton can be difficult to remove from protein samples and may affect subsequent chromatographic steps aimed toward purifying the endonuclease.

4. If the samples are to be stored at this point, they should be dialyzed against 30 m*M* Tris-HCl (pH 7.4), 2 m*M* dithiothreitol, 0.5 m*M* PMSF, and 15% (v/v) glycerol, and stored frozen at -80°.

Preparation of Salt-Extracted Polysomes

Most of the polysomal RNases identified to date can be recovered in the soluble fraction after extraction of polysomes with 0.4 or 0.5 *M* KCl. Such extracts can be made in the following manner from polysomes isolated as described above.

1. Add a stock solution of 4 *M* KCl dropwise with gentle mixing to polysomes suspended in 30 m*M* Tris-HCl, 2 m*M* dithiothreitol, 1 m*M* $MgCl_2$ at 4° until the desired concentration is attained.

2. Gently mix the suspension at 4° on a rocking platform for 2 hr, and then centrifuge at 4°, $125{,}000g_{max}$ for 60 min in a fixed-angle rotor to separate the polysome salt extract from the remaining ribosomes. Both the supernatant and ribosome-containing pellet should be retained.

3. After determining the amount of recovered protein in the sample the supernatant is dialyzed at 4° against 30 m*M* Tris (pH 7.4), 2 m*M* dithiothreitol, 15% (v/v) glycerol, and stored at -80°. The salt-extracted ribosomes can be stored as a frozen pellet or resuspended in an equal volume of buffer and dialyzed as described above before being stored frozen.

Simple Method for Separating Messenger Ribonucleoprotein and Polysome Fractions

McGrew *et al.*[11] developed a simple method to fractionate mRNP and polysome complexes from postmitochondrial extracts which we have found useful. Postmitochondrial tissue extract is layered onto a step gradient of 10% (w/v) sucrose (2 ml) and 35% (w/v) sucrose (2.5 ml) prepared in 30 m*M* Tris-HCl (pH 7.4), 2 m*M* $MgCl_2$, 2 m*M* dithiothreitol, and centrifuged at $147{,}000g_{max}$ for 2 hr at 4° in a Beckman SW50.1 rotor. Polysomes pellet to the bottom of the tube whereas mRNP complexes band at the interface between the 10 and 35% (w/v) sucrose layers. The

[11] L. L. McGrew, E. Dworkin-Rastl, M. B. Dworkin, and J. D. Richter, *Genes Dev.* **3,** 803 (1989).

polysome-containing pellet is dissolved in the buffer used for the sucrose gradients, and both the polysome and mRNP fractions are dialyzed against 30 m*M* Tris-HCl (pH 7.4), 2 m*M* $MgCl_2$, 2 m*M* dithiothreitol, 15% (v/v) glycerol at 4°, and aliquots stored at −80°. The absorbance at 260 nm can be used to estimate the amount of recovered material.

Rapid Method for Preparing Polysome Salt Extracts

Because of the rapid clearance rate of a tabletop ultracentrifuge polysomes and salt extract polysome-associated proteins can be prepared quickly. The following is adapted for 5–7.5 g of liver tissue.

1. Postmitochondrial extract is centrifuged at 4° for 50 min at $165{,}000g_{max}$ in a Beckman TL100.3 rotor, TL100 ultracentrifuge.
2. The pellet obtained from this procedure is carefully resuspended in 1 ml of homogenization buffer, using a B pestle of a Dounce homogenizer. The resuspended material is loaded onto a 15–40% (w/v) sucrose gradient [30 m*M* Tris-HCl (pH 7.4), 2 m*M* $MgCl_2$, 2 m*M* dithiothreitol] and centrifuged for 3.5 hr at $225{,}000g_{max}$ in an SW41 rotor.
3. The recovered pellet is then suspended in 2 ml of 30 m*M* Tris-HCl (pH 7.4), 2 m*M* dithiothreitol, 12% (v/v) glycerol. The solution is adjusted to 0.4 *M* KCl and incubated as described above. The extract is centrifuged at 4° for 50 min at $165{,}000g_{max}$ in the TL100.3 rotor and the resulting supernatant is dialyzed against several hundred volumes of 30 m*M* Tris-HCl (pH 7.4), 2 m*M* dithiothreitol, 15% (v/v) glycerol and stored in aliquots at −80°. Extracts prepared in this manner typically have a protein concentration of 0.25–1 mg/ml.

Considerations When Assaying Endonuclease Activity *in Vitro*

The protocols described above were designed to obtain the most likely subcellular fractions in which to look for mRNA endonuclease activity. Once these are in hand the process becomes more empirical. Before embarking on the *in vitro* analysis and purification of an mRNA endonuclease like PMR-1 it is vital to identify one or more mRNAs that are not subject to degradation by the identified enzyme. This can take several forms. For example, we used ferritin and β-globin mRNAs, both of which are expressed in liver but are not subject to estrogen regulation, and neither of which are good substrates for PMR-1 *in vitro*. Alternatively, it might be possible to identify a particular region in an mRNA that has a single endonuclease cleavage site that can be mapped with high definition. A mutant RNA that is resistant to cleavage by the identified endonuclease can then be used as a control substrate.

The most straightforward way to evaluate an endonuclease activity like PMR-1 is to use a uniformly ^{32}P-labeled substrate RNA and assay both the disappearance of input transcript and the generation of specific cleavage products. The latter are best visualized by using a denaturing polyacrylamide gel. To determine kinetic parameters at this point it is preferable to assay the disappearance of substrate, because endonuclease cleavage products may be susceptible to further degradation. A more detailed treatment of this may be found in Schoenberg and Cunningham.[12] Because PMR-1 came from a cold-blooded vertebrate the *in vitro* activity assays were routinely performed at 23°. Wang and Kiledjian[13] have reported that performing reactions at a lower temperature allowed them to identify degradation intermediates for an endonuclease activity that cleaves within the α-globin mRNA 3′ untranslated region (UTR). Initially we used a capped, uniformly labeled full-length (2-kb) transcript, and the identification of unique degradation intermediates with this substrate allowed us to subsequently reduce the size of the RNA substrate to a 160-nucleotide (nt) RNA bearing multiple PMR-1 cleavage sites.

The next step after the initial assay is to examine the subcellular fractions obtained above for enzymatic activity. In the case of PMR-1 this involved assaying for selective degradation of an albumin mRNA substrate transcript versus ferritin mRNA, and the concurrent generation of a doublet cleavage product from albumin mRNA produced by cleaving at overlapping APyrUGA elements in the 5′ coding region of the message.[9] A variety of approaches to this are described in detail in an article by Ross,[14] and those we have used successfully are summarized here. Postmitochondrial extract, polysomes, mRNPs, S100, or polysome salt extracts are incubated for varying times in buffer containing an assortment of reagents that may serve as cofactors for a putative mRNA endonuclease. A typical buffer will contain 30 m*M* Tris-HCl (pH 7.4), 2 m*M* $MgCl_2$, 30 m*M* NaCl, and 10 m*M* ATP, and 2 m*M* dithiothreitol (DTT). Subsequent work showed that neither ATP nor Mg^{2+} was required for PMR-1 activity. It proved important to incorporate placental ribonuclease inhibitor (RNasin) into the buffers used in the initial experiments. Because this protein binds to and inhibits activity of members of the superfamily of RNase A-related proteins it both reduces nonspecific RNase activities in crude extracts and indicates whether the endonuclease under study belongs to that family of enzymes. Because PMR-1 is a member of the peroxidase gene family its activity was unaffected by this reagent. Another important parameter to test early on is the pH dependence of the endonuclease activity. PMR-1 activity is highly pH dependent, with maximal activity achieved at pH 7.5, and no activity at pH 7.0 or 8.0. Clearly, the temperature dependence of Tris-based buffer pH can have a major effect on whether assaying either PMR-1 or related enzymes with similar pH dependence is successful. In addition to these parameters we found that adding a large excess of total liver RNA to the reaction mixtures reduced nonspecific nuclease activity. For our experiments this involved addition of 10 μg of total liver RNA.

[12] D. R. Schoenberg and K. S. Cunningham, *Methods* **17,** 60 (1999).

[13] Z. Wang and M. Kiledjian, *EMBO J.* **19,** 295 (2000).

[14] J. Ross, *in* "RNA Processing: A Practical Approach" (S. J. Higgins and B. D. Hames, eds.), p. 107. Oxford University Press, New York, 1994.

Assays for Polysomal Ribonuclease 1 Activity

1. PMR-1 activity is assayed by using a transcript corresponding to the first 470 bp of albumin mRNA, or a 160-nt transcript derived by removing the 5′ and 3′ sequences flanking a stem–loop structure bearing the major PMR-1 cleavage sites in the 5′ portion of albumin mRNA.[15] The basic assay is similar to that described previously,[9] where the fractions to be analyzed are mixed on ice with 500 pg of uniformly ^{32}P-labeled substrate transcript and 10 μg of liver RNA in a 20-μl reaction mixture containing 30 m*M* Tris (pH 7.4) and 2 m*M* DTT. Reactions are usually carried out for 30 min at 23°.

2. Initially both the amount of protein added and the time of incubation are varied. It is best to work under conditions in which approximately 50% degradation of input transcript is achieved. Tips for using this approach to determine the kinetic parameters of an endonuclease are presented in Schoenberg and Cunningham.[12] Also, if specific degradation products are detected at this stage it may be desirable to optimize incubation conditions to enhance their appearance.

3. A number of biochemical parameters should be examined to optimize the *in vitro* assay. Ionic strength influences both the stability of the enzyme in solution and the structure of the RNA substrate. PMR-1 was found to be inhibited by salt concentrations in excess of 150 m*M* NaCl. It should be noted that Tris is a metal ion-binding buffer, and so inactivation of RNase activity by EDTA or EGTA is necessary to indicate involvement of a divalent metal ion in catalysis.

Purification of Polysomal Ribonuclease 1

Considerations before Purification

A considerable amount of qualitative information about PMR-1 was obtained in the process of its purification. Protein extracted from polysomes had an apparent molecular mass of 60,000 Da by gel filtration. This information indicated both the expected size of the final purified protein and that the material extracted from polysomes was not part of a larger complex. In the course of testing various column matrices before embarking on the large-scale purification of PMR-1 we discovered that the enzymatic activity that generated the characteristic albumin mRNA degradation product did not bind to either DEAE cellulose or Mono Q, thus indicating PMR-1 was a basic protein.[16] When the final product purified by the method described below was analyzed by two-dimensional gel electrophoresis it separated into a series of basic isoforms with isoelectric points of 9.6, 9.7, and 9.8.[9]

The following protocol is adapted for 30 adult male frogs generating approximately 75–100 g of liver. Frogs are injected in the dorsal lymph sac with

[15] E. Chernokalskaya, R. E. Dompenciel, and D. R. Schoenberg, *Nucleic Acids Res.* **25,** 735 (1997).
[16] R. L. Pastori and D. R. Schoenberg, *Arch. Biochem. Biophys.* **305,** 313 (1993).

1 mg of 17β-estradiol in 0.1 ml of propylene glycol–dimethyl sulfoxide (DMSO) (9 : 1, v/v) 24 hr before harvest. They are anesthetized with 0.1% (w/v) 3-aminobenzoic acid ethyl ester (tricane methanesulfonate) for 30 min and then livers are removed with standard sterile tools and techniques. All steps are carried out at 4°. Livers are perfused with sterile 1× SSC (0.15 *M* NaCl, 0.015 *M* sodium citrate, pH 7.0) before excision, then cut into small pieces, and homogenized with 2.5 volumes of homogenization buffer per gram of liver [40 m*M* Tris-HCl (pH 7.5), 10 m*M* $MgCl_2$, 7% (w/v) sucrose, 2 m*M* dithiothreitol, and protease inhibitors: 0.2 m*M* PMSF, leupeptin (0.5 μg/ml), pepstatin A (0.7 μg/ml), aprotinin (2 μg/ml)] in a Teflon–glass homogenizer.

The first steps in purifying PMR-1 are identical to the protocols outlined above for the preparation of post-mitochondrial extract and preparation of polysomes with a few modifications. After the 125,000*g* spin of the post-mitochondrial supernatant, it can be seen that the pellet is composed of two parts. The topmost thin flocculent layer (pellet 1) contains the majority of PMR-1 activity, and is composed primarily of RNPs and membrane-bound polysomes. The larger layer at the bottom of the tube (pellet 2) contains free polysomes and ribosomal subunits. Using five strokes of a Dounce homogenizer, pellet 1 is resuspended in a few milliliters of 40 m*M* Tris-HCl (pH 7.5), 2 m*M* dithiothreitol, and protease inhibitors. EDTA is added to a final concentration of 50 m*M*, KCl is added to a final concentration of 500 m*M*, and the mixture is gently mixed for 2 hr on ice.

The salt-extracted preparation is centrifuged at 106,000*g* for 60 min and protein is recovered from the supernatant (S106) by slowly adding ammonium sulfate to 80% (w/v) final concentration with gentle stirring, and the mixture is further stirred on ice for 2 hr. Protein is recovered by centrifugation at 10,000*g* for 10 min at 4° and the pellet is resuspended in 3 ml of 40 m*M* Tris-HCl (pH 7.5), 2 m*M* dithiothreitol and dialyzed against >1000 volumes of 40 m*M* Tris-HCl (pH 7.5), 2 m*M* dithiothreitol, 50 m*M* NaCl, and 0.2 m*M* PMSF.

PMR-1 is first fractionated from polysome salt extract by fast protein liquid chromatography (FPLC) on a connected series of three 5-ml Econo-Pac Q (QAE) cartridges (Bio-Rad, Hercules, CA), with the highly basic ribonuclease recovered in the flow-through fraction. The flow-though fraction is then dialyzed overnight against 50 m*M* sodium phosphate (pH 6.75), 2 m*M* DTT, and 0.2 m*M* PMSF, and the dialysate is then applied to a series of three 5-ml Econo-Pac S (SE) cartridges (Bio-Rad). The SE column is developed with a gradient of 0.01–0.5 *M* NaCl, with PMR-1 activity eluting between 200 and 350 m*M* NaCl. These fractions are pooled and dialyzed against 50 m*M* sodium phosphate (pH 6.75), 2 m*M* DTT, and 0.2 m*M* PMSF overnight. The dialysate is applied to a single 5-ml column of Econo-Pac OHT-II (hydroxylapatite; Bio-Rad) and eluted with a linear gradient of 0.01–0.5 *M* NaCl.

At all steps in the purification each of the column fractions is assayed for PMR-1 activity as described above, using a uniformly radiolabeled [α-^{32}P]albumin substrate transcript as described above. PMR-1 activity is scored both by the disappearance of the input RNA and the appearance of the characteristic doublet cleavage product that is generated by cleavage within overlapping APyrUGA elements in the 470- or 160-nt transcripts prepared from the 5′ end of albumin mRNA as noted above. Figure 1 presents the results obtained using this assay to identify PMR-1 containing fractions in the final purification step. Figure 1 (top) shows the elution profile for enzymatic activity, with the doublet cleavage product of the 470-nt albumin substrate transcript indicated by an open arrow on the autoradiogram. Figure 1 (middle) shows the OD_{206} profile of the column and the salt gradient used for eluting bound protein. PMR-1 activity is present only in the small leading peak of protein eluting in fractions 9 and 10. A silver-stained sodium dodecyl sulfate (SDS)–polyacrylamide gel of each of the column fractions is shown in Fig. 1 (bottom). In agreement with gel-filtration data obtained with crude polysome salt extract, the major product appears as a doublet of ~60 kDa, and there is a lesser amount of a 40-kDa peptide. The hydroxylapatite column was the only one found to adequately separate PMR-1 from an abundant 14-kDa peptide that eluted at a higher salt concentration. Protease mapping showed that the proteins constituting the 60-kDa doublet and 40-kDa moiety are related, and the variable appearance of the 40-kDa peptide suggests it is a breakdown product. The final fractions are pooled, concentrated with a Centricon (Amicon, Danvers, MA) centrifugal concentrator, dialyzed against 40 m*M* Tris-HCl (pH 7.5), 2 m*M* dithiothreitol, 20% (v/v) glycerol, 0.2 m*M* PMSF, aliquoted, and stored at −80° until needed. A typical purification from 30 frogs yields approximately 12.5 μg of protein or 5000 units of PMR-1, where 1 unit has been defined as the amount of PMR-1 that completely degrades 1 ng of albumin 5′ transcript RNA in 30 min at 25°, with an approximate 235-fold purification of PMR-1 from the starting polysome salt extract.

Reconstitution Assay: In-Gel Assay for Ribonuclease Activity

To evaluate the 62/64- and 40-kDa polypeptides for ribonuclease activity, an "in-gel" assay was performed. Briefly, a denaturing SDS–polyacrylamide gel with approximately 200 μg of unlabeled albumin transcript per milliliter of gel mix is prepared and 2 μg of purified nuclease is denatured by incubation for 3 min at 37° in 6.5 m*M* Tris-HCl (pH 6.8), 2 m*M* EDTA, 10% (v/v) glycerol, 0.12 *M* 2-mercaptoethanol (2-ME), and 1% (w/v) SDS and subsequently electrophoresed with the standard Laemmli buffer at 4°. The gel is washed twice for 15 min in 25% (v/v) 2-propanol, 40 m*M* Tris-HCl (pH 7.5), followed by three washes in 40 m*M* Tris-HCl, pH 7.5, for 10 min each. The gel is then incubated at 37° for 1.5 hr in 40 m*M* Tris-HCl (pH 7.5), 1 m*M* EDTA, washed once more in the same buffer for 10 min, and then stained with 0.2% (w/v) toluidine blue for 10 min. The higher temperature is necessary to obtain sufficient clearing of the degraded transcript from regions of the gel containing the nuclease. The gel is destained

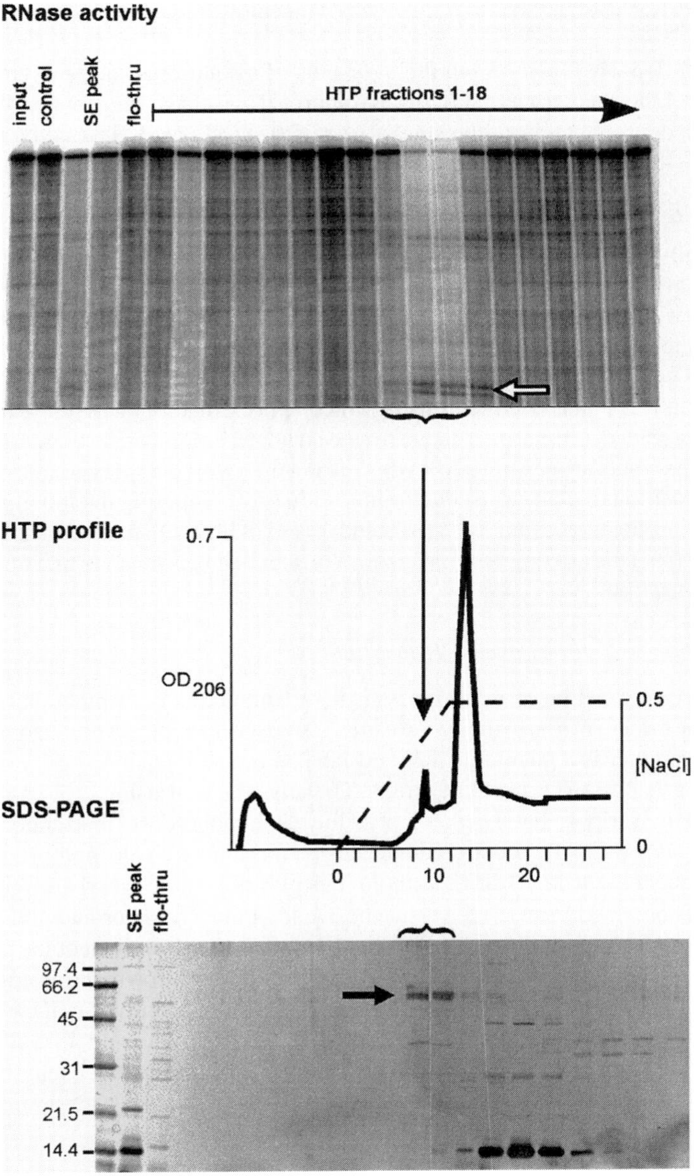

FIG. 1. Purification of PMR-1 by chromatography on hydroxylapatite. Fractions from an SE column containing PMR-1 activity were pooled, dialyzed, and applied to an Econo-Pac HTP hydroxylapatite column as described in text. One-milliliter fractions were collected and were assayed as described in text. *Top:* An activity assay, with the 194-nt doublet product from cleavage in the 5′ 470-nt albumin substrate transcript indicated by an open arrow. *Middle:* The protein elution profile at OD_{206} and the linear NaCl elution gradient. *Bottom:* A silver-stained SDS–polyacrylamide gel of the proteins present in each fraction. PMR-1 is identified by the filled arrow.

in 40 m*M* Tris-HCl, pH 7.5, until clear bands of ribonuclease activity are visible against a blue background. Using this approach we have shown that both the ~60-kDa doublet band and the single band of 40 kDa have RNase activity.

Mapping *in Vitro* Endonuclease Cleavage Sites

Initially we used primer extension to map the *in vitro* PMR-1 cleavage sites,[9] and the basic assay used is described below for mapping *in vivo* cleavage sites. The techniques described here are those we found most readily applied for mapping *in vitro* cleavage sites. Transcripts are prepared that are labeled at either the 5′ or 3′ end as described below. These are incubated with crude or purified PMR-1 for 30 min at 23° under conditions in which approximately one-half of the input RNA is degraded as described above for the standard PMR-1 assay conditions. The reaction products are separated on a denaturing 6% (w/v) polyacrylamide–urea gel and the position of cleavage is determined either by comparing the mobility of the cleavage products with known size standards or with a DNA sequencing ladder prepared from the starting plasmid, using a primer that begins at the first nucleotide of the transcript.

Preparation of 5′-End-Labeled Transcript

Transcripts can be made *in vitro,* using a commercially available kit such as the Megascript kit from Ambion (Austin, TX). Because the first nucleotide from the T7 promoter is a guanidine a 5′-end-labeled transcript can be easily prepared by adding [γ-^{32}P]GTP to the reaction. The advantage of using a 5′-end-labeled substrate transcript is that it makes it possible to unambiguously map endonuclease cleavage sites when the products are electrophoresed next to an appropriately prepared sequencing ladder from a homologous plasmid, and it clearly differentiates fragments obtained from the 5′ versus the 3′ end of the RNA. For mapping PMR-1 cleavage sites 1 μg of linearized plasmid containing a 160-nucleotide fragment from the 5′ end of the albumin gene is added to a transcription reaction containing 50 μCi of [γ-P^{32}]GTP (6000 Ci/mmol) and a 10 μM concentration of each of the respective unlabeled nucleotide triphosphates. Transcription with T7 polymerase is performed for 1 hr at 30°. Transcript prepared in this manner may be clean enough to use as is, but should be gel purified if it contains a high proportion of smaller fragments.

Preparation of 3′-End-Labeled Transcript

Formal proof that a particular RNase is an endonuclease requires the ability to identify products both 5′ and 3′ to the cleavage site. A variety of methods have been used successfully to label the 3′ end of the 160-nt albumin substrate transcript. To label using terminal deoxynucleotidyltransferase, 5 pmol of transcript is incubated with 10 U of terminal transferase in reaction buffer containing 100 mM sodium cacodylate (pH 7.0), 1 mM $CoCl_2$, 0.1 mM DTT, 20 μM [α-^{32}P]dCTP, bovine serum albumin (BSA, 50 μg/ml) for 30 min at 37°. The reaction is stopped by adding 2 μl of 0.5 M EDTA and phenol–chloroform–isoamyl alcohol extraction. Radiolabeled RNA is recovered by ethanol precipitation, using a one-half volume of 3 M ammonium acetate, pH 5.2, and 2 volumes of ethanol. The advantage of this method is that it provides highly labeled transcript because approximately 10 residues are polymerized onto the end of each transcript. However, this is a disadvantage if a cleavage site is to be mapped with nucleotide resolution. In the latter case it is preferable to 3′-end label the transcript, using T4 RNA ligase plus 5′-[^{32}P]pCp, or poly(A) polymerase plus [α-^{32}P]cordycepin triphosphate.[17]

Test for Cleavage of Double-Stranded RNA

Two approaches are employed to test whether PMR-1 (or another endonuclease) is specific for single-stranded RNA, or whether it can cleave double-stranded RNA. In the first approach 1 ng of uniformly labeled albumin substrate transcript is mixed with 5 ng of unlabeled antisense transcript prepared from the same plasmid. The mixture is heated for 5 min at 80°, and then hybridized overnight at 52° in 80% (v/v) formamide, 40 mM 1,4-piperazinediethanesulfonic acid (PIPES, pH 6.4), 400 mM NaCl, 1 mM EDTA. Under these conditions all of the radiolabeled RNA becomes double stranded. After ethanol precipitation the duplex RNA is either used directly in an PMR-1 activity assay or gel purified if the presence of unlabeled antisense RNA in the reaction is to be avoided. The second approach requires more detailed information about the sequence of the PMR-1 cleavage site and the RNA secondary structure of the site. As noted above, the characteristic doublet cleavage product of PMR-1 results from cleavage of overlapping APyrUGA elements in a single-stranded loop of a stem–loop structure.[15] When one-half of the loop is mutated to extend the structure to a double-stranded hairpin, the duplexed APyrUGA element becomes resistant to cleavage. In contrast, mutations that do not alter the structure of the loop do not interfere with cleavage, thus confirming the requirement for single-stranded structure for cleavage by PMR-1.

Determining Phosphorylation State of Polysomal Ribonuclease 1 Cleavage Products with Snake Venom Phosphodiesterase

Members of the RNase A superfamily of enzymes generate cleavage products bearing 3′-phosphate termini that result from hydrolysis of a 2′,3′-cyclic phosphate intermediate. Such products pose a challenge for mRNA decay, because a 3′-phosphate blocks degradation by known 3′ → 5′-exonucleases. Although

[17] J. Lingner and W. Keller, *Nucleic Acids Res.* **21,** 2917 (1993).

T4 kinase can be used to test for the presence of a 5′-hydroxyl on a given cleavage product, this type of analysis can give a false positive because phosphatase activity in the reaction mixture generates fragments with 5′-hydroxyl groups that can then become labeled. To overcome this issue we instead relied on the ability of a 3′-phosphate to block the activity of snake venom phosphodiesterase.

A 5′-end-labeled albumin substrate transcript is diluted in RNase T1 buffer [2.5 m*M* Tris-HCl (pH 7.2), 2.5 m*M* $MgCl_2$, 25 m*M* KCl], S1 nuclease buffer [30 m*M* sodium acetate (pH 4.6), 50 m*M* NaCl, 10 m*M* zinc acetate, and 5% (v/v) glycerol], or PMR-1 buffer [40 m*M* Tris-HCl (pH 7.5), 2 m*M* DTT, and 5 m*M* $MgCl_2$], heated at 50° for 5 min, and cooled slowly to 25°. RNase T1 and S1 nuclease are controls that generate products with 3′-phosphates or 3′-hydroxyl groups, respectively. The individual tubes receive either RNase T1 (0.05 U/μl), S1 nuclease (0.05 U/μl), or PMR-1 (10 ng/μl). Each reaction also has yeast tRNA (0.25 μg/μl). Twenty-microliter reactions are incubated at 23° for 10 min (30 min for PMR-1), and stopped by addition of 15 μl of 7.5 *M* ammonium acetate and 115 μl of H_2O. The cleaved RNA is recovered by extraction with pheno–chloroform–isoamyl alcohol (25 : 24 : 1), and precipitated with 300 μl of ethanol. The pellets are then dissolved in 15 μl of 100 m*M* Tris-HCl (pH 8.7), 100 m*M* NaCl, and 14 m*M* $MgCl_2$, and 3-μl portions are mixed with 1-μl portions of snake venom phosphodiesterase (Roche Diagnostics/Boehringer Mannheim, Indianapolis, IN) containing 10^{-5} to 10^{-1} units of activity. Digestion is performed for 10 min at 23°, and the reaction is stopped by addition of 4 μl of formamide loading buffer and heating for 5 min at 70°. The products are analyzed by electrophoresis on a 6% (w/v) polyacrylamide–urea gels. For nucleases like S1 and PMR-1, which leave a 3′-hydroxyl group, it is observed that discrete products in control lanes smear to smaller sizes on digestion with increasing amounts of snake venom phosphodiesterase. In contrast, the products generated by RNase T1 are resistant to exonuclease digestion.

Identifying *in Vivo* Endonuclease Cleavage Products

Whereas it is relatively straightforward to identify *in vitro* endonuclease cleavage sites, the identification of such sites *in vivo* is much more difficult for most mRNAs. The reason for this is that these products are substrates for subsequent degradation, for example, by 3′ → 5′-exonucleases. The first comprehensive study of *in vivo* degradation products was performed by Binder *et al.* for both the chicken apo-VLDL II[18] and transferrin mRNA.[19] We have modified their primer extension and S1 mapping approaches for mapping *in vivo* cleavage of albumin mRNA after estrogen stimulation, and present these below.

[18] R. Binder, S. P. Hwang, R. Ratnasabapathy, and D. L. Williams, *J. Biol. Chem.* **264,** 16910 (1989).

[19] R. Binder, J. A. Horowitz, J. P. Basilion, D. M. Koeller, R. D. Klausner, and J. Harford, *EMBO J.* **13,** 1969 (1994).

Primer Extension

Primer extension is performed in a manner similar to that described previously.[19] Ten micrograms of total liver RNA is ethanol precipitated with 1–2 × 10^5 dpm of 5′-end-labeled primer. The primer CACTCAGGAGTTTTGTCATTAA, which is complementary to nucleotides 280–301 of the 74-kDa albumin mRNA, is used to map the cleavage sites. The precipitated RNA and primer are dissolved in 10 μl of annealing buffer [50 m*M* Tris-HCl (pH 8.7), 0.54 *M* KCl, 1 m*M* EDTA], heated at 65° for 10 min, and slowly cooled to 25°. Each tube then receives a mixture of dATP, dCTP, dGTP, and dTTP (0.9 m*M* each), 50 m*M* Tris-HCl (pH 8.3), 13 m*M* $MgCl_2$, 7 m*M* dithiothreitol, and 200 units of moloney murine leukemia virus reverse transcriptase to a total volume of 40 μl. The reaction mixture is incubated for 1.5 hr at 42° and stopped by the addition of 260 μl of stop solution [0.3 *M* sodium acetate, 10 m*M* Tris-HCl (pH 8.0), 1 m*M* EDTA], followed by the addition of 600 μl of cold ethanol. The precipitated product is recovered by centrifugation after 30 min at −70°. The final pellets are dissolved in 6 μl of formamide loading buffer and electrophoresed on a 6% (w/v) polyacrylamide–urea gel.

The positions of nuclease cleavage sites are determined relative to a sequencing ladder prepared from the cloned cDNA, using the same primer, or a primer that begins at the 5′ end of the transcribed RNA. It is also important to include a control sample prepared from an *in vitro* transcript to be able to differentiate sites of polymerase pausing from actual nuclease cleavage sites. The biggest drawback of this technique is sensitivity. To be able to detect even the intermediates of a high-abundance transcript may require scaling up the reaction volume and using exposure times approaching 1 month for X-ray film, and at least 1 week for a storage phosphor screen.

S1 Nuclease Protection Assay

S1 nuclease protection is the easiest way to map cleavage sites *in vivo* and a positive result here is definitive evidence of endonucleolytic cleavage. For PMR-1 this was done with a 500-bp DNA probe complementary to the 5′ end of albumin mRNA made by asymmetric polymerase chain reaction (PCR), using 200 ng of 5′-end-labeled primer 5′-TTTCATTCACCAGCTTAGATAA corresponding to region 255–234. The probe included a portion of the plasmid multiple cloning site in order to differentiate the larger undigested probe from the final protected product after electrophoresis on a denaturing 6% (w/v) polyacrylamide–urea gel.

Probe (5 × 10^5 dpm) is ethanol precipitated with 10 μg of total RNA isolated from either control male *Xenopus* or animals that have been injected with 1 mg of estradiol 12 hr before sacrifice. The recovered pellet is rinsed with 70% (v/v) ethanol and resuspended in 30 μl of hybridization buffer containing 40 m*M* PIPES (pH 6.4), 1 m*M* EDTA, 0.4 *M* NaCl, 80% (v/v) formamide. The mixture is denatured at 85° for 5 min and immediately transferred to 37° and incubated overnight.

The next morning 300 μl of ice-cold S1 solution consisting of 0.28 *M* NaCl, 50 m*M* sodium acetate (pH 4.6), 4.5 m*M* $ZnSO_4$, 6 μg of denatured salmon sperm DNA, and 75 units of S1 nuclease is added. This is then incubated at 25° for 30 min and chilled on ice followed by the addition of 80 μl of stop solution consisting of 4 *M* ammonium acetate, 0.1 *M* EDTA, yeast tRNA (50 μg/μl). The reaction mixture is extracted with phenol–chloroform–isoamyl alcohol (25 : 24 : 1), and the nucleic acid is precipitated from the aqueous phase by addition of one-half volume of 7.5 *M* ammonium acetate and 2 volumes of ethanol. The pellet recovered after a 30-min centrifugation at 4° in a microcentrifuge, is dissolved in 6 μl of loading buffer containing 90% (v/v) formamide, 4 m*M* NaOH, 2 m*M* EDTA, 0.1% (w/v) each bromphenol blue and xylene cyanol, heated in a boiling water bath for 5 min, and immediately loaded onto a polyacrylamide–urea gel. As noted for primer extension, the position of the cleavage sites can be determined relative to a DNA sequencing ladder for T and C residues, using the primer employed in the synthesis of the S1 probe. Results showing the estrogen-induced appearance of PMR-1 cleavage sites in albumin mRNA using this assay can be seen in Schoenberg and Cunningham.[12] As with the primer extension protocol, it can take between 2 and 4 weeks to see a signal because the products being looked for are so rare.

Isolating Polysomal Ribonuclease 1-Containing Ribonucleoprotein Complexes

We have determined that PMR-1 associates with substrate mRNA *in vivo* in an RNP complex.[20] Two methods were used to recover these complexes; selection of poly(A)-containing complexes with oligo(dT)–cellulose, and immunoprecipitation of PMR-1-containing complexes with a polyclonal antibody to PMR-1.

Recovery of Messenger Ribonucleoprotein Complexes on Oligo(dT)-Cellulose

The protocol described below was created for use with approximately 5–10 g of tissue. It can be adjusted for the amount of sample.

1. Diced tissue is placed into 2.5 volumes (v/w) of ice-cold homogenization buffer [40 m*M* Tris-HCl (pH 7.5), 10 m*M* $MgCl_2$, 7% (w/v) sucrose, 2 m*M* dithiothreitol, and protease inhibitors: 0.2 m*M* PMSF, leupeptin (0.5 μg/ml), pepstatin A (0.7 μg/ml), aprotinin (2 μg/ml)], and homogenized with a Teflon–glass homogenizer.
2. The homogenate is filtered through a polyamide nylon mesh (Nitex 27621, 0.5 mm; Tetko) or a double layer of cheescloth, and centrifuged at 4°, $1000g_{max}$ for 15 min in a swinging bucket rotor. The supernatant is then centrifuged at 4° for 15 min at $15{,}000g_{max}$ in a swinging bucket rotor.

[20] K. S. Cunningham, M. N. Hanson, and D. R. Schoenberg, *Nucl. Acids. Res.* **29,** 1156 (2001).

3. The recovered postnuclear supernatant is adjusted to 0.5 *M* LiCl and 0.5% (w/v) SDS. Oligo(dT) cellulose (200 mg) equilibrated in 25 m*M* Tris-HCl (pH 7.5), 0.5 *M* LiCl, and 0.5% (w/v) SDS is added to the postnuclear extract and the suspension is rocked for 2 hr at 4°. The mixture is then centrifuged at 1000*g* for 2 min and the supernatant is removed, leaving the oligo(dT) matrix bearing bound poly(A)-containing mRNPs. The oligo(dT) cellulose is washed five times in 10 volumes of 25 m*M* Tris-HCl (pH 7.5), 0.5 *M* LiCl, and 0.5% (w/v) SDS and centrifuged at 1000*g* between each wash. Five bed volumes of diethylpyrocarbonate (DEPC)-treated sterile H_2O is added to resuspend oligo(dT) matrix, and to elute the bound RNP complexes, and the supernatant is recovered after centrifugation is removed to a new tube. The RNP complexes are recovered by addition of tRNA (10 μg/ml) and sodium acetate, pH 5.2, to 0.3 *M*, followed by an equal volume of cold 2-propanol. After precipitating overnight at −20° the complexes are recovered by centrifugation at 100,000*g* for 30 min at 4° in a tabletop ultracentrifuge.

4. The recovered RNA–protein complexes are dissolved in 20 μl of 10 m*M* Tris-HCl, pH 8.0, and the RNA present in the sample is removed by digestion with 5 μg of RNase A for 10 min at 37°. Protein recovered in this preparation is assayed by Western blot for recovery of PMR-1.

Immunoprecipitation of Albumin Messenger Ribonucleoproteins

1. A postnuclear supernatant (S15) is generated as described above from a nonestrogenized frog. One hundred micrograms of a polyclonal antibody to PMR (not crude serum) is added to the S15 extract and the mixture is rocked at 4° for 2 hr to allow sufficient time for antibody binding to mRNP complexes.

2. Fifty milligrams of protein A–Sepharose (Pharmacia, Piscataway, NJ) equilibrated in homogenization buffer containing protease inhibitors is added, followed by gentle rocking for another 60 min at 4°. The mixture is centrifuged at 1000*g* for 2 min and the S15 supernatant is removed. The beads are washed for 5 min in 10 volumes of ice-cold homogenization buffer and recovered by centrifugation at 1000*g* for 2 min at 4° for a total of five washes. The washed beads are then washed once more in buffer lacking protease inhibitors before the cDNA synthesis step (see below).

3. The beads are suspended in an equal volume of sterile H_2O and oligo $(dT)_{12-18}$ is added to 0.5–1 μM. The mixture is heated at 65° for 5 min and then cooled quickly on ice. The beads are removed by centrifugation at 1000*g* for 2 min at 4° and the RNA-containing supernatant is used for cDNA synthesis. cDNA synthesis is performed with the Qiagen (Valencia, CA) Sensiscript kit for low levels of RNA (<50 ng) according to the manufacturer recommended protocol. mRNA recovered by immunoprecipitation with antibody to PMR-1 is detected by PCR, using gene-specific primers located near the 3′ end of albumin mRNA to minimize loss of signal due to reverse transcription artifacts or degradation of

the mRNA during handling. The PCR products are visualized either by using a 5′-end ^{32}P–labeled primer for the reaction or by adding an [α-^{32}P]dNTP to the reaction. The final products are visualized by autoradiography or PhosphorImager (Molecular Dynamics, Sunnyvale, CA) analysis after electrophoresis on a 6% (w/v) polyacrylamide–urea gel.

Concluding Remarks

PMR-1 is the first mRNA endoribonuclease to be purified and cloned. Because we have yet to inactivate mRNA decay by inactivating PMR-1 it cannot formally be termed a "messenger RNase" at this time. However, there is convincing evidence of the appearance of PMR-1 cleavage sites in albumin mRNA coincident with the estrogen-induced initiation of albumin mRNA decay.[12,21] Thus, the protocols developed for this mRNA endonuclease should be generally applicable to the characterization of other mRNA endonucleases as they are identified.

Acknowledgments

Our work on PMR-1 is supported by Grant GM38277 from the National Institutes of Health. K.S.C. was supported by a Presidential Fellowship from Ohio State University. K.S.C. and M.N.H. contributed equally to this work.

[21] M. N. Hanson and D. R. Schoenberg, *J. Biol. Chem.* **276,** 12331 (2001).

[5] Liver Perchloric Acid-Soluble Ribonuclease

By TATSUYA SAWASAKI, TATSUZO OKA, and YAETA ENDO

Introduction

Oka *et al.*[1] were the first to report the occurrence of perchloric acid-soluble protein (L-PSP), a 136-amino acid protein from rat liver that inhibits protein synthesis. It was later reported that 14-kDa translational inhibitor proteins, remarkably similar to L-PSP, are also present in human monocytes and mouse liver.[2–4] A homology search revealed that these proteins belong to a new group of small proteins,

[1] T. Oka, H. Tsuji, C. Noda, K. Sakai, Y.-M. Hong, I. Suzuki, S. Munoz, and Y. Natori, *J. Biol. Chem.* **270,** 30060 (1995).

0076-6879/01 $35.00

named the YER057c/YJGF family,[2] which is of unknown physiological function. The protein sequences of these family members are highly conserved in prokaryotes (including cyanobacteria), fungi, and eukaryotes, suggesting that the proteins may be involved in a basic cellular process. Indeed, the mRNA of the translational inhibitor p14.5, the human homolog of L-PSP, becomes significantly upregulated with the induction of differentiation to macrophages.[3] In addition, the synthesis of K-PSP in the rat kidney increases from fetal day 17 to postnatal week 4, and then enters a steady state level.[5] In contrast, the expression of K-PSP in renal tumor cells is downregulated.[5]

Schmiedeknecht *et al.* have identified the functional promoter of the human p14.5 translational inhibitor gene.[6] They reported a head-to-head orientation of the p14.5 gene with the gene for the protein subunit hPOP1 of RNase P and RNase MRP ribonucleoproteins. The promoter region between p14.5 and hPOP1 acts as a bidirectional promoter.[6] As bidirectional transcription units commonly encode proteins that are different in structure but have similar biological function,[7] the authors suggested that the p14.5–hPOP1 cluster may encode functionally related proteins as well.

Although Oka *et al.* suggested that L-PSP has a mode of inhibition of translation similar to that of the heme-regulated eukaryotic initiation factor 2α kinase (HRI),[1] it is now clear that the protein is a ribonuclease and that this activity is responsible for the inhibition of translation.[8]

Methods

Purification of Perchloric Acid-Soluble Protein from Rat Liver

Livers from male Wistar rats are homogenized in 2 volumes of cold 0.25 *M* sucrose in buffer [50 m*M* Tris-HCl (pH 7.5), 25 m*M* KCl, and 10 m*M* $MgCl_2$] with a Potter–Elvehjem type homogenizer.[1] The homogenate is centrifuged at 10,000*g* for 30 min at 4°. The supernatant is made 5% (v/v) with respect to perchloric acid (PCA) by the addition of 60% (w/v) PCA and centrifuged at 10,000*g* for 15 min at 4°. The supernatant is then made 25% (v/v) with respect to trichloroacetic acid, and the precipitate is collected by brief centrifugation. The precipitate is washed three

[2] G. Schmiedeknecht, C. Kerkhoff, E. Orso, J. Stohr, C. Aslanidis, G. M. Nagy, R. Knuechel, and G. Schmitz, *Eur. J. Biochem.* **242,** 339 (1996).

[3] S. J. Samuel, S.-P. Tzung, and S. A. Cohen, *Hepatology* **25,** 1213 (1997).

[4] F. Ceciliani, L. Faotto, A. Negri, I. Colombo, B. Berra, A. Bartorelli, and S. Ronchi, *FEBS Lett.* **393,** 147 (1996).

[5] A. Asagi, T. Oka, K. Arao, I. Suzuki, M. K. Thakur, K. Izumi, and Y. Natori, *Nephron* **79,** 80 (1998).

[6] G. Schmiedeknecht, C. Buchler, and G. Schmitz, *Biochem. Biophys. Res. Commun.* **241,** 59 (1997).

[7] S. Chen, P.-L. Nagy, and H. Zalkin, *Nucleic Acids Res.* **25,** 1809 (1997).

[8] R. Morishita, A. Kawagoshi, T. Sawasaki, K. Madin, T. Ogasawara, T. Oka, and Y. Endo, *J. Biol. Chem.* **274,** 20688 (1999).

times with 50 ml of acetone to remove trichloroacetic acid, and then the washed precipitate is collected by brief centrifugation and dried under vacuum. The dried material is resuspended in 0.9% (w/v) acetic acid, and dialyzed extensively against 0.1 *M* sodium phosphate buffer (pH 7.5). After clarification by a 10-min centrifugation at 10,000*g* at 4°, the proteins in the dialysate are fractionated with ammonium sulfate. The precipitate formed between 0 and 20% saturation is collected by centrifugation. The precipitate is suspended in 0.1 *M* sodium phosphate buffer (pH 7.5) and dialyzed against the same buffer. The dialysate is applied to a column (2.5 × 30 cm) of CM-Sephadex C-25. L-PSP is eluted in the flow-through fraction and appears homogeneous on sodium dodecyl sulfate (SDS)–polyacrylamide gels. Almost all the L-PSP is recovered in the postmitochondrial supernatant fraction and about 30 mg of the protein is obtained from 100 g of rat liver.

Preparation of Recombinant Rat Liver Perchloric Acid-Soluble Protein

To prepare recombinant L-PSP,[9] complementary DNA encoding L-PSP is inserted into an inducible bacterial expression vector, pGEX-4T-1 (Amersham Pharmacia Biotech, Piscataway, NJ). The resulting construct, pGEX-PSP, encodes a fusion protein of glutathione *S*-transferase (GST) and L-PSP. The plasmid is propagated in *Escherichia coli* BL21 (DE3). Transformants are grown at 30° in L-broth containing ampicillin (50 μg/ml). The expression of the fusion protein is initiated by the addition of 1 m*M* isopropylthio-β-D-galactopyranoside (IPTG) and the protein is produced during continuous aerobic incubation at 37°. The product is purified by means of affinity chromatography, using glutathione–Sepharose 4B (Amersham Pharmacia Biotech). After incubation of the purified recombinant protein with thrombin, the L-PSP is separated from the GST moiety by passing it through the glutathione–Sepharose 4B column a second time. The yield of L-PSP is about 0.3 mg per liter of culture. The amino-terminal amino acid of the recombinant L-PSP was confirmed to be the expected serine, like the N-terminal serine of the authentic L-PSP, which is acetylated. The translation inhibitory activity of the protein is usually about 10 times lower compared with that of authentic L-PSP isolated from rat liver.

Determination of Amino Acid Sequence of Rat Liver Perchloric Acid-Soluble Protein

The N- and C-terminal peptides of L-PSP are isolated after the digestion of the protein by tolylsulfonyl phenylalanyl chloromethyl ketone (TPCK)-treated trypsin, using an anhydrotrypsin–agarose column (TaKaRa Shuzo, Kyoto, Japan), which specifically adsorbs peptides that have an arginine or lysine at their C terminus.

[9] T. Oka, Y. Nishimoto, T. Sasagawa, H. Kanouchi, Y. Kawasaki, and Y. Natori, *Cell. Mol. Life Sci.* **55,** 131 (1999).

The N terminus of L-PSP is insensitive to both automated Edman degradation and aminopeptidase M. Analysis by electrospray ionization (ESI) mass spectrometry of the N-terminal peptides isolated after extensive digestion of anhydrotrypsin–agarose column-bound peptides, together with deblocking by treatment with the acyl amino acid-releasing enzyme (Boehringer Mannheim, Indianapolis, IN), shows that the N-terminal amino acid sequence of L-PSP is acetyl-Ser-Ser-Ile-Ile-Arg. The C-terminal peptide of the PSP is recovered in the unbound fraction of the anhydrotrypsin–agarose column of the TPCK-treated trypsin digest. The amino acid sequence is determined by automated Edman degradation and also by ESI mass spectrometry and identified as the C terminus of L-PSP; its sequence is Ile-Glu-Ile-Glu-Ala-Ile-Ala-Val-Gln-Gly-Pro-Phe-Thr-Thr-Ala-Gly-Leu. The complete amino acid sequence of L-PSP is determined by cDNA cloning and sequencing. A cDNA encoding L-PSP was obtained by immunoscreening of a cDNA expression library constructed in λgt11. One clone, named pPSP-1 and obtained by screening of the library with the anti-L-PSP antisera, contains an insert of 0.9 kb. The sequence of cDNA pPSP-1 has an open reading frame of 411 base pairs encoding a 136-amino acid protein; the deduced amino acid sequence fully agrees with the sequences of 9 peptides (including the above-described N- and C-terminal peptides; Fig. 1). Thus L-PSP is a 136-amino acid protein with a molecular mass of 14,149 Da; interestingly, it lacks histidine.

Protein Synthesis Inhibition Assay

The standard cell-free protein synthesis system derived from rabbit reticulocyte lysate[10] contains in a final volume of 30 μl : 15 μl of micrococcal nuclease-treated lysate (containing 25 μM hemin), 25 mM HEPES (pH 7.6), 2 mM dithiothreitol (DTT), 1 mM ATP, 20 μM GTP, 8 mM creatine phosphate, 1.2 μg of creatinine phosphokinase, a 25 μM concentration of each of the 20 amino acids, 0.1 μCi of L-[U-^{14}C]leucine (13.3 Ci/mol), 90 mM potassium acetate, 1 mM magnesium acetate, 0.6 mM spermidine, and 2 μg of mRNA. Endogenous globin synthesis is done under the same conditions as described above but with lysate that is not treated with micrococcal nuclease.

Incubation at 30° of micrococcal nuclease-treated rabbit reticulocyte lysate, containing exogenous capped mRNA encoding dihydrofolate reductase, and all other components necessary for protein synthesis, leads to the incorporation of [^{14}C]leucine into newly synthesized proteins. Addition of L-PSP to the system results in inhibition of protein synthesis in a concentration-dependent manner, with an estimated median inhibitory concentration (IC_{50}) of 51 nM (Fig. 2A). An L-PSP concentration of 1 μM almost completely abolishes the capacity of the cell-free system to support protein synthesis. However, when similar experiments are

[10] R. J. Jackson and T. Hunt, *Methods Enzymol.* **96,** 50 (1983).

```
                                                                         60
      CGGGATCCTCCAGAGGGAAGGATTAGC ATG TCG TCA ATA ATC AGA AAA GTG ATC AGC ACT
                                  Met Ser Ser Ile Ile Arg Lys Val Ile Ser Thr
                                               I                      II  10

                                                                        120
TCA AAA GCC CCG GCG GCC ATT GGT GCC TAC AGC CAA GCT GTG CTA GTG GAC AGG ACC ATT
Ser Lys Ala Pro Ala Ala Ile Gly Ala Tyr Ser Gln Ala Val Leu Val Asp Arg Thr Ile
                                                                         30

                                                                        180
TAC GTT TCT GGA CAG ATA GGC ATG GAT CCT TCC AGT GGA CAA CTT GTG CCA GGA GGG GTA
Tyr Val Ser Gly Gln Ile Gly Met Asp Pro Ser Ser Gly Gln Leu Val Pro Gly Gly Val
                                    III                                  50

                                                                        240
GCA GAA GAA GCT AAA CAA GCT CTT AAA AAC CTG GGT GAG ATT CTG AAG GCT GCA GGG TGT
Ala Glu Glu Ala Lys Gln Ala Leu Lys Asn Leu Gly Glu Ile Leu Lys Ala Ala Gly Cys
                        IV                  V                            70

                                                                        300
GAC TTC ACT AAT GTG GTG AAG ACA ACT GTT TTA CTG GCT GAC ATA AAT GAC TTT GGC ACT
Asp Phe Thr Asn Val Val Lys Thr Thr Val Leu Leu Ala Asp Ile Asn Asp Phe Gly Thr
  VI                                                          VII        90

                                                                        360
GTC AAT GAG ATC TAC AAA ACA TAT TTC CAG GGT AAC CTT CCT GCC AGG GCT GCT TAC CAA
Val Asn Glu Ile Tyr Lys Thr Tyr Phe Gln Gly Asn Leu Pro Ala Arg Ala Ala Tyr Gln
                                                        VIII            110

                                                                        420
GTC GCC GCT TTA CCC AAA GGA AGT CGA ATT GAA ATC GAA GCC ATC GCT GTC CAG GGG CCT
Val Ala Ala Leu Pro Lys Gly Ser Arg Ile Glu Ile Glu Ala Ile Ala Val Gln Gly Pro
                                                   IX                   130

                                                                        492
TTC ACC ACA GCA GGA CTG  TAAGGAGCCATGCTGATGCTGACTCTGGAGTCTTTACAATGTCTCTCACACTTT
Phe Thr Thr Ala Gly Leu  ***
                    136

                                                                        572
AAATTTTACAAATGATTCTGGGAATGCACACATGAGCCGTGTGAAGTGATCGTGGAAATGATCATACATGTGGAGATTTG
                                                                        652
ACATTAATTGGGGATTATTCAGTACAGTGGCTAGTGCTGTAATTTCACTTATGTGGCTGGGTGTGAGAGAAAGGCACATA
                                                                        732
GCTGCTAACATGCAGAGCAGAATCAAAGACGAAACAGTAAAGAGCACACCTAATTAAATCATATCAATTTGCTTGTGTGG
                                                                        812
AATACTTAGTTCTCATGTTAGATATCAGAAACTAAATTAAAATATTCAAACTTTAATAGTACCAGGAGAAAAAAGTGGAC
                                                      869
CAAAATCTTATAAGATAATACTGCCTTAATGGAATAAAAAACAGATGTGTAACTTTC
```

FIG. 1. Nucleotide sequence and deduced amino acid sequence of L-PSP. Amino acids representing the sequenced peptides are underlined. The cDNA sequence is deposited in EMBL/DDBJ/GenBank under accession number D49363.

done with endogenous globin mRNA, inhibition by L-PSP appears to be about 10 times less efficient (IC_{50}, 0.68 μM) (Fig. 2B). The observed difference in the IC_{50} can be ascribed to the characteristics of the two translation systems: in the latter system, a large population of ribosomes is already engaged with globin mRNA (as polysomes) at the start of incubation, while practically no mRNA is associated with

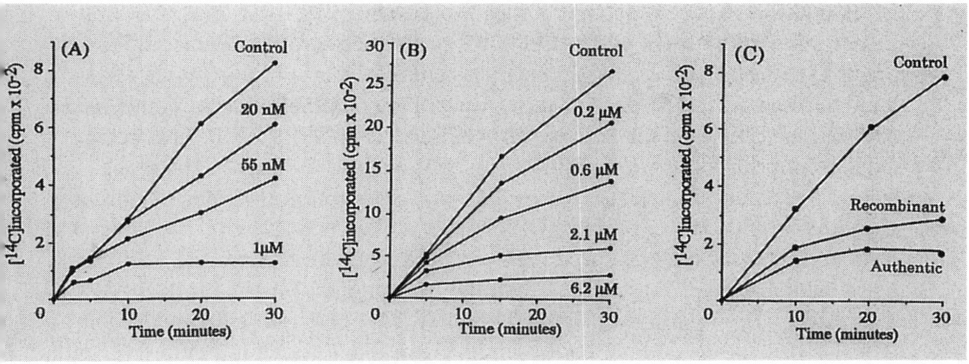

FIG. 2. Inhibition of cell-free protein synthesis by L-PSP., Exogenous mRNA-programmed lysate (A) or globin-synthesizing lysate (B) is incubated without or with the indicated concentration of L-PSP. In (C), 30 μg of mRNA in 30 μl is incubated first with 20 n*M* L-PSP or recombinant L-PSP for 3 hr and aliquots containing the same amounts of mRNA are added to the mRNA-programmed lysate. Under these conditions, the final concentration of L-PSP is about 1.3 n*M*. Reactions are done at 30°, and 5-μl aliquots were removed at the specified times, spotted on paper filters, and assayed for [^{14}C]leucine incorporation into hot trichloroacetic acid-insoluble material. [From R. Morishita, A. Kawagoshi, T. Sawasaki, K. Madin, T. Ogasawara, T. Oka, and Y. Endo, *J. Biol. Chem.* **274,** 20688 (1999), with permission.]

ribosomes in the former system. Thus the results suggest that the lysate containing a higher proportion of polysomes is less susceptible to L-PSP.

An important observation here is that the kinetics of the inhibition is monophasic in both systems rather than biphasic as reported earlier.[1] Inhibition of translation initiation by proteins such as HRI shows biphasic kinetics.[11] However, this typical biphasic shape, in which protein synthesis proceeds at the initial rate for several minutes before an abrupt decline occurs, can only be expected when the globin-synthesizing system is used because only this system has an initial rate (runoff of polysomes). Obviously, inhibitors of initiation show monophasic kinetics of translation inhibition only in the exogenous mRNA-programmed system, which has no initial rate because there is no initial protein synthesis.

In experiments that use higher amounts of mRNA than the standard reaction, the inhibitory effect on protein synthesis tends to be less (data not shown). To gain more insight into the mode of action of this small protein, the effect of preincubation of mRNA with L-PSP can be examined. For these experiments, mRNA is first incubated with a low amount of L-PSP for a prolonged time, and then the activity of the mRNA of a small portion of the preincubation reaction mixture is tested in the standard translation system. This strategy minimizes the effect of L-PSP on the translation system because the concentration of L-PSP in

[11] J. J. Chen, J. S. Crosby, and I. M. London, *Biochimie (Paris)* **76,** 761 (1994).

the translation reaction mixture is low. As shown in Fig. 2C, preincubation of mRNA with L-PSP results in a significant decrease in protein synthesis. Recombinant L-PSP affects the template quality of mRNA in a similar manner, showing that the observed inactivation of mRNA is not due to contaminants in the L-PSP preparation. However, the activity on mRNA of authentic L-PSP differs significantly from that of recombinant PSP, as had been found earlier.[9] Although we cannot be certain, the difference is likely to be due to a loss of activity because of the expression of recombinant PSP in a prokaryotic system. A similar but even larger discrepancy has been reported for the human homolog of PSP, p14.5: the authentic form is more active than the recombinant form by three orders of magnitude.[2] The inhibition of protein synthesis by both authentic and recombinant L-PSP is not prevented by the ribonuclease inhibitor from human placenta, which tightly binds and inhibits the activity of the ribonuclease A family (data not shown).

Assay for Ribonucleolytic Activity

The results indicate that the main target of the protein is mRNA rather than the translation system. There are at least two alternative mechanisms for mRNA inactivation: L-PSP might specifically modify mRNA at its 5′-untranslated region, including the cap structure (m^7GpppG), which is nearly essential for the initiation reaction; or the protein might degrade mRNA nucleolytically.

The alternative possibilities are tested experimentally by the use of sucrose density gradient centrifugation to investigate polysome profiles in the presence of cycloheximide. Incubation of the globin-synthesizing lysate system with L-PSP results in a disaggregation of polysomes into 80S ribosomes (Fig. 3A), supporting the above-described results. Addition of a ribonuclease inhibitor from human placenta at the start of incubation did not prevent polysome disaggregation. An important feature of this experiment is that both the incubation and the analysis are performed in the presence of a low concentration of cycloheximide, at which the antibiotic freezes the elongation reaction but not initiation. Polysome disaggregation in the presence of L-PSP demonstrates that the protein does not affect the initiation process. Rather, the results strongly suggest that L-PSP disintegrates polysomes through fragmentation of polysomal mRNA because of ribonuclease activity.

Ribonucleolytic Activity of Rat Liver Perchloric Acid-Soluble Protein

An mRNA having the cap structure is synthesized with [α-^{32}P]UTP and the four nucleotide triphosphates, using phage SP6 RNA polymerase and linearized plasmid pSP65 containing the gene encoding *E. coli* form 1 dihydrofolate reductase as template.[12] The transcript is 1079 nucleotides long, and consists of a

[12] Y. Endo, S. Otsuzuki, K. Ito, and K. Miura, *J. Biotechnol.* **25,** 221 (1992).

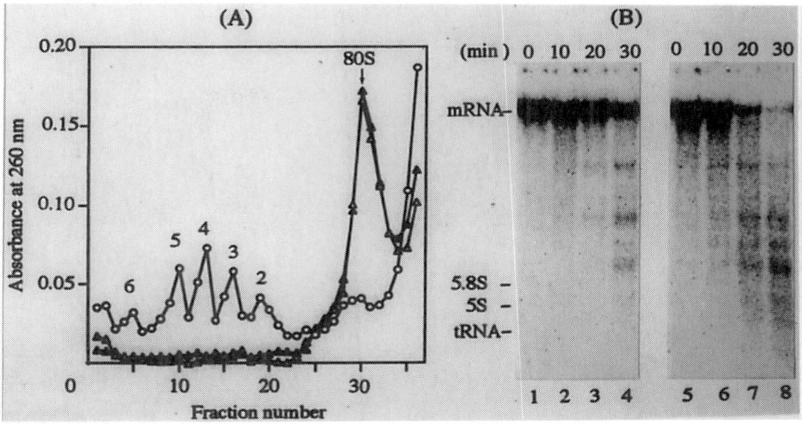

FIG. 3. Effect of L-PSP on polysome profiles and degradation of mRNA. (A) Translation mixtures (30 μl) prepared with endogenous globin-synthesizing lysate are incubated in the presence of cycloheximide (0.25 μM) at 30° for 15 min without (○) or with L-PSP (2.1 μM) (△), or with L-PSP together with human ribonuclease inhibitor (266 units) (▲). Samples are analyzed on a linear 10–45% sucrose gradient and 20 drops are collected from the bottom of the tube and used for absorbance measurements at 254 nm. (B) Translation mixtures (30 μl) composed of micrococcal nuclease-treated lysate, [^{32}P]mRNA (0.5 μg, 10,000 cpm), and 266 units of human ribonuclease inhibitor were incubated in the absence (lanes 1–4) or in the presence (lanes 5–8) of L-PSP (2.1 μM) at 30° for the indicated periods of time. RNA is extracted and separated on a 4% (w/v) polyacrylamide gel containing 7 *M* urea, and an autoradiograph of the gel was prepared. [From R. Morishita, A. Kawagoshi, T. Sawasaki, K. Madin, T. Ogasawara, T. Oka, and Y. Endo, *J. Biol. Chem.* **274,** 20688 (1999), with permission.]

coding region of 477 nucleotides close to the 5′ end and a 3′ noncoding region of 565 nucleotides. To obtain direct evidence, micrococcal nuclease-treated rabbit reticulocyte lysate is incubated as described above with ^{32}P-labeled dihydrofolate reductase mRNA in the presence of both the ribonuclease inhibitor and L-PSP. After incubation for various periods of time (Fig. 3B), RNA is extracted and separated on polyacrylamide gels. The autoradiograph shows intensive digestion of mRNA into small fragments in the presence of L-PSP, whereas the mRNA is fairly stable in the control sample. A similar digestion pattern of the mRNA is observed when the incubation is carried out in the absence of the ribonuclease inhibitor (data not shown). The amount of the ribonuclease inhibitor (266 units) used in Fig. 3A and B inhibits the activity of 1.33 μg of ribonuclease A by 50%, whereas the amount of L-PSP is 0.89 μg in a 30-μl reaction mixture. The result excludes the possibility that the inhibition by L-PSP is due to contamination with ribonuclease A or other A-like ribonucleases, RNases known to be abundant in both animal tissues and *E. coli*. These results show that L-PSP is a ribonuclease that hydrolyzes phosphodiester bonds of RNA in an endonucleolytic fashion.

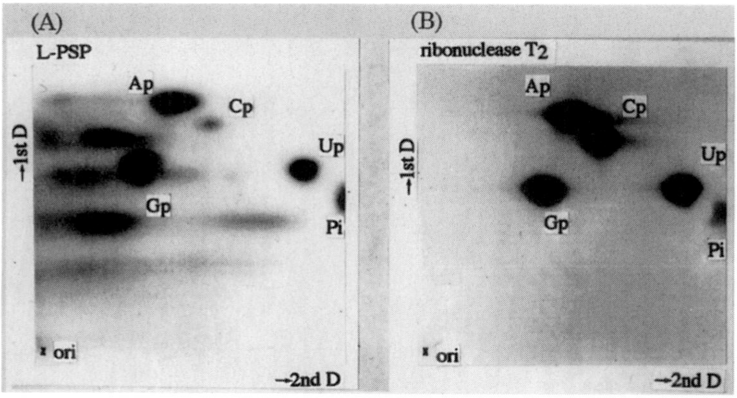

FIG. 4. Ribonucleotide specificity of L-PSP. [^{32}P]mRNA (1 μg, 50,000 cpm) is incubated in 10 μl with 6.2 μM L-PSP (A) or with ribonuclease T2 (100 units/ml) (B) in 10 mM Tris-HCl, pH 7.6, at 37° for 60 min. A portion from each reaction was chromatographed on cellulose plates in two dimensions [isobutyric acid–0.5 M ammonium hydroxide (5 : 3, v/v) for the first dimension and 2-propanol–HCL–H_2O (70 : 15 : 15, v/v/v) for the second dimension] and autoradiographs were made. [From R. Morishita, A. Kawagoshi, T. Sawasaki, K. Madin, T. Ogasawara, T. Oka, and Y. Endo, *J. Biol. Chem.* **274,** 20688 (1999), with permission.]

Determination of Substrate Specificity

Our results demonstrating the substrate specificity of the enzyme are shown in Fig. 4A. The [^{32}P]mRNA is incubated with L-PSP and then the sample is separated by thin-layer chromatography.[13] Subsequent autoradiography shows the production of the four 3′ nucleotide monophosphates as identified by comparison with standard nucleotides obtained from ribonuclease T2 digests (Fig. 4B). The results show that L-PSP cleaves the phosphodiester bonds of all four nucleotides, yielding 3′-AMP, 3′-GMP, 3′-UMP, and 3′-CMP. In addition to the four discrete spots, some streaks are also seen in Fig. 4A, which may represent residual small oligomers accumulating under the digestion conditions. Practically the same digestion pattern of the RNA was obtained when using recombinant L-PSP (data not shown). By measuring the radioactivity of the spots, the relative catalytic activity of L-PSP for the various substrates is determined: $Np/A^{32}p/U = Np/G^{32}p/U = Np/U^{32}p/U > Np/C^{32}p/U$ (slashes mark the sites of hydrolysis). The mechanism of cleavage probably involves 2′,3′-cyclic phosphate intermediates. L-PSP has little activity when DNA is the substrate.

L-PSP is further characterized in terms of specificity for single- or double-stranded regions of RNA. Rat liver 5S rRNA is a good substrate for the test because (1) the secondary structure is known and (2) the RNA has a highly ordered

[13] K. Randerath, and E. Randerath, *Methods Enzymol.* **12,** 323 (1967).

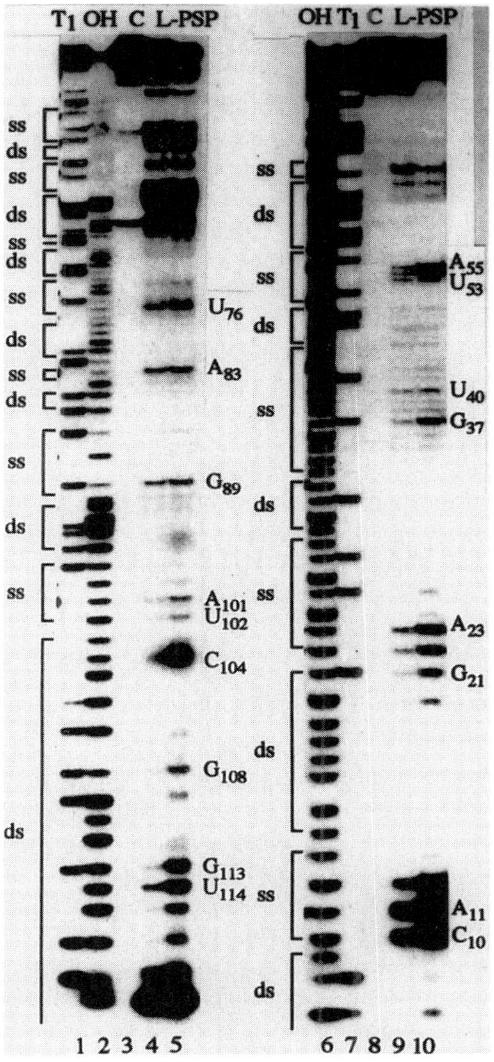

FIG. 5. Effect of L-PSP on 5S rRNA. Rat liver 5S rRNA is made radioactive at either the 5′ end (gels on the right) or 3′ end (gels on the left). The RNA is dissolved in 50 m*M* Tris-HCl (pH 7.6)–50 m*M* KCl and the 3′-labeled RNA is incubated with recombinant PSP (8.5 μM for lane 4 and 17 μM for lane 5). The 5′-labeled RNA is incubated with authentic L-PSP (2.1 μM for lane 9 and 3.5 μM for lane 10). After 20 min, the products are analyzed on polyacrylamide sequencing gels. Lanes 2 and 6, alkaline hydrolysates; lanes 1 and 7, ribonuclease T1 digest; lanes 3 and 8, incubation without the protein. *Abbreviations:* ds, Double-stranded regions of the nucleic acid; ss, single-stranded regions. [From R. Morishita, A. Kawagoshi, T. Sawasaki, K. Madin, T. Ogasawara, T. Oka, and Y. Endo, *J. Biol. Chem.* **274,** 20688 (1999), with permission.]

structure.[14] 5S rRNA is labeled at its 3′ end with [5′-^{32}P]pCp or at its 5′ end with [γ-^{32}P]ATP, and the sample is treated under nondenaturing conditions with either authentic PSP from rat liver or recombinant PSP. An alkaline digest as well as T1 digests are analyzed together with the products of the protein treatment on a 10% (w/v) polyacrylamide sequencing gel (Fig. 5). Both L-PSP preparations cleave only in single-stranded regions of the molecule.

It may be worth mentioning here that besides the lack of significant sequence homology with other ribonuclease, L-PSP also lacks histidine residues.[1] Histidine is known to be the indispensable general acid in the catalytic mechanism of other ribonucleases.[15,16] Another ribonuclease that lacks histidine has been reported[17]: The C-terminal domain of the bacteriocin, colicin E5, inhibits bacterial protein synthesis *in vitro* by cleaving several tRNAs at a specific site, the 3′ side of the queosine nucleotide in the anticodon loop. The cytotoxin recognizes the same site even if unmodified tRNA is the substrate, in which case cleavage occurs on the 3′ side of the guanosine nucleotide, that is, at the queosine position. The proposed mechanism of cleavage involves a 2′,3′-cyclic phosphate intermediate, but the exact enzymatic mechanism is unknown. A computer-aided homology search between L-PSP and the colicin E5 peptide reveals a short sequence, NDFGTV (amino acids 87 to 92) in L-PSP and NDPATV (positions 84 to 89 of the 115-amino acid C-terminal domain) in E5, but the functional significance of this homology remains to be seen.

Summary of Enzymatic Properties of Rat Liver Perchloric Acid-Soluble Protein

Rat liver perchloric acid-soluble protein is a unique ribonuclease, and this activity is responsible for the inhibition of translation. The inhibition appears to be due to an endoribonucleolytic activity because both authentic and recombinant L-PSP (1) directly affect mRNA template activity, and (2) induce disaggregation of the reticulocyte polysomes into 80S ribosomes even in the presence of cycloheximide. Analysis by thin-layer chromatography of [α-^{32}P]UTP-labeled mRNA incubated with the protein shows production of the ribonucleoside 3′-monophosphates Ap, Gp, Up, and Cp, providing direct evidence that the protein is an endoribonuclease. When either 5′- or 3′-^{32}P-labeled 5S rRNA is the substrate, L-PSP cleaves phosphodiester bonds only in the single-stranded regions of the molecule.

[14] Y. Endo, P. W. Huber, and I. G. Wool, *J. Biol. Chem.* **258,** 2662 (1983).

[15] L. Zegers, A. F. Haikal, R. Palmers, and L. Wyns, *J. Biol. Chem.* **269,** 127 (1994).

[16] J. C. Fontecilla-Camps, R. de Llorens, M. H. le Du, and C. M. Cuchillo, *J. Biol. Chem.* **269,** 21526 (1994).

[17] T. Ogawa, K. Tomita, T. Ueda, K. Watanabe, T. Uozumi, and H. Masaki, *Science* **283,** 2097 (1999).

[6] *Escherichia coli* Ribonuclease G

By MASAAKI WACHI and KAZUO NAGAI

Discovery of *cafA* Gene

The *cafA* gene, a name used previously for the *rng* gene encoding *Escherichia coli* RNase G, was originally found as a putative open reading frame (ORF) of unknown function during analysis of mecillinam-resistant shape determination mutants.[1] The mecillinam-resistant Δ*mre-678* mutant has a large deletion encompassing *mreB, mreC, mreD, orfE,* and *cafA* at 71 min on the *E. coli* genetic map.[2] The Δ*mre-678* mutant transformed with a plasmid carrying *mreB, mreC, mreD,* and *orfE* did not show any apparent phenotype.[3] Therefore, its function had remained to be elucidated. We found that extensive overproduction of the CafA protein caused formation of chained cells and minicells in *E. coli.* We observed long axial filament bundles, termed cytoplasmic axial filaments, running through the center of their cytoplasms[4] (Fig. 1). These findings led us to the idea that this filament may have a role as a cytoskeletal element either in cell division or chromosome segregation.

Genetic Interaction between RNase E and CafA

McDowall *et al.* reported significant sequence similarity between the CafA protein and the N-terminal half of RNase E, which is encoded by the *rne* gene at 24 min on the *E. coli* genetic map[5] (Fig. 2). RNase E is an essential *E. coli* endoribonuclease involved in the maturation of 5S rRNA and the decay of many mRNAs.[6] The *rne-1* (also called *ams-1*) temperature-sensitive mutation was initially described as increasing the chemical half-life of total RNA.[7] Another independently isolated mutation, mapped to the same gene, *rne-3071,* leads to the accumulation of 9S rRNA, a precusor of 5S rRNA.[8,9] These two mutations were

[1] M. Wachi, M. Doi, T. Ueda, M. Ueki, K. Tsuritani, K. Nagai, and M. Matsuhashi, *Gene* **106,** 135 (1991).

[2] M. Wachi, M. Doi, S. Tamaki, W. Park, S. Nakajima-Iijima, and M. Matsuhashi, *J. Bacteriol.* **169,** 4935 (1987).

[3] M. Wachi, M. Doi, Y. Okada, and M. Matsuhashi, *J. Bacteriol.* **171,** 6511 (1989).

[4] Y. Okada, M. Wachi, A. Hirata, K. Suzuki, K. Nagai, and M. Matsuhashi, *J. Bacteriol.* **176,** 917 (1994).

[5] K. J. McDowall, R. G. Hernandez, S. Lin-Chao, and S. N. Cohen, *J. Bacteriol.* **175,** 4245 (1993).

[6] S. N. Cohen and K. J. McDowall, *Mol. Microbiol.* **23,** 1099 (1997).

[7] M. Kuwano, M. Ono, H. Endo, K. Hori, K. Nakamura, Y. Hirota, and Y. Ohnishi, *Mol. Gen. Genet.* **154,** 279 (1977).

[8] B. K. Ghora and D. Apirion, *Cell* **15,** 1055 (1978).

0076-6879/01 $35.00

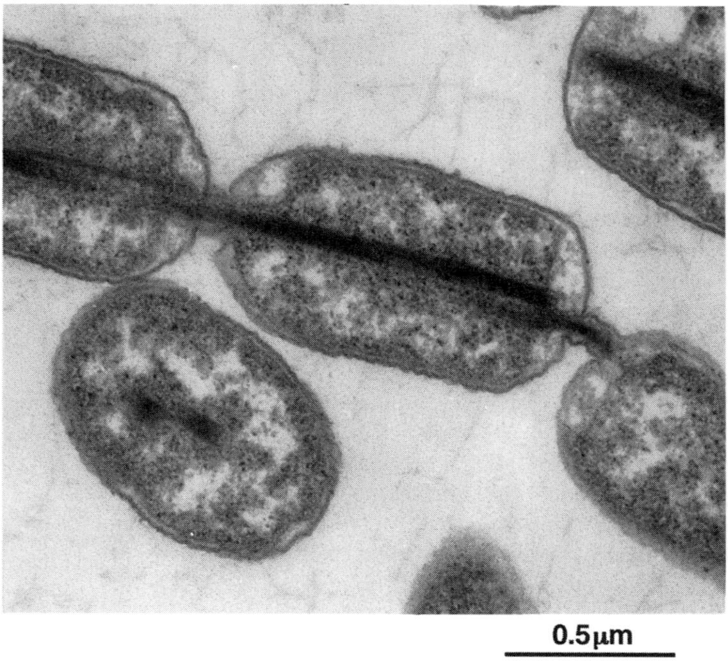

FIG. 1. Electron micrograph of ultrathin sections of CafA-overproducing cells. Bar: 0.5 μm. [micrograph kindly provided by A. Hirata, University of Tokyo.]

caused by amino acid substitutions in the highly homologous region between RNase E and RNase G.[5] As the first step in analyzing the function of the CafA protein, we examined the functional relationship between RNase E and CafA. To investigate the possible genetic interactions between *rne* and *cafA,* we constructed a *cafA*::*cat* insertional mutant strain. Introduction of the *cafA*::*cat* mutation into the *rne-1* mutant strain enhanced its temperature sensitivity (decreased the upper limit temperature for growth). On the other hand, introduction of the wild-type *cafA* allele on a high copy number plasmid pBR322 partially suppressed the temperature sensitivity of *rne-1*.[10] These results suggested that the CafA protein possesses some of the essential functions of RNase E in *E. coli*. The finding that the *rne-1* and *rne-3071* alleles mapped in the N-terminal half of RNase E,[5] which is sufficient for ribonuclease activity[11] and which has

[9] E. A. Mudd, H. M. Krisch, and C. F. Higgins, *Mol. Microbiol.* **4,** 2127 (1990).
[10] M. Wachi, G. Umitsuki, and K. Nagai, *Mol. Gen. Genet.* **253,** 515 (1997).
[11] K. J. McDowall and S. N. Cohen, *J. Mol. Biol.* **255,** 349 (1996).

```
E.col-RNaseE     ---------------MKRMLINATQQEELRVALVDGQRLYDLDIESPGHEQKKANIYKG  44
E.col-RNaseG     ---------------MTAELLVNVTPSETRVAYIDGGILQEIHIEREARRGIVGNIYKG  44
H.inf-RNaseE     MRFILRFFKYLIRELTMKRMLINATQKEELRVALVDGQRLFDLDIESPGHEQKKANIYKG  60
H.inf-RNaseG     --------------MDAVELLMNVTPNETRIALVETGMLREVHIERQAKRGIVGNIYKG  45
N.men-RNaseE     ---------------MKRMLFNATQAEELRVAIVDGQNLLDLDIETLGKEQRKGNIYKG  44
N.men-RNaseG     ---MLSGLPIPKDIARPPETILVNITPQETRVAVLEENNICELHIERNSEHSLVGNIYLG  57
                                   ::      .* *:* ::     : ::.**   ...   .*** *

E.col-RNaseE     KITRIEPSLEAAFVDYGAERHGFLPLKEIAREYFPANYSAHGRP---NIKDVLREGQEVI 101
E.col-RNaseG     RVSRVLPGMQAAFVDIGLDKAAFLHASDIMPHTECVAGEEQKQFTVRDISELVRQGQDLM 104
H.inf-RNaseE     KITRVEPSLEAAFVDYGAERHGFLPLKEIAREYFPDDYVFQGRP---NIRDILVEGQEVI 117
H.inf-RNaseG     RVTRVLPGMQSAFVDIGLEKAAFLHAADIVSHTECVDENEQKQFKVKSISELVREGQDIV 105
N.men-RNaseE     IITRIEPSLEACFVDYGTDRHGFLPFKEVSRSYFQDYEGGRAR-----IQDVLKEGMEVI  99
N.men-RNaseG     VVRRVLPGMQSAFIDIGLERAAFLHIVDVLEQRRNP-EETQR------IEHMLFEGQSVL 110
                 .:*:.*.:::.*:* * ::.**     ::             :         * .:: :* .::
                    HSR1
E.col-RNaseE     VQIDKEERGNKGAALTTFISLAGSYLVLMPNNPRAGGISRRIEGDDRTELKEALASLELP 161
E.col-RNaseG     VQVVKDPLGTKGARLTTDITLPSRYLVFMPGASHVGVSQRIESESERERLKKVVAEYCDE 164
H.inf-RNaseE     VQVNKEERGNKGAALTTFVSLAGSYLVLMPNNPRAGGISRRIEGDERTELKEALSSLDVP 177
H.inf-RNaseG     VQVVKEPLGTKGARLTTDITLPSRHLVFMPENSHVGVSQRIESEEERARLKALVEPFCDE 165
N.men-RNaseE     VQVEKDERGNKGAALTTFISLAGRYLVLMPNNPRGGGVSRRIEGEERQELKAAMAQLDIP 159
N.men-RNaseG     VQVIKDPINTKGARLSTQISLAGRFLVHLPQEDHIGVSQRIEDDAERSSLRERLDKLLPE 170
                 **: *:   ..*** *:* ::*.. .** :*    : *  .*  .   :*  *:  :

E.col-RNaseE     -EGMGLIVRTAGVGKSAEALQWDLSFRLKHWEAIKKAAESRPAPFLIHQESNVIVRAFRD 220
E.col-RNaseG     --QGGFIIRTAAEGVGEAELASDAAYLKRVWTKVMERKKRPQTRYQLYGELALAQRVLRD 222
H.inf-RNaseE     -DGVGLIVRTAGVGKSPEELQWDLKVLLHHWEAIKQASQSRPAPFLIHQESDVIVRAIRD 236
H.inf-RNaseG     --LGGFIIRTATEGASEEELRQDAEFLKRLWRKVLERKSKYPTKSKIYGEPALPQRILRD 223
N.men-RNaseE     -NGMSIIARTAGIGRSAEELEWDLNYLKQLWQAIEEAGKAHHDPYLLFMESSLLIRAIRD 218
N.men-RNaseG     NACRGYIIRTNAENATDEQLQSDIDYLTKVWEHIQEQAKIRPPETLLYQDLPLSLRVLRD 230
                   . * **    .      *   *      : *  : :  .         :.  :   :   * :**

E.col-RNaseE     YLRQDIGEILIDNPKVLELARQHIAALGRPDFSSKIKLYTGEIPLFSHYQIESQIESAFQ 280
E.col-RNaseG     FADAELDRIRVDSRLTYEALLEFTSEYI-PEMTSKLEHYTGRQPIFDLFDVENEIQRALE 281
H.inf-RNaseE     YLRRDIGEILIDSPKIFEKAKEHIKLVR-PDFINRVKLYQGEVPLFSHYQIESQIESAFQ 295
H.inf-RNaseG     FIGTNLEKIRIDSKLCFGEVKEFTDEFM-PELSDKLVLYSGNQPIFDVYGVENAIQTALD 282
N.men-RNaseE     YFRPDIGEILVDNQEVYDQVAEFMSYVM-PGNIGRLKLYEDHTPLFSRFQIEHQIESAFS 277
N.men-RNaseG     MVGCDTQKILVDSTVNHGRMTRFAEQYV-HGALGRIELFKGERPLFETHNVEQEISRALQ 289
                     :  .* :*.         ..          .::  : ..  *:*.  . :*  *.  *:.

E.col-RNaseE     REVRLPSGGSIVIDSTEALTAIDINSARATRGGDIEETAFNTNLEAADEIARQLRLRDLG 340
E.col-RNaseG     RKVELKSGGYLIIDQTEAMTTVDINTGAFVGHRNLDDTIFNTNIEATQAIARQLRLRNLG 341
H.inf-RNaseE     REVRLPSGGSIVIDVTEALTAIDINSARSTRGGDIEETALNTNLEAADEIARQLRLRDLG 355
H.inf-RNaseG     KRVNLKSGGYLIIEQTEAMTTIDINTGAFVGHRNLEETIFNTNIEATKAIAHELQLRNLG 342
N.men-RNaseE     RSVSLPSGGAIVIDHTEALVSIDVNSARATRGADIEDTAFKTNMEAAEEVARQMRLRDLG 337
N.men-RNaseG     PRVNLNFGSYLIIESTEAMTTIDVNTGGFVGARNFDETIFRTNLEACHTIARELRLRNLG 349
                   * *  *.  ::*:  ***:.::*:*:.   .       ::::*  :.**:**  .:*::::**:**
                                                              HSR2
E.col-RNaseE     GLIVIDFIDMTPVRHQRAVENRLREAVRQDRARIQISHISRFGLLEMSRQRLSPSLGESS 400
E.col-RNaseG     GIIIIDFIDMNNEDHRRRVLHSLEQALSKDRVKTSVNGFSALGLVEMTRKRTRESIEHVL 401
H.inf-RNaseE     GLVVIDFIDMTPIRHQREVENRIRDAVRPDRARIQISRISRFGLLEMSRQRLSPSLGESS 415
H.inf-RNaseG     GIIIIDFIDMQTDEHRNRVLQSLCDALSKDRMKTNVNGFTQLGLVEMTRKRTRESLEHVL 402
N.men-RNaseE     GLVVIDFIDMENPKHQRDVENVLRDALKKDRARVQMGKLSRFGLLELSRQRLKPALGESS 397
N.men-RNaseG     GIIIIDFIDMAQESHREAVLQELAKALAFDRTRVTLHGFTSLGLVELTRKRSRENLNQVL 409
                 *::::******        *:.  *  :  .*:  **  :   :   :: :**:*::*:*      : .

E.col-RNaseE     HHVCPRCSGTGTVRDNESLSLSILRLIEEEALKENTQEVHAIVPVPIASYLLNEKRSAVN 460
E.col-RNaseG     CNECPTCHGRGTVKTVETVCYEIMREIVRVHHAYDSDRFLVYASPAVAEALKGEESH-SL 460
H.inf-RNaseE     HHICPRCQGTGKVRDNESLSLSILRLLEEEALKENTKQVHTIVPVQIASYLLNEKRKAIS 475
H.inf-RNaseG     CDECPTCHGRGRVKTVETVCYEIMREIIRVYHLFSSEQFVVYASPAVSEYLINEESHGLL 462
N.men-RNaseE     HVACPRCAGTGVIRGIESTALHVLRIIQEEAMKDNTGEVHAQVPVDVATFLLNEKRAELF 457
N.men-RNaseG     CEPCPSCQGRGRLKTPQTVCYEIQREIVREARRYDAESFRILAAPNVIDLFLDEESQ-SL 468
                    ** * * * ::  ::.   : * : .        .:   .    ..   :    : .*:

E.col-RNaseE     AIETRQDGVRCVIVPNDQMETPHYHVLRVRKGEETPTLSYMLPKLHEEAMAL....... 1061
E.col-RNaseG     AEVEIFVGKQVKVQIEPLYNQEQFDVVMM---------------------------  489
H.inf-RNaseE     NIEKRHN-VDIIVAPNEAMETPHFSVFRLRDGEEVNELSYNLAKIHCAQDEN.......  951
H.inf-RNaseG     PEVEMFIGKRVKVKTEQFYNQEQFDVVVM---------------------------  491
N.men-RNaseE     AMEERLD-VNVVLIPNIHLENPHYEINRIRTDDVEEDGEPSYKRVAEPEENE.......  919
N.men-RNaseG     AMLIDFIGKPISLAVETAYTQEQYDIVLM---------------------------  497
                                 :  :        :: :  :
```

FIG. 2. Multialignment of RNase E and RNase G of several bacteria. Underlines indicate high-similarity regions (HSRs) 1 and 2. E. col, *Escherichia coli;* H. inf, *Haemophilus influenzae;* N. men, *Neisseria meningitidis*. Alignment was done by the ClustalW program (http://spiral.genes.nig.ac.jp/homology/clustalw-e.shtml).

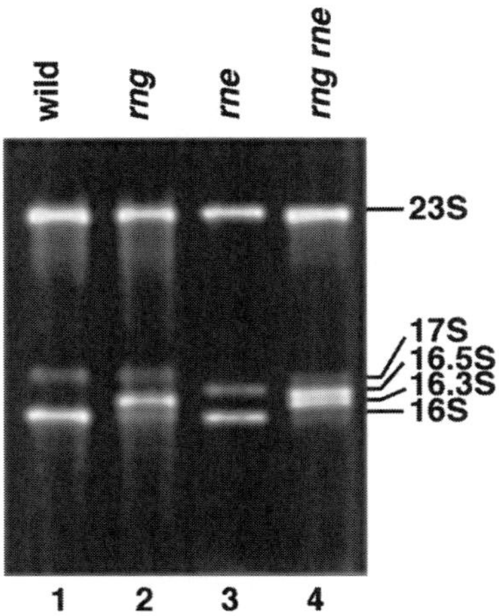

FIG. 3. Total cellular RNA was extracted and analyzed by the quick extraction method and modified agarose gel electrophoresis as described previously.[13] The wild-type strain GW10 (lane 1) and the *rng*::*cat* (*cafA*::*cat*) strain GW11 (lane 2) were grown at 30°. The *rne-1* mutant strain GW20 (lane 3) and the *rng*::*cat rne-1* double mutant strain GW21 (lane 4) were grown at 44°.

high sequence similarity with CafA, suggested that *cafA* also encodes RNase activity.

Accumulation of Precursor Molecule of 16S rRNA in *cafA* :: *cat* Mutant

We investigated the effect of *cafA* mutation on cellular RNA metabolism. Total cellular RNA extracted from a *cafA*::*cat* mutant was analyzed by modified agarose gel electrophoresis.[12] We found that a precursor molecule of 16S rRNA accumulated in the *cafA*::*cat* mutant cells (Fig. 3, lanes 1 and 2). Because the precursor molecule disappeared when wild-type *cafA* gene was introduced on a plasmid, the CafA protein appeared to be involved in the maturation of 16S rRNA.[12]

Escherichia coli contains seven rRNA gene clusters organized in a similar way: promoter–16S rRNA–tRNA–23S rRNA–5S rRNA–terminator, although there are some variations. The process of maturation is similar in all seven gene clusters.[13] RNase III introduces double cleavages in each of two stems, producing precursors of the 16S rRNA and 23S rRNA. Less well known is how the mature termini of 16S and 23S rRNA are generated from RNase III-cleaved intermediates.

[12] M. Wachi, G. Umitsuki, M. Shimizu, A. Takada, and K. Nagai, *Biochem. Biophys. Res. Commun.* **259,** 483 (1999).

[13] D. Apirion and A. Miczak, *BioEssays* **15,** 113 (1993).

BUMMER Mutant

In 1978, Dahlberg *et al.* reported an *E. coli* mutant, called BUMMER, which accumulated a precursor form of 16S rRNA, named 16.3S rRNA, which has 66 extra nucleotides at the 5′ end.[14] Extracts from the parental strain catalyzed conversion of 16.3S rRNA to mature 16S rRNA *in vitro*. We compared this precursor in the *cafA*::*cat* with that in the BUMMER mutant strain. The precursor molecule in the *cafA*::*cat* mutant strain migrated to the same position as the 16.3S precursor in the BUMMER strain. Moreover, accumulation of the precursor form in the BUMMER strain was suppressed by introduction of a plasmid carrying wild-type *cafA*, as in the case of *cafA*::*cat*. These results strongly suggested that the BUMMER strain has a mutation in the *cafA* gene. DNA sequencing analysis of the mutant *cafA* gene revealed that the BUMMER mutant allele has an 11-bp deletion at nucleotides 729–739 in the coding region of *cafA*. This results in a frame shift and formation of a truncated CafA protein with an extra 23 amino acid residues.[12]

These results indicated that the *cafA* gene encodes an RNase activity involved in the 5′-end maturation of 16S rRNA, which is defective in the BUMMER mutant. We proposed that this activity was provided by a new enzyme we named RNase G. The gene was renamed as *rng*. This was confirmed by Tock *et al.*, using highly purified protein.[15] Hereafter we use "RNase G" and "*rng*" instead of "CafA" and "*cafA*," respectively.

Two-Step Maturation of 5′ End of 16S rRNA by RNase E and RNase G

The mature form of 16S rRNA is formed in *rng*::*cat* mutant cells (Fig. 3, lane 2). In a rifampicin-chase experiment, a 16.3S precursor form is converted to the mature 16S form at a half-life of about 15 min. This means that *E. coli* has additional activity for converting 16.3S to 16S other than RNase G. Is this residual activity due to RNase E? In the *rng*::*cat rne-1* double mutant strain, almost no mature form of 16S rRNA is detected at the temperature restrictive for *rne-1* (44°) (Fig. 3, lane 4). A rifampicin-chase experiment shows that the 16.3S precursor form is stable in the double mutant. These results clearly indicate that the residual activity is indeed due to RNase E.[12] High-resolution analysis reveals that, in addition to the mature 16S, a product with four or five additional 5′-nucleotide residues is produced in the *rng*::*cat* mutant cells, indicating that processing by RNase E at the 5′ end is less accurate than RNase G.[16]

In the *rne-1* mutant and the *rng*::*cat rne-1* double mutant, a novel precursor form of 16S rRNA, which migrates a little slower than 16.3S (tentatively named

[14] A. E. Dahlberg, J. E. Dahlberg, E. Lund, H. Tokimatsu, A. B. Rabson, P. C. Calvert, F. Reynolds, and M. Zahalak, *Proc. Natl. Acad. Sci. U.S.A.* **75,** 3598 (1978).

[15] M. R. Tock, A. P. Walsh, G. Carrol, and K. J. McDowall, *J. Biol. Chem.* **275,** 8726 (2000).

[16] Z. Li, S. Pandit, and M. P. Deutscher, *EMBO J.* **18,** 2878 (1999).

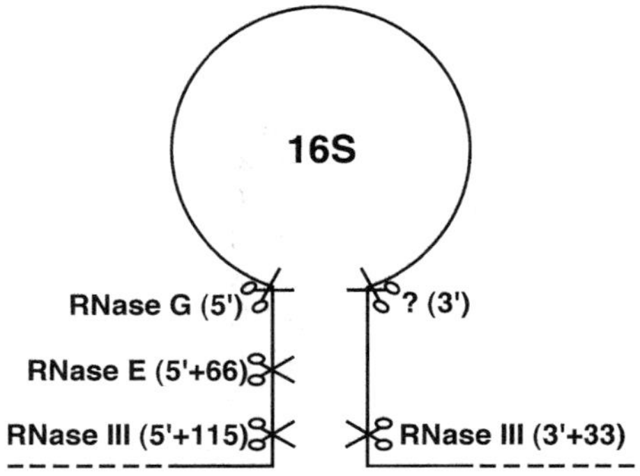

FIG. 4. A model for the maturation of 16S rRNA. From the primary transcripts of *rrn* operons, the 17S precursor is generated by cleavages with RNase III. The mature 3′ end is formed by an unknown enzyme (indicated as "?") to produce the 16.5S precursor. The 16.5S precursor is processed by RNase E to produce the 16.3S molecule. Finally the mature 16S rRNA is generated by the action of RNase G.

16.5S), is found (Fig. 3, lanes 3 and 4). This indicates that RNase E is involved in the step converting 16.5S to 16.3S. Dahlberg *et al.* have also reported that they occasionally detected a precursor molecule migrating between 16.3S and 17S, which contained the 5′ end of 17S and the mature 3′ end.[14] A rifampicin-chase experiment shows that 16.5S is still converted to 16S at a half-life of about 60 min in the *rne-1* single mutant. On the other hand, in the *rng::cat rne-1* double mutant, 16.5S is as stable as 16.3S. These results indicate that RNase G can also catalyze the reaction that converts 16.5S to 16S, although less efficiently.[12]

From our results[12] and those of Li *et al.*,[16] a possible mechanism of maturation of 16S rRNA can be proposed (Fig. 4). A 17S precursor is generated from primary transcripts from *rrn* operons by cleavage with RNase III. The mature 3′ end is formed by an unknown enzyme (indicated as "?" in Fig. 4) to produce the 16.5S precursor. The 16.5S precursor is processed by RNase E to produce 16.3S. Finally, the mature 16S rRNA is generated by action of RNase G. The reaction by RNase E may precede the 3′ maturation in wild-type cells, because the 3′ maturation is significantly slowed in *rne-1* mutant cells.[16]

RNase G: A 5′-End-Dependent Ribonuclease

The endoribonucleolytic activity of RNase G was confirmed by Tock *et al.* by using highly purified protein and synthetic oligoribonucleotide substrates *in vitro*.[15] A synthetic decaribonucleotide, 5′-ACAGUAUUUG-3′, which corresponds in sequence to the single-stranded segment at the 5′ end of RNA I and which is a good substrate of RNase E, is efficiently cleaved by RNase G endoribonucleolytically. The action of RNase G is dependent on the nature of the 5′ end of the substrates. RNase G efficiently cleaves 5′-monophosphorylated substrates but not 5′-hydroxylated substrates. RNase G action is also blocked by a 5′-triphosphate group. Interestingly, a 5′-monophosphate group may only stimulate cleavage at sites present on the same RNA molecule. Considering these characteristics, Tock *et al.* suggest that RNase G prefers to cut to completion 5′-monophosphorylated decay or processing intermediates rather than initiate the decay of intact 5′- triphosphorylated RNAs. In contrast, the 3′-phosphorylation status does not affect the rate of cleavage by RNase G.[15]

Intact RNA I is as efficiently cleaved by RNase G at the RNase E site as are the synthetic oligosubstrates. In addition to RNA I, the 5′-untranslated region (UTR) of *ompA* mRNA, which is a substrate of RNase E, is cleaved by RNase G at the proper RNase E sites, although more slowly than RNase E. In contrast, RNase G does not cleave at an RNase E site of 9S rRNA (which is also a substrate of RNase E). However, a synthetic decanucleotide that corresponds in sequence to the RNase E site is efficiently cleaved by RNase G. This indicates that the context of the clevage site, not its sequence, blocks cleavage by RNase G.[15]

They also found that RNase G has only weak poly(A) nuclease activity *in vitro* relative to RNase E.[15]

Role of RNase G in mRNA Degradation *in Vivo*

From the results of *in vitro* experiments, Tock *et al.* suggested that RNase G contributes in some manner to the decay of RNAs in *E. coli*.[15] Our genetic evidence, showing that functions of RNase E and RNase G overlap, although partially,[10] also supports this idea.

We have found that *rng*::*cat* mutants overproduce a protein of about 100 kDa. N-terminal amino acid sequencing of this protein shows that it is identical to the fermentative alcohol dehydrogenase, the product of the *adhE* gene located at 28 min on the *E. coli* genetic map.[17] The level of *adhE* mRNA is significantly higher in the *rng*::*cat* mutant strain than in its parental strain; such differences are not seen with other genes examined. A rifampicin-chase experiment reveals that the half-life of *adhE* mRNA is 2.5 times longer in the *rng*::*cat* disruptant than in the wild type. These results indicate that, in addition to rRNA processing, RNase G is involved in *in vivo* mRNA degradation (our unpublished data, 2000).

RNase E/RNase G Family in Bacteria

Homologs of the N-terminal endoribonucleolytic domain of RNase E or RNase G are found in all Gram-negative and some Gram-positive bacteria that have been

[17] P. E. Goodlove, P. R. Cunningham, J. Parker, and D. P. Clark, *Gene* **85,** 209 (1989).

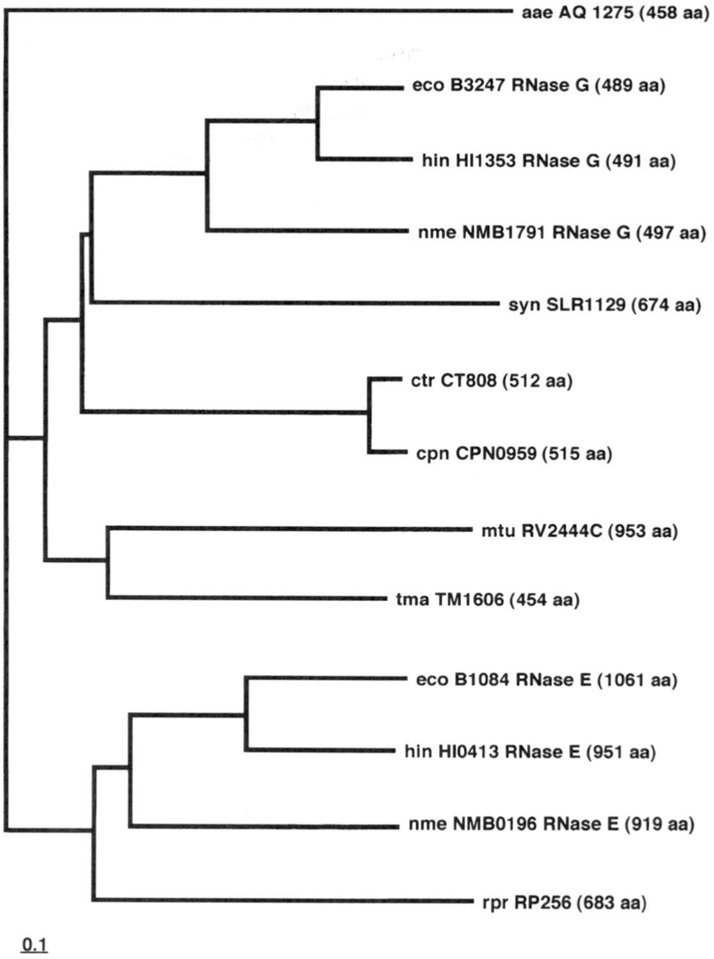

FIG. 5. Phylogenetic tree of RNase E/RNase G family. Bar: 0.1 substitutions per site. aae, *Aquifex aeolicus;* cpn, *Chlamydia pneumoniae;* ctr, *Chlamydia trachomatis,* eco, *Escherichia coli;* hin, *Haemophilus influenzae,* mtu, *Mycobacterium tuberculosis*; nme, *Neisseria meningitidis;* rpr, *Rickettsia prowazekii;* syn, *Synechocystis* sp.; tma, *Thermotoga maritima.*

completely sequenced (Fig. 5).[18] *Escherichia coli, Haemophilus influenzae,* and *Neisseria meningitidis* have both RNase E-type and RNase G-type enzymes, but others only have one homolog. The *Rickettsia prowazekii* enzyme seems to be

[18] V. R. Kaberdin, A. Miczak, J. S. Jakobsen, S. Lin-Chao, K. J. McDowall, and A. von Gabain, *Proc. Natl. Acad. Sci. U.S.A.* **95,** 11637 (1998).

more related to RNase E, although it does not have a long C-terminal domain. The others seem to belong to G-type enzymes, judging from sequence homology and their sizes, although the *Mycobacterium tuberculosis* enzyme has an extra N-terminal domain. In these homologs, extensive sequence similarity is found in two regions: residues 39–68 and 314–350 in *E. coli* RNase E (Fig. 2). The former region, designated high-similarity region 1 (HSR1), contains a consensus motif for nucleotide binding[5] and is a part of the S1-like RNA-binding domain.[19] Two *rne* mutations, *rne-1* (G66S) and *rne-3071* (L68F), are mapped in this region.[5] The latter region, designated HSR2, appears to be required for endoribonucleolytic activity,[11] but its precise function is still unknown. The high degree of sequence similarity of these regions suggests that the mechanism for sensing the presence of monophosphate group on the 5′ end of RNAs may be ancient and evolutionarily conserved.

Because RNase G can make cleavages that resemble those of RNase E, careful assignment of contribution of these two enzymes in RNA decay or processing is required in organisms that possess both enzymes. It is also interesting to investigate the role of RNase G or E in those organisms that have only one homolog.

Possible Role of RNase E and RNase G as Cytoskeletal Elements

The biological role, if any, of the cytoplasmic axial filament formed by the RNase G protein remains to be determined. When histidine-tagged RNase G proteins were constructed for the purpose of purification, we observed that some of the constructs did not form cytoplasmic axial filaments by overproduction, but still could complement the defect of 16S rRNA maturation in the *rng*::*cat* mutant. This suggests that RNase activity and the ability to form filaments are distinguishable (our unpublished data, 2000). Interestingly, it has also been reported that the RNase E protein reacts with monoclonal antibodies against yeast myosin heavy chain and that anti-RNase E antibody reacts with nonmuscle myosins.[20] Tock *et al.* observed that the purified RNase G protein tends to aggregate at high concentrations *in vitro*.[15] They also pointed out that there is a precedent in eukaryotes for associations between RNA and cytoskeletal filaments.[21] Therefore, the possible cytoskeletal functions of RNase E and RNase G became still more attractive to us.

Acknowledgments

We thank Moselio Schaechter for critical reading of the manuscript. We also thank Aiko Hirata for providing the electron micrograph.

[19] M. Bycroft, T. J. P. Hubbard, M. Proctor, S. M. V. Freund, and A. G. Murzin, *Cell* **88,** 235 (1997).
[20] S. Casarégola, A. Jacq, D. Laoudj, G. McGurk, S. Margarson, M. Tempête, V. Norris, and I. B. Holland, *J. Mol. Biol.* **228,** 30 (1992).
[21] R. P. Jansen, *FASEB J.* **13,** 455 (1999).

[7] *Escherichia coli* Transcript Cleavage Factors GreA and GreB: Functions and Mechanisms of Action

By SERGEI BORUKHOV, OLEG LAPTENKO, and JOOKYUNG LEE

Biological Role of Gre Factors

Two homologous prokaryotic proteins, GreA and GreB,[1,2] and a eukaryotic analog TFIIS (SII)[3] comprise a novel class of transcription factors that are involved in endonucleolytic hydrolysis of nascent RNA in ternary complexes (TCs) of RNA polymerase (RNAP). The RNA cleavage typically occurs 2–18 bases upstream from the 3′ terminus, followed by dissociation from the TC of the 3′-proximal fragment carrying the phosphate group.[4] The 5′-proximal fragment with the free 3′-hydroxyl group remains in the TC, and can be extended in the presence of rNTPs. GreA was shown to induce hydrolysis of only di- and trinucleotides (type "A" cleavage activity), whereas SII and GreB induce cleavage of 2- to 18-nucleotide-long RNA fragments (type "B" cleavage activity).[1–4]

Unlike classic RNases, SII and Gre proteins do not display any nucleolytic activity toward free RNA or RNA–DNA duplex[2,4] and generally require formation of a stable TC. In the absence of factors, the intrinsic endonucleolytic activity of *Escherichia coli* RNAP and eukaryotic RNAP II can be stimulated by mild alkaline pH[5] and pyrophosphate,[6] respectively. Both RNA synthesis and RNA hydrolysis reactions require at least two Mg^{2+} ions with similar K_m values. Moreover, the two reactions are activated and inhibited by similar groups of divalent ions, such as Mn^{2+}, Fe^{2+}, and Co^{2+}, and Pb^{2+}, Cu^{2+}, and Zn^{2+}, respectively. These results suggest that the same active center of RNAP is involved in both reactions. These and other data indicate that Gre and SII are not nucleases per se, but rather cofactors that activate the intrinsic RNase activity of RNAP. This intrinsic activity has proved to be an evolutionarily conserved function among all multisubunit RNAPs.[4]

The transcript cleavage occurs when RNAP in a TC engages in "backtracking," which is characterized by reverse translocation of RNAP together with the transcription bubble along the DNA template.[7–9] Backtracking of RNAP results in

[1] S. Borukhov, A. Polyakov, V. Nikiforov, and A. Goldfarb, *Proc. Natl. Acad. Sci. U.S.A.* **89,** 8899 (1992).
[2] S. Borukhov, V. Sagitov, and A. Goldfarb, *Cell* **72,** 459 (1993).
[3] M. G. Izban and D. S. Luse, *Genes Dev.* **6,** 1342 (1992).
[4] S. M. Uptain, C. M. Kane, and M. J. Chamberlin, *Annu. Rev. Biochem.* **66,** 117 (1997).
[5] M. Orlova, J. Newlands, A. Das, A. Goldfarb, and S. Borukhov, *Proc. Natl. Acad. Sci. U.S.A.* **92,** 4596 (1995).
[6] M. D. Rudd, M. G. Izban, and D. S. Luse, *Proc. Natl. Acad. Sci. U.S.A.* **91,** 8057 (1994).
[7] T. C. Reeder and D. K. Hawley, *Cell* **87,** 767 (1996).

0076-6879/01 $35.00

the displacement of the 3′-terminal portion of the nascent RNA from DNA–RNA hybrid and its extrusion from the transcription bubble, with simultaneous repositioning of the RNAP catalytic center from the 3′ terminus to an internal site of the transcript. Backtracking has been demonstrated both *in vitro* and *in vivo*[10]; however the nature of the signals that cause it (besides destabilization of the DNA–RNA hybrid in the TC[11]) is poorly understood.

The nucleolytic reaction is believed to be an integral part of the *in vivo* mechanism that confers proofreading capability to RNAP.[12] It may also allow RNAP to overcome obstacles encountered during elongation, such as pausing and arresting sites in the DNA sequences, nucleosomes, or DNA-binding proteins and drugs.[1–4] *In vitro* studies have shown that GreB can rescue arrested TCs in which RNAP has irreversibly backtracked. The cleavage of the extruded portion of RNA induced by GreB allows RNAP to resume transcription from the newly generated 3′ terminus, thus reactivating the arrested TC ("antiarrest activity").[2] Unlike GreB, GreA is unable to induce cleavages in preformed arrested TCs and reactivate them.[2,13] However, both GreA and GreB can prevent elongating TCs from falling into an arrested conformation ("readthrough activity"), presumably by inducing RNA cleavages at the onset of backtracking and reengaging the active center with the RNA 3′ terminus.[1,2] The functional selectivity of GreA may be related to its ability to induce cleavage only of short RNAs. Backtracking of RNAP, and subsequent Gre-induced hydrolysis of RNA, were also observed when nucleotide analogs that disrupt RNA–DNA hybrids were incorporated into RNA in *in vitro* transcription assays.[9,11]

Transcript cleavage reactions may also play an important role during transcription initiation. It was shown *in vitro* and *in vivo* that on select promoters, GreA and GreB facilitate the transition of RNAP from initiation to the elongation stage.[14,15] During initiation Gre factors were observed to induce substantial cleavages in nascent RNAs 2 to 9 nucleotides in length.[13–15] It was proposed that by inducing cleavage, GreA and GreB prevent RNAs from dissociating from the open promoter complex, increasing the time of RNA occupancy in the initiation complex. This leads to an increased rate of promoter clearance by RNAP and synthesis of the full-length RNA.[16]

[8] N. Komissarova and M. Kashlev, *Proc. Natl. Acad. Sci. U.S.A.* **94,** 1766 (1997).

[9] E. Nudler, A. Mustaev, E. Lukhtanov, and A. Goldfarb, *Cell* **89,** 33 (1997).

[10] F. Toulme, M. Guerin, N. Robichon, M. Leng, and A. R. Rahmouni, *EMBO J.* **18,** 5052 (1999).

[11] E. Nudler, *J. Mol. Biol.* **288,** 1 (1999).

[12] D. A. Erie, O. Hajiseyedjavadi, M. C. Young, and P. H. von Hippel, *Science* **262,** 867 (1993).

[13] G. H. Feng, D. N. Lee, D. Wang, C. L. Chan, and R. Landick, *J. Biol. Chem.* **269,** 22282 (1994).

[14] M. H. Hsu, N. V. Vo, and M. J. Chamberlin, *Proc. Natl. Acad. Sci. U.S.A.* **92,** 11588 (1995).

[15] A. Das, Unpublished data (1995).

[16] D. Kulish, J. Lee, I. Lomakin, B. Nowicka, A. Das, S. Darst, K. Normet, and S. Borukhov, *J. Biol. Chem.* **275,** 12789 (2000).

Structure–Function Relationship of Gre Factors

Escherichia coli GreA and GreB are homologous polypeptides of 158 amino acids.[2] According to the established three-dimensional (3-D) structures, the two factors have similar structural organization and surface charge distribution, and are made of two domains: an N-terminal extended coiled-coil domain (NTD) and a C-terminal globular domain (CTD).[17,18] Amino acid sequence alignment of Gre homologs from more than 30 different prokaryotic organisms indicates that the spatial organization of the two domains and the salient structural features of *E. coli* GreA and GreB (including the location of basic patch residues) are highly conserved.[16–18] The NTD is responsible for the induction of type-specific nucleolytic activity by Gre factors as well as readthrough and antiarrest activities [18,19] whereas CTD is responsible for the high-affinity binding of Gre to RNAP.[18–20] The charge distribution analysis of 3-D structures revealed a cluster of positively charged residues on the NTD of GreA and GreB that form a distinct structural feature, described as a basic patch, on the side of the protein that is presumed to face RNAP in the TC.[17] Genetic and biochemical analyses revealed that basic patch residues are responsible for interaction of Gre factors with nascent RNA in the TC.[16]

The exact molecular mechanism by which Gre factors activate the catalytic center of RNAP and induce RNA cleavage in the TC is unknown. The current model of Gre action involves the following three essential steps. (1) The CTD of Gre protein binds to RNAP near the opening of the secondary channel visible on the 3-D crystal structure of *Thermus aquaticus* RNAP core enzyme.[21] This binding not only tethers Gre factor to the TC but also causes conformational changes in RNAP and provides the Gre NTD efficient contacts with RNA and RNAP; (2) the NTD attracts through its basic patch the 3′-terminal portion of the nascent RNA extruded from backtracked TC. The specific electrostatic interactions with the exposed 3′-terminal portion of the nascent transcript confer a proper orientation and alignment of an internal phosphodiester bond with respect to the RNAP catalytic center; and (3), the NTD, through unidentified residues, activates, the catalytic site of RNAP by affecting the environment in the immediate vicinity of the RNAP active center, in particular within the conserved regions D, F, and G of the β' subunit.

[17] C. E. Stebbins, S. Borukhov, M. Orlova, M. Polyakov, A. Goldfarb, and S. A. Darst, *Nature (London)* **373,** 636 (1995).

[18] D. Koulich, M. Orlova, A. Malhotra, A. Sali, S. A. Darst, and S. Borukhov, *J. Biol. Chem.* **272,** 7201 (1997).

[19] D. Koulich, V. Nikiforov, and S. Borukhov, *J. Mol. Biol.* **276,** 379 (1998).

[20] A. Polyakov, C. Richter, A. Malhotra, D. Koulich, S. Borukhov, and S. A. Darst, *J. Mol. Biol.* **281,** 262 (1998).

[21] G. Zhang, E. A. Campbell, L. Minakhin, C. Richter, K. Severinov, and S. A. Darst, *Cell* **98,** 811 (1999).

The purification and *in vitro* transcriptional assays of GreA and GreB have been described in detail.[22] Here, we describe three methods useful for biochemical and structure–function analyses of Gre factors: the Gre–RNAP binding assay, the Gre–RNA photo-cross-linking assay, and Fe^{2+}-induced hydroxyl radical mapping of Gre interactions with the active center of RNAP.

Gre–RNA Polymerase Binding Assay

Both GreA and GreB competitively bind to the RNAP core, holoenzyme, or TC with an apparent K_d of 20–30 and 0.1–0.2 μM, respectively.[18] These values were estimated by size-exclusion and affinity chromatography of ^{35}S-labeled Gre factors in complexes with RNAP and TCs. Loisos and Darst[23] have described a direct GreB–RNAP binding assay using a native gel-shift electrophoresis of ^{32}P-labeled GreB, tagged with a phosphorylation site for heart muscle kinase (HMK), in complex with core RNAP. Under conditions of native polyacrylamide gel electrophoresis (PAGE) at pH 8.3, negatively charged GreB–RNAP complexes and free RNAP migrate into the gel, thereby separating themselves from free GreB, which is partially positively charged and does not enter the gel. After PAGE, the GreB–RNAP complexes are visualized and quantified by Phospho-rlmager (Molecular Dynamics-Pharmacia, Sunnyvale, CA). The apparent K_d for the GreB–RNAP complex calculated by this method was ~87 nM, which is in agreement with our data reported earlier. The gel-shift method is fast and simple and allows direct measurements of the apparent binding affinity of GreB for RNAP. However, this method is not suitable for GreA, which is negatively charged and migrates closely with RNAP during native PAGE. Also, under conditions of low ionic strength and in the absence of carrier proteins, GreB forms nonspecific complexes with RNAP, which appear on the autoradiogram as a smear or multiple bands.

To complement the native gel-shift binding assay and to overcome some of its limitations we developed an indirect, semiquantitative Gre–RNAP binding assay based on the ability of RNAP to remain transcriptionally active after it specifically binds to an immobilized Gre protein [e.g., hexahistidine-tagged GreA (His_6–GreA) or GreB attached to Ni^{2+}–nitrilotriacetic acid (NTA)–agarose beads].[18] In the presence of DNA template (*E. coli* ribosomal operon *rrnB* P1 promoter fragment with an initial transcribed sequence of CACCACUGACACGG), dinucleotide initiator CpA, ATP, and radiolabeled CTP, RNAP forms a stable TC carrying the hexanucleotide RNA product CpA**pC**pC**pA**pC (6C-TC) (here and elsewhere in this article, boldface type denotes a radioactive phosphate).[22] After 6C-TC is formed on immobilized Gre inside the pores of the beads, its diffusion is restricted because

[22] S. Borukhov and A. Goldfarb, *Methods Enzymol.* **274,** 315 (1996).

[23] N. Loizos and S. A. Darst, *J. Biol. Chem.* **274,** 23378 (1999).

of the large size of the TC. Thus, the trapped 6C-TC can be washed to remove unincorporated radioactive CTP and the remaining 6C-RNA can be quantified after separation by denaturing PAGE and analyses by phosphorimaging. Several washings of beads reduce the background and substantially increase the sensitivity of the assay. Optimally, about 1 fmol of 6C-TC can be detected by this method.

Because neither GreA nor GreB affects the formation and the stability of the TC, the amount of radiolabeled transcript synthesized by RNAP on the beads is proportional to the amount of RNAP molecules adsorbed to immobilized Gre and, therefore, reflects the binding affinities of Gre factors to RNAP.[18] The apparent K_d values obtained by this assay for GreA–RNAP and GreB–RNAP complexes were consistent with K_d values obtained by other methods. In control experiments in which Ni^{2+}–NTA–agarose is used for the assay without immobilized Gre factors, no radioactive 6C-transcript could be detected.[18] These data indicate that the formation of 6C-TC results from specific binding of RNAP to immobilized Gre proteins and depends only on their ability to retain RNAP on agarose beads.

Procedure 1

Step 1. Gre factor (5 μg) carrying a COOH-terminal hexahistidine tag (His_6–Gre) is incubated for 10 min at 4° with 5 μl of Ni^{2+}-chelating NTA–agarose beads (Qiagen, Valencia, CA) in 15 μl of buffer A [40 m*M* Tris-HCl (pH 7.5), 30 m*M* KCl, 0.1 m*M* EDTA, and 0.1 m*M* dithiothreitol (DTT)] containing 0.8 *M* NaCl. The beads are briefly centrifuged (10 sec) at low speed (1000 rpm), the supernatant is aspirated (without drying the beads), and the beads are washed three times with 500 μl of the initial buffer. Typically, ~95% of the initial Gre protein is immobilized on the Ni^{2+}–agarose.

Step 2. Specific amounts of RNAP holoenzyme (1–50 μg) in 20 μl of buffer A containing 0.1 *M* NaCl, 10 m*M* $MgCl_2$, and bovine serum albumin (BSA, 0.5 mg/ml) are added to the beads and incubated with gentle shaking for 15 min. Free RNAP is removed by centrifugation and the beads are quickly washed with 100 μl of buffer A. In the control experiment, RNAP is incubated with Ni^{2+}–NTA–agarose beads and washed as described above but in the absence of immobilized Gre factors. Note that for all procedures only GreA/GreB-free RNAP should be used. The isolation and purification of RNAP from *E. coli greA$^-$ greB$^-$* strain AD8571 is performed as described.[24]

Step 3. The beads carrying adsorbed RNAP are suspended in 20 μl of transcription buffer (buffer A, supplemented with BSA at 0.5 mg/ml and 10 m*M* $MgCl_2$) containing 0.5 μg of *rrnB* P1 DNA template (202-bp fragment, end points −152

[24] S. Borukhov and A. Goldfarb, *Protein Express. Purif.* **4,** 503 (1993).

and +50, relative to the +1 transcription start site[25]). CpA (0.5 m*M*; Sigma, St. Louis, MO), 2 μM ATP, and 1 μM [α-^{32}P]CTP (600 Ci/mmol; ICN, Costa Mesa, CA). After incubation at 37° for 10 min, the beads are washed four times with 200 μl of buffer A containing BSA (0.2 mg/ml) and 0.2 *M* NaCl. The TC formed by adsorbed RNAP and carrying hexameric transcript CpA**p**C**p**CpA**p**C (6C-TC) is eluted from the beads with 50 μl of denaturing buffer containing formamide (80%, v/v) and 20 m*M* EDTA in buffer A after incubation for 2 min at 95°. A 20-μl aliquot of the eluate is analyzed by denaturing 23% (w/v) PAGE in the presence of 8 *M* urea.[22] Gels are visualized by PhosphorImager (Storm; Molecular Dynamics), and the intensities of 6C-TC are quantitated using ImageQuant software (Molecular Dynamics) and plotted against the initial concentrations of RNAP used for binding to Gre. The apparent K_d values are calculated graphically as described.[23]

The indirect binding assay described above was successfully used for identifying domains within Gre factors responsible for high-affinity binding to RNAP and for comparing the specific binding affinities of various GreA and GreB mutant proteins.[16,18,19] However, because the immobilized Gre molecules are not equally accessible to RNAP and only about 20% are available for specific interactions with RNAP, this method provides only an approximate estimation of the Gre–RNAP specific binding affinity.

Gre–RNA Photo-Cross-Linking Assay

To probe the interactions of GreA and GreB with the nascent transcript in TCs, we developed an RNA–protein cross-linking assay using photoactive analogs of AMP (8-N_3-AMP) or UMP (4-thio-UMP) incorporated into nascent RNA 3′ terminus. Both photoactive probes have short spacer arms (2–3 Å) connecting the photoactive group with the nucleotide base, thus permitting cross-linking of target proteins at positions that lie in the immediate vicinity of the transcript. Unlike 4-thio-UTP, 8-N_3-ATP is a terminating substrate for RNAP, and can be used only as a 3′-terminal probe.[26] The two probes may also display slightly different reactivities toward nucleophilic amino acid side chains in proteins.[26,27] The TCs used in the assay carry a radiolabeled heptameric transcript, CpA**p**C**p**CpA**p**CpU$_{4S}$ (7U*-TC), or a nonameric transcript, CpA**p**C**p**CpA**p**CpUpGpA$_{N_3}$ (9A*-TC). The 7U*- and 9A*-TCs were prepared by a stepwise extension of the radiolabeled 6C-TC obtained with the *rrnB* P1 promoter DNA in the presence of appropriate NTPs (Fig. 1A). The two TCs differ by the extent of RNAP backtracking: 2 nucleotides (nt) in 7U*-TC and up to 4 nt in 9A*-TC.

[25] S. Borukhov, V. Sagitov, C. A. Josaitis, R. Gourse, and A. Goldfarb, *J. Biol. Chem.* **268,** 23477 (1993).

[26] M. M. Hanna, *Methods Enzymol.* **180,** 383 (1989).

[27] N. Riehl, P. Remy, J. P. Ebel, and B. Ehresmann, *Eur. J. Biochem.* **128,** 427 (1982).

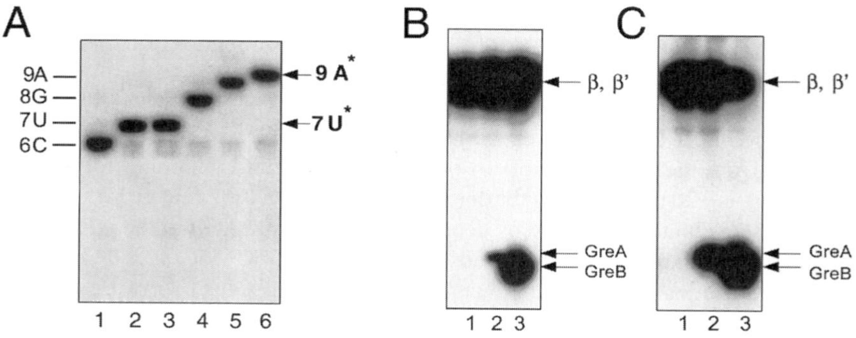

FIG. 1. (A) The initial 6C-TC formed on the *rrnB* P1 promoter fragment (lane 1) is incubated with UTP (lane 2), 4-thio-UTP (lane 3), UTP + GTP (lane 4), UTP + GTP + ATP (lane 5), or UTP + GTP + 8-N_3ATP (lane 6) for a stepwise extension of the transcript. The products are separated by urea–23% (w/v) PAGE and visualized by autoradiography. In each reaction, the initial 6C-TC is quantitatively converted to a specific TC carrying RNA of expected size: 7-, 8-, or 9-mer. The incorporation of 8-N_3ATP at position +9 results in reduced electrophoretic mobility of the 9A*-TC relative to normal 9A-TC (compare lanes 5 and 6). The concentrations of 4-thio-UTP, 8-N_3ATP, and NTPs are 20, 50, and 5 μM, respectively. (B) Fifty femtomoles of 7U*-TC carrying a photoactive probe 4-thio-UTP at the RNA 3′ terminus is subjected to UV irradiation alone (lane 1), or in the presence of 2 μg of GreA (lane 2) or 2 μg of GreB (lane 3). The results are analyzed by SDS–14% (w/v) PAGE and autoradiography. In the absence of factors, the radiolabeled RNA specifically cross-links to the β' subunit of RNAP, appearing on the autoradiogram as a high molecular weight radioactive species (lane 1). In the presence of Gre factors, additional low molecular weight radioactive species become visible in the gel, corresponding to cross-linked adducts of GreA and GreB (lanes 2 and 3). The efficiency of RNA–GreB cross-linking is about 20 times higher than that of RNA–GreA cross-linking, reflecting the higher binding affinity of GreB toward the TC. (C) Fifty femtomoles of 9A*-TC, which carries the photoactive probe 8-N_3ATP at the RNA 3′ terminus, is subjected to UV irradiation as in (B). Similar to 7U*-TC, the radiolabeled RNA cross-links specifically to the β' subunit of RNAP in the absence of Gre (lane 1), and to GreA (lane 2) and GreB (lane 3) in the presence of factors. However, cross-linking in the presence of GreB also results in a substantial decrease in the β' cross-link (lane 3). This implies that in 9A*-TC, which is more prone to backtrack than 7U*-TC, the interaction between the 3′ terminus of the transcript and the catalytic site of RNAP is sterically altered because of the presence of GreB.

In the absence of Gre factors, UV irradiation of 7U*- and 9A*-TCs induces cross-linking of RNA to the β' subunit of RNAP (the yield is ~15 and 8% of the initial TC, respectively) (Fig. 1B and C, lane 1). The addition of GreB to TCs results in efficient cross-linking of this factor to both photoprobes, with a similar yield of approximately 18% (Fig. 1B and C, lane 3). In the case of 9A*-TC this reaction is also accompanied by a substantial (3-fold) decrease in β' cross-linking (Fig. 1B, lane 3). Compared with GreB, GreA cross-links to RNA in 7U*-TC and 9A*-TC with a much lower efficiency, with a yield of 1 and 5%, respectively (Fig. 1B and C, lane 2), and does not affect cross-linking to the β' subunit. The lower yield of GreA cross-linking is attributed to its low binding affinity to RNAP, and to the small size of its basic patch, which precludes GreA from efficient interaction with RNA.[16] Using this photo-cross-linking technique in combination with chemical and enzymatic mapping procedures, we have demonstrated that the 3′ terminus of the transcript in 7U*-TC and 9A*-TC contacts the β' subunit of RNAP near the conserved regions G and F, and contacts both Gre proteins at a site near the tip of their N-terminal domains.[17,18] The specificity of the cross-linking reactions was confirmed by the observation that under the same experimental conditions, the carrier proteins BSA or lysozyme that were present in excess did not cross-link to RNA.

Procedure 2

Step 1. The initial radiolabeled 6C-TC is prepared by incubation of 0.9 μg (7.3 pmol) of the DNA fragment with 12 μg (30 pmol) of GreA/GreB-free RNAP, BSA (1 mg/ml), 0.5 m*M* CpA (Sigma), 5 μM ATP, and 1 μM [α-^{32}P]CTP (1000 Ci/mmol; ICN) in 35 μl of standard transcription buffer [40 m*M* Tris–acetate (pH 7.9), 30 m*M* KCl, and 10 m*M* $MgCl_2$] for 10 min at 37°. The 6C-TC is further purified by gel filtration on a Quick-Spin G-50 column. The typical yield of 6C-TC is ~3 pmol (~40% based on the initial amount of DNA used in the reaction).

Step 2. (*Note:* Steps 2 and 3 are performed in the dark or under reduced light.) The radiolabeled 7U*-TC and 9A*-TC are obtained by incubation of ~1 pmol of 6C-TC for 5 min at 30° in 30 μl of transcription buffer (see above) in the presence of, respectively, 20 μM 4-thio-UTP (Amersham-Pharmacia Biotech, Piscataway, NJ), or 5 μM UTP, 5 μM GTP, and 50 μM 8-azido-ATP (Sigma). The reactions are stopped by addition of 2 μl of 0.5 *M* EDTA and kept on ice. The identity and the quantity of the resulting RNA products are verified by analyzing the 0.5-μl aliquots of each transcription reaction by urea–23% (w/v) PAGE, followed by autoradiography and quantitation by PhosphorImager. The yield of 7U*- and 9A*-TC depends on the purity of NTPs and is typically in the range of 70–90% of the theoretical value.

Step 3. One-microliter aliquots (20–30 fmol) of the resulting 7U*-TC or 9A*-TC are diluted into 10 μl of buffer B [40 m*M* Tris-acetate (pH 7.9), 200 m*M* NaCl, and 10 m*M* EDTA] containing BSA (0.2 mg/ml), supplemented with different amounts of purified GreA or GreB (1–2000 ng, 0.055–110 pmol), placed into the wells of a standard 96-well microtiter plate, and incubated for 5 min on ice.

Step 4. The samples in the microtiter plate (kept on ice) are UV irradiated for 10 min, using a Spectroline UV lamp (302 nm, 6 W) placed at a distance of 1 cm from the plate surface. During irradiation, the plate should be rotated every 3 min. The cross-linking reaction is stopped by addition of 1 μl of 0.5 *M* DTT and 10 μl of the 2× SDS-PAGE sample-loading buffer to each sample. The material is withdrawn from the wells, transferred to a standard 1.5-ml tube, and boiled for 2 min. The samples are separated by 4–20% gradient SDS-PAGE (Novex, San Diego,

CA), and the gel is Coomassie stained, dried, and subjected to autoradiography and quantitation by phosphorimaging.

To detect the presence of Gre in complex with TC irrespective of its interaction with RNA, we used a photoactive 4-azido-2-nitrophenyl derivative of 5-aminoallyl-UMP (Sigma) (N_3-NPA-UMP), which is incorporated into internal position +7 of the transcript in 9A**-TC.[16,19] Unlike 8-N_3-AMP or 4-thio-UMP, the photoactive azido group of N_3-NPA-UMP is connected to the nucleotide base through a long spacer (~12 Å). Therefore, Gre proteins that are capable of binding to RNAP in TC but unable to contact the nascent RNA can easily be identified with this probe. The 9A**-TC is obtained similarly to 9A*-TC, by extension of radiolabeled 6C-TC in the presence of 50 μM N_3-NPA-UTP, and 5 μM GTP and ATP, and the cross-linking reactions are carried out as described above in step 4. The yields of Gre and RNAP cross-linking to 9A**-TC are comparable to those of the 7U*-TC. Another photoactive analog of N_3-NPA-UTP, 5-[(4-azidophenacyl)thio]UTP[26] (5-APAS-UTP; Designer Genes, Phoenix, AZ) can be used with comparable cross-linking efficiencies for a similar purpose.

The usefulness of using both short and long spacearm probes in the cross-linking assay is illustrated by the results obtained with mutant Gre proteins lacking the basic patch. These proteins do not cross-link to short spacearm probe in TC, but cross-link efficiently to the long spacearm probe in 9A**-TC,[16] thus providing direct evidence that basic patch is important for Gre–RNA interaction, but not necessarily for Gre–RNAP interaction. Also, comparative Gre–RNA cross-linking experiments with wild-type Gre proteins and their individual domains using short and long spacearm photoprobes demonstrated that the N-terminal domain of Gre factors may occupy two alternative sites in TC, proximal and distal, with respect to nascent RNA.[19]

Fe^{2+}-Induced Hydroxyl Radical Mapping of Gre Interactions with RNA Polymerase Active Center

To identify the sites of Gre–RNAP interaction near the enzyme active center we applied Fe^{2+}-induced hydroxyl radical mapping analysis. This method is based on the ability of chelated Fe^{2+} ion, such as in Fe^{2+}–EDTA complexes, to generate in aqueous solutions highly reactive hydroxyl radicals in the presence of molecular oxygen or hydrogen peroxide, and a reducing agent, 2-mercaptoethanol or dithiothreitol. The hydroxyl radicals cause cleavages in biopolymers with a diffusion-limited rate, in the distance range of 10 Å. The process is known as the Fenton reaction[28] and is used for general footprinting of protein–protein and

[28] M. A. Price and T. D. Tullius, *Methods Enzymol.* **212,** 194 (1992).

nucleic acid–protein complexes. This approach has been successfully applied in studies of the intersubunit interface in *E. coli* RNAP[29] and RNAP binding sites on GreB.[23]

A modification of this method was developed to carry out highly localized hydroxyl radical mapping experiments. By exchanging the catalytic Mg^{2+} ion in the active center of RNAP with Fe^{2+}, specific cleavages of protein, DNA, and nascent RNA can be introduced in the immediate vicinity of the catalytic center.[30] Four major sites of Fe^{2+}-induced cleavage were observed in the RNAP β' subunit: one lies in the evolutionarily conserved region D, near three aspartate residues that coordinate metal ion in the active center; two lie within conserved region F; and one lies in region G[31] (see also Fig. 2). According to the 3-D crystal structure of *T. aquaticus* RNAP,[21] these conserved regions comprise a deeply buried catalytic pocket of RNAP and could serve as effective targets (direct or allosteric) for GreA and GreB action.

To detect possible perturbations within the RNAP active center caused by interactions with Gre factors, we subjected the core RNAP to localized Fe^{2+}-induced hydroxyl radical mapping in the presence and absence of GreA or GreB (Fig. 2). To visualize protein cleavages and facilitate mapping, the β' subunit of RNAP is N-terminally tagged with a phosphorylation site for heart muscle kinase (HMK)[32] and is radiolabeled with [γ-^{33}P]ATP. As shown in Fig. 2, specific and reproducible differences in the cleavage pattern of the β' subunit can be observed in correlation with the presence and absence of GreB. These differences (marked in Fig. 2 by asterisks) include the appearance of a double band in regions G (site I) and D (site IV) and the selective/efficient protection of site III in region F. Importantly, the cleavage site II in region F (Fig. 2) remains unchanged and serves as an internal control that shows the specificity of GreB effects on Fe^{2+}-induced cleavages in RNAP. Similar patterns of protein cleavages were also observed for GreA, but not for individual domains of GreA or for GreA mutants unable to bind to RNAP.[33] The identities of protein fragments are determined with molecular weight markers generated by specific cleavages of the radiolabeled β' subunit at methionine, cysteine, or tryptophan residues (only the CNBr cleavages at methionine are shown in lane 8 of Fig. 2). The established positions of the cleavage sites that are visibly protected or altered by GreB are identified within residues 466–484, 725–743, and 922–932 of regions D, F, and G, respectively. The control experiments carried out in the

[29] T. Heyduk, E. Heyduk, K. Severinov, H. Tang, and R. H. Ebright, *Proc. Natl. Acad. Sci. U.S.A.* **93,** 10162 (1996).

[30] E. Zaychikov, E. Martin, L. Denissova, M. Kozlov, Markovtsov, M. Kashlev, H. Heumann, V. Nikiforov, A. Goldfarb, and A. Mustaev, *Science* **273,** 107 (1996).

[31] A. Mustaev, M. Kozlov, V. Markovtsov, E. Zaychikov, L. Denissova, and A. Goldfarb, *Proc. Natl. Acad. Sci. U.S.A.* **94,** 6641 (1997).

[32] Z. Keiman, V. Naktinis, and M. O'Donnell, *Methods Enzymol.* **262,** 430 (1995).

[33] O. Laptenko, J. Lee, and S. Borukhov, in preparation (2001).

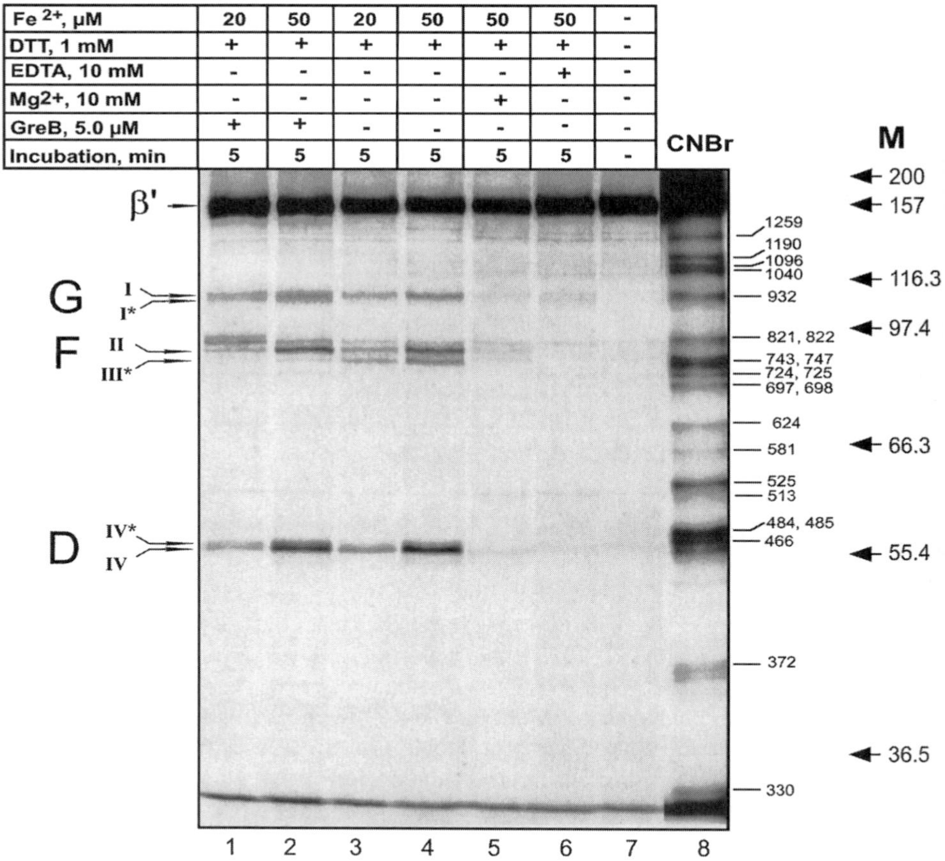

FIG. 2. Fe^{2+}-induced hydroxyl radical mapping of β' subunit in RNAP in complex with GreB, shown as an autoradiogram of SDS–7% PAGE. The initial core RNAP radiolabeled at the N terminus of the β' subunit is stripped of bound divalent metal ions by treatment with 10 m*M* EDTA followed by gel filtration through a QuickSpin G-50 column (Boehringer Mannheim) equilibrated in chelated HEPES buffer (lane 7). The purified RNAP (0.25 μM) is incubated in the presence of 5 μM BSA with or without 5 μM GreB and treated with Fe^{2+} and DTT to induce localized cleavages of the ^{33}P-labeled β' subunit. In the absence of GreB (lanes 3 and 4), specific cleavages occur at four major sites in the β' subunit that are spatially adjacent to the catalytic center of RNAP (marked on the left by arrows and Roman numerals). In the presence of GreB (lanes 1 and 2), site III* is effectively protected while two additional sites (sites I* and IV*) become susceptible to Fe^{2+}-induced cleavages. Cleavages at sites I, II, and IV are unaffected by the presence of Gre, demonstrating the specificity of the cleavage reactions. In the control experiments, both EDTA and Mg^{2+} protect the β' subunit from all Fe^{2+}-induced cleavages (lanes 5 and 6). These data imply that binding of Gre causes a subtle but detectable rearrangement of the β' polypeptide within the catalytic center of RNAP. Lane 8 shows the size markers, products of chemical degradation of radiolabeled β' at methionine by CNBr.[31] The positions of methionine residues in the sequence ladder of β' are indicated next to lane 8. The cleavage products that are altered in the presence of GreB are marked by asterisks. The capital letters next to Roman numerals refer to the conserved regions in β' where the cleavages occur. Molecular weight standards ($\times 10^{-3}$, Mark 12; Novex) are shown at the right-hand side letter M.

presence of competitors, EDTA, and Mg^{2+} (Fig. 2, lanes 5 and 6, respectively) demonstrate the specificity of Fe^{2+}-induced cleavages.

Procedure 3

Step 1. HMK-tagged RNAP core enzyme is purified as described[5] from *E. coli* strain CAG3016 transformed with plasmid pTRC-N-HMKrpoC that expresses the β' subunit with an N-terminal His_6 tag and HMK site (LRRASV)[32] under the control of the isopropyl-β-D-thiogalactopyranoside (IPTG)-inducible *trc* promoter. pTrcN-HMKrpoC was constructed by excising the *rpoC* gene with double tags from initial expression plasmid pET16b-β'-PKA[34] with *Nco*I and *Hin*dIII, and inserting the resulting fragment into *Nco*I/*Hin*dIII-linearized expression vector pTRC99A (Pharmacia). *Escherichia coli* CAG3016 strain-recipient[34] carries a chromosomal RifD18 mutation in the *rpoB* gene[35] encoding the β subunit of RNAP, which renders *E. coli* cells resistant to the antibiotic rifampicin. In the RifD18 allele, serine at position 531 is substituted by phenylalanine, which eliminates a strong intrinsic phosphorylation site for HMK.[36] The RNAP purified from this strain can be selectively and specifically phosphorylated by HMK at the N terminus of the β' subunit (see Fig. 2, lane 7).

Step 2. Purified HMK-tagged RNAP core enzyme is ^{33}P-end labeled in a 20-μl reaction containing 20 μg (1.8 μM) of HMK-RNAP, 5 units of reconstituted HMK (Sigma), 0.8 μM [γ-^{33}P]ATP (3000 Ci/mmol; ICN), 20 mM Tris-HCl (pH 7.9), 1 mM DTT, 0.2 M NaCl, and 10 mM $MgCl_2$. After the incubation (20 min at 37°), the radiolabeled enzyme is supplemented with 10 mM EDTA and BSA (0.5 mg/ml) and passed through a QuickSpin G-50 desalting column (Roche Molecular Biochemicals) to remove free ATP and any remaining divalent metal ions. The column should be preequilibrated with chelated[30] 20 mM Na-HEPES buffer, pH 7.5, containing 50 mM NaCl. The typical yield of radiolabeled RNAP is 13–15 μg, 5–7 μCi. Gre (3 mg/ml) and BSA (3 mg/ml) are purified similarly with a QuickSpin G-25 column equilibrated in the same HEPES buffer, but containing 0.8 M NaCl.

Step 3. Fe^{2+}-mapping reactions are performed at 25° for 5 min in 10-μl reactions containing 2 μg of purified radiolabeled RNAP, 5–7 μg of BSA with or without 2 μg of Gre protein, 20 mM Na-HEPES buffer (pH 7.5), 0.3 M NaCl, 20–50 μM $(NH_4)_2Fe(SO_4)_2$ (Sigma), and 1 mM DTT. Reactions are initiated by adding 1 μl of freshly made 10× $(NH_4)_2Fe(SO_4)_2$ solution in Milli-Q water and 1 μl of 10 mM DTT to an 8-μl mixture of RNAP with BSA alone or with Gre protein preincubated on ice for 10 min. In the control experiments, inhibitors (10 mM $MgCl_2$, or 10 mM EDTA) should be included to verify the specificity of Fe^{2+}-induced cleavage reaction.

[34] Plasmid pET16b-β'-PKA and *E. coli* CAG3016 strain were gifts of K. Severinov.

[35] D. J. Jin and C. A. Gross, *J. Mol. Biol.* **202,** 45 (1988).

[36] K. Severinov, unpublished data (2000).

Step 4. Reactions are terminated by addition of 1 μl of 0.5 *M* EDTA, and 10 μl of 2× SDS gel sample loading buffer containing 10% (v/v) 2-mercaptoethanol, followed by boiling for 2 min. The products of Fe^{2+}-induced cleavage reactions are resolved by Tris–glycine/SDS–7% (w/v) PAGE,[37] visualized by autoradiography, and analyzed by PhosphorImager (Storm); Molecular Dynamics). Molecular weight markers for the β' subunit are generated as described[31] by residue-specific cleavages of the radiolabeled RNAP at cysteine, methionine, and tryptophan under limited cleavage conditions.

The method described above provides a unique experimental tool with which to assess the structural consequences of Gre–RNAP interaction at the molecular level, with a special emphasis on the microenvironment of the catalytic center of RNAP. Specifically, it allowed us to identify the sites in the immediate vicinity of the RNAP catalytic center that undergo conformational changes on binding of Gre. The polypeptide regions that are located within the same distance (less than 15 Å) from the source of hydroxyl radicals are nonetheless differentially protected by Gre, demonstrating the high degree of sensitivity of this method. Indeed, the main advantage of this method is that it allows us to detect subtle changes in highly localized, internal sites of the target protein that may not be readily accessed by other, conventional methods of detection. However, it should be noted that this method is indirect, and does not reveal whether the altered protection pattern of RNAP peptides in the presence of Gre reflects the conformational changes in RNAP resulting from direct or allosteric effect of Gre binding to RNAP.

Results obtained by Fe^{2+}-induced hydroxyl radical mapping suggest that the binding site for Gre on RNAP lies near the ~12-Å-wide opening of the "secondary channel" described in the structure of *T. aquaticus* RNAP as the probable site of substrate entry.[21] On the one hand, the direct interactions of Gre factor with the catalytic residues of RNAP are unlikely on structural grounds.[16,17] Therefore, the observed alterations near the catalytic site in region D (site IV*) are apparently due to an allosteric effect of GreB. On the other hand, the nascent RNA 3′ end may be able to interact with Gre proteins only if it protrudes through the secondary channel, as was suggested to occur in backtracked TCs.[16,21] This interaction is possible if the NTD of Gre occupies a corridor, near the opening of the secondary channel, formed by an extended coiled-coil subdomain of β' subunit (beginning of region β'-E) and parts of conserved regions β'-F and β'-G. In this case, the basic patch of NTD will be in close proximity to region β'-G, where cross-linking to the nascent RNA 3′ terminus occurs in backtracked TCs.[11] This model is consistent with the observation that GreB efficiently protects one segment of region F (site III*) from hydroxyl radical cleavages and alters the protein cleavages in region G (site I*).

[37] U. K. Laemmli, *Nature (London)* **227,** 680 (1970).

[8] *Escherichia coli* Ribonuclease P

By LEIF A. KIRSEBOM

Introduction

Ribonuclease P is an endoribonuclease responsible for the maturation of the 5′ termini of the majority of all known tRNAs in all cell types studied to date. In *Escherichia coli,* RNase P consists of an RNA subunit, M1 RNA, and a small basic protein, C5. The catalytic activity is associated with M1 RNA, and cleavage of various substrates *in vitro* does not require the presence of the C5 protein.[1] However, C5 is essential for RNase P activity *in vivo*. Although this review focuses on *E. coli* RNase P, other bacterial RNase P systems such as that of *Bacillus subtilis* have also been studied in detail (Altman and Kirsebom,[2] and references therein). This review has been divided into two parts: the first part describes protocols for *in vitro* studies of RNase P activity and the second part discusses an experimental system that can be used to study tRNA processing with an emphasis on RNase P *in vivo*.

Preparation of Substrates and RNase P RNA

In *E. coli,* as in bacteria in general, there is a single RNase P enzyme species that must interact with a large number of different tRNA precursors in the cell. Approximately 80 tRNA genes have been identified in *E. coli* and these are transcribed (1) as monomeric or multimeric tRNA precursors (i.e., precursors carrying more than one tRNA), or (2) together with rRNA precursors, or (3) together with mRNAs. In addition, several non-tRNA precursors have been identified as substrates for RNase P. These include the precursor to transfer mRNA (tmRNA), certain plant viral RNAs, the precursors to 4.5S RNA and bacteriophage M3 RNA, and the phage-derived antisense C4 RNA. These precursors possess tRNA-like structures or stem–loop structures harboring an RCC(A/C) motif at their 3′ termini. It has also been suggested that a stem–loop structure in mRNA functions as a target for RNase P *in vivo*. This latter suggestion is in keeping with the finding that various *in vitro*-generated model substrates consisting of a hairpin–loop structure are cleaved efficiently by bacterial RNase P. In fact, the smallest RNase P substrate consists of 3 base pairs, an RCCA motif at the 3′ end, and an unpaired 5′ residue. For references see Altman and Kirsebom.[2]

[1] C. Guerrier-Takada, K. Gardiner, T. Marsh, N. Pace, and S. Altman, *Cell* **38,** 849 (1983).

[2] S. Altman and L. A. Kirsebom, *in* "RNA World II" (R. F. Gesteland, T. R. Cech, and J. F. Atkins, eds.), 2nd Ed., pp. 351–380. Cold Spring Harbor Laboratory Press, Cold Spring Harbor, New York, 1999.

0076-6879/01 $35.00

We and others use monomeric tRNA precursors to study cleavage by M1 RNA and by the holoenzyme. In addition, *in vitro*-generated model hairpin–loop substrates are commonly used. These substrates are generated with *in vitro*-synthesized gene constructs behind the phage T7 promoter and cloned into an appropriate plasmid. These constructs usually carry a specific restriction cleavage site downstream of the gene to be transcribed. A convenient enzyme to use is *Fok*I, because it allows the generation of transcripts with specific 3′ ends that mimic *in vivo* precursors. Other enzymes that we have used are *Sfa*NI and *Bsp*MI. After cleavage of the plasmid carrying the gene construct with *Fok*I and transcription with T7 DNA-dependent RNA polymerase according to Milligan *et al.*,[3] the transcript is purified by gel electrophoresis on denaturing polyacrylamide gels, with the percent polyacrylamide depending on the size of the transcript. The transcript is visualized by autoradiography or UV shadowing and eluted either by electroelution (we use a BIOTRAP; Schleicher & Schuell, Keene, NH), or by crushing and soaking in TE buffer [10 m*M* Tris-HCl (pH 7.9), 1 m*M* EDTA] or crush and soak buffer [10 m*M* Tris-HCl (pH 7.5), 1 m*M* EDTA, 0.1 *M* Sodium acetate (pH 5.2), and 0.25% (w/v) sodium dodecyl sulfate (SDS)] followed by ethanol precipitation. As a guideline the yield is approximately 1 μg of transcript per microliter transcription volume. To label the RNA we carry out parallel transcription reactions using a final UTP concentration of 0.2 m*M* (instead of 2 m*M*) and [α^{32}-P]UTP, resulting in a final specific activity of 5 μCi/mmol in the transcription mixture.

Shorter RNA substrates such as model hairpin loops can be chemically synthesized and after synthesis these RNA molecules must be purified according to the procedure described above. We are currently using chemically synthesized, 45-nucleotide-long RNA substrates and find that these are cleaved with the same kinetic rate constants as the same substrate transcribed by T7 RNA polymerase. The chemically synthesized substrates are labeled at the 5′ end according to standard procedures using T4 polynucleotide kinase. Given the large number of different substrates cleaved by RNase P it is important to study the cleavage of different categories of substrates in order to elucidate how RNase P operates.

To generate RNase P RNA for *in vitro* studies the same construction, transcription and purification procedures are in principle applicable. However, to date no complete RNase P RNA with a natural bacterial sequence has yet been chemically synthesized.

Genetics is a powerful tool with which to understand the function of an enzyme and this is also applicable in studies of the function of RNase P. Thus, a large number of mutant substrates and M1 RNA derivatives have been generated. To construct genes encoding mutant substrates we have used the following strategies. First, we identify two unique restriction sites within the gene encoding the substrate.

[3] J. F. Milligan, D. R. Groebe, G. W. Whiterell, and O. C. Uhlenbeck, *Nucleic Acids Res.* **15,** 8783 (1987).

The plasmid carrying the gene encoding the substrate is then cut with these restriction enzymes and the fragment is replaced by a fragment carrying mutation(s) generated by polynucleotide chain reaction (PCR), using an oligonucleotide carrying the sequence we wish to introduce. Alternatively, we use chemical synthesis to make double-stranded DNA harboring the mutation. To change nucleotides at specific positions in M1 RNA we use the same strategy as described above, with restriction sites and fragment replacement. However, if there is no unique restriction site close to the position we want to change we use an oligonucleotide (not longer than 30 nucleotides) carrying the mutation in the middle and an *in vitro* mutagenesis kit according to the manufacturer instructions. As template we have been using, for example, the wild-type M1 RNA gene behind the T7 promoter, but any M1 RNA gene variant can be used. We routinely use the Pharmacia (Piscataway, NJ) Biotech USE mutagenesis kit.

Generation of Substrate and M1 RNA Molecules Carrying Modified Nucleotides at Specific Positions

To introduce modified nucleotides at specific positions either in the substrate or in M1 RNA different approaches can be taken. The chemical synthesis of RNA has improved such that small RNase P model substrates carrying a modified residue at a specific position can now be purchased from one of the companies devoted to chemical synthesis of RNA (see, e.g., Perrault and Altman[4,5]). When we and others have studied the catalytic performance of RNase P as a function of a modified nucleotide in either a tRNA precursor or in M1 RNA the ligation protocol according to Moore and Sharp[6] has been used. This approach is particularly suitable to generate larger RNA molecules, such as a tRNA precursor or M1 RNA, carrying modified nucleotides.[7–9]

To make tRNA precursors harboring modified nucleotides a chemically synthesized RNA oligonucleotide carrying the modification (at, e.g., the cleavage site) is radioactively labeled at the 5′ end. The length of the RNA oligonucleotide can vary; however, in our example the 3′ end corresponds to the residue at position 14 in the D-loop. The second RNA fragment covering the remainder of the tRNA molecule starting at position G15 in the D-loop is generated with T7 RNA polymerase[3] except that transcription is performed in the presence of 5 m*M* GMP and 1 m*M* GTP, to ensure incorporation of monophosphate at the 5′ end of the

[4] J.-P. Perreault and S. Altman, *J. Mol. Biol.* **226,** 399 (1992).

[5] J.-P. Perreault and S. Altman, *J. Mol. Biol.* **230,** 750 (1993).

[6] M. J. Moore and P. A. Sharp, *Science* **256,** 992 (1992).

[7] J. Kufel and L. A. Kirsebom, *Proc. Natl. Acad. Sci. U.S.A.* **93,** 6085 (1996).

[8] J. Kufel and L. A. Kirsebom, *RNA* **4,** 777 (1998).

[9] J. M. Warnecke, J. P. Fürste, W.-D. Hardt, V. E. Erdmann, and R. K. Hartmann, *Proc. Natl. Acad. Sci. U.S.A.* **93,** 8924 (1996).

RNA.[10,11] According to the ligation protocol a monophosphate at the 5′ end (i.e., at the point of ligation) is a requirement for ligation. Our experience is that ligation between positions 14 and 15 in the D-loop is efficient (>50%). We have also tried to ligate in the T-loop but with less success. These RNA molecules are annealed to a DNA oligonucleotide (~20-mer; defined as the splint), complementary to the junction region in the D-loop, in 50 m*M* KCl, 50 m*M* Tris-HCl (pH 7.9) by heating at 95° for 5 min, followed by cooling on ice. Ligation is performed in ligation buffer [50 m*M* Tris-HCl (pH 7.5), 10 m*M* $MgCl_2$, 20 m*M* dithiothreitol (DTT), bovine serum albumin (BSA, 50 μg/ml), and polyethylene glycol (PEG) 6000 to a final concentration of 12–14% (w/v)] supplemented with RNA Guard (0.4 unit/μl; Pharmacia) and 1 m*M* ATP (final concentration), using T4 DNA ligase (1 unit/μl; Pharmacia). After incubation overnight at 16° the products are separated on a denaturing polyacrylamide gel and purified as described above. Precursors generated in this way are subsequently used as substrates for RNase P as described above.

To introduce a modified nucleotide at a specific position in M1 RNA, for example replacement of the R_p- or S_p-oxygen 3′ of G293, the following approach can be taken.[8] Two M1 RNA fragments are made using T7 DNA-dependent RNA polymerase, one containing residues 1–291 and the other, starting with a monophosphate (GMP), containing residues 300–377. The remaining fragment comprising residues 292–299 with and without substitutions is chemically synthesized. The three RNA fragments are then ligated according to Moore and Sharp[6] as described above, using a DNA splinter of appropriate length. The ligated M1 RNA derivatives are gel purified (see above). With this protocol the ligation efficiency is approximately 1%. We have tried other ligation points but so far we have not been able to identify a ligation point that gives a higher efficiency of ligation than 1% when three pieces of RNA are ligated simultanously.

The fact that T7 RNA polymerase can use other nucleotides[10,12] (e.g., GMP, dGMP, IMP, and m^7GMP) to initiate transcription can be exploited to introduce a modified G at a specific position in M1 RNA. By following a modification of the protocol described above two RNA fragments are generated, one starting at position 1 and ending at position 299 and the other starting at position 300 (wild-type M1 RNA carries a G at position 300). The transcription of the second RNA molecule is done in the presence of a modified G (monophosphate), resulting in molecules with a modified G at the 5′ end. These two pieces of RNA are then ligated as described above. RNA molecules that carry a triphosphate at the 5′ end will not be used as substrate during the ligation step because ligation requires the presence of a 5′-monophosphate. Here our experience is that the efficiency can be >1%; however, this appears to depend on the modified nucleotide as well as the

[10] J. R. Sampson and O. C. Uhlenbeck, *Proc. Natl. Acad. Sci. U.S.A.* **85,** 1033 (1988).

[11] J. R. Wyatt, M. Chastain, and J. D. Puglisi, *BioTechniques* **11,** 764 (1991).

[12] C. Pitulle, R. G. Kleineidam, B. Sproat, and G. Krupp, *Gene* **112,** 101 (1992).

point of ligation. The concentrations of RNA fragments and DNA splinter have been approximately 1 μM irrespective of ligation.

To incorporate modified nucleotides into small RNA molecules we have also used a modified transcription protocol described by Conrad *et al.*[13] In this protocol three of the unmodified NTPs and one of the corresponding modified nucleotides [dUTP (Sigma) or NTPαS (Amersham, Arlington Heights, IL)], are added to the reaction mixture at a final concentration of 1 mM. Noteworthy is that in the transcription reactions with dUTP, $MgCl_2$ was substituted with 2.5 mM $MnCl_2$. The RNA obtained contains modified bases at several positions depending on the sequence of the template. In certain experiments these types of modified RNA can be useful.[8]

Preparation of C5 Protein and Holoenzyme

The RNase P protein subunit, C5, is essential for catalytic activity *in vivo*. To investigate its role in the cleavage reaction catalyzed by RNase P the catalytic performance of the holoenzyme or of the reconstituted holoenzyme is studied. These data are compared with data obtained by studying cleavage by M1 RNA alone. Different protocols have been established for preparing purified holoenzyme and C5 protein. The purified C5 protein and *in vitro*-transcribed M1 RNA are then used to generate reconstituted holoenzyme. The most commonly used protocols for purification of either RNase P holoenzyme or C5 protein have been described in detail in a previous volume in this series.[14] We have followed this protocol to obtain highly active C5 protein. Briefly, C5 protein is overexpressed in a BL21(DE3) strain carrying the C5 gene behind the T7 promoter, the cells are disrupted, and an S30 extract is prepared in which the protein remains in the supernatant. From the S30 extract an S100 extract is prepared and the C5 protein sticks to the ribosomal pellet. The pellet is washed with 1 M NH_4Cl, the suspension is centrifuged at 100,000g at 4° and the supernatant containing the C5 protein is collected for further purification by dialysis and fractionation on a CM-Sephadex C-50 column. The purified C5 protein is stored at −80° in small aliquots at a concentration of 0.2 mg/ml or higher in a buffer containing 7 M urea. Note that diluted solutions of C5 are unstable. This procedure, with some modifications, has also been used to purify mutant C5 proteins.[14,15] The modifications are the following: (1) instead of crushing the cells in alumina the cells are disrupted by sonication; and (2) to promote C5 dimer formation a reducing agent such as DTT is omitted. The idea with the latter modification is that this facilitates purification of mutant variants. An advantage

[13] F. Conrad, A. Hanne, R. K. Gaur, and G. Krupp, *Nucleic Acids Res.* **23,** 1845 (1995).

[14] M. F. Baer, J. G. Arnez, C. Guerrier-Takada, A. Vioque, and S. Altman, *Methods Enzymol.* **181,** 569 (1990).

[15] V. Gopalan, A. D. Baxevanis, D. Landsman, and S. Altman, *J. Mol. Biol.* **267,** 818 (1997).

is that while the wild-type C5 protein remains in the supernatant S30, the mutant C5 proteins are found to be associated with the pellet, P30, and the distribution between the S30 fraction and the pellet varies depending of C5 variant. This facilitates the purification of mutant C5 protein even in the presence of wild-type C5 protein.

Baer *et al.*[14] provide a protocol for purification of the RNase P holoenzyme. However, here a protocol developed by Lazard and Meinnel[16] is briefly mentioned. The source for RNase P in their purification scheme is a strain that overproduces both M1 RNA and the C5 protein, because of the presence of a plasmid carrying the genes encoding both RNase P subunits. It is claimed that this strain overproduces RNase P holoenzyme 15-fold. After cultivation, harvest, and ammonium sulfate precipitation an RNase P fraction is obtained that is further purified by gel filtration and anion-exchange chromatography. This is followed by concentration of the fractions containing RNase P activity and chromatography twice on a Superdex 200 (Pharmacia) column. The RNase P holoenzyme is stored in 55% (v/v) glycerol and is claimed to be >95% pure as judged by SDS–polyacrylamide gel electrophoresis (PAGE). The authors also claim that the yield is on the order of 0.8 mg/liter of bacterial culture and that RNase P holoenzyme prepared in this way retains full activity after storage at −30° for several months.

Reaction Conditions

Buffer Conditions

M1 RNA alone and the holoenzyme have different buffer requirements for optimal cleavage activity, such that a higher concentration of Mg^{2+} is required for cleavage in the absence of the C5 protein (Guerrier-Takada *et al.*[1]; see also below). The standard buffer conditions we have used for cleavage by M1 RNA alone are described by Guerrier-Takada *et al.*[17] Buffer A contains 50 m*M* Tris-HCl, 100 m*M* NH_4Cl, 100 m*M* $MgCl_2$, and PEG 6000 (5%, w/v) with a final pH at 37° of 7.2. In the presence of C5 the same buffer is used with the exception that the concentration of Mg^{2+} is reduced to 10 m*M* (buffer B). Other buffer conditions are used in other laboratories and these function equally well. To give a few examples: 50 m*M* Tris-HCl (pH 8), 0.1% (w/v) SDS, 0.05% (w/v) Nonidet P-40, and high concentrations (≤3 *M*) of NH_4Cl[18]; 50 m*M* morpholineethanesulfonic acid (MES), 0.1 m*M* EDTA, 1 *M* ammonium acetate, 15 m*M* Magnesium acetate with a final pH of 6.0 at 37°[9] and 50 m*M* Tris-HCl

[16] M. Lazard and T. Meinnel, *Biochemistry* **37,** 6041 (1998).

[17] C. Guerrier-Takada, A. van Belkum, C. W. A. Pleij, and S. Altman, *Cell* **53,** 267 (1988).

[18] E. S. Haas, J. W. Brown, C. Pitulle, and N. R. Pace, *Proc. Natl. Acad. Sci. U.S.A.* **91,** 2527 (1994).

(final pH at 37° of 7.2), 10 m*M* spermidine, 10 m*M* $MgCl_2$, 1 m*M* NH_4Cl, and 5% (w/v) PEG 6000.[19] The choice of buffer conditions is of course dependent on the experiment. However, when the task is to compare the catalytic performance of different M1 RNA variants it is important that cleavage by the variants as well as by the wild type is performed under identical buffer conditions. This is also relevant when the aim is to assess the function of the C5 protein in the RNase P-catalyzed reaction. A problem here is the fact that an increase in the concentration of Mg^{2+} (starting at 10 m*M*) results in a reduced activity for the reconstituted holoenzyme while cleavage by M1 RNA alone is stimulated. To investigate the role of C5 we have determined the kinetic parameters for cleavage both in the absence and in the presence of C5 using a "compromise" buffer: Buffer IB [50 m*M* Tris-HCl (pH 7.5), 100 m*M* NH_4Cl, 20 m*M* $MgCl_2$ and 5% (w/v) PEG 6000].[20]

Assay Conditions

The concentrations of M1 RNA and holoenzyme vary according to the type of experiment that is performed; however, the following protocol can be used as a guideline. In the reaction with M1 RNA alone, typically 20 n*M* M1 RNA is preincubated before the addition of substrate for at least 6 min at 37° in buffer A. This step is necessary to allow proper folding of the RNA.[21] This is followed by the addition of substrate that has been incubated at 37° for at least 2 min. The final concentration of substrate is dependent on the experiment and on which substrate is used. Samples are withdrawn at different time points and stopped by adding four to five times the volume of warm stop solution (10 *M* urea, 10 m*M* EDTA, bromphenol blue, and xylene cyanol blue). When kinetic constants are determined conditions are adjusted so that the measurements are carried out in the linear portion of the cleavage reaction for different concentrations of substrate added. The extent of cleavage is quantified with a PhosphorImager (Molecular Dynamics, Sunnyvale, CA) and the kinetic constants are determined by Eadie–Hofstee plots.

Generally cleavage in the presence of C5 requires less M1 RNA, typically 1.6 n*M*. The amount of C5 needed to generate reconstituted holoenzyme depends on the activity of the C5 preparation. To determine this we carry out a titration with C5 to find the amount required to obtain optimal activity. In these assays we mix M1 RNA and C5 on ice and preincubate this mixture for at least 6 min at 37° before the addition of the prewarmed substrate (see above). The reaction is terminated as described above. A variant to this protocol is first to preincubate M1 RNA alone in buffer B, add prewarmed substrate, and continue to incubate this mixture for

[19] N. E. Mikkelsen, M. Brännvall, A. Virtanen, and L. A. Kirsebom, *Proc. Natl. Acad. Sci. U.S.A.* **96,** 6155 (1999).
[20] A. Tallsjö and L. A. Kirsebom, *Nucleic Acids Res.* **21,** 51 (1993).
[21] C. Guerrier-Takada, K. Haydock, L. Allen, and S. Altman, *Biochemistry* **25,** 1509 (1986).

2 min (under these buffer conditions no cleavage by M1 RNA alone is detected). This preincubation is followed by the addition of C5 and samples are withdrawn at different time points and cleavage is stopped as outlined above. The result will be a lag, which is due in part to reconstitution of the holoenzyme. If a mutant M1 RNA that carries a change at a position important for C5 interaction is used this lag is increased.[22] Thus, this qualitative assay is useful to obtain an indication of whether a specific residue or a structural element in M1 RNA is important for binding of the RNase P protein.

As a guideline, under our standard optimal reaction conditions for cleavage of a precursor to $tRNA^{His}$ by M1 RNA in the absence and in the presence of C5 we have determined k_{cat} to be 2 and 14 min^{-1}, respectively, and K_m to be 95 n*M* and 53 n*M*, respectively.[22]

Determination of Cleavage Site

To understand the function of RNase P and its catalytic RNA subunit it is necessary to understand the mechanism of cleavage site recognition, that is, to understand the accuracy of cleavage by RNase P. We have taken this approach and used genetics as a tool to identify residues in M1 RNA that are important in this process. To achieve this methods to determine the cleavage site are crucial.

Cleavage by RNase P generates a 5′-phosphate and 3′-hydroxyl. Hence, the cleavage site can be determined (or verified) in the following way. For example, a tRNA precursor substrate in which the nucleotide at position +1 in the 5′ matured tRNA is a G can be labeled with [α-^{32}P]GTP in the transcription reaction. After cleavage with RNase P or M1 RNA the reaction products are separated on an 8% (w/v) denaturing polyacrylamide gel containing 7 *M* urea in TEB buffer [90 m*M* Tris–borate (pH 8.5), 2.5 m*M* EDTA]. Here it is important to stop the reaction before the tRNA precursor is completely processed. The 5′ matured tRNA is excised and eluted from the gel, and is digested to completion with different RNases. To detect the 5′-phosphate group-carrying nucleotide the digestion products are analyzed by two-dimensional thin-layer chromatography (TLC) according to Nishimura.[23] The first dimension solvent is NH_4OH [25% (w/v) NH_3]–propanol–H_2O (30 : 60 : 10, v/v/v); and the second dimension solvent is HCl (concentrated)–2-propanol–H_2O (17.6 : 68 : 14.4, v/v/v). The TLC plates we have used are purchased from Merck (Rahway, NJ) and coated with cellulose. To visualize the separated nucleotides we have used either autoradiography or a PhosphorImager. An alternative nucleotide that can be labeled is the nucleotide at

[22] A. Tallsjö, S. G. Svärd, J. Kufel, and L. A. Kirsebom, *Nucleic Acids Res.* **21,** 3927 (1993).

[23] S. Nishimura, *Prog. Nucleic Acids Res. Mol. Biol.* **12,** 50 (1972).

the +2 position. Because it is the α-phosphate that is labeled, the ^{32}P label will be linked to the pNp nucleotide as a result of RNase digestion (see also Tallsjö *et al.*[24]).

A complementary way to determine the cleavage site is to compare the mobility of 5′ cleavage fragments on a denaturing polyacrylamide gel (see above) with the size of a known RNase P cleavage fragment. The percentage of polyacrylamide is dictated by the size of the expected 5′ cleavage fragment. Furthermore, it is well documented that certain precursors that are miscleaved in the 5′ leader generate a product that is subsequently used as a substrate and is cleaved a second time (Tallsjö *et al.* [24] and references therein). Thus, determination of the cleavage site by TLC (see above) as well as by the size of the 5′ cleavage fragment will show whether this is the case or not. An advantage of analyzing the size of the 5′ cleavage fragment is that information about the recognition of the initial cleavage site is obtained. This is particularly informative when a precursor is cleaved initially at more than one position.

As an alternative method, primer extension has been used to determine the cleavage site.[25] However, a disadvantage with this approach is that it is impossible to determine whether the 5′ matured tRNA carries a phosphate at its 5′ terminus.

Metal Ion-Induced Cleavage of M1 RNA

The RNase P RNA-catalyzed reaction requires the presence of divalent metal ions, with Mg^{2+} promoting the most efficient cleavage to date (2). The functions of Mg^{2+} are to (1) ensure correct folding of the RNA, (2) promote interaction with the substrate, and (3) participate in the chemistry of cleavage resulting in cleavage at the correct site. It has been suggested that the true substrate for RNase P is a substrate with a Mg^{2+} bound close to the cleavage site.[5] Taken together, to understand the mechanism of RNase P cleavage it is necessary to identify the functionally important Mg^{2+} ions and where they are located in RNase P RNA in the absence as well as in the presence of the substrate. Different approaches have been explored to identify nucleotides in an RNA molecule that are involved in Mg^{2+} binding.

Divalent Metal Ion Cleavage of RNase P RNA and Its Substrate

The information that can be extracted by divalent metal ion cleavage of RNase P RNA and its substrate suggests that a divalent metal ion(s) is positioned close to the metal ion-induced cleavage site, but this does not give any information about how the metal ion is coordinated. However, on the basis of the suggestion that

[24] A. Tallsjö, J. Kufel, and L. A. Kirsebom, *RNA* **2,** 299 (1996).
[25] C. J. Green and B. S. Vold, *J. Biol. Chem.* **263,** 652 (1988).

the 2′-hydroxyl immediately 5′ of the scissile bond is actively involved in the chemistry of cleavage gives some structural constraints for the positioning of the divalent metal ion(s).[26] Here two protocols used to cleave M1 RNA with divalent metal ions are described.

1. Pb^{2+}-induced cleavage of M1 RNA (and/or its substrate): The protocol we have followed was first described by Ciesiolka *et al.*[27] When labeled M1 RNA is used the RNA is usually labeled at the 3′ end with [^{32}P]pCp, using standard procedures,[28] and purified on a 6% (w/v) denaturing polyacrylamide gel as described above. The RNA is renatured by heating for 5 min at 55°. Approximately 20,000 to 30,000 cpm of labeled M1 RNA is mixed with ~2.5 pmol of unlabeled M1 RNA and preincubated in 50 m*M* Tris-HCl (pH 7.5), 100 m*M* NH_4Cl, and 10 m*M* $MgCl_2$ for 10 min at 37°. Available data suggest that M1 RNA is completely folded under these conditions (see Altman and Kirsebom[2]). Cleavage is initiated by the addition of freshly prepared lead acetate to a final concentration of 0.5 m*M*. However, depending on the nature of the experiment we have used other concentrations of lead acetate. The final volume of the reaction is 10 μl and the reaction is terminated after 10–15 min by the addition of 2 volumes of stop solution [9 *M* urea, 25 m*M* EDTA, 0.1% (w/v) bromphenol blue]. The cleavage products are separated on an 8% (w/v) denaturing polyacrylamide gel. The Pb^{2+}-induced cleavage sites are mapped by primer extension analysis, using primers complementary to specific positions in M1 RNA. For this purpose we use unlabeled M1 RNA, and 15–20 nucleotide primers. An increase in the concentration of Mg^{2+} (or some other divalent metal ion such as Mn^{2+}) results in a suppression of the Pb^{2+}-induced cleavage of M1 RNA, suggesting that Mg^{2+} and Pb^{2+} bind, if not to the same site, at least to overlapping sites. In combination with the use of genetics (i.e., by using M1 RNA variants) or by studying cleavage of the M1 RNA–substrate complex, it is also possible to use the Pb^{2+}-induced cleavage to probe for structural changes in M1 RNA. Note that the formation of M1 RNA–substrate complex requires a higher concentration of Mg^{2+} ($\geq$20 m*M*). Therefore, an increased concentration of Pb^{2+} is needed to detect cleavage.

2. M1 RNA is also cleaved by other divalent metal ions such as Mg^{2+}.[29] However, the Mg^{2+}-induced cleavage of M1 RNA is less efficient compared with the Pb^{2+}-induced cleavage and in order to detect cleavage the reaction must be performed at a higher pH and in the presence of 10% (v/v) ethanol. The conditions we have used are 50 m*M* 2(*N*-cyclohexylaminoethane sulfonic acid (CHES) buffer

[26] R. S. Brown, J. C. Dewan, and A. Klug, *Biochemistry* **24,** 4785 (1985).

[27] J. Ciesiolka, W.-D. Hardt, J. Schlegl, V. A. Erdmann, and R. K. Hartmann, *Eur. J. Biochem.* **219,** 49 (1994).

[28] T. E. England, A. G. Bruce, and O. C. Uhlenbeck, *Methods Enzymol.* **65,** 65 (1980).

[29] S. Kazakov and S. Altman, *Proc. Natl. Acad. Sci. U.S.A.* **88,** 9193 (1991).

(pH 9.5), 100 m*M* NH_4Cl, 10 m*M* $MgCl_2$ (a higher concentration of Mg^{2+} can be used), and 10% (v/v) ethanol.[29] The reaction is incubated at 37° for 6 hr, and terminated and the cleavage products are separated and characterized as described above.

3. Modification–interference approaches have also been used to identify *pro*-R_p-oxygens used as ligands for functionally important Mg^{2+} in M1 RNA.[30,31] Briefly stated, the idea is that replacement of a *pro*-R_p-oxygen with a sulfur can result in an inactive M1 RNA in the presence of Mg^{2+}, because Mg^{2+} has a significantly lower affinity for sulfur than for oxygen (see below). The next step is to perform a rescue experiment by adding a more thiophilic metal ion, for example, Mn^{2+} or Cd^{2+}, to regain activity.

4. An approach we have used to map Mg^{2+} binding sites in a functionally important region of M1 RNA is to generate a small model RNA molecule representing the M1 RNA region.[8] Different model R_p-phosphorothioate-modified RNA variants carrying modified nucleotides at specific positions were generated using T7 RNA polymerase as described above. The idea is that sulfur is a poor coordination partner for Mg^{2+}, whereas Pb^{2+} is much more thiophilic.[32] As a consequence, coordination of Mg^{2+} to the ligands at the positions substituted with sulfur will be significantly decreased, whereas Pb^{2+} will retain or even increase its binding capacity. Thus, by following the Pb^{2+}- and Mg^{2+}-induced cleavage patterns of the R_p-phosphorothioate-modified RNA variants it is possible to identify *pro*-R_p-oxygen ligands for both metal ions. Cleavage with the Pb^{2+} and Mg^{2+} is performed as described above. To be able to apply the information obtained using small model RNA molecules it is important to demonstrate that the folding of the domain is similar, irrespective of whether it is part of the full-length M1 RNA or part of the model RNA.

5. Phosphothioates and deoxy-modified nucleotides have also been introduced at the cleavage site in different substrates in order to map ligands that are important for Mg^{2+} binding at the cleavage site.[4,5,7,9,24,33] In this context see also the work by Christian *et al.*[34]

Cross-Linking Protocols

Cross-linking can provide information about nucleotides in M1 RNA that are in close proximity to residues in the substrate. We have followed two different protocols: (1) direct UV-induced cross-linking as described by Guerrier-Takada

[30] W.-D. Hardt, J. M. Warnecke, V. A. Erdmann, and R. K. Hartmann, *EMBO J.* **14,** 2935 (1995).
[31] M. E. Harris and N. R. Pace, *RNA* **1,** 210 (1995).
[32] E. K. Jaffe and M. Cohn, *J. Biol. Chem.* **254,** 10839 (1979).
[33] Y. Chen, X. Li, and P. Gegenheimer, *Biochemistry* **36,** 2425 (1997).
[34] E. L. Christian, N. M. Kaye, and M. E. Harris, *RNA* **6,** 511 (2000).

et al.[35] and Kufel and Kirsebom,[36] and (2) a specific UV-induced cross-linking protocol in which we used substrates carrying a 4-thiouridine (4-thioU) at the cleavage site.[7,37] Other cross-linking procedures have also been used in the laboratories of Pace and Harris (see, e.g., Harris *et al.*[38] and Christian and Harris[39]).

Direct UV-Induced Cross-Linking

To allow M1 RNA–substrate complex formation, precursor tRNA (~40 μg) is incubated with M1 RNA (~40 μg) in buffer A for 10 min at 37°. This mixture is UV irradiated at 254 nm on a plastic microtiter plate at a distance of ~1 cm from the UV source. As a UV source we have used a short-wave UV lamp model UVG-54 (Ultra Violet Products, San Gabriel, CA). After 10 min the reaction is terminated by adding 2 volumes of stop solution (see above) and the cross-linked products are separated on a 6% (w/v) denaturing polyacrylamide gel and cross-linked M1 RNA–substrate complexes are eluted as described above. Identification of cross-linked residues is performed by primer extension analysis using short oligonucleotides complementary either to M1 RNA or to the substrate (see above).

Specific UV-Induced Cross-Linking

A precursor carrying a 4-thioU at a specific position (e.g., at the cleavage site) is prepared according to the protocol described above. Approximately 20 μg of the modified precursor is incubated with a 2.5-fold molar excess of M1 RNA in buffer A for 10 min at 37°. This mixture is UV irradiated on ice for 30 min, using a long-wave UV lamp (366 nm), followed by precipitation and gel separation. The cross-linked species are identified, eluted, and primer extension analysis is used to map cross-linked residues in M1 RNA. The advantage of using the 4-thioU-modified substrate is that it is known from the start that the 4-thioU is the cross-linking partner in the substrate, making it possible to map nucleotides in M1 RNA that are in close contact with this specific residue.

In Vivo Studies

Most of what is currently known about RNase P processing in relation to substrate recognition comes from studies performed *in vitro*. To investigate RNase P processing *in vivo* an experimental system has been devised in which plasmids carrying different tRNA precursor gene constructs are placed under the control of

[35] C. Guerrier-Takada, N. Lumelsky, and S. Altman, *Science* **286,** 1578 (1989).
[36] J. Kufel and L. A. Kirsebom, *J. Mol. Biol.* **244,** 511 (1994).
[37] J. Kufel and L. A. Kirsebom, *J. Mol. Biol.* **263,** 685 (1996).
[38] M. E. Harris, A. V. Kazantsev, J.-L. Chen, and N. R. Pace, *RNA* **3,** 561 (1997).
[39] E. L. Christian and M. E. Harris, *Biochemistry* **38,** 12629 (1999).

the T7 promoter. These tRNA precursors are expressed *in vivo* in various *E. coli* BL21(DE3) derivatives.[40,41] The BL21(DE3) strain harbors the T7 RNA polymerase gene on the chromosome under the control of the *lacUV5* promoter. It is also possible to use an *E. coli* strain carrying a plasmid that encodes the T7 RNA polymerase gene under the control of an inducible promoter. An advantage with the *lacUV5* promoter, however, is that it is leaky, resulting in low basal expression even in the absence of an inducer gene construct with a T7 promoter. Basal expression is usually sufficient for *in vivo* studies; however, expression can be increased by the addition of isopropyl-β-D-thiogalactopyranoside (IPTG). Depending on the experiment we have used final concentrations of IPTG ranging between 0 and 2 m*M*. In our studies we have used the precursor to tRNA ^{Tyr}Su3 (pSu3) as a model precursor. Processing of this precursor generates a tRNA nonsense (UAG) suppressor that can easily be detected *in vivo* (see below). Thus, to investigate tRNA processing *in vivo*, in particular with respect to RNase P processing, various tRNA precursor gene constructs are cloned behind the T7 promoter in the multiple cloning site of pUC19. An advantage with these constructs is that they can be used to generate tRNA precursor substrates for *in vitro* analysis, as well as to generate novel substrate variants using various *in vitro* mutation protocols (see above). Below is described the use of pSu3 as a model substrate but any substrate can be analyzed using this experimental setup.

Processing of pSu3 generates a functional tRNA nonsense (UAG) suppressor, allowing us to study whether changes in the tRNA domain, the 5′ leader, and/or the 3′ trailer of the precursor affect the efficiency of suppression. Naturally, factors influencing suppression efficiency are not restricted to the processing event. However, in most of our studies we have used pSu3 constructs that were chosen on the basis of *in vitro* studies and demonstrated to influence RNase P cleavage. This might indicate that those precursor constructs resulting in a change in the efficiency of suppression are processed less efficiently also *in vivo*. To investigate that this is actually the case, total RNA is extracted from cells harboring the different precursor constructs. The RNA is subjected to both Northern blot analysis and primer extension analysis. Northern blot analysis can reveal whether there is a correlation between the level of matured tRNA and the efficiency of suppression, whereas primer extension is a way to map the RNase P cleavage site (however, see also the discussion above).

Determination of Efficiency of Suppression

The efficiency of suppression is determined with the fused *lacIlacZ* system described by Miller *et al.*[42] In this system the F′ derivative F′122 carries an amber

[40] F. W. Studier and B. A. Moffatt, *J. Mol. Biol.* **189,** 113 (1986).

[41] L. A. Kirsebom and S. G. Svärd, *Nucleic Acids Res.* **20,** 425 (1992).

[42] J. H. Miller, C. Coulondre, and P. J. Farabaugh, *Nature (London)* **274,** 770 (1978).

(UAG) stop codon at position 189 in the *lacI* part of the hybrid gene while F′Δ14 harbors the corresponding wild-type *lacIlacZ* hybrid gene. In addition, these F′ factors carry a *pro*$^+$ gene. Different tRNA suppressors cloned behind the T7 promoter in pUC19 plasmids are transformed into a BL21(DE3) derivative harboring a deletion of the *lac* operon, Δ(*proBlac*), and transformants are grown on LB plates supplemented with ampicillin (50 μg/ml). F′Δ14 and F′122 are introduced into the resulting BL21(DE3)Δ(*proBlac*) derivative by selection for the Pro$^+$ property associated with the F′ factor, with resistance to tetracycline (Tet^R) used for counterselection [the BL21(DE3)Δ(*proBlac*) derivative carries a Tn*10* on the chromosome, which gives Tet^R]. The resulting derivatives are grown to midexponential phase at 37° in M9 minimal medium supplemented with 0.2% (w/v) glucose, ampicillin (50 μg/ml), and recommended concentrations of amino acids, except proline.[43] The cell cultures are cooled on ice and kept on ice until tested for ß-galactosidase activity. Preparation of cell extracts and ß-galactosidase activity measurements are performed as described in detail elsewhere.[44] The efficiency of suppression is given as a percentage value, calculated by dividing the specific activity of ß-galactosidase for a strain carrying F′ 122 with the ß-galactosidase activity as determined in the same strain carrying the wild-type F′Δ14.

RNA Preparations for Northern Analysis and Primer Extension Analysis

The BL21(DE3)Δ(*proBlac*)/F′ factor derivatives harboring different tRNA precursors are grown in minimal medium as described above. At an OD_{540} of ~0.5, 500 μl is transferred to an RNase-free 1.5-ml Eppendorf tube containing 200 μl of stop mix [95% (v/v) ethanol, 5% (v/v) phenol] and snap-frozen in liquid nitrogen. At this stage the cells can be stored at −70° for at least 3 months if kept in stop mix and not thawed. To prepare total RNA cells are thawed on ice and centrifuged for 5 min at 13,000 rpm at 4°. The cells are lysed by the addition of 400 μl of lysis buffer [100 m*M* Tris-HCl (pH 7.5), 40 m*M* EDTA, 200 m*M* NaCl, 0.5% (w/v) SDS] and heated to 60°. RNA is isolated by two extractions with warm (60°) TE-saturated phenol [TE: 10 m*M* Tris-HCl (pH 8.0), 1 m*M* EDTA] followed by one extraction with chloroform–isoamyl alcohol (24 : 1, v/v). The RNA is precipitated at −80° for 20 min (or overnight) by the addition of glycogen (Boehringer, Mannheim, Germany), final concentration 0.1 mg/ml, and 2.5 volumes of 95% (v/v) ethanol. This is followed by centrifugation at 13,000 rpm for 20 min and the precipitate is washed twice in cold 75% (v/v) ethanol. The resulting RNA is dissolved in 50 μl of double-distilled H_2O. RNA prepared in this way can be stored at −20° for at least 1 year. The concentration of total RNA in each preparation is determined by measuring the OD_{260}.

[43] F. C. Neidhardt, P. L. Bloch, and D. F. Smith, *J. Bacteriol.* **119,** 736 (1974).

[44] D. I. Anderson, K. Bohman, L. A. Isaksson, and C. G. Kurland, *Mol. Gen. Genet.* **187,** 467 (1982).

Northern Blot Analysis

Total RNA (10–20 μg) is separated on 8% (w/v) denaturing polyacrylamide gels and blotted onto Zeta-Probe blotting membranes (Bio-Rad, Hercules, CA) with a Bio-Rad electroblotting apparatus according to the manufacturer instructions. Filters are air dried for 30 min followed by UV cross-linking. Prehybridization is performed at 60° for 2 hr in 6× SSC, 3× Denhardt, 0.5% (w/v) SDS, salmon sperm DNA (100 μg/ml). For preparation of SSC buffer and Denhardt's solution, see Sambrook *et al.*[45] Hybridization with labeled probes is performed at 37° in prehybridization solution without salmon sperm DNA. Oligonucleotides are labeled with [γ-^{32}P]ATP, using T4 polynucleotide kinase according to the manufacturer instructions. Our experience is that the probes should be 20–25 nucleotides long and we have found that the most efficient and specific hybridization is achieved when the probe is complementary to the variable loop and the anticodon in the tRNA. The filters are washed in 6× SSC at 37–60° depending on the required stringency.

Primer Extension Analysis

The primer extension experiments are performed in principle as described by Stirling *et al.*[46] Total RNA is treated with DNase I before the primer extension analysis. Here it is important to heat the RNA (10–20 μg) at 85° for 5 min together with the labeled primer under hybridization conditions and then let it cool slowly to 42°. Our experience indicates that the best result is obtained when the primer extension reaction is performed at 42°. The tertiary structure of the matured tRNA can sometimes generate problems due to premature termination during the reaction. This problem is overcome by using an oligodeoxyribonucleotide complementary to the D-loop of the tRNA.[41] To identify the RNase P cleavage site a sequencing ladder with the same oligodeoxyribonucleotide used in the primer extension reaction is generated. The 5′ end of an RNA molecule can also be mapped by 5′ rapid amplification of cDNA ends (RACE) or reverse ligation-mediated (RLM)-RACE.[47]

One problem is that a change in the tRNA precursor that results in low amounts of matured tRNA might reflect degradation of either the precursor and/or the processed tRNA. To address this we have developed a protocol to study the processing of a specific precursor *in vivo*.[48] Here we take advantage of the fact that *E. coli*

[45] J. Sambrook, E. F. Fritsch, and T. Maniatis, "Molecular Cloning: A Laboratory Manual," 2nd Ed. Cold Spring Harbor Laboratory Press, Cold Spring Harbor, New York, 1989.

[46] D. A. Stirling, C. S. J. Hulton, L. Waddell, S. F. Park, G. S. A. B. Stewart, I. R. Booth, and C. F. Higgins, *Mol. Microbiol.* **3,** 1025 (1989).

[47] B. C. Schaefer, *Anal. Biochem.* **227,** 255 (1995).

[48] L. A. Kirsebom and S. G. Svärd, *in* "Ribozymes" (G. Krupp and R. K Gaur, eds.). Eaton Publishing, Natick, Massachusetts, 2000 (in press).

RNA polymerase is inhibited by rifampin whereas T7 RNA polymerase is not sensitive to this antibiotic. Addition of rifampin will therefore shut down transcription of genes that are transcribed by *E. coli* RNA polymerase, whereas a gene construct with a T7 promoter will still be transcribed by the T7 RNA polymerase. A tRNA gene construct, for example, pSu3, carrying a T7 promoter as well as a T7 terminator downstream of the tRNA domain of the gene is transferred into an *E. coli* strain harboring the T7 RNA polymerase [e.g., BL21(DE3)]. The rationale with the T7 transcription terminator is that its presence will lead to tRNA precursor accumulation and consequently allow us to obtain an estimate of the rate of tRNA processing *in vivo* as a function of changes in the precursor by following the disappearance of the precursor and the accumulation of the resulting matured tRNA. For details see Kirsebom and Svärd.[48]

In Vivo Labeling Protocol

A strain harboring the T7 RNA polymerase gene, for example, BL21(DE3) and a tRNA precursor gene construct (see above) is grown to an OD_{540} of ~0.5 in 10 ml of phosphate-starvation medium [0.1 *M* Tris-HCl (pH 7.4), 30 m*M* KCl, 2.5 m*M* $MgCl_2$, 1 m*M* $CaCl_2$, 0.2% (w/v) glucose, 1 m*M* $FeCl_3$, 0.3 m*M* Na_2PO_4, 0.4 m*M* Na_2SO_4, vitamin B_1 (1 μg/ml), ampicillin (50 μg/ml)]. At an OD_{540} of ~0.5, IPTG is added to a final concentration of 2 m*M*. Rifampin is added to a final concentration of 200 μg/ml at time =25 min, and 10 μCi of ortho[^{32}P]phosphate is added 45 min after the addition of IPTG. We define this time as 0 min. Samples (500 μl) are withdrawn at different time intervals (5–60 min) after addition of ortho[^{32}P]phosphate and total RNA is extracted as described above. Equal amounts of radioactively labeled total RNA are added to each lane and the result is analyzed on 8% (w/v) denaturing polyacrylamide gels.

The appearance of mature tRNA is the result of cleavage by RNase P as well as other processing enzymes. Therefore, to obtain information about RNase P cleavage the data generated by the various *in vivo* approaches should be taken into account as well as *in vitro* data where cleavage of the same tRNA precursors has been analyzed. In addition, the stringency of the *in vivo* analysis can be improved by using *E. coli* strains carrying mutations in the gene encoding either M1 RNA or the C5 protein.

Acknowledgments

Dr. D. Hughes is acknowledged for critical reading of the manuscript. The ongoing work in the laboratory of L.A.K. is supported by the Swedish Natural Science Research Council and the Foundation for Strategic Research.

[9] Human Ribonuclease P

By NAYEF JARROUS and SIDNEY ALTMAN

Introduction

Ribonuclease P is a ribonucleoprotein nuclease required for the site-specific cleavage of the 5′ leader sequence of precursor tRNAs. In eubacteria, the RNA subunit of RNase P is the catalytic moiety and is capable of processing precursor tRNA in the presence of divalent metal ions.[1] The single protein subunit of bacterial RNase P acts as a cofactor.[2,3] In contrast, the RNA subunits of eukaryotic RNase P (human and yeast) do not exhibit enzymatic activity *in vitro,* an indication that their protein components are required for the catalytic reaction. In this article, we describe a purification procedure for human RNase P from HeLa cells that facilitated the characterization of nine of its protein subunits. This procedure also allows the separation of RNase P from the structurally related RNase MRP nuclease.

Purification of RNase P from HeLa Cells

Preparation of S100 Extracts of HeLa S3 Cells

An outline of a purification procedure of human RNase P from HeLa cells is shown in Fig. 1. For purification of sufficient RNase P for analysis of protein subunits,[4] frozen pellets of HeLa S3 cells [Cell Culture Center, National Center for Research Resources (NCRR), National Institutes of Health (NIH), Bethesda, MD] harvested from 120 liters of cell culture at a density of 6×10^5 cells/ml are used. The estimated copy number of RNase P is $\sim 2 \times 10^5$/cell.[5] Cell pellets are placed on ice for 1–2 hr for thawing before addition of 5 volumes of lysis buffer [10 m*M* Tris-HCl (pH 8.0), 2.5 m*M* $MgCl_2$, 5 m*M* KCl, 1 m*M* dithiothreitol (DTT), 0.2 m*M* Pefabloc protease inhibitor] made with sterilized Milli-Q water. Resuspended cells are left on ice for 30 min before their complete disruption by 10–15 strokes in a Dounce homogenizer. Cell debris is removed from the crude extract by centrifugation in a Sorvall (Newtown, CT) SS34 rotor at

[1] C. Guerrier-Takada, K. Gardiner, T. Marsh, N. Pace, and S. Altman, *Cell* **35,** 849 (1983).

[2] A. Vioque and S. Altman, *Proc. Natl. Acad. Sci. U.S.A.* **83,** 5904 (1986).

[3] V. Gopalan, H. Kuhne, R. Biswas, H. Li, G. W. Brudvig, and S. Altman, *Biochemistry* **38,** 1705 (1999).

[4] P. S. Eder, R. Kekuda, V. Stolc, and S. Altman, *Proc. Natl. Acad. Sci. U.S.A.* **91,** 1101 (1997).

[5] Y.-T. Yu, E. C. Scharl, C. M. Smith, and J. A. Steitz, *in* "The RNA World" (R. F. Gesteland, T. R. Cech, and J. F. Atkins, eds.), p. 487. Cold Spring Harbor Laboratory Press, Cold Spring Harbor, New York, 1999.

0076-6879/01 $35.00

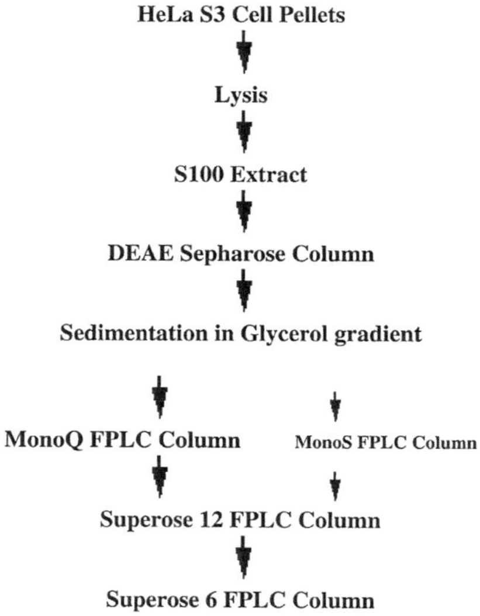

FIG. 1. An outline of purification of human nuclear RNase P from HeLa S3 cells. Large arrows show main path of purification of the enzyme for protein analysis, as described in text.

7000 rpm for 30 min at 4°. This step is followed by ultracetrifugation of the extract at 40,000 rpm in a Beckman (Fullerton, CA) Ti60 rotor at 2° for 2 hr to obtain S100 extract. The specific activity of RNase P in the S100 extracts is determined (see assay below), and then KCl is added to a final concentration of 100 m*M*.

Fractionation of RNase P by DEAE Anion-Exchange Chromatography

DEAE Sepharose is a fast flow, weak anion exchanger suitable for the initial purification of ribonucleoproteins from cell crude extracts. A slurry of 500 ml of DEAE Sepharose Fast Flow beads (Amersham Pharmacia Biotech, Piscataway, NJ) is degassed under vacuum for 30 min and packed in a column at 4°. The DEAE column is prewashed with 5 liters of equilibration buffer [10 m*M* Tris-HCl (pH 8.0), 100 m*M* KCl, 2.5 m*M* $MgCl_2$, 1 m*M* DTT] before loading the S100 extract described above. The column is washed with 5 liters of equilibration buffer after the sample is loaded and bound RNase P is eluted with a linear gradient from 0.1 to 0.5 *M* KCl in 4 liters of equilibration buffer. Fractions (20 ml) are assayed for RNase P activity, which is typically eluted between 0.2 and 0.4 *M* KCl. Fractions containing the peak of RNase P activity are pooled and the specific activity of the enzyme is determined (Table I; purification is about 30-fold over the S100 extract). A nonspecific nuclease activity can be detected in fractions preceding

TABLE I
PURIFICATION PROFILE OF HUMAN RNase P FROM HeLa CELLS[a]

Purification step	Volume (ml)	Protein (mg)	Total activity ($U \times 10^5$)	Specific activity ($U/mg \times 10^5$)	Purification (fold)
S-100	450	6,300	9,000	1.4	—
DEAE	180	740	32,000	43	30
Glycerol	80	140	24,000	170	120
Mono Q	7	3	2,100	720	510
Superose 12	2	0.2	640	3,200	2,300
Superose 6	ND	ND	ND	ND	ND

Abbreviation: ND, Not determined.
[a] P. S. Eder, R. Kekuda, V. Stolc, and S. Altman, *Proc. Natl. Acad. Sci. U.S.A.* **91,** 1101 (1997).

those enriched with RNase P. Therefore, these fractions should be excluded from further analysis.

Velocity Sedimentation of RNase P in Glycerol Density Gradients

The pool of the DEAE-purified RNase P described above is concentrated to 6 ml by centrifugation at 3500 rpm in Centriprep-10 concentrators (Amicon, Danvers, MA). Samples of 1 ml each are layered onto 11 ml glycerol density gradients [15–30% (v/v) glycerol in 10 m*M* Tris-HCl (pH 8.0), 2.5 m*M* $MgCl_2$, 200 m*M* KCl, 1 m*M* DTT, 0.1 m*M* Pefabloc protease inhibitor] prepared in Beckman Ultra-Clear centrifuge tubes ($9/16 \times 3\frac{1}{2}$ in.) and centrifuged at 2° for 26 hr in a Beckman SW41 rotor at 40,000 rpm. Fractions of 0.4 ml are collected from the bottom of the tube by inserting an 18-gauge needle while the top of the tube is sealed in a gradient fraction collector. Fractions are assayed for RNase P and the specific activity of the peak is determined. The purification of RNase P at this step is ~120-fold (Table I).[4]

Human nuclear RNase P has a sedimentation coefficient of ~15S in glycerol gradients.[6] Its sedimentation profile partially overlaps with that of the structurally related ribonucleoprotein RNase MRP, an RNA-processing nuclease found in nucleoli and mitochondria.[7,8]

Fractionation of RNase P by Mono Q Ion-Exchange Chromatography

For further purification of RNase P, we use a Mono Q column in a fast protein liquid chromatography (FPLC) system (Pharmacia, Piscataway, NJ). A Mono Q

[6] M. Bartkiewicz, H. Gold, and S. Altman, *Genes Dev.* **3,** 488 (1989).
[7] J. N. Topper, J. L. Bennett, and D. A. Clayton, *Cell* **70,** 16 (1992).
[8] T. Kiss and W. Filipowicz, *Cell* **70,** 11 (1992).

HR 10/10 column (Pharmacia) is a strong anion exchanger and it has a large binding capacity for RNase P. The column is equilibrated with 10 column volumes of MQ buffer [200 m*M* KCl, 10 m*M* $MgCl_2$, 10 m*M* Tris-HCl (pH 7.5), 2 m*M* DTT] before the RNase P sample (10 ml) from the glycerol gradient, diluted in MQ buffer and then concentrated so that the glycerol concentration is less than 5%, is injected onto the column with a 10-ml Superloop (Pharmacia). After washing with 5 column volumes of MQ buffer (at a flow rate of 0.5 ml/min), bound RNase P is eluted with a gradient of 0.2–0.7 *M* KCl. Eluted fractions (1 ml each) with peak RNase P activity are pooled (~6 ml).

Although the recovered activity of RNase P is significantly reduced at this step of fractionation, the purification factor of the enzyme is >500-fold (Table I).[4] In addition, the Mono Q FPLC column facilitates the separation of nuclear RNase P from RNase MRP RNA,[9] as has also been demonstrated for their counterparts in yeast.[10]

Gel Filtration of RNase P

RNase P can be separated from smaller protein contaminants by gel-filtration chromatography. Gel filtration of RNase P on a Superose 12 HR 10/30 (Pharmacia) FPLC column, with a fractionation range of 1×10^3 to 3×10^5 Da, is an efficient step for the purification of the enzyme as an intact ribonucleoprotein complex. Therefore, the pool of Mono Q purified RNase P (~6 ml) described above is concentrated to 0.4 ml by centrifugation at 3500 rpm in Centriprep-10 concentrators and then injected by syringe onto a Superose 12 FPLC column prewashed with S12 buffer [10 m*M* Tris-HCl (pH 8.0), 0.5 *M* KCl, 10 m*M* $MgCl_2$, 3 m*M* DTT]. RNase P is excluded from the column during washing with the S12 buffer at a flow rate of 0.5 ml/min. Fractions of 3 ml each are collected and assayed for enzyme activity.

After this step of purification, RNase P is purified >2300-fold (Table I).[4] Separation of proteins in the S12 fractions that contain RNase P by sodium dodecyl sulfate (SDS)/12–16% (w/v) polyacrylamide gel electrophoresis (PAGE), followed by silver staining, shows at least nine proteins that copurify with the enzyme activity(Table II).[4,9,11,12] Eight of these proteins, Rpp14, Rpp20, Rpp21, Rpp25, Rpp29, Rpp30, Rpp38, and Rpp40, are relatively small polypeptides (Table II). A larger protein of 115 kDa, hPop1, was previously identified on the basis of its homology to a yeast RNase P protein subunit.[13] Except for Rpp25, these proteins have been proved to be genuine subunits of nuclear RNase P.[4,9,11–13]

[9] N. Jarrous, P. S. Eder, C. Guerrier-Takada, C. Hoog, and S. Altman, *RNA* **4,** 407 (1998).

[10] Z. Lygerou, C. Allmang, D. Tollervey, and B. Seraphin, *Science* **272,** 268 (1996).

[11] N. Jarrous, P. S. Eder, D. Wesolowski, and S. Altman, *RNA* **5,** 153 (1999).

[12] N. Jarrous, R. Reiner, D. Wesolowski, H. Mann, C. Guerrier-Takada, and S. Altman, *RNA* **7,** in press (2001).

[13] Z. Lygerou, H. Pluk, W. J. van Venrooij, and B. Seraphin, *EMBO J.* **15,** 5936 (1996).

TABLE II
SUBUNIT COMPOSITION OF HUMAN NUCLEAR RNase P

Subunit	cDNA	Molecular mass: Theoretical/p*I*	Molecular mass: In gel (kDa)	Recombinant Rpp	Antibodies
Rpp14[a]	+	14/7.6	14	+	+
Rpp20[b]	+	15.6/8.7	20	+	+
Rpp21[c]	+	17.6/9.6	21	+	+
Rpp25[c]	−	—	25	−	−
Rpp29[a]	+	25.4/10.2	29	+	+
Rpp30[d]	+	29.5/9.3	30	+	+
Rpp38[d]	+	32/9.9	38	+	+
Rpp40[b]	+	34.5/5.3	40	+	+
hPop1[e]	+	115/9.4	115	−	+
H1 RNA[f]	+	102 (340 nucleotides)			

[a] N. Jarrous, P. S. Eder, D. Wesolowski, and S. Altman, *RNA* **5,** 153 (1999).
[b] N. Jarrous, P. S. Eder, C. Guerrier-Takada, C. Hoog, and S. Altman, *RNA* **4,** 407 (1998).
[c] N. Jarrous, R. Reiner, D. Wesolowski, H. Mann, C. Guerrier-Takada, and S. Altman, *RNA* **7,** in press (2001).
[d] P. S. Eder, R. Kekuda, V. Stolc, and S. Altman, *Proc. Natl. Acad. Sci. U.S.A.* **91,** 1101 (1997).
[e] Z. Lygerou, H. Pluk, W. J. van Venrooij, and B. Seraphin, *EMBO J.* **15,** 5936 (1996).
[f] M. Bartkiewicz, H. Gold, and S. Altman, *Genes Dev.* **3,** 488 (1989).

The large protein contaminants (>100 kDa) that coeluted with RNase P from the Superose 12 column[4] can be separated from the enzymatic activity by a second gel-filtration chromatography using a Superose 6 HR 10/30 FPLC column (Pharmacia). The Superose 6 column has a higher resolution capacity with a wider fractionation range (5×10^3–5×10^6 Da) compared with that of the Superose 12. The nine Rpp polypeptides described above remain tightly associated with RNase P after elution from the Superose 6 column, as determined by silver staining.[12] The same buffer and conditions described above for the use of the Superose 12 column are employed for the Superose 6 column.

Cation-Exchange Chromatography

A strong cation exchanger, such as Mono S, can also be exploited for RNase P purification. However, fractionation of Mono Q- or Superose 12-purified RNase P on a Mono S FPLC column usually results in a considerable decrease in or even complete loss of the activity of the enzyme. Therefore, we recommend loading on the Mono S column a preparation of RNase P that has been purified through the glycerol gradient step and subsequent to this column to proceed to the Superose 12 gel filtration step (Fig. 1). However, for protein subunit analysis of RNase P, we follow the main path of purification (Fig. 1, large arrows) without the Mono S step, as explained below.

Mono S HR 10/10 (Pharmacia) is equilibrated with MS buffer [10 mM Tris-HCl (pH 8.0), 10 mM $MgCl_2$, 30 mM KCl, 2 mM DTT] and the RNase P preparation is loaded with a 10-ml Superloop as described above for the Mono Q FPLC column. After washing of the column with MS buffer (20 ml; 0.5 ml/min), bound RNase P is eluted with a linear gradient of 0.03–0.5 M KCl. Fractions (1 ml each) collected are assayed for enzyme activity.

Silver staining of proteins in the Mono S fractions showed that Rpp38 is found in fractions eluted after those that contained the peak of RNase P activity (and the H1 RNA).[4,9,12] This observation was confirmed by Western blot analysis (see below).[4,12] Accordingly, it appears that Rpp38 can be stripped from the RNase P complex without complete inactivation of enzyme function.[4] This result also corroborates that RNase P binds to the cation-exchange Mono S column through its positively charged protein subunits, as would be expected.

Technical Considerations in Preparation and Storage of Active RNase P

Highly purified RNase P is an unstable complex, sensitive to both heat and to prolonged storage at 4°. Accordingly, storage of active enzyme is recommended in 30% (v/v) glycerol at −20°.

Because RNase P is a ribonucleoprotein, special caution must be undertaken during the purification procedure to avoid nuclease and protease contamination. In addition to the full-length 340-nucleotide H1 RNA, two smaller derivatives of H1 RNA of 160–170 nucleotides are detected in RNase P preparations as judged by Northern blot hybridization analysis with a specific probe against the H1 RNA.[6] These smaller RNAs are probably generated as a result of breakdown of intact H1 RNA.

Assay of RNase P Activity

The only reliable assay for measuring RNase P activity is a PAGE analysis of the processing of the 5′ leader sequence of precursor tRNA. The specific activity of RNase P is measured in units of enzyme per milligram protein (Table I). One unit of RNase P is defined as the amount of enzyme required to process 1 pmol of precursor tRNA in 1 min at 37°.[4] A standard buffer for testing active human RNase P in processing of a ^{32}P-labeled precursor tRNA is composed of 10 mM $MgCl_2$, 50 mM Tris-HCl (pH 7.5), 100 mM NH_4Cl, and 2 mM DTT. Although human RNase P is quite active in some other buffers, including those used for testing RNase MRP[7] or *E. coli* RNase P[1] activity, the presence of divalent ions in the assay buffer is critical for enzyme function. RNase P cleavage products of the ^{32}P-labeled precursor tRNA are separated in an 8% (w/v) polyacrylamide denaturing gel and then visualized and quantitated by autoradiography or by using a PhosphorImager (Fuji, Tokyo, Japan).

RNase P preparations with little activity can be partially reactivated by addition of reducing reagents, such as 0.1–0.5 m*M* 2-mercaptoethanol or 5 m*M* DTT, to the assay buffer. When crude extracts of cells are tested for enzyme function, the addition of nonspecific RNA, such as 1–10 μg of poly(I : C), and RNasin (10–50 units; Promega; Madison, WI) can minimize the nonspecific degradation of precursor tRNA.

Characterization of Protein Subunits of RNase P

RNase P purified through the gel-filtration steps (main pathway in Fig. 1) described above was separated in a preparative protein gel, followed by Coomassie blue staining.[4,12] Copurifying polypeptides (>100 ng) were excised from the gel and subjected to microsequencing of tryptic peptides (W. M. Keck Facility at Yale University).[4,9] The peptide sequences obtained were used for searching sequence databases (GenBank and EMBL) for their corresponding cDNA clones (expressed sequence tags), or for designing degenerate primers for cDNA cloning using the reverse transcription-polymerase chain reaction.[4,9,11]

The molecular cloning of the Rpp cDNAs facilitated the overexpression of their corresponding recombinant proteins in *Escherichia coli* strain BL21 (DE3) using the pHTT7K vector.[9,11] The overexpressed Rpp polypeptides contain a hexahistidine tag at their amino terminus to enable their purification by nickel (Ni^{2+}) chelate affinity chromatography. His-Bind Resin (Novagen, Madison, WI) or HiTrap FPLC columns (Pharmacia) charged with Ni^{2+} ions were utilized for affinity purification, according to the manufacturer instructions. Rpp polypeptides were purified under nondenaturing or denaturing conditions (6 *M* urea) depending on their solubility in *E. coli*. Highly purified (>90%) recombinant Rpp polypeptides (Table II) were reproducibly obtained after a single affinity purification step. For some Rpp polypeptides, such as Rpp29 and Rpp38, the predicted molecular weights deduced from their cDNA sequences were smaller than their sizes as determined by mobility in SDS–PAGE (Table II). These size discrepancies are due to anomalous migration rather than to extensive posttranslation modifications, as judged by the migration of the recombinant (bacterial) Rpp proteins.[9,11]

Recombinant Rpp polypeptides and/or synthetic Rpp peptides were used to raise polyclonal rabbit antibodies (Pocono Rabbit Farm, Canadensis, PA)[9,11] that were affinity purified through Affi-Gel 10 supports (Bio-Rad, Hercules, CA). The purified antibodies were then tested by Western blot analysis to confirm their antigenic specificity, using RNase P preparations from HeLa cells. Anti-Rpp antibodies precipitate RNase P activity and its RNA subunit from HeLa cells. In addition, anti-hPop1 and anti-Rpp antibodies bring down RNase MRP RNA from crude extracts of HeLa cells.[9,11,13,14] Therefore, these two human ribonucleoprotein nucleases share several protein subunits, a conclusion that is supported by genetic and immunobiochemical studies in lower eukaryotes, for example, in *Saccharomyces cerevisiae*.[15] However, the protein composition of human RNase MRP needs to be elucidated by biochemical purification.[16]

Conclusions

The purification procedure described above shows that nuclear RNase P from HeLa cells has at least nine protein subunits in association with a single RNA species, H1 RNA. Nuclear RNase P in other eukaryotes, such as *S. cerevisiae*[15] and *Aspergillus nidulans,*[17] also consists of multiple protein subunits, some of which are evolutionarily conserved,[11,13,15] and an RNA subunit. The precise function of these subunits in tRNA processing is not yet understood. Reconstitution of the activity of RNase P *in vitro* may reveal the role of these subunits in catalysis.

Finally, RNase P activity is also found in the mitochondria of human cells.[18,19] However, the composition of this mitochondrial nuclease and its structural relationship to the nuclear RNase P and its subunits, which are mainly localized in the nucleoli,[20,21] remain in question.

Acknowledgments

We thank Craig Crews (Yale University) and Dieter Söll (Yale University) for the use of FPLC apparatus and columns, and our colleagues Cecilia Guerrier-Takada and Donna Wesolowski for assistance and comments on this manuscript. This work was supported by U.S. Public Health Service Grant GM-19422 and Human Frontiers Science Program Grant RG O2N1 1997M to S. Altman. N. Jarrous is the recipient of a Kahanoff Foundation fellowship and supported by the Leszynski Fund for Advanced Research.

[14] H. Pluk, H. van Eenennaam, S. A. Rutjes, G. J. Pruijn, and W. J. van Venrooij, *RNA* **5,** 512 (1999).

[15] J. R. Chamberlain, Y. Lee, W. S. Lane, and D. R. Engelke, *Genes Dev.* **12,** 1678 (1998).

[16] R. Karwan, J. L. Bennett, and D. A. Clayton, *Genes Dev.* **5,** 1264 (1991).

[17] S. J. Han, B. J. Lee, and H. S. Kang, *Eur. J. Biochem.* **251,** 244 (1998).

[18] C. J. Doersen, C. Guerrier-Takada, S. Altman, and G. Attardi, *J. Biol. Chem.* **260,** 5942 (1985).

[19] W. Rossmanith and R. Karwan, *Biochem. Biophys. Res. Commun.* **247,** 234 (1998).

[20] E. Bertrand, F. Houser-Scott, A. Kendall, R. H. Singer, and D. R. Engelke, *Genes Dev.* **12,** 2463 (1998).

[21] N. Jarrous, J. S. Wolenski, D. Wesolowski, C. Lee, and S. Altman, *J. Cell Biol.* **146,** 559 (1999).

[10] *Saccharomyces cerevisiae* Nuclear Ribonuclease P: Structure and Function

By FELICIA HOUSER-SCOTT, WILLIAM A. ZIEHLER, and DAVID R. ENGELKE

Introduction

The study of macromolecules that cleave and process ribonucleic acids has been an area of intense study. The majority of these enzymes utilize protein-based mechanisms to achieve this goal (reviewed extensively in this volume of *Methods in Enzymology*). In the case of ribonuclease P (RNase P), cleavage of the 5′ leader sequence of tRNA precursors (pre-tRNA) in a cell is primarily based on an RNA-catalyzed mechanism. Well-known examples of RNA-catalyzed reactions in which intramolecular cleavage of the RNA catalyst occurs include the self-splicing group I and II introns, the hammerhead, hairpin, and hepatitis delta ribozymes. The most common mechanism of intramolecular recognition and self-cleavage is mediated through Watson–Crick base pairing between regions of complementarity within the same RNA chain. RNase P is unique in that the RNA substrate is present *in trans* and substrate recognition and cleavage do not occur through Watson–Crick base pairing but through recognition of common features in the three-dimensional structure of the pre-tRNA substrates. Thus, the question of how RNase P uses a RNA-based mechanism to recognize and cleave a substrate *in trans* is an interesting evolutionary question.

Multiple Forms of RNase P

Ribonuclease P (RNase P; EC 3.1.26.5) is a ribonucleoprotein endoribonuclease that cleaves pre-tRNA molecules to produce mature 5′ termini (for review, see Refs. 1–4). The enzyme is conserved in all three kingdoms—bacteria, archaea, and eukarya—with mitochondria and chloroplasts having activities separate from the nucleus in eukaryotes. With the exception of RNase P from spinach chloroplast[5,6] and possibly human mitochondria,[7] most RNase P activities studied to date are composed of a single RNA subunit and one or more protein subunits.

[1] N. R. Pace and J. W. Brown, *J. Bacteriol.* **177,** 1919 (1995).

[2] D. N. Frank and N. R. Pace, *Annu. Rev. Biochem.* **67,** 153 (1998).

[3] S. Altman, L. Kirsebom, and S. J. Talbot, *in* "tRNA: Structure, Biosynthesis, and Function" (D. Söll and U. L. RajBhandary, eds.), p. 67. American Society of Microbiology, Washiington, D.C., 1995.

[4] J. R. Chamberlain, A. J. Tranguch, E. Pagán-Ramos, and D. R. Engelke, *in* "Progress in Nucleic Acid Research and Molecular Biology" (K. Moldave and W. Cohn, eds.), p. 87. Academic Press, San Diego, California, 1996.

[5] B. C. Thomas, X. Li, and P. Gegenheimer, *RNA* **6,** 545 (2000).

0076-6879/01 $35.00

RNase P has been studied extensively in bacteria, particularly in *Escherichia coli* and *Bacillus subtilis* (reviewed in Refs. 1, 2, 8, and 9). The bacterial holoenzyme is composed of a single RNA subunit (~400 nucleotides) and a single protein subunit (~14 kDa). The bacterial RNA subunit is catalytic *in vitro* under high Mg^{2+} and monovalent salt conditions, but both the RNA and protein subunits are required *in vivo*.[8,10–12] A subset of archaeal RNase P RNAs has been shown to be catalytic *in vitro* as well.[13] In contrast, no eukaryotic RNase P RNA subunits have been shown to be catalytic in the absence of protein.

By density measurements, the bacterial protein subunit comprises less than 10% of the holoenzyme,[14,15] whereas protein accounts for ~50% of the eukaryotic and archaeal holoenzymes.[16–21] Despite the increased protein content in the eukaryotic and archaeal holoenzymes, the conserved RNA core structure supports the idea that the RNA still comprises the catalytic core of the enzyme.

While bacteria and archaea appear to contain only one form of cellular RNase P, eukaryotes contain multiple forms of the enzyme—one in the nucleus and a second in mitochondria—and in the case of plants there is a third in chloroplasts. Nuclear RNase P is responsible for processing nuclear-encoded pre-tRNAs whereas mitochondrial RNase P and chloroplast RNase P are responsible for processing pre-tRNA encoded in the mitocondrial and chloroplast genomes, respectively. In *Saccharomyces cerevisiae,* the mitochondrial RNase P RNA component is encoded by the mitochondrial *RPM1* gene and is ~500 nucleotides long.[22–24] The mitochondrial RNase P RNAs from several closely related yeast have also

[6] M. J. Wang, N. W. Davis, and P. A. Gegenheimer, *EMBO J.* **7,** 1567 (1988).
[7] W. Rossmanith and R. M. Karwan, *Biochem. Biophys. Res. Commun.* **247,** 234 (1998).
[8] S. Altman, L. Kirsebom, and S. Talbot, *FASEB J.* **7,** 7 (1993).
[9] L. A. Kirsebom and A. Vioque, *Mol. Biol. Rep.* **22,** 99 (1995).
[10] R. Kole and S. Altman, *Biochemistry* **20,** 1902 (1981).
[11] C. Guerrier-Takada, K. Gardiner, T. Marsh, N. Pace, and S. Altman, *Cell* **35,** 849 (1983).
[12] C. Guerrier-Takada and S. Altman, *Science* **223,** 285 (1984).
[13] J. A. Pannucci, E. S. Haas, T. A. Hall, J. K. Harris, and J. W. Brown, *Proc. Natl. Acad. Sci. U.S.A.* **96,** 7803 (1999).
[14] E. Akaboshi, C. Guerrier-Takada, and S. Altman, *Biochem. Biophys. Res. Commun.* **96,** 831 (1980).
[15] K. Gardiner and N. R. Pace, *J. Biol. Chem.* **255,** 7507 (1980).
[16] N. Lawrence, D. Wesolowski, H. Gold, M. Bartkiewicz, C. Guerrier-Takada, W. H. McClain, and S. Altman, *Cold Spring Harb. Symp. Quant. Biol.* **52,** 233 (1987).
[17] L. Kline, S. Nishikawa, and D. Söll, *J. Biol. Chem.* **256,** 5058 (1981).
[18] A. K. Knap, D. Wesolowski, and S. Altman, *Biochimie* **72,** 779 (1990).
[19] G. P. Jayanthi and G. C. Van Tuyle, *Arch. Biochem. Biophys.* **296,** 264 (1992).
[20] M. Doria, G. Carrara, P. Calandra, and G. P. Tocchini-Valentini, *Nucleic Acids Res.* **19,** 2315 (1991).
[21] S. C. Darr, B. Pace, and N. R. Pace, *J. Biol. Chem.* **265,** 12927 (1990).
[22] D. L. Miller and N. C. Martin, *Cell* **34,** 911 (1983).
[23] M. J. Morales, C. A. Wise, M. J. Hollingsworth, and N. C. Martin, *Nucleic Acids Res.* **17,** 6865 (1989).
[24] C. Wise and N. C. Martin, *Nucleic Acids Res.* **19,** 4773 (1991).

been identified.[25] All of these RNAs are AU rich and range in size from 140 to 500 nucleotides. A single 105-kDa protein subunit, encoded by the nuclear *RPM2* gene, has been shown to be required for mitochondrial RNase P activity in *S. cerevisiae*[26,27] as well as involved in the biogenesis of the RNase P RNA subunit, *RPM1* RNA.[28] RPM2 is also essential for normal cell growth under conditions that do not require mitochondrial protein synthesis.[29] The essential nuclear function is still unknown, but RPM2 does not appear to affect nuclear RNase P activity.

In this review, we focus primarily on the structure and function of the nuclear RNase P holoenzyme from *S. cerevisiae* because much progress has been made in the characterization of this enzyme. The nuclear RNase P RNA subunit is encoded by the nuclear *RPR1* gene and is ~369 nucleotides long.[30] The *RPR1* gene is essential for both life and nuclear pre-tRNA processing. Nuclear RNase P RNAs from several yeasts have been identified and in some cases can substitute functionally for the *S. cerevisiae* gene.[31] Biochemical purification of the *S. cerevisiae* RNase P holoenzyme identified nine protein subunits (ranging in size from 15 to 100 kDa) that copurify with the *RPR1* RNA.[32] A more detailed discussion of the *RPR1* RNA and protein subunits is presented below.

Substrate Recognition

The RNase P reaction is unique among ribozymes in that substrate pre-tRNAs are recognized by tertiary structure and not via Watson–Crick base pairing. This is a requisite feature of RNase P because of the enormous sequence diversity present in the tRNA gene repertoire of a cell (or organelle). The primary determinants for substrate recognition are located in the mature tRNA portion of the substrate, as determined by deletion studies and the use of synthetic model substrates. The 5′ leader can be reduced to a single nucleotide and still be processed efficiently by both bacterial and eukaryotic enzymes.[33,34] Mature tRNA is also a strong inhibitor of the RNase P reaction, indicating the 5′ leader does not contribute a major portion of the binding energy. Minimization of the mature tRNA domain of the substrate

[25] C. A. Wise and N. C. Martin, *J. Biol. Chem.* **266,** 19154 (1991).

[26] M. J. Morales, Y. L. Dang, Y. C. Lou, P. Sulo, and N. C. Martin, *Proc. Natl. Acad. Sci. U.S.A.* **89,** 9875 (1992).

[27] Y. L. Dang and N. C. Martin, *J. Biol. Chem.* **268,** 19791 (1993).

[28] V. Stribinskis, G. J. Gao, P. Sulo, Y. L. Dang, and N. C. Martin, *Mol. Cell. Biol.* **16,** 3429 (1996).

[29] C. K. Kassenbrock, G. J. Gao, K. R. Groom, P. Sulo, M. G. Douglas, and N. C. Martin, *Mol. Cell. Biol.* **15,** 4763 (1995).

[30] J. Y. Lee, C. E. Rohlman, L. A. Molony, and D. R. Engelke, *Mol. Cell. Biol.* **11,** 721 (1991).

[31] E. Pagán-Ramos, A. J. Tranguch, D. W. Kindelberger, and D. R. Engelke, *Nucleic Acids Res.* **22,** 200 (1994).

[32] J. R. Chamberlain, Y. Lee, W. S. Lane, and D. R. Engelke, *Genes Dev.* **12,** 1678 (1998).

[33] C. K. Surratt, B. J. Carter, R. C. Payne, and S. M. Hecht, *J. Biol. Chem.* **265,** 22513 (1990).

[34] W. H. McClain, C. Guerrier-Takada, and S. Altman, *Science* **238,** 527 (1987).

has further shown that the TΨC and acceptor stems contain the primary recognition determinants. Synthetic substrates composed of these two helices are processed by bacterial RNase P.[35–37] As with the bacterial RNase P, multiple studies have shown that the tertiary structure of substrates, rather than specific sequences, is important for substrate recognition by eukaryotic RNase P.[38–43] Unlike the bacterial ribozyme, however, the yeast nuclear holoenzyme does not easily recognize simple substrates that mimic only the coaxial aminoacyl/TΨC stems.[34,44,45] This suggests additional contacts in the D or anticodon arms might be required for efficient recognition.

Eukaryotic pre-tRNAs contain 5′ leader and 3′ trailer sequences, which must be removed to form mature tRNA molecules that are competent for aminoacylation. Synthesis of pre-tRNAs by RNA polymerase III is terminated by a polyuridine stretch at the end of the transcript, which becomes part of the 3′ trailer structure. In yeast, the steady state pool of nuclear precursors suggests that RNase P cleavage precedes removal of the 3′ trailer by exo- or endonuclease activities.[46–48] Previous studies have demonstrated Watson–Crick base-pairing interactions between the purine-rich leader sequences and the polyuridine sequence of the 3′ trailer. The more stable stems between the leader and trailer sequences in naturally occurring yeast pre-tRNAs typically have mismatched bulges that prevent the formation of continuous extensions of the aminoacyl stem past the $-1/+1$ position. The -1 mismatch is important for RNase P activity, as minor alterations that cause uninterrupted extension of the aminoacyl acceptor stem severely inhibit processing of the pre-tRNA *in vitro* and *in vivo*.[48] There are likely to be mechanisms that antagonize the stem structure *in vivo,* thereby making the separated 5′ and/or 3′ strands available for interaction with RNase P. One possible mechanism could be the binding of La protein to the 3′ poly(U) trailer.[49,50]

[35] J. Schlegl, W. D. Hardt, V. A. Erdmann, and R. K. Hartmann, *EMBO J.* **13,** 4863 (1994).
[36] W. D. Hardt, J. Schlegl, V. A. Erdmann, and R. K. Hartmann, *Biochemistry* **32,** 13046 (1993).
[37] S. Gunnery, Y. Ma, and M. B. Mathews, *J. Mol. Biol.* **286,** 745 (1999).
[38] E. S. Haas, J. W. Brown, C. Pitulle, and N. R. Pace, *Proc. Natl. Acad. Sci. U.S.A.* **91,** 2527 (1994).
[39] C. J. Green and B. S. Vold, *J. Biol. Chem.* **263,** 652 (1988).
[40] D. Kahle, U. Wehmeyer, and G. Krupp, *EMBO J.* **9,** 1929 (1990).
[41] D. L. Thurlow, D. Shilowski, and T. L. Marsh, *Nucleic Acids Res.* **19,** 885 (1991).
[42] L. A. Kirsebom and S. G. Svard, *Nucleic Acids Res.* **20,** 425 (1992).
[43] P. S. Holm and G. Krupp, *Nucleic Acids Res.* **20,** 421 (1992).
[44] J. Schlegl, J. P. Furste, R. Bald, V. A. Erdmann, and R. K. Hartmann, *Nucleic Acids Res.* **20,** 5963 (1992).
[45] J. Y. Lee and D. R. Engelke, unpublished data (1990).
[46] D. R. Engelke, P. Gegenheimer, and J. Abelson, *J. Biol. Chem.* **260,** 1271 (1985).
[47] J. P. O'Connor and C. L. Peebles, *Mol. Cell. Biol.* **11,** 425 (1991).
[48] Y. Lee, D. W. Kindelberger, J. Y. Lee, S. McClennen, J. Chamberlain, and D. R. Engelke, *RNA* **3,** 175 (1997).
[49] S. L. Wolin and A. G. Matera, *Genes Dev.* **13,** 1 (1999).
[50] C. J. Yoo and S. L. Wolin, *Cell* **89,** 393 (1997).

In bacteria, both the 5′ leader and 3′ trailing sequences are thought to interact with the RNase P holoenzyme. The 3′-CCA that is encoded in many bacterial pre-tRNAs immediately downstream of the aminoacyl stem interacts with the P15 internal loop of the RNase P RNA subunit, helping to stabilize binding in the ribozyme reaction.[51,52] In addition, the protein subunit of bacterial RNase P has been shown to possess a single-stranded RNA-binding cleft that recognizes the unpaired 5′ leader of bacterial pre-tRNAs.[53–55] In eukaryotic nuclear pre-tRNAs the 3′–terminal CCA is not normally contained in the primary transcript, and the complementary sequence in the RNase P RNA P15 loop is also missing.[2,56] If there is pairing between any 3′ trailing sequences in nuclear pre-tRNAs and the nuclear RNase P RNA subunit it is not obvious and has not been shown to have a detectable positive effect on substrate recognition. Even in the bacterial enzyme, the effects of the –CCA/RNA subunit pairing are more pronounced in the ribozyme reaction (RNA only) than when the holoenzyme is used.[51]

In the absence of enzyme–substrate base pairing, it seems likely that the nuclear holoenzyme has the capacity to interact with multiple regions of the substrate. The yeast enzyme has significantly more protein subunits associated with the RNA subunit than are found in the eubacterial enzymes (see below), and they are quite positively charged.[32] Thus, the nuclear holoenzyme could potentially interact with the 5′ leader and 3′ trailer separately, although sites of contact between the pre-tRNA substrate and the nuclear RNase P holoenzyme are currently unclear. It has been shown that the single-stranded 3′ trailer plays a substantial role in binding substrate and that the enzyme can also interact with the 5′ leader to improve binding by *S. cerevisiae* nuclear RNase P.[57] The second-order rate constant, k_{cat}/K_m, for pre-tRNA cleavage is at or near the diffusion-controlled limit at $1.3 \times 10^8 M^{-1}\text{sec}^{-1}$.[58] This indicates that the association rate constant for substrate binding to enzyme is likely rate limiting under these conditions, and up to 10-fold even more efficient than the already very efficient *B. subtilis* RNase P.[59–61] *Saccharomyces cerevisiae* nuclear RNase P appears to have binding sites for both the tRNA structure and single-stranded RNAs, including the 3′ trailer structure.

[51] B. K. Oh and N. R. Pace, *Nucleic Acids Res.* **22,** 4087 (1994).

[52] L. A. Kirsebom and S. G. Svard, *EMBO J.* **13,** 4870 (1994).

[53] C. Spitzfaden, N. Nicholson, J. J. Jones, S. Guth, R. Lehr, C. D. Prescott, L. A. Hegg, and D. S. Eggleston, *J. Mol. Biol.* **295,** 105 (2000).

[54] T. Stams, S. Niranjanakumari, C. A. Fierke, and D. W. Christianson, *Science* **280,** 752 (1998).

[55] S. Niranjanakumari, T. Stams, S. M. Crary, D. W. Christianson, and C. A. Fierke, *Proc. Natl. Acad. Sci. U.S.A.* **95,** 15212 (1998).

[56] M. P. Deutscher, *Crit. Rev. Biochem.* **17,** 45 (1984).

[57] W. Ziehler, J. Day, C. Fierke, and D. R. Engelke, *Biochemistry* **39,** 9909 (2000).

[58] S. H. Northrup and H. P. Erickson, *Proc. Natl. Acad. Sci. U.S.A.* **89,** 3338 (1992).

[59] J. C. Kurz, S. Niranjanakumari, and C. A. Fierke, *Biochemistry* **37,** 2393 (1998).

[60] J. A. Beebe and C. A. Fierke, *Biochemistry* **33,** 10294 (1994).

[61] D. Smith, A. B. Burgin, E. S. Haas, and N. R. Pace, *J. Biol. Chem.* **267,** 2429 (1992).

Catalysis

The RNase P cleavage reaction has been characterized in both bacteria and eukaryotes. Bacterial studies have primarily been performed with synthetic ribozymes, and to a lesser extent with reconstituted bacterial holoenzyme. Initial characterization of the eukaryotic enzyme reaction was accomplished with partially purified enzyme from yeast.[61,62]

Cleavage of pre-tRNA by RNase P is accomplished in three separate steps. First, the substrate is bound, followed by hydrolysis of the scissile phosphodiester bond, leaving a 3′-hydroxyl group on the −1 nucleotide (first nucleotide upstream of cleavage) and a 5′ phosphate on the +1 nucleotide (first position downstream of cleavage). The 5′ leader and mature tRNA products then dissociate, regenerating enzyme for another reaction cycle. The rate constants for the separate steps of the reaction pathway have been measured for the bacterial RNA reaction, using both presteady state and steady state conditions.[60,63–66] The rate of phosphodiester hydrolysis under single-turnover conditions is between 180 and 360 min^{-1} for the *E. coli* and *B. subtilis* ribozymes.[60,64] Multiple substrate turnover is approximately 100 times slower (2–3 min^{-1}) because of the limiting release of product. The catalytic rate of the eukaryotic holoenzyme under multiple turnover conditions is approximately 80 min^{-1},[57] although the rate can vary significantly with different pre-tRNAs.[48] The RNase P reaction absolutely requires divalent metal cofactors for catalysis. However, tRNA binding can occur in the absence of metals, evident from photoaffinity cross-linking between tRNA and the bacterial ribozyme in the absence of divalent metal ions.[63] Interestingly, the cross-linking patterns are identical in the presence and absence of divalent ions, suggesting that the RNase P RNA structure is not significantly altered by divalent ions. However, substrate binding is enhanced by their presence.[33,63,67,68]

Magnesium is the preferred divalent metal cofactor. Manganese and calcium can substitute in the bacterial ribozyme, but cause a decrease in the catalytic activity, ranging from a slight defect for Mn^{2+} to a 1000-fold reduction for Ca^{2+}.[60] The *S. cerevisiae* RNase P holoenzyme is intolerant of changes in divalents,[69] and is irreversibly inactivated by the removal of Mg^{2+}.[70] It is estimated that approximately three metal ions (Hill coefficient of 3.2) are necessary for maximal

[62] E. Pagán-Ramos, Y. Lee, and D. R. Engelke, *RNA* **2,** 1100 (1996).
[63] D. Smith and N. R. Pace, *Biochemistry* **32,** 5273 (1993).
[64] W. D. Hardt, J. Schlegl, V. A. Erdmann, and R. K. Hartmann, *J. Mol. Biol.* **247,** 161 (1995).
[65] J. A. Beebe, J. C. Kurz, and C. A. Fierke, *Biochemistry* **35,** 10493 (1996).
[66] A. Tallsjo and L. A. Kirsebom, *Nucleic Acids Res.* **21,** 51 (1993).
[67] T. Pan, *Biochemistry* **34,** 902 (1995).
[68] P. P. Zarrinkar, J. Wang, and J. R. Williamson, *RNA* **2,** 564 (1996).
[69] T. Pfeiffer, A. Tekos, J. M. Warnecke, D. Drainas, D. R. Engelke, B. Séraphin, and R. K. Hartmann, *J. Mol. Biol.* **298,** 559 (2000).
[70] J. Y. Lee, University of Michigan, Ann Arbor, Michigan (1991).

enzyme activity for the *E. coli* enzyme.[63,71] The specificity for these metal ions further suggests that the active site forms a highly ordered metal-binding pocket, designed to accommodate a particular atomic size and coordination geometry. It is hypothesized that one of the metal ions required for RNase P catalysis serves to direct an attacking nucleophilic hydroxide ion to the scissile phosphodiester bond.[72]

Yeast Nuclear RNase P RNA Structure and Function

RPR1 Gene Organization and Expression

The *S. cerevisiae* RNase P RNA gene, *RPR1*, is an essential single-copy gene transcribed by RNA polymerase III.[30,73] The 369-nucleotide mature RNA is processed from a primary transcript (pre-RPR1 RNA) containing an 84-nucleotide 5′ leader sequence and ~28 nucleotides of extra sequence at the 3′ terminus, upstream of the RNA polymerase III poly(U) terminator. The 84-nucleotide leader sequence contains a tRNA-like internal promoter with "A block" and "B block" elements for RNA polymerase III transcription. The 5′ leader sequence appears to be cleaved off in a single step after assembly of the RNase P holoenzyme by an uncharacterized activity, and does not accumulate stably in the cell.[30,74] This arrangement of a transcription unit in which an internal RNA polymerase III promoter is lost by processing of the primary transcript appears to be unique to the *S. cerevisiae RPR1* gene. This type of internal promoter is not found in either the *Schizosaccharomyces pombe* or human nuclear RNase P RNA gene.[75–78] Current data suggest that the extra 3′-terminal nucleotides are removed by a series of cleavage events,[30] including exoribonucleases.[79]

Phylogenetic Studies

Initial phylogenetic structure analysis of RNase P RNAs from several *Schizosaccharomyces* species identified some conserved features of the RNA subunits, but the degree of evolutionary divergence among the RNAs was low.[77] Although preliminary efforts to generate a *RPR1* RNA secondary structure using computer

[71] J. M. Warnecke, J. P. Furste, W. D. Hardt, V. A. Erdmann, and R. K. Hartmann, *Proc. Natl. Acad. Sci. U.S.A.* **93,** 8924 (1996).
[72] J. P. Perreault and S. Altman, *J. Mol. Biol.* **226,** 399 (1992).
[73] J. Y. Lee, C. F. Evans, and D. R. Engelke, *Proc. Natl. Acad. Sci. U.S.A.* **88,** 6986 (1991).
[74] J. R. Chamberlain, R. Pagan, D. W. Kindelberger, and D. R. Engelke, *Nucleic Acids Res.* **24,** 3158 (1996).
[75] B. Cherayil, G. Krupp, P. Schuchert, S. Char, and D. Söll, *Gene* **60,** 157 (1987).
[76] G. Krupp, B. Cherayil, D. Frendewey, S. Nishikawa, and D. Söll, *EMBO J.* **5,** 1697 (1986).
[77] S. Zimmerly, V. Gamulin, U. Burkard, and D. Söll, *FEBS Lett.* **271,** 189 (1990).
[78] G. J. Hannon, A. Chubb, P. A. Maroney, G. Hannon, S. Altman, and T. W. Nilsen, *J. Biol. Chem.* **266,** 22796 (1991).
[79] A. van Hoof, P. Lennertz, and R. Parker, *EMBO J.* **19,** 1357 (2000).

folding programs and sequence similarity to the bacterial consensus were also partially successful,[80] no general secondary structure for yeast RNase P RNA was identified. By analyzing the RNase P RNA genes of a number of closely to distantly related *Saccharomyces* and *Schizosaccharomyces* species, a consensus secondary structure model for yeast nuclear RNase P was determined[81] and later confirmed by structure-sensitive RNA footprinting, using a variety of enzymatic and chemical reagents.[82] Several conclusions were drawn from the RNA footprinting studies of the yeast RNase P holenzyme versus the RNase P RNA alone. First, as supported by the buoyent density measurements, the proteins of the yeast holoenzyme protect a much larger percentage of the RNA subunit than the bacterial protein subunit (Fig. 1). Second, the local secondary structures of the *RPR1* RNA are maintained in the deproteinized RNA; however, the tertiary structure is lost.[82] These observations suggest that one function of the protein(s) in the yeast holoenzyme is to maintain the tertiary structure of the *RPR1* RNA. This might account for the lack of catalytic activity of the yeast RNA subunit alone.

Comparison of the yeast secondary structure with the bacterial consensus structure, as well as with higher eukaryotic RNase P RNAs, revealed a number of similarities, suggesting that RNase P from evolutionarily diverse sources might share a core of conserved elements. Subsequently, a universally conserved core structure of RNase P RNA that has been maintained throughout evolution was identified.[83] The core structure is characterized by five distinct regions containing conserved nucleotide sequences (CR): CRI–CRV as well as several helical elements that vary in sequence, but occur at similar positions in different RNAs. Only 21 of ~200 nucleotides in the core structure of RNase P are conserved across phylogenetic groups.[2] Secondary structure homologs of helices P1, P2, P3, P4, and P10/P11 exist in bacteria, archaea, and eukaryotes. Eukaryotic RNase P RNAs differ from the bacterial structure in the P3 and P15 regions. The eukaryotic P3 region is more complex than the bacterial or archaeal counterpart. The eukaryotic P3 helix contains an internal loop that has been implicated in subcellular localization of the RNase P holoenzyme, which is consistent with its decreased importance in bacteria and archaea because organellar localization is less likely to be an issue.[84,85] In contrast, the P15 region is significantly reduced or missing altogether in eukaryotes. Important functions of the P15 loop in bacteria include binding the 3′-CCA that is encoded in the primary transcripts of bacterial tRNAs.[51,52] The P15 loop

[80] J. Y. Lee and D. R. Engelke, *Mol. Cell. Biol.* **9,** 2536 (1989).

[81] A. J. Tranguch and D. R. Engelke, *J. Biol. Chem.* **268,** 14045 (1993).

[82] A. J. Tranguch, D. W. Kindelberger, C. E. Rohlman, J. Y. Lee, and D. R. Engelke, *Biochemistry* **33,** 1778 (1994).

[83] J. L. Chen and N. R. Pace, *RNA* **3,** 557 (1997).

[84] M. R. Jacobson, L.-G. Cao, Y.-L. Wang, and T. Pederson, *J. Cell Biol.* **131,** 1649 (1995).

[85] M. R. Jacobson, L.-G. Cao, K. Taneja, R. H. Singer, Y. L. Wang, and T. Pederson, *J. Cell Sci.* **110,** 829 (1997).

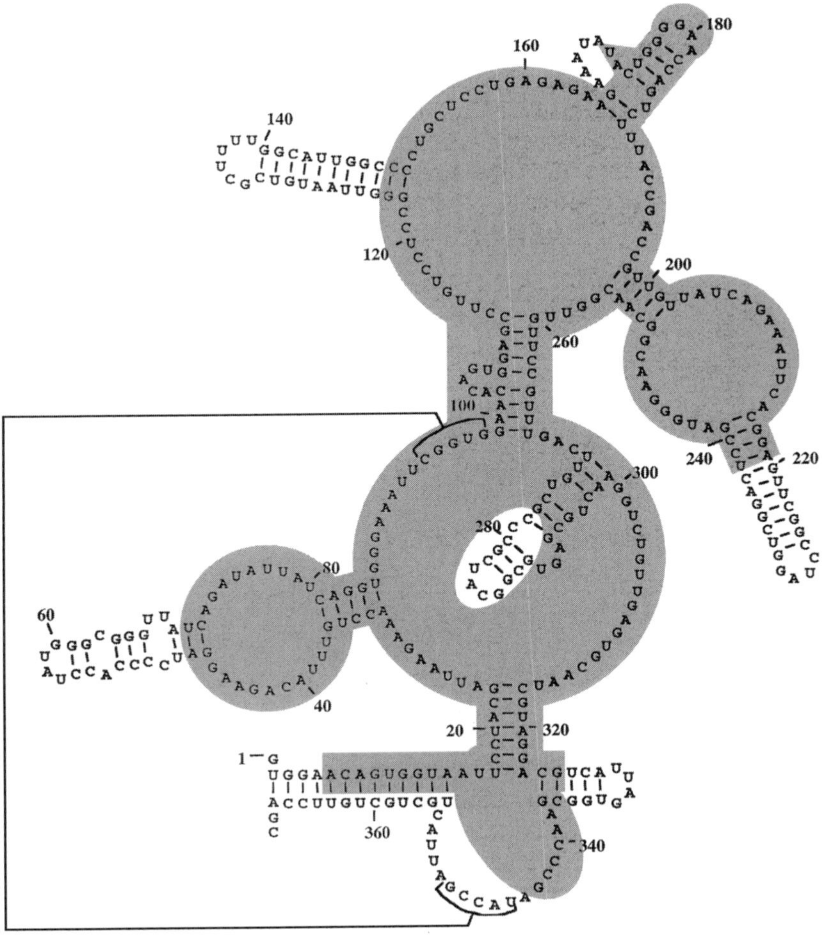

FIG. 1. Regions of *RPR1* RNA protected from structure-probing reagents in the holoenzyme. Nucleotides within the shaded regions become accessible to modification or cleavage on deproteinization of the holoenzyme. [Adapted with permission from A. J. Tranguch, D. W. Kindelberger, C. E. Rohlman, J.-Y. Lee, and D. R. Engelke, *Biochemistry* **33,** 1778 (1994). Copyright 1994 American Chemical Society.]

is also thought to coordinate catalytic metal ions in bacterial enzyme.[86] Several helices such as P5/7, P8, and P9 are not supported by phylogenetic data in yeast, but they do appear in some eukaryotes. CRI–CRV are indicated in the *RPR1* RNA secondary structure shown in Fig. 2.

[86] J. Kufel and L. A. Kirsebom, *J. Mol. Biol.* **263,** 685 (1996).

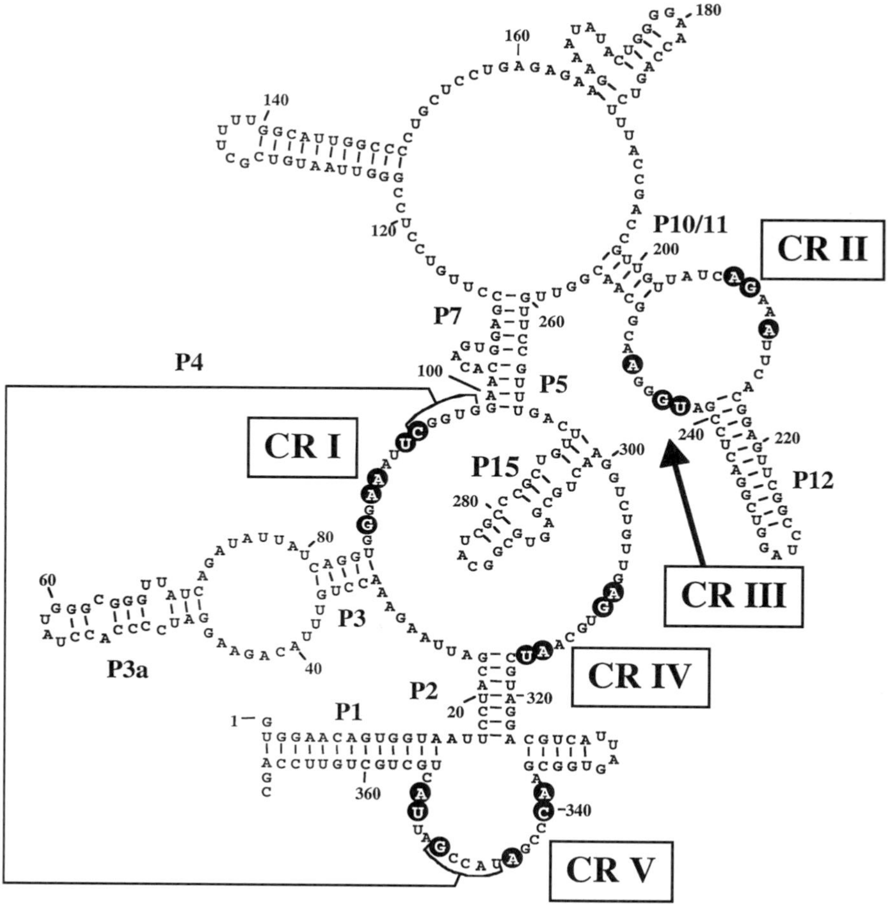

FIG. 2. Secondary structure of RNase P RNA from *Saccharomyces cerevisiae* (*RPR1* RNA). Regions of homology with the bacterial RNA consensus are labeled "P#" for the corresponding helix. The P4 helix is marked with brackets connected by a line. Locations of the universally conserved critical regions are denoted by white nucleotides on black circles and labeled "CR" followed by a roman numeral. [CRI–CRV designated according to J. L. Chen and N. R. Pace, *RNA* **3,** 557 (1997).]

Functional Analysis of RNA Subdomains

CRI, IV, and CRV were the first regions identified as conserved core elements. The P4 helix, which is composed of a long-distance interaction between CRI and CRV (Fig. 2), is the most conserved structural feature in all RNase P RNAs. Eleven of the 21 invariant nucleotides of RNase P RNA occur within or adjacent to the P4 helix.[2] The P4 helix is thought to constitute the catalytic center of the

RNase P enzyme. CRIV is next to the active site[87] and may participate in active site formation along with the P4 helix.

Site-directed mutagenesis, deletional analysis, and phosphorothioate substitution studies have demonstrated that alterations in the P4 helix significantly reduce or abolish catalysis by bacterial RNase P.[38,88–93] Site-directed mutagenesis of conserved nucleotides in CRI, CRIV, and CRV in the *S. cerevisiae* holoenzyme gave mixed phenotypes. Only 2–4% of the possible sequence combinations at the highly conserved positions gave viable phenotypes. Even so, most of the positions tolerated sequence changes in some sequence contexts without severely affecting enzyme function.[94] Both K_m and k_{cat} values of the RNase P holoenzyme were affected by mutations in these regions, suggesting that, as in the bacterial enzyme, they might interact with the substrates. Mutations in CRI, CIV, and CV did not affect RNase P assembly, as measured by the ability to produce mature *RPR1* RNA.

CRII and CRIII were initially identified as a conserved structural domain in yeast RNase P RNAs,[81,95] and later refinements to the bacterial consensus structure revealed the presence of a more complex, but similar, domain, containing stems P10, P11, and P12.[83] Tertiary structural models of the *E. coli* RNase P RNA place the P10/11–P12 domain near the tRNA cleavage site.[88,95,96] The bulk of the data from bacteria and yeast suggests that the P10/11–P12 domain is involved in magnesium utilization by RNase P. In bacteria, phosphorothioate substitution and manganese ion rescue experiments showed that binding of tRNA substrate requires divalent metal ion-mediated interactions with the conserved internal loop.[97] Moreover, the domain is sensitive to magnesium and lead-induced cleavages in bacteria.[93,98]

Deletion studies of yeast RNA have demonstrated that the P12 helix is dispensible for function *in vivo,* whereas the loop region containing CRIII and the conserved sequence CA<u>G</u>AA<u>A</u> in CRII is essential for viability.[31] Randomization

87 A. B. Burgin and N. R. Pace, *EMBO J.* **9,** 4111 (1990).
88 M. E. Harris, J. M. Nolan, A. Malhotra, J. W. Brown, S. C. Harvey, and N. R. Pace, *EMBO J.* **13,** 3953 (1994).
89 M. E. Harris and N. R. Pace, *RNA* **1,** 210 (1995).
90 S. C. Darr, K. Zito, D. Smith, and N. R. Pace, *Biochemistry* **31,** 328 (1992).
91 S. Altman, D. Wesolowski, and R. S. Puranam, *Genomics* **18,** 418 (1993).
92 C. Guerrier-Takada and S. Altman, *Biochemistry* **32,** 7152 (1993).
93 S. Kazakov and S. Altman, *Proc. Natl. Acad. Sci. U.S.A.* **88,** 9193 (1991).
94 E. Pagán-Ramos, Y. Lee, and D. R. Engelke, *RNA* **2,** 441 (1996).
95 E. Pagán-Ramos, A. J. Tranguch, J. M. Nolan, N. R. Pace, and D. R. Engelke, *Nucleic Acids Symp. Ser.* **33,** 89 (1995).
96 E. Westhof and S. Altman, *Proc. Natl. Acad. Sci. U.S.A.* **91,** 5133 (1994).
97 W. D. Hardt, J. M. Warnecke, V. A. Erdmann, and R. K. Hartmann, *EMBO J.* **14,** 2935 (1995).
98 J. Ciesiolka, W. D. Hardt, J. Schlegl, V. A. Erdmann, and R. K. Hartmann, *Eur. J. Biochem.* **219,** 49 (1994).

mutagenesis of this sequence followed by kinetic analysis of the defective holoenzymes showed that mutations primarily affected the catalytic rate (k_{cat}), with minor changes in apparent substrate binding.[62] The underlined G and A positions were particularly intolerant of mutations. As with the other critical regions, mutations in CRII had no obvious effect on holoenzyme assembly. Several mutant enzymes required increased magnesium concentrations, suggesting magnesium utilization was poor for these mutants. Structural analysis of the P10/11–P12 domain of yeast confirms that the structure is sensitive to magnesium.[99] On the basis of these observations it is feasible that the P10/11–P12 domain coordinates magnesium at the active site.

The P3 domain of RNase P RNA is a conserved structural feature of eukaryotes, and is distinct from the P3 stem found in bacteria. The eukaryotic P3 domain is composed of a helix–internal loop–helix that contains several conserved nucleotides. Deletional analysis of yeast RNase P RNA has shown that the P3 helix is essential for enzyme function *in vivo*.[31] More recent studies in yeast have shown that the P3 domain of RNase P is functionally equivalent to the P3 domain of the RNase MRP RNA subunit.[100] RNase MRP is a closely related ribosomal RNA-processing enzyme that appears to share eight of nine protein subunits with RNase P. Each enzyme has at least one distinctive protein subunit (Rpr2p in RNase P, Snm1p in RNase MRP) and a different (although related) RNA subunit. The RNase P and MRP RNA subunits share several secondary structural features, including at least CRI, CIV, CRV, and P3.[101,102] Phylogenetic analysis suggests that, unlike the critical regions, the P3 domains of RNases MRP and P are more highly conserved within a species than RNase P P3 domains are conserved between species. This observation led to the hypothesis that the P3 domain might be the binding site of one of the proteins shared between MRP and RNase P and that the P3 domain coevolved with the protein in the RNase P and MRP RNA subunits. This would be consistent with the P3 domain being the binding site for common protein antigens in human RNases P and MRP.[103,104] In addition, recognition of the P3 domain in human MRP and RNase P has been implicated in nucleolar localization of these RNAs.[84,85] Yeast RNase P RNA has also been localized to the nucleolus.[105] It is unknown whether the yeast localization requires the P3 domain, but mutagenesis results suggest that P3 may be required for overall holoenzyme assembly.[106]

[99] W. A. Ziehler, J. Yang, A. V. Kurochkin, P. O. Sandusky, E. R. Zuiderweg, and D. R. Engelke, *Biochemistry* **37,** 3549 (1998).

[100] L. Lindahl, S. Fretz, N. Epps, and J. M. Zengel, *RNA* **6,** 653 (2000).

[101] A. C. Forster and S. Altman, *Cell* **62,** 407 (1990).

[102] M. E. Schmitt, J. L. Bennett, D. J. Dairaghi, and D. A. Clayton, *FASEB J.* **7,** 208 (1993).

[103] P. S. Eder, R. Kekuda, V. Stolc, and S. Altman, *Proc. Natl. Acad. Sci. U.S.A.* **94,** 1101 (1997).

[104] M. H. Liu, Y. Yuan, and R. Reddy, *Mol. Cell. Biochem.* **130,** 75 (1994).

[105] E. Bertrand, F. Houser-Scott, A. Kendall, R. H. Singer, and D. R. Engelke, *Genes Dev.* **12,** 2463 (1998).

[106] W. A. Ziehler, J. Morris, F. H. Scott, C. Millikin, and D. R. Engelke, *RNA* **7,** 565 (2001).

TABLE I
SUBUNIT COMPOSITION OF *Saccharomyces cerevisiae* NUCLEAR RNase P

Yeast gene	Subunit type	Molecular mass (kDa)	Isoelectric point (p*I*)	Ref.
RPR1	RNA	120	—	*a*
POP8	Protein	15.5	4.57	*b*
POP7[c]	Protein	15.8	9.34	*b, d*
RPR2	Protein	16.3	9.99	*b*
POP6	Protein	18.2	9.28	*b*
POP5	Protein	19.6	7.79	*b*
POP3	Protein	22.6	9.57	*e*
RPP1[c]	Protein	32.2	9.76	*f*
POP4[c]	Protein	32.9	9.26	*g*
POP1[c]	Protein	100.5	9.84	*h*

[a] J. Y. Lee, C. E. Rohlman, L. A. Molony, and D. R. Engelke, *Mol. Cell. Biol.* **11,** 721 (1991).
[b] J. R. Chambertain, Y. Lee, W. S. Lane, and D. R. Engelke, *Genes Dev.* **12,** 1678 (1998).
[c] Human ortholog of yeast protein subunit is known. See text for details.
[d] V. Stolc, A. Katz, and S. Altman, *Proc. Natl. Acad. Sci. U.S.A.* **95,** 6716 (1998).
[e] B. Dichtl and D. Tollervey, *EMBO J.* **16,** 417 (1997).
[f] V. Stolc and S. Altman, *Genes Dev.* **11,** 2414 (1997).
[g] S. Chu, J. M. Zengel, and L. Lindahl, *RNA* **3,** 382 (1997).
[h] Z. Lygerou, P. Mitchell, E. Petfalski, B. Seraphin, and D. Tollervey, *Genes Dev.* **8,** 1423 (1994).

Temperature-sensitive mutants showed an accumulation of the precursor form of *RPR1* RNA with a marked decreased in the mature form of the RNA. Pop1 protein appears to be the yeast subunit that interacts specifically with the P3 domain.[106]

Characterization of RNase P Proteins

Physical Characteristics

Several protein subunits of yeast RNase P were initially identified by genetic and immunoprecipitation studies as possible integral subunits of both RNase P and RNase MRP (Pop1p, Pop3p, Pop4p, and Rpp1p; see Table I).[107–110] However, the complete subunit composition remained elusive for many years because of

[107] Z. Lygerou, P. Mitchell, E. Petfalski, B. Séraphin, and D. Tollervey, *Genes Dev.* **8,** 1423 (1994).
[108] S. Chu, J. M. Zengel, and L. Lindahl, *RNA* **3,** 382 (1997).
[109] B. Dichtl and D. Tollervey, *EMBO J.* **16,** 417 (1997).
[110] V. Stolc and S. Altman, *Genes Dev.* **11,** 2414 (1997).

the low abundance of nuclear RNase P (~200–400 copies/cell). The 400,000-fold biochemical purification of the nuclear holoenzyme from *S. cerevisiae* to apparent homogeneity confirmed the presence of the previously suspected protein subunits and identified five additional polypeptides (Pop5p, Pop6p, Pop7p, Pop8p, and Rpr2) that copurify with the *RPR1* RNA.[32] At about the same time, Pop7p was also independently identified on the basis of its sequence similarity to the human protein Rpp20 and given the alternative name of Rpp2p.[111] Thus, the subunit composition of yeast RNase P is much more complex than that of the bacterial enzyme, containing one RNA subunit and nine protein subunits. All of the subunits are essential for life and for RNase P activity.[32] Of the nine protein subunits, seven of the proteins are highly basic with isoelectric points (p*I*) greater than 9, Pop8p is the only acidic protein with a p*I* of 4.5, whereas Pop5p has a p*I* of 7.7.

There is limited information on the function of the individual protein subunits. Attempted alignment of all nine protein sequences with the bacterial RNase P proteins does not produce convincing similarities. Protein database searches have allowed the identification of the human counterparts of several of the yeast RNases P and MRP protein subunits (see below), while attempts to identify putative functions and/or activities based on sequence similarities to proteins of known function have been unproductive so far. Pairwise alignments with known RNA-processing and -modifying enzymes have also not revealed any close relationships.

Six of the nine protein subunits exhibit RNA-binding activity *in vitro,* but it has not yet been possible to determine the specificity of these interactions because of the low solubility of most of the proteins when expressed in bacteria.[112] Experiments using the two-hybrid system to examine specific protein–protein interactions reveal extensive contacts among the protein subunits, with an especially large number of contacts between Pop4p and the other subunits.[113] In addition, experiments using the three-hybrid system to examine RNA–protein interactions suggest that Pop1p and Pop4p interact specifically with the *RPR1* RNA subunit. The specific regions of the RNA subunit in contact with the individual protein subunits are under investigation. Because of experiments described in previous sections, it is considered highly likely that one or both of these proteins interact specifically with the P3 domain to nucleate ribonucleoprotein particle formation. Taken together, these data are providing us with a low-resolution map of the architecture of *S. cerevisiae* RNase P.[113] The RNase P RNA from *S. cerevisiae* has not yet been shown to be catalytic in the absence of protein, and our attempts

[111] V. Stolc, A. Katz, and S. Altman, *Proc. Natl. Acad. Sci. U.S.A.* **95,** 6716 (1998).

[112] F. Houser-Scott, C. Milliken, and D. R. Engelke, unpublished data (1999).

[113] F. Houser-Scott, C. Milliken, W. A. Ziehler, and D. R. Engelke, in preparation (2000).

to reconstitute the activity *in vitro* from the RNA and protein components have been unsuccessful to date. This, in part, may be due to technical difficulties in soluble protein expression, but it is not uncommon to have severe, if not insurmountable, difficulty in correctly reconstituting multisubunit enzymes and RNP complexes.

Relationship to RNase P Protein Subunits from Other Organisms

There is no striking sequence homology between the protein subunits of bacterial, mitochondrial, and eukaryotic nuclear RNases P. Even between eukaryotes, sequence conservation in the protein subunits is not strong, but several of the human protein homologs of the yeast RNase P/MRP proteins have been identified by partial characterization of RNase P from HeLa cells. Human RNase P consists of a single RNA species, H1 RNA,[114] and at least seven polypeptides that copurify with the enzymatic activity as determined on the basis of silver-stained gels.[103] These polypeptides were originally identified as Rpp14p, Rpp20p, Rpp25p, Rpp29p, Rpp30p, Rpp38p, and Rpp40p. A Pop1p homolog was the first human RNase P protein to be characterized on the basis of its homology to a yeast RNase P subunit.[115] Like its yeast counterpart, it is approximately 100 kDa and also part of human RNase MRP. Putative Pop1p homologs have also been identified in *Caenorhabditis elegans, Drosophila,* and *Xenopus.*[116,117] The product of the yeast *POP4* gene also has an ortholog in human RNases P and MRP, Rpp29p.[117] The human homolog of Rpp1p is Rpp30p,[110] whereas the human homolog of Pop7p is Rpp20p.[111] Most of the human homologs have been identified by sequence database searches, although the amino acid sequence similarity between the human and yeast proteins is low. Once the putative human homologs were identified, they were shown to be integral parts of the human RNase P and MRP complexes by immunoprecipitation with antibodies raised against the human proteins. Several proteins have also been identified as part of the human MRP complex by UV cross-linking, but their exact relationship to the yeast protein subunits is still unclear.[118]

Relationship to RNase MRP

RNase MRP, which appears to be an exclusively eukaryotic enzyme, is an endoribonuclease involved in pre-rRNA processing. The RNA subunits of RNase

[114] M. Bartkiewicz, H. Gold, and S. Altman, *Genes Dev.* **3,** 488 (1989).
[115] Z. Lygerou, H. Pluk, W. J. van Venrooij, and B. Séraphin, *EMBO J.* **15,** 5936 (1996).
[116] S. Xiao and D. R. Engelke, unpublished data (2000).
[117] N. Jarrous, P. S. Eder, C. Guerrier-Takada, C. Hoog, and S. Altman, *RNA* **4,** 407 (1998).
[118] H. Pluk, H. van Eenennaam, S. A. Rutjes, G. J. Pruijn, and W. J. van Venrooij, *RNA* **5,** 512 (1999).

MRP and RNase P share common structural features.[101,102] In addition, eight of the nine protein subunits of RNase P appear to be shared with RNase MRP.[32,107–111] The extensive protein subunit overlap between RNases P and MRP, combined with the clear evolutionary relatedness of the large RNA subunits, strongly suggests that nuclear RNase P and RNase MRP are simply variant forms of a single ancestral enzyme, with RNase MRP having developed an altered substrate specificity for rRNA transcripts at positions that have no detectable tRNA-like structure. It is not known whether changes in the RNA subunit, the single different protein subunits, or both are responsible for differences in substrate specificity between RNases P and MRP. No similarity in substrate structure or sequence is apparent at the cleavage site in pre-rRNA and pre-tRNA. Localization studies suggest that both enzymes might function in the same subcellular compartment, the nucleolus.[105] Interestingly, a mutation in the *RPR1* RNA subunit has minor effects on ribosomal processing.[74]

Conclusions and Future Directions

Despite progress that has been made in understanding the complexity of nuclear RNase P in *S. cerevisiae,* a large number of questions remain to be addressed. Although the integral subunits have been identified and the beginnings of an overall structure is emerging, nearly nothing is known about the functions of the protein subunits. The need for the presence of the subunits to activate an otherwise inactive RNA subunit suggests that some functions might have been transferred from the RNA subunit in the prokaryotic enzyme to one or more protein subunits in the eukaryotic holoenzyme. It is unlikely, however, that much of the 18-fold increase in the eukaryotic protein content is devoted to taking over essential functions of the RNA. All of the highly conserved "critical regions" from the bacterial RNA structure are still contained in the yeast RNA subunit. This, combined with mutagenesis studies of the RNA subunit suggesting these regions still contribute to substrate binding and catalysis, strongly indicate that the RNA subunit has simply become dependent on protein for correct folding, and possibly for additional protein contacts that help discriminate substrates.

What, then, do all of these subunits do? It is possible that some, at least, make contact with the leader, trailer, and anticodon regions of the pre-tRNA substrates. Kinetic experiments suggest that at least the termini make contacts with the enzyme and contribute to substrate recognition.[57] Because seven of the nine proteins are highly basic, there are a large number of candidate sites for nucleic acid interaction. It is possible that the nuclear enzyme has significantly more potential for interacting with single-stranded regions of RNAs, as judged by the ability of the nuclear holoenzyme, but not the bacterial holoenzyme, to tightly bind and be inhibited by RNA homopolymers. Site-specific chemical cross-linking could sort out where in the holoenzyme structure contacts with pre-tRNA substrates are made.

Increased RNA recognition by proteins is likely to be only part of the story for the increase in protein subunit composition. Although none of the subunits are closely related to any of the RNA-processing enzymes examined to date, it is possible that additional enzymatic activities are carried along by the RNase P holoenzyme. The discovery that RNase P, like its sibling enzyme RNase MRP, is largely nucleolar[105] invites the speculation that some number of the subunits common to both enzymes are used for correctly positioning the holoenzyme in a spatially ordered pathway. The ribosomal RNA-processing pathway has long been known to be ordered and spatially organized in the nucleolus. It is possible that pre-tRNAs, which also undergo a large number of nuclear processing events before export to the cytoplasm, also have at least their early processing enzymes organized into discrete sites within the nucleus. The *in situ* hybridization signals for several pre-tRNAs and RNase P were found in several nucleoplasmic foci, in addition to the nucleolus. It remains to be seen whether the tRNA genes are also clustered at these sites, as the ribosomal RNA genes are in the nucleolus, or whether pre-tRNA primary transcripts need to be transported to these processing centers. In either case, a large part of the function of the RNase P protein subunits might be to position the enzyme correctly with respect to other enzymes, the processing center, or the cell. Experiments are currently under way to address these issues.

Acknowledgments

We thank Christopher Millikin for experimental collaboration on this work, Paul Good for always lending a helpful hand, Shaohua Xiao for sharing information on protein alignments, and Lasse Lindahl and Carol Fierke for sharing results before publication.

[11] Cyanelle Ribonuclease P: Isolation and Structure–Function Studies of an Organellar Ribonucleoprotein Enzyme

By CHRISTIAN HEUBECK and ASTRID SCHÖN

I. Introduction

Ribonuclease P (RNase P, EC 3.1.26.5) is the endonuclease required for generating the mature 5′ end of tRNA in all organisms; in bacteria, it is also involved in maturation of 4.5S RNA, transfer-messenger RNA (tmRNA), and several other small RNAs (for reviews, see, e.g., Refs. 1–3). The ribonucleoprotein character of this enzyme has now been proved in most organisms and organelles; exceptions, however, are still the chloroplasts, plant nuclei, and metazoan mitochondria, where no associated RNAs have been detected to date. In contrast to the known RNA subunits, which are fairly well conserved in size (about 400 nucleotides) and structure among diverse phylogenetic groups,[4] the protein contribution to the holoenzyme is highly variable in size and number of the individual components. In bacteria, a single small, basic protein of about 15 kDa is associated with the RNA; in contrast, the yeast mitochondrial enzyme contains a large (100-kDa) protein subunit, and the eukaryotic nuclear enzymes are composed of nine proteins in addition to the essential RNA.[5–7] Although all photosynthetic organelles are phylogenetically related to cyanobacteria, RNase P from different types of plastids shows striking diversity: whereas no RNA subunit can be detected in the chloroplast enzyme from green algae or higher plants, the plastids of some primitive algae encode an RNase P RNA of the cyanobacterial type. The photosynthetic organelle (cyanelle) of the glaucocystophyte *Cyanophora paradoxa* combines properties of free-living cyanobacteria and true chloroplasts, which makes it an ideal model system to study the evolution, structure, and function of RNase P from plastids. Here we describe the methods used to grow this organism in amounts sufficient for enzyme preparation; the isolation of the organellar RNase P holoenzyme, detection of its activity, and analysis of the RNase P RNA structure; and the preparation of RNA and protein components from this and related organisms for use in homologous and heterologous reconstitution experiments.

[1] L. A. Kirsebom and A. Vioque, *Mol. Biol. Rep.* **22,** 99 (1996).
[2] D. N. Frank and N. R. Pace, *Annu. Rev. Biochem.* **67,** 153 (1998).
[3] A. Schön, *FEMS Microbiol. Rev.* **23,** 391 (1999).
[4] J. W. Brown, *Nucleic Acids Res.* **27,** 314 (1999).
[5] Y. L. Dang and N. C. Martin, *J. Biol. Chem.* **26,** 19791 (1993).
[6] J. R. Chamberlain, Y. Lee, W. S. Lane, and D. R. Engelke, *Genes Dev.* **12,** 1678 (1998).
[7] N. Jarrous, P. S. Eder, D. Wesolowski, and S. Altman, *RNA* **5,** 153 (1999).

0076-6879/01 $35.00

II. Materials, Equipment, and General Procedures

A. *Buffers and Media*

Buffer A: 6 *M* Guanidinium hydrochloride (Gu-HCl), 100 m*M* KH_2PO_4, 10 m*M* Tris-HCl (pH 8.0), 20 m*M* NaCl, 25 m*M* imidazole

Buffer B: 8 *M* urea, 100 m*M* KH_2PO_4, 10 m*M* Tris-HCl (pH 8.0), 20 m*M* NaCl, 25 m*M* imidazole

Buffer C: Same as buffer B but pH 6.3

Buffer D: Same as buffer B but pH 4.5

Buffer E: Same as buffer B but 250 m*M* imidazole and pH 4.5

Buffer F: 20 m*M* Tris-HCl (pH 7.0), 130 m*M* KCl, 15 m*M* $MgCl_2$

Buffer P: 50 m*M* Tris-HCl (pH 7.0), 2 *M* NH_4Cl, 250 m*M* $MgCl_2$

RNA elution buffer: 500 m*M* Tris-HCl (pH 7.0), 1 m*M* ethylenediaminetetraacetic acid disodium salt (Na_2EDTA), 1% (w/v) sodium dodecyl sulfate (SDS)

SDS gel loading buffer (2× concentrated): 500 m*M* Tris-HCl (pH 6.8), 4% (w/v) SDS, 17% (w/v) glycerol, 0.8 *M* 2-mercaptoethanol, 0.1% (w/v) bromphenol blue

T7 buffer (10× concentrated): 400 m*M* Tris-HCl (pH 8.0), 50 m*M* dithiothreitol, 10 m*M* spermidine, 120 m*M* $MgCl_2$

TBE: 89 m*M* Tris, 89 m*M* boric acid, 2 m*M* Na_2EDTA

TE: 10 m*M* Tris-HCl (pH 7.5), 1 m*M* Na_2EDTA

Buffer TM: 20 m*M* Tris-HCl (pH 8.0), 5 m*M* $MgCl_2$

Urea gel loading buffer: 8 *M* urea, 0.03% (w/v) bromphenol blue, 0.03% (w/v) xylene cyanol

Buffer Xa: 20 m*M* Tris-HCl (pH 8.0), 100 m*M* NaCl, 1 m*M* $CaCl_2$

Buffers I to IV used for structure probing: Described in the relevant sections

LB medium: 1% (w/v) tryptone, 0.5% (w/v) yeast extract (both from Difco, Detroit, MI), 0.5% (w/v) NaCl; sterilize by autoclaving. Antibiotics, 1-isopropyl-β-D-1-thiogalactopyranoside (IPTG), and 5-bromo-4-chloro-3-indolyl-β-D-galactopyranoside (X-Gal) are filter sterilized and added after cooling

B. *Enzymes and Cloning Procedures*

All cloning procedures are performed as described[8] and conform to the German Federal Law for Biological Safety; *Escherichia coli* DH5α, JM109, and SG 13009 are used throughout.

[8] J. Sambrook, E. F. Fritsch, and T. Maniatis, "Molecular Cloning: A Laboratory Manual," 2nd Ed. Cold Spring Harbor Laboratory Press, Cold Spring Harbor, New York, 1989.

Restriction DNases and other enzymes used in cloning procedures are purchased from Roche Molecular Biochemicals (formerly Boehringer Mannheim) or New England BioLabs (Beverly, MA) if not stated otherwise. *Tth* and *Pfu* DNA polymerases used for polymerase chain reaction (PCR) are from Epicentre (Madison, WI) and Stratagene (La Jolla, CA), respectively. Sequences of all genetic constructs are determined either manually with the Sequenase 2.0 kit (Amersham-USB, Cleveland, OH) or automatically by an ABI-Prism sequencer (PE Biosystems, Foster City, CA). Factor Xa protease is from Roche Molecular Biochemicals. T7 RNA polymerase is prepared according to Zawadzki and Gross.[9]

Reagents. All reagents are of analytical or reagent grade and purchased from Roche Molecular Biochemicals (formerly Boehringer Mannheim), New England BioLabs, or Merck (Darmstadt, Germany) if not stated otherwise.

C. Equipment and General Procedures

Work with RNA requires special precautions to avoid RNase contaminations. Thus, all plasticware and aqueous solutions are sterilized by autoclaving if the stability of components permits; heat-labile solutions are prepared from sterile stable components. Glassware is baked at 150° for several hours.

Separation of nucleic acids from complex proteinaceous mixtures is achieved by extraction with TE-saturated phenol. Nucleic acids are precipitated by addition of 0.1 volume of 2 *M* sodium acetate (pH 6.0) and 2.5 volumes of ice-cold ethanol, incubation at −20° for 30 min, and centrifugation at a relative centrifugal force (RCF) of 10,000 to 15,000 for 30 min at 4°. Polyacrylamide gels for nucleic acid separation are generally 0.3 to 0.4 mm thick and consist of TBE, 8 *M* urea, and the specified polymer concentration.[8]

III. Methods

A. Preparation and Functional Analysis of Native Cyanelle Ribonuclease P

a. Cell Growth and Preparation of Cyanelle S100

Cyanophora paradoxa (strain PCC C.7201; Pasteur Culture Collection) cells are grown by aeration with sterile filtered air in Stanier's medium,[10] in a light–dark cycle of 16 : 8 hr with 1000 lux measured at the flask surface. For the large-scale preparation of cyanelles, cultures are grown to about 6×10^9 cells/ml in 3- or 5-liter bottles filled halfway with medium, harvested by centrifugation at an RCF of 10,000 for 10 min at 4°, suspended in TE, and lysed by addition of Triton

[9] V. Zawadzki and H. J. Gross, *Nucleic Acids Res.* **19,** 1948 (1991).

[10] M. Herdman and R. Y. Stanier, *FEMS Microbiol. Lett.* **1,** 7 (1977).

X-100 to 1% (v/v) final concentration.[11] Cyanelles are collected by centrifugation, subjected to several washes with TE, shock frozen in liquid N_2, and stored at $-80°$.

All steps in the enzyme isolation are performed on ice or at 4°; all solutions contain 0.2 m*M* phenylmethylsulfonyl fluoride (PMSF) to reduce proteolysis. Protein concentration is determined according to Bradford[12] with bovine serum albumin (BSA) as standard. For preparation of extracts active in tRNA processing, frozen cyanelles are ground with an equal volume of sterile sea sand in a small volume of extraction buffer [50 m*M* Tris-HCl (pH 7.5), 10 m*M* $MgCl_2$, 20 m*M* KCl, 2 m*M* Na_2EDTA, 10 m*M* 2-mercaptoethanol]. The sand is removed by low-speed centrifugation and a postribosomal supernatant (S100) is prepared by centrifugation for 1 h at $150{,}000g_{max}$ in a Beckman (Fullerton, CA) 80 Ti or 60 Ti rotor. The extract is extensively dialyzed against buffer TM (or buffer F) containing 10 m*M* 2-mercaptoethanol and 20% (v/v) glycerol, shock frozen in liquid N_2 in small aliquots, and kept at $-80°$.

b. Purification of Native Cyanelle Ribonuclease P

i. Ion-Exchange Chromatography. Endogenous tRNA present in the postribosomal supernatant may interfere with activity assays and RNA analysis. In addition, most proteins present in the cyanelle are eluted from anion-exchange resins at lower ionic strength than ribonucleoprotein particles such as RNase P. Thus, fractionation of the crude extract by anion-exchange chromatography is an important step toward RNase P purification. Although a variety of chromatography media are available for this purpose, in our hands a trimethylaminoethyl-derivatized material (TMAE) gives the most reproducible results. However, the enzyme activity is inhibited by high concentrations of monovalent cations; in addition, high ionic strength conditions irreversibly inactivate cyanelle RNase P. Thus, it is essential that the KCl present in the gradient be removed by dialysis into buffer TM containing 20% (v/v) glycerol before the fractions are assayed and stored.

For fractionation of the cyanelle extract, 30 to 50 mg of protein is applied to a 20-ml column of TMAE Fractogel-650(M) (Merck) equilibrated in buffer TM plus 10% (v/v) glycerol. After washing off unbound material, a 150-ml gradient from 0 to 500 m*M* KCl in buffer TM is developed and 5-ml fractions are collected. The protein content of the eluate is monitored by absorption at 280 nm, and determined in individual fractions by dye binding.[12] Ten microliters of each dialyzed fraction is then used for RNase P assays; cyanelle RNase P typically elutes at about 400 m*M* KCl.

ii. Gel Filtration. Size-exclusion chromatography can be used for analytical purposes (size determination of RNase P) or on a preparative scale during the

[11] M. Baum, A. Cordier, and A. Schön, *J. Mol. Biol.* **257,** 43 (1996).

[12] M. M. Bradford, *Anal. Biochem.* **72,** 248 (1976).

purification of larger amounts of the enzyme. Although it has the advantage that enzyme activity is not destroyed because of high-salt conditions as in ion-exchange chromatography or equilibrium gradient centrifugation, the inevitable dilution during the separation process requires that the pooled fractions be concentrated after this step.

Chromatography is performed on a Superdex 200 HR gel-filtration column (120 ml, 1.6×60 cm; Pharmacia, Uppsala, Sweden) on a Pharmacia Fast Performance Liquid Chromatography (FPLC) apparatus. The column is equilibrated with buffer F at a flow rate of 1 ml/min at 4° and calibrated with the following standards (all from Boehringer Mannheim/Roche): carbonic anhydrase (29 kDa), ovalbumin (45 kDa), BSA (68 kDa), alcohol dehydrogenase (150 kDa), ferritin (450 kDa), and blue dextran (2400 kDa). Four milliliters of cyanelle S100 extract (14 mg of protein) is then loaded onto the column and the chromatography is carried out under conditions identical to the calibration run. Fractions can be assayed for RNase P activity without further treatment.

iii. Equilibrium Density Gradient Centrifugation. The difference in density between nucleic acids and proteins provides not only a convenient means for separation, but also for the identification, purification, and characterization of nucleoprotein particles such as ribosomes or RNase P. Density gradients can be conveniently formed by centrifugation of concentrated salt solutions; because of their good solubility, the chlorides or sulfates of cesium or rubidium are usually employed for this purpose.[13,14] Although cyanelle RNase P is sensitive to high ionic strength, isopycnic density gradient centrifugation can be used on an analytical scale, provided that the salt is removed by dialysis immediately after centrifugation.

For determination of the buoyant density of cyanelle RNase P, a solution of Cs_2SO_4 in buffer TM (starting density $\rho = 1.32$ g/ml) and 500 μg of S100 protein is prepared in a total volume of 10 ml. Density gradients are formed by centrifugation of this solution in a Beckman 80 Ti rotor for 70 hr at 35,000 rpm and 15°. Fractions of 0.5 ml are collected from the bottom of the tube and their density is determined from the refractive index and a calibration curve. Cs_2SO_4 is removed from the fractions by extensive dialysis against buffer TM before testing for RNase P activity.

c. Determination of Cyanelle RNase P Activity

A unique feature of plastids and most cyanobacteria is the presence of an unusually structured T-stem in $tRNA^{Glu}$, the tRNA required for porphyrin biosynthesis. It differs from the tRNA consensus by the A53-U61 base pair at the distal

[13] W. Szybalski, L. Grossman, and K. Moldave, *Methods Enzymol.* **12,** 330 (1968).

[14] M. G. Hamilton, L. Grossman, and K. Moldave, *Methods Enzymol.* **20,** 512 (1971).

end of the T-stem.[15,16] Another phenomenon distinct from most bacteria is the absence of the terminal 3′-CCA end in tRNA genes from members of this phylogenetic lineage.[17] To analyze the effect of these features on pre-tRNA cleavage by RNase P from plastids and cyanobacteria, we have employed a barley chloroplast pre-$tRNA^{Glu}$ and two mutants that restore the tRNA consensus to the T-domain (pre-$tRNA^{Glu}$-GC), or lack the 3′-terminal CCA (pre-$tRNA^{Glu}$-ΔCCA).

i. Preparation of Pre-tRNA Substrate. Plasmids allowing transcription of barely chloroplast pre-$tRNA^{Glu}$ containing the natural 5′-flank and terminating either with the mature CCA end or with the discriminator base A73, have been described; a construct in which the unique A–U pair has been converted to the canonical G53–C61 is also available.[18] An additional frequently used RNase P substrate is *E. coli* pre-$tRNA^{Tyr}$.[19] All constructs contain a T7 promotor and a *Fok*I restriction site located appropriately for correct and efficient runoff transcription. Templates are prepared by *Fok*I cleavage of these plasmids, followed by phenol extraction, ethanol precipitation and adjustment of DNA concentration to 0.1 μg/μl.

For the synthesis of ^{32}P-labeled transcripts, 6 μl of the restricted template is mixed in a total volume of 20 μl with 2 μl of 10× T7 buffer, 2 μl of rNTP mix (rATP, rCTP, rUTP, 10 m*M* each), 2 μl of 2 m*M* rGTP, 2 μl of [α^{32}P]rGTP (10 μCi/μl), 1 μl of T7 RNA polymerase (0.37 μg/μl), and double distilled H_2O. The reaction mixture is incubated at 37° for 2 hr. After addition of 80 μl of 25 m*M* EDTA to stop transcription, the reaction mixture is extracted with phenol and the aqueous layer is precipitated. The dried pellet is resuspended in 10 μl of urea loading buffer; the solubility of the transcript can be increased by the addition of 0.02% (w/v) Triton X-100. The RNA is purified by electrophoresis on a 8% (w/v) polyacrylamide gel and localized by autoradiography. The desired band is cut out, 300 μl of RNA elution buffer is added, and the vial is incubated at −20° until the contents are frozen. The RNA is eluted by vigorous shaking for 6–12 hr, the buffer is recovered, and the RNA is precipitated with 700 μl of ethanol. The resulting RNA pellet is resuspended in TE buffer to give a concentration of 10× 10^3 cpm/μl.

ii. Assay and Quantitation of Holoenzyme Activity. RNase P activity is assayed by monitoring the cleavage of the 5′-leader sequence from different pre-tRNA substrates. Correct cleavage of *E. coli* pre-$tRNA^{Tyr}$ produces a 43-nucleotide 5′ flank, whereas the pre-$tRNA^{Glu}$ variants yield a 33-nucleotide flank. Standard processing reactions are performed in buffer F at 30° and contain 1000 to 7500 cpm

[15] A. Schön, G. Krupp, S. Gough, S. Berry-Lowe, C. G. Kannangara, and D. Söll, *Nature (London)* **322,** 281 (1986).

[16] G. P. O'Neill, D. M. Peterson, A. Schön, M.-W. Chen, and D. Söll, *J. Bacteriol.* **170,** 3810 (1988).

[17] M. Sprinzl, C. Horn, M. Brown, A. Ioudovitch, and S. Steinberg, *Nucleic Acids Res.* **26,** 148 (1998).

[18] W. R. Hess, C. Fingerhut, and A. Schön, *FEBS Lett.* **431,** 138 (1998).

[19] L. A. Kirsebom, M. F. Baer, and S. Altman, *J. Mol. Biol.* **204,** 879 (1988).

(4 to 30 fmol) of α-^{32}P-labeled substrate and up to 3.5 μg of RNase P preparation in a total volume of 20 μl. The reaction is stopped by adding 80 μl of ice-cold double distilled H_2O, and the sample is phenol extracted and precipitated with ethanol. Before electrophoresis on a 10% (w/v) polyacrylamide gel, the pellet is resuspended in 10 μl of urea loading buffer and denatured at 95° for 2 min. The RNase P cleavage reaction is analyzed by autoradiography or quantitated with a PhosphorImager (Molecular Dynamics, Sunnyvale, CA).

iii. Identification of Cleavage Site. For all RNA processing experiments, correct cleavage of the substrate must be verified. Primer extension is routinely used to localize the cleavage site; however, when a newly identified enzyme such as the cyanelle RNase P is investigated, the direct identification of the 5′-phosphate group in the reaction product is of equal importance.

For primer extension analysis, processing reactions with RNase P holoenzymes or ribozymes (see below) are scaled up to 50 μl. To obtain a strong signal, the ^{32}P-labeled pre-tRNA should be supplemented with unlabeled substrate to a total amount of approximately 1 pmol. In a preliminary experiment, reaction time and enzyme concentration should be optimized so that maximum turnover is achieved. The reaction product is purified by phenol extraction (this step is not necessary for ribozyme reactions), concentrated by ethanol precipitation, and dissolved in 9 μl of H_2O. After addition of 1 μl of oligonucleotide (1 pmol), the mixture is heated to 95° for 2 min to denature the pre-tRNA, and slowly cooled to 42°. Three microliters of avian myeloblastosis virus (AMV) buffer (5× concentrated), 2 μl of dNTP solution (5 m*M* each), 0.5 μl of [α-^{32}P]dATP (10 μCi/μl), and 3 U of reverse transcriptase (AMV-RT; Promega, Madison, WI) is added and the reaction is incubated for 1 hr at 42–48°. After supplementation with 1 μl of RNase A (DNase free, 0.5 mg/ml) and 2 μl of Na_2EDTA (250 m*M*), the radioactive RNA is degraded for 10 min at 37°. The products of the primer extension are then analyzed on a 6 or 8% (w/v) denaturing polyacrylamide gel. To allow the identification of the cleavage site, sequencing reactions are performed with the same primer and the uncleaved plasmid encoding the pre-tRNA; they are run in adjacent lanes. It should be pointed out that, although other reverse transcriptases may be used, the stable secondary structure of tRNAs can lead to strong stops of elongation at incubation temperatures below 42°; thus, AMV-RT is the enzyme of choice because it is active between 42 and 48°. The choice of primer is also important for the successful reverse transcription of tRNAs: the start of elongation should be at least 20 bases downstream from the mature 5′ end to allow incorporation of enough radioactive label for a strong signal; in addition, the primer annealing region in the tRNA should not exhibit a stable secondary structure.

The direct identification of the 5′ end group of the matured tRNA (nearest neighbor analysis) requires that the nucleotide adjacent to the 5′ terminus provide the radioactive label. Thus, if this second nucleotide is not a guanosine, the corresponding [α-^{32}P]rNTP must be used in the transcription reaction (see Section III,A,c,i) and the mixture of unlabeled nucleotides must be adjusted accordingly. At least 10^6 cpm of pre-tRNA is used in a preparative-scale processing reaction as described above; the reaction products are separated by gel electrophoresis, and the band corresponding to the mature tRNA is eluted from the gel. The RNA precipitate is dissolved in 6 μl of H_2O; 2 μl of 50 mM ammonium acetate (pH 4.5) and 1 μl of RNase mix [RNase T2 (50 mU/μl) and RNase A (0.1 mg/ml)] are added and the hydrolysis is incubated for 5 hr at 37°. Both RNases produce nucleoside 3′-phosphates; the only 5′,3′-nucleoside diphosphate resulting from this reaction corresponds to the 5′ end of mature tRNA. The end group can then be readily identified by two-dimensional thin-layer chromatography.[20]

d. Identification and Functional Characterization of Ribonuclease P RNA

i. Identification of Ribonuclease P-Associated RNA. For the identification of novel RNase P RNA sequences, the two short "conserved" regions within helix P4 of all RNase P RNAs are used to design amplification primers for PCR. For phylogenetic reasons, these oligonucleotides should be derived from cyanobacterial sequences if plastids are to be analyzed. It is crucial that the cultures in question be free from contaminating bacteria; if possible, nucleic acid preparations from isolated and nuclease-treated organelles should be used. Total plastid RNA is obtained by phenol extraction of the organelles in the presence of 25 mM Na_2EDTA and 1% (w/v) SDS, followed by ethanol precipitation. These preparations contain only traces of plastome DNA.[21] The sequence of the RNase P RNA core is determined by RT-PCR. First-strand cDNA synthesis is performed by annealing 20 pmol of 3′ primer (complementary to the 3′ strand of P4) to 5 μg of total plastid RNA and extension with AMV-RT as described in Section III,A,c,iii. One-tenth of this cDNA is then amplified with this 3′ primer and a 5′ primer corresponding to the 5′ strand of P4 under conditions suggested by the manufacturer of the thermostable polymerase. The PCR product is then cloned and sequenced to allow the design of internal primers for the determination of the full-length sequence. The 5′ and 3′ termini of the native RNase P RNA can then be easily obtained by rapid amplification of cDNA ends (RACE) with a specific nested primer set and commercially available RACE kits (e.g., Roche; GIBCO-BRL, Gaithersburg, MD).

ii. Functional Verification of Ribonuclease P-Associated RNA. The identity of any newly identified RNA as an integral and essential RNase P component should be proved by several independent methods. For bacterial RNase P RNAs, a test for ribozyme activity can be used; nuclear and organellar RNase P RNAs, however, do not exhibit catalytic activity without their protein complement. Copurification

[20] H. Beier and H. J. Gross, "Sequence Analysis of RNA in Essential Molecular Biology—A Practical Approach" (T. A. Brown, ed.), Vol. II, p. 221. IRL Press, Oxford, 1991.

[21] M. Baum and A. Schön, *FEBS Lett.* **382,** 60 (1996).

of an RNA with RNase P activity over consecutive isolation steps is an unequivocal proof of its identity as enzyme component. This is most easily done by Northern blotting of RNA prepared from enzyme fractions by phenol extraction and ethanol precipitation.[11] The samples are separated on a denaturing 8% (w/v) polyacrylamide gel, electroblotted onto a Zetaprobe membrane (Bio-Rad, Hercules, CA), and hybridized with a gene-specific antisense primer that has been radioactively labeled at the 5′ end with T4 polynucleotide kinase (PNK) and [γ-^{32}P]ATP.[8] Appropriate standards (e.g., *in vitro* transcripts of known RNase P RNAs) should be included on the gel to allow an estimate of RNA sizes.

An additional proof for an essential RNA component is nuclease treatment of RNase P. This is most conveniently performed with micrococcal nuclease in the presence of $CaCl_2$. The nuclease is then inactivated by chelating Ca^{2+} with EGTA before RNase P activity is assayed. The data obtained with this type of experiment should, however, be interpreted with caution, as "substrate masking" by inactivated micrococcal nuclease may mimic loss of RNase P function.[22]

B. Determination of Cyanelle Ribonuclease P RNA Structure in Vitro and in Situ

To evaluate whether misfolding might be a reason for the lack of ribozyme activity in an RNase P RNA, or to elucidate the RNA structure within the RNase P holoenzyme, or to identify RNA regions that are protected by the protein in the holoenzyme, structure analysis of the RNase P RNA is performed. Specifically, the two-dimensional RNA structure is determined for a corresponding *in vitro* transcript, and a structural footprint analysis is performed for the native RNA in the holoenzyme. The construction of plasmid pT7CyRPR allowing efficient runoff transcription of cyanelle RNase P RNA, and the *in vitro* transcription with T7 RNA polymerase, have been described (see Section III,C,a,i and Refs. 11 and 23). Here, we describe how the methods of RNA structure analysis previously developed for yeast nuclear RNase P[24] have been adapted to probe this RNA, both as a transcript and *in situ*.

a. Structure Analysis of in Vitro-Synthesized Cyanelle Ribonuclease P RNA

i. Probing of RNA Base Accessibility to Chemical Reagents. Unpaired nucleotides in cyanelle RNase P RNA are modified with dimethyl sulfate (DMS) or 1-cyclohexyl-3-[2-(*N*-methylmorpholino)ethyl]carbodiimide *p*-toluenesulfonate (CMCT). The modified positions (N1 of adenine and N3 of cytosine for DMS; N3 of uracil and rarely N1 of guanine for CMCT) can then be detected by primer

[22] M. J. Wang and P. Gegenheimer, *Nucleic Acids Res.* **18,** 6625 (1990).

[23] A. Cordier and A. Schön, *J. Mol. Biol.* **289,** 9 (1999).

[24] A. J. Tranguch, D. W. Kindelberger, C. E. Rohlman, J.-Y. Lee, and D. R. Engelke, *Biochemistry* **33,** 1778 (1994).

extension.[25] Each reaction contains 200 ng of cyanelle RNase P RNA transcript (see Section III,C,a,i) in a total volume of 200 μl; final concentrations of the reagents are 33 m*M* for CMCT and 50 m*M* for DMS. To mimic the conditions under which most RNase P ribozymes are functioning, high ionic strength buffers are used for probing native cyanelle RNase P RNA structure. For CMCT modifications, buffer I (20 m*M* Tris-HCl, 10 m*M* $MgCl_2$, 500 m*M* NH_4Cl, pH 7.5) is employed for native conditions and buffer II (20 m*M* Tris-HCl, 1 m*M* Na_2-EDTA, 500 m*M* NH_4Cl, pH 7.5) is employed for semidenaturing and denaturing conditions. For DMS modifications, buffer III (50 m*M* sodium borate, 10 m*M* $MgCl_2$, 500 m*M* NH_4Cl, pH 7.8) is used for native conditions and buffer IV (50 m*M* sodium borate, 1 m*M* Na_2-EDTA, 500 m*M* NH_4Cl, pH 7.5) is used for semidenaturing and denaturing conditions. Reactions under native or semidenaturing conditions are incubated at 30° for 10 min; incubation under denaturing conditions is at 95° for 30 sec. Reactions are stopped by ethanol precipitation, and pellets are washed twice with 70% (v/v) ethanol and dried before primer extension analysis. Controls are treated in an identical manner, but with omission of the chemical reagent. Primer extension for the detection of modified positions is essentially performed as described in Section III,A,c,iii, except that the unlabeled oligonucleotide and [α-^{32}P]dATP are replaced by a primer that carries a 5′-^{32}P label.[8]

ii. Probing RNA Backbone Accessibility with Structure-Sensitive Enzymatic Probes. Phosphodiester bonds accessible to enzymatic cleavage are probed with RNase ONE (RNase 1 from *E. coli*) and RNase T1 (both are single-strand specific; RNase T1 cleaves only after guanosines), and RNase V1 (specific for double-stranded and stacked regions). For each *in vitro* structure probing reaction, 600 ng of cyanelle RNase P RNA transcript is digested with RNase V1 (Pharmacia, 5 mU) or RNase ONE (Promega, 5 mU) in buffer I at 30° in a total volume of 200 μl. Digestions with RNase T1 (CalBiochem, La Jolla, CA) are performed in buffer I at 30° (native) and in buffer II at 30° (semidenaturing), using 300 mU of enzyme; under denaturing conditions, 5 U of T1 is incubated at 50° in T1-mix (7 *M* urea, 1 m*M* Na_2-EDTA, 20 m*M* trisodium-citrate, pH 5.0). At 2, 5, and 15 min, aliquots are withdrawn, the reactions are stopped by phenol–chloroform extraction, and the RNA is precipitated and dried before primer extension analysis (Section III,B,a,i). Controls are treated in the same way, except that the respective enzyme is omitted.

For *in situ* probing of RNA accessibility in the holoenzyme, 595 μl of S100 extract in buffer F (2 mg of protein) is used per reaction. Digestions with RNase V1 (10 mU), RNase ONE (2 mU), or RNase T1 (300 mU) are performed at 30° in a total volume of 600 μl. At 2, 5, and 15 min, 200-μl aliquots are withdrawn, the reactions are terminated, and the RNA is analyzed by primer extension.

[25] S. Stern, D. Moazed, and H. Noller, *Methods Enzymol.* **164,** 481 (1988).

C. Reconstitution of Ribonuclease P from Pure Recombinant Subunits and Functional Analysis of Reconstituted Holoenzymes

a. Preparation and Functional Analysis of Recombinant Ribonuclease P RNA

The construction of plasmids pT7CyRPR, pT7PholRPR, and pT7PmRPR allowing efficient run-off transcription of RNase P RNAs from cyanelles and the two cyanobacteria *Prochlorothrix hollandica* and *Prochlorococcus marinus,* respectively, have been described.[11,18,26] The RNA component of *Synechocystis* 6803 RNase P (SynRPR) is often used as a functional control when other cyanobacterial-type RNase P RNAs are assayed; the construction of a plasmid allowing preparation of this RNA *in vitro* has been published.[27] Here we describe the construction and use of a high-efficiency transcription clone for this RNA as an example for all RNase P RNAs mentioned in this article.

i. Construction of Transcription Template for Ribonuclease P RNA and Preparation of Transcript. The *rnpB* gene is amplified by PCR from *Synechocystis* PCC 6803 genomic DNA, using the primers ET7Syn5′ (GCGAATTC*TAATACGACTCACTATAGGG*AGAGTTAGGGAGG) and Syn3′Bam (GCGGATCCGACGCAT*CTCGAG*AGTTAGTCGTAAG). These two primers match the 5′ and 3′ ends of the RNase P RNA coding region, respectively. ET7Syn5′ includes the T7 RNA polymerase promoter (italics) such that transcription begins with the first of the three terminal guanosines; the last of these three nucleotides corresponds to the 5′ end of native *Synechocystis* RNase P RNA. An *Eco*RI restriction site (underlined) is added at the 5′ end for cloning purposes. Syn3′Bam contains restriction sites for *Bam*HI (underlined) and *Xho*I (italics) for cloning and subsequent run off transcription, respectively. The PCR product is cloned into the *Eco*RI and *Bam*HI sites of pUC19 to give pT7SynRPR. Templates are prepared by cleavage with *Xho*I (or appropriate restriction enzymes for other RNase P RNA genes) and further processed as in Section III,A,c,i.

In vitro transcription of RNase P RNAs is essentially performed as described.[28] T7 RNA polymerase is prepared according to Zawadzki and Gross[9]; the optimum amount of enzyme required for efficient transcription must be determined empirically for each batch. Alternatively, commercial preparations may be used. The reaction mixture for preparative-scale transcription consists of 125 μl (12.5 μg) of restricted plasmid, 25 μl of 10$\times$ T7 buffer, 25 μl of rNTP mix (rATP, rCTP, rGTP, rUTP, 20 m*M* each), 2.5 μl of T7 RNA polymerase (3.7 μg/μl), and double-distilled H_2O to a total volume of 250 μl. The reaction is incubated for 2 hr at 37° and Na_2EDTA is added to a final concentration of 20 m*M*. It complexes Mg^{2+}, which is essential for the T7 RNA polymerase, and avoids formation of

[26] C. Fingerhut and A. Schön, *FEBS Lett.* **428,** 161 (1998).

[27] A. Vioque, *Nucleic Acids Res.* **20,** 6331 (1992).

[28] U. Weber, H. Beier, and H. J. Gross, *Nucleic Acids Res.* **24,** 2212 (1996).

an insoluble precipitate. The reaction mixture is phenol extracted, and the nucleic acids are precipitated with ethanol, resuspended in 100 μl of urea loading buffer, and purified on a denaturing 6% (w/v) polyacrylamide gel. The transcript is detected by UV shadowing as follows: The gel is carefully sandwiched between two layers of household plastic wrap and placed on a fluorescence-modified thin-layer plate (e.g., Merck F254). The RNA is then visualized under a UV lamp (λ 254 nm) as a dark band on a bright green background, cut out, eluted, and precipitated. The dried pellet is resuspended in TE buffer and the RNA concentration is determined by measuring the absorbance at 260 nm.

b. Preparation of Recombinant Ribonuclease P Protein

The sequence of the RNase P protein gene from the cyanobacterium *Synechocystis* PCC 6803, and the construction of an overproducing clone, have been described.[29] To facilitate large-scale purification of the *Synechocystis* RNase P protein (SynPP), we have developed a genetic construct for the overexpression of this protein containing an affinity tag. Because the functional properties of the small, charged RNase P protein might be changed by the addition of six histidine residues to either end, we have also constructed a clone in which the hexahistidine (His_6) sequence is followed by a protease cleavage site. This permits the removal of the engineered N terminus and thus allows the evaluation of any possible effect of this affinity tag on RNase P function.

i. Construction of Clone Allowing Overproduction of Affinity-Tagged Cyanobacterial Ribonuclease P Protein. The *Synechocystis rnpA* gene encoding the protein component of RNase P is isolated by PCR amplification, using genomic DNA as template. In all eubacteria analyzed so far, the *rnpA* gene is downstream and closely linked to the *rpmH* gene, which encodes ribosomal protein L34. In *Synechocystis* PCC 6803 this gene organization is conserved. To avoid problems of nonspecific priming, the *rpmH-rnpA* operon is first amplified from genomic DNA by using oligonucleotides complementary to the 5′ and 3′ ends of the two reading frames, respectively. *Hin*dIII restriction sites (underlined) are included in the amplification primers (SynL34Hind, 5′-GCGC<u>AAGCTT</u>ATGACTCAACGAAC-3′; SynPPHind, 5′-GCGC<u>AAGCTT</u>CTCATAATAAGTGG-3′) to facilitate cloning. The amplified DNA is cloned into vector pUC19 to give the plasmid pSynL34PP.

In a subsequent round of PCR, the two histidine-tagged *rnpA* variants are amplified from the template pSynL34PP. The 5′-primers SynPPBam (5′-CGC<u>GGATCC</u>ATGGGACTACCCAAA-3′) and SynPPXa (5′-CGC<u>GGATCC</u>*AUCGAAGGACGA*ATGGGACTACCCAAA-3′; factor Xa cleavage site in italics), the latter of which contains an additional sequence coding for the factor Xa protease recognition site, both contain a *Bam*HI restriction site (underlined) for cloning. The 3′ primer SynPPHind is as described above. The amplified *rnp*A gene variants are cloned

[29] A. Pascual and A. Vioque, *Eur. J. Biochem.* **241,** 17 (1996).

into the *Bam*HI and *Hin*dIII sites of the vector pQE30 (Qiagen, Hilden, Germany), which adds a series of six histidines at the amino terminus of the RNase P protein, thus allowing efficient purification by chromatography on a nickel chelate column. In addition, this vector combines a phage T5-derived promoter with a repression system to allow tight regulation of protein production by means of the *lac* operator/*lac* repressor. The resulting plasmids are called pSynPP-Xa and pSynPP-His.

ii. Overproduction of Synechocystis Ribonuclease P protein in Escherichia coli. For overexpression of the protein, the plasmids pSynPP-Xa and pSynPP-His are transformed into *E. coli* SG13009, which carries the pREP4 repressor plasmid, thus reducing "leaky" expression before induction with IPTG. Cells are grown overnight in 5 ml of LB medium supplemented with ampicillin (0.1 mg/ml) and kanamycin (0.05 mg/ml) to select for the presence of both plasmids. A 2-liter flask containing 500 ml of LB medium and antibiotics is then inoculated with the overnight culture at a 1 : 1000 dilution. The culture is incubated with vigorous shaking at 34° until the absorbance at 550 nm reaches 0.6–0.7. At this point IPTG is added to a final concentration of 1 m*M* and the incubation is continued. After 4 hr the cells are harvested by centrifugation (Beckman J-6B, 4000 rpm, 10 min, 4°). Protein yield is not increased by longer incubation. The cell pellet is recovered, weighed, and frozen in liquid nitrogen. About 2.5 g of wet cell mass is usually recovered from 500 ml of induced culture. The frozen cells can be stored at −80°. It is essential to check the degree of overexpression of *Synechocystis* RNase P protein by analyzing a small aliquot of total cellular protein on a 12.5% (w/v) polyacrylamide–SDS gel.[30] It is worth mentioning that Gu-HCl, used for preparative cell lysis, will form a precipitate when combined with SDS loading buffer. Therefore, proteins should be separated from the Gu-HCl suspension by trichloroacetic acid (TCA) precipitation or, more easily, a small aliquot of the culture is lysed directly by addition of SDS loading buffer and vortexing. An intense band of *Synechocystis* RNase P protein corresponding to a molecular mass of 15.3 kDa for SynPP-His or 15.8 kDa for SynPP-Xa, respectively, will be visible if the overexpression has been successful.

iii. Purification of Recombinant Synechocystis Ribonuclease P Protein. All operations using denaturing buffers are performed at room temperature. The purification procedure for SynPP containing a histidine tag (SynPP-His) and the protein with a histidine tag and factor Xa cleavage site (SynPP-Xa) are identical except where the differences are pointed out.

The cell pellet is resuspended in lysis buffer A (5 ml/1-g cell pellet). To complete cell lysis the suspension is agitated for 1 hr. After centrifugation (Beckman J6-B, 4000 rpm, 10 min, 4°) to remove cell debris, the protein is recovered in the supernatant and the pellet is discarded. Nickel nitrilotriacetic acid agarose (Ni^{2+}-NTA, 2.5 ml; Qiagen), equilibrated in buffer A, is added to the supernatant.

[30] U. K. Laemmli, *Nature* (*London*) **227,** 680 (1970).

The resulting suspension is gently agitated in a 15-ml polypropylene test tube for 30 min before being placed in a 15-ml column (1.6 × 7.5 cm). This batch procedure promotes efficient binding of the histidine-tagged protein, especially when the protein in the lysate is present at a low concentration. The column is washed successively with 20 ml of buffers B, C, and D at a rate of 1.0 ml/min. The urea contained in buffers B, C, and D keeps RNA from binding to the protein. SynPP is then eluted with 5 ml of buffer E at a rate of 0.5 ml/min. Fractions of 0.5 ml are collected. The presence of SynPP and the completeness of elution are checked by dye binding.[12] The protein composition in each fraction is analyzed by gel electrophoresis followed by staining with Coomassie blue or silver.[31] The fractions containing the RNase P protein are pooled and either processed immediately, or stored at −20°. Endogenous *E. coli* RNase P protein (C5) or holoenzyme will not bind to the column and thus will not copurify with the recombinant protein because the basis for the Ni^{2+}-NTA column purification requires the histidine-tagged terminal sequence.

The *Synechocystis* RNase P protein obtained after the Ni^{2+}-NTA column treatment is further purified by DEAE-cellulose chromatography to remove any remaining traces of nucleic acids that survived the first step. The flow-through contains the basic SynPP while RNA contaminations bind to the anion exchanger. This purification step is carried out by mixing the protein with 3 ml of DEAE-cellulose (DE52; Whatman, Clifton, NJ) equilibrated with buffer B, in a 50-ml polypropylene test tube; complete binding of RNA is ensured by agitation for 1 hr. The DEAE-cellulose suspension is then poured into a 15-ml column. The protein concentration of the eluate is checked by dye binding or by measuring the absorbance at 280 nm, using an extinction coefficient of 7953 M^{-1} cm^{-1} based on the amino acid composition of this protein.[32] Protein concentrations determined by the extinction coefficient differ by <10% from those based on the dye-binding method. The resulting material is then loaded onto a carboxymethyl-cellulose column (Whatman CM23) of the same dimensions and eluted with a 30-ml gradient from 0 to 500 m*M* NaCl. The basic RNase P protein, but not the RNA contaminants, is expected to bind to this cation exchanger. This method has been described for the *E. coli* wild-type[33] and recombinant[34] C5 protein. In our case, this purification step did not result in higher purity but in lower yields. As aliquots of the SynPP preparation that are directly renatured after the DEAE column treatment do not show any RNase P activity, the presence of RNA contaminants can be excluded at this point. Therefore an additional CM-cellulose column treatment is not recommended for the large-scale purification procedure.

[31] H. Blum, H. Beier, and H. J. Gross, *Electrophoresis* **8,** 93 (1987).

[32] S. C. Gill and P. H. von Hippel, *Anal. Biochem.* **182,** 319 (1989).

[33] M. F. Baer, J. G. Arnez, C. Guerrier-Takada, A. Vioque, and S. Altman, *Methods Enzymol.* **181,** 569 (1990).

[34] C. J. Green, R. Rivera-Leon, and B. S. Vold, *Nucleic Acids Res.* **24,** 1497 (1996).

Up to the last step the purification procedure is the same for both recombinant proteins, SynPP-His and SynPP-Xa. At this point SynPP-His is concentrated with an Centricon 10 concentrator (Amicon, Danvers, MA), followed by dialysis against buffer F containing 50% (v/v) glycerol and 0.01 m*M* PMSF. SynPP-Xa is dialyzed against buffer Xa and concentrated with an Centricon 10 concentrator (Amicon). The pure renatured proteins can be kept at −20° at concentrations of 0.5–1.0 mg/ml. A total yield of about 3 mg of highly pure histidine-tagged RNase P protein per liter of induced *E. coli* SG13009 culture can be expected.

The serine protease factor Xa, used to remove the His_6 tag from SynPP-Xa, is irreversibly inhibited by PMSF; the inhibitor is thus omitted from this buffer. Factor Xa cleaves after the arginine residue in its preferred cleavage site (Ile-Glu-Gly-Arg). Sometimes it recognizes other basic residues, depending on the conformation of the protein substrate.[35] As SynPP has a basic character, it is necessary to modify the usual protease reaction parameters suggested by the manufacturer. The proteolysis is carried out at 19° with a molar ratio of 1 : 500 (enzyme vs substrate). After 15 min the digestion is stopped by adding ethylene glycol tetraacetate (EGTA) to a final concentration of 10 m*M*. RNase P protein without a histidine tag is separated from the histidine-peptide and uncleaved protein by Ni^{2+}-NTA column chromatography as described above. Protein without a histidine tag does not bind to the column and can be collected as flow-through. Uncleaved protein that remains on the column can be recovered by elution as described above. Factor Xa (43 kDa), which also does not bind to the column, is separated from the eluted protein by a Centricon 30 concentrator. The RNase P protein passes through the filter and is collected as filtrate. For storage, the protein is dialyzed against buffer F containing 50% (v/v) glycerol and 0.01 m*M* PMSF and can be kept at −20° for several months. Efforts to purify the cleaved RNase P protein by electroelution after preparative polyacrylamide gel electrophoresis were not successful due to irreversible binding of SDS to the protein. Therefore this procedure is not recommended. Purity and concentration of the preparation are determined as described for the histidine-tagged protein; usually about 200 μg of pure and active RNase P protein-Xa per liter of induced *E. coli* SG13009 culture can be recovered.

c. Functional Analysis of Cyanelle Ribonuclease P Reconstituted from Heterologous Components

In contrast to bacterial RNase P RNAs [e.g., M1 RNA, *Bacillus subtilis* P RNA, *P. marinus* RNase P RNA, or *Synechococcus* RNase P RNA (SynRPR)], or the *C. paradoxa* RNase P RNA alone does not show catalytic activity under the high-salt conditions typical for bacterial RNase P RNA.[11] Neither restoration of the bacterial consensus sequence at the positions where the plastid RNA deviates, nor heterologous reconstitution with the protein subunit from *E. coli* RNase P, can

[35] K. Nagai and H. C. Thogersen, *Nature (London)* **309,** 810 (1984).

restore cleavage activity of cyanelle RNase P RNA.[23] However, a cyanobacterial protein subunit can restore holoenzyme activity to this RNase P RNA.[36] The same is true for recombinant histidine-tagged SynPP and the variant devoid of the extra hexapeptide. Holoenzyme reconstitution and analysis of processing function are described for the bacterial RNase P RNAs (M1 and SynRPR), and for the plastid RNase P RNA from the *C. paradoxa* cyanelle.

i. Homologous and Heterologous Reconstitution of Ribonuclease P Holoenzymes with Cyanobacterial Protein Subunit. Reconstitution of RNase P holoenzymes from homologous or heterologous components is performed by direct mixing of the subunits. SynPP must be used in molar excess over RNase P RNA to maximize the yield of active holoenzyme after reconstitution, as in most protein preparations only a fraction is capable of promoting holoenzyme formation. To determine the optimal molar ratio between SynPP and the respective RNA subunit, a titration of protein against a given amount of RNA (10 n*M*) must be performed for any given preparation. In general, reconstitution is achieved by incubating the RNA subunit with a 5-fold molar excess of the renatured protein subunit in buffer F at 37° for 10 min. This preincubation allows conformational changes necessary to obtain the maximally active holoenzyme.[37]

ii. Functional Assay of Reconstituted Ribonuclease P. An aliquot (9 μl/reaction) of the reconstituted holoenzyme in buffer F is incubated with 1 μl of ^{32}P-labeled substrate (5000 cpm) at 37°. Time courses must be taken for each set of reconstituted components; usually, the 5-min time point is well within the linear range of the kinetics, and total turnover is reached after 30 min. The reaction is stopped by adding 40 μl of ice-cold doubly distilled H_2O, and the sample is processed and analyzed as described (see Section III,A,c,ii). Under these conditions, no cleavage activity is observed for the RNA subunits in the absence of SynPP. Similarly, the recombinant proteins SynPP-His and SynPP-Xa show no cleavage activity when assayed without RNase P RNA, indicating that the preparations are completely free of cellular (*E. coli*) RNase P. Cleavage activity of SynRP or M1 RNA alone may be tested as ribozyme control cleavage reactions in a similar manner, using the high-salt buffer P. Our results indicate that the presence of the hexahistidine sequence at the amino terminus of the recombinant SynPP-His does not interfere with protein function in the holoenzyme (C. Heubeck and A. Schön, unpublished data, 2001). This is in accordance with observations made with *E. coli* RNase P reconstituted with histidine-tagged C5 protein, and also conforms to the results that have been obtained with other affinity-tagged proteins.[38,39]

[36] A. Pascual and A. Vioque, *FEBS Lett.* **442,** 7 (1999).

[37] C. Guerrier-Takada, K. Haydock, L. Allen, and S. Altman, *Biochemistry* **25,** 1509 (1986).

[38] R. Rivera-Leon, C. J. Green, and B. Vold, *J. Bacteriol.* **177,** 2564 (1995).

[39] R. Janknecht, G. de Martynoff, J. Lou, A. Hipskind, and A. Nordheim, *Proc. Natl. Acad. Sci. U.S.A.* **88,** 8972 (1991).

IV. Conclusions

The unusual variability of RNase P in the phylogenetic lineage leading to and including modern chloroplasts necessitates the analysis of different members of this group of organisms and requires different approaches and methodologies. *Cyanophora paradoxa* is a primitive alga with a ribonucleoprotein-type plastid RNase P. This organism can be grown in amounts sufficient for enzyme purification, allowing studies of holoenzyme structure and function. The identification of an RNA component in other primitive algae is greatly facilitated by PCR, which permits the analysis of organisms that are difficult to grow in sufficient amounts for more traditional biochemical approaches. Structural and functional studies can be performed with *in vitro*-synthesized RNase P RNAs, which are easily obtained in large quantities by T7 transcription. Of particular interest with respect to substrate recognition by the plastid enzymes is their possible adaptation to unusually structured substrates, for example, to the "multifunctional" pre-tRNA^{Glu} typical of plastids and cyanobacteria; these substrates are also prepared by *in vitro* transcription. As for all processing enzymes, the verification of correctly cleaved ends is of crucial importance for the assessment of enzyme identity and cleavage mechanism.

Protein components of RNase P have not yet been identified for any of these organisms; however, the *C. paradoxa* RNase P RNA can be reconstituted to a functional holoenzyme with a cyanobacterial protein subunit. The purification of overproduced protein subunits is greatly facilitated by attaching a terminal affinity tag. As no functional impairment by this modification is observed in the case of RNase P proteins, functional studies of these *in vitro*-assembled "hybrid" holoenzymes can be performed.

Acknowledgments

Research on RNase P in the authors' laboratory is supported by grants from the Deutsche Forschungsgemeinschaft.

[12] Characterization of Ribonuclease MRP Function

By TI CAI and MARK E. SCHMITT

Introduction

Ribonuclease (RNase) MRP is a ribonucleoprotein endoribonuclease that cleaves an RNA sequence in a site-specific manner.[1] The enzyme has two known RNA substrates: one in the mitochondria, where it processes a primer RNA for the initiation of DNA replication,[2] and a second in the nucleus, where it processes the rRNA precursor to help generate the 5.8S rRNA.[3] The exact substrate for a third function in cell cycle control has not yet been determined.[4]

RNase MRP has been isolated from various organisms including human, mouse, rat, cow, *Xenopus,* yeast, *Arabidopsis,* and tobacco (see Reilly and Schmitt[5] for review). In yeast, *Saccharomyces cerevisiae,* the enzyme complex is known to consist of an RNA component of 340 nucleotides in length and nine protein components, although the full protein complement is as yet undefined. The RNA component of the enzyme was found to be structurally and evolutionally related to RNase P RNA, the ribonucleoprotein endoribonuclease that processes the 5′ end of tRNA, and several protein components are shared between the two enzyme complexes.[6] The data suggest that the enzymes, while ancient, have been highly conserved over time to perform essential cellular RNA-processing functions both in the nucleus and the mitochondria.

The use of mutagenesis in yeast has become an increasingly powerful tool for studying RNase MRP.[4,7] We describe here methodologies for testing various yeast mutants for endoribonuclease activity both *in vivo* and *in vitro.* In addition, a method for testing for assembly of the individual RNase MRP constituents with the complex is described.

In Vitro Assay for Ribonuclease MRP Endoribonuclease Activity

The protocol given here describes the *in vitro* assay of RNase MRP on its mitochondrial substrate. RNA transcribed from the mitochondrial origin is the most well-characterized substrate for RNase MRP. A similar assay can been used

[1] D. D. Chang and D. A. Clayton, *EMBO J.* **6,** 409 (1987).
[2] L. L. Stohl and D. A. Clayton, *Mol. Cell. Biol.* **12,** 2561 (1992).
[3] Z. Lygerou, C. Allmang, D. Tollervey, and B. Seraphin, *Science* **272,** 268 (1996).
[4] T. Cai, T. R. Reilly, M. E. Cerio, and M. E. Schmitt, *Mol. Cell. Biol.* **19,** 7857 (1999).
[5] T. H. Reilly and M. E. Schmitt, *Mol. Biol. Rep.* **22,** 87 (1995).
[6] J. R. Chamberlain, Y. W. S. Lane, and D. R. Engelke, *Genes Dev.* **12,** 1678 (1998).
[7] G. S. Shadel, G. A. Buckenmeyer, D. A. Clayton, and M. E. Schmitt, *Gene* **245,** 175 (2000).

0076-6879/01 $35.00

with minor modifications for human, mouse, and yeast enzymes. We describe the assay that we use for the yeast, *S. cerevisiae.*

Partial Purification of Yeast Ribonuclease MRP

Partial purification by the method described here is highly efficient and has been used for the purification of both wild-type enzymes and enzymes with mutated protein or RNA subunits. For wild-type purification we regularly use the strain LSY413-8D-rho^{o} [*Mat***a** *sep1::URA3 pep4::LEU2 nuc1::LEU2 leu2-3,112 trp1-1 his3-11,15 can1-100 ura3-1 rho*^{o}] (gift of L. Symington, Columbia University, New York, NY). The deletions of *sep1, nuc1,* and *pep4* remove most of the endogenous nucleases and proteases that interfere with the purification of intact enzyme and subsequent assay. The use of a *rho*^{o} strain helps to eliminate many of the protein contaminants in the final sample. RNase MRP can also be assayed directly from crude extracts (see below, Preparation of Yeast Whole Cell Soluble Protein Extracts), but this can provide inconsistent results.

1. Yeast cells are grown in 10 liters of YPD [1% (w/v) yeast extract, 2% (w/v) peptone, 2% (w/v) dextrose] with vigorous aeration to early saturation, 1×10^8 cells/ml. Cells are harvested by centrifugation at 4000*g* for 5 min. The yeast cells can be frozen at -70° and thawed before subsequent use or used fresh.
2. All purification procedures are performed at 4° on ice unless otherwise stated. Harvested cells (about 100–150 g) are washed once with 400 ml of doubly distilled H_2O and once in 400 ml of breaking buffer A-100 [50 m*M* Tris-HCl (pH 8.0), 100 m*M* NaCl, 5 m*M* EDTA, 10% (v/v) glycerol, 0.1% (v/v) Tween 20, 1 m*M* dithiothreitol (DTT), 1 m*M* phenylmethylsulfonyl fluoride (PMSF)]. The cells are then resuspended in 150 ml of buffer A-100 and placed in a 400-ml BeadBeater chamber (BioSpec, Bartlesville, OK) with 100 g of glass beads (425–600 μm in size). The chamber is filled to the brim with buffer A-100 (400-ml total volume) and the mixture is beaten (ten 1-min pulses alternating with 1 min of cooling).
3. After cell breakage, the mixture is brought to 500 ml with buffer A-100 and allowed to stir for 5 min. The cells and glass beads are removed by two sequential centrifugations at 4000*g* for 5 min. The supernatant, ~350 ml, is centrifuged at 100,000*g* for 100 min. The clarified extract is carefully removed to avoid the loose pellet and the fatty upper layer. This is critical for subsequent purification.
4. The clarified extract (~240 ml at 10–15 mg of protein per milliliter) is batch absorbed to 100 ml of DEAE-Sephacel (Pharmacia, Piscataway, NJ) that has been preequilibrated in buffer A-100. This is best performed in a 350-ml medium sinter funnel. The matrix is subsequently washed three times with 200 ml of buffer A-100. The enzyme is then eluted from the resin with 100 ml of buffer A-250 (A-100 with 250 m*M* NaCl instead of 100 m*M* NaCl).
5. The eluted sample (~100 ml at 1.2–2 mg of protein per milliliter) is mixed with 10 g of Bio-Rex 70 resin (Bio-Rad, Hercules, CA) that had been preequilibrated

with buffer A-250. The mixture is stirred for 15 min, and then poured into a 2.5 × 10 cm column. The column is washed with 10 column volumes of buffer A-250, and the RNase MRP activity is eluted with a step of buffer A-500 (A-100 with 500 m*M* NaCl instead of 100 m*M* NaCl). Fractions are monitored at 280 nm, and the protein peak is pooled (about 2–5 ml at 1 mg of protein per milliliter).

6. The sample is then concentrated 5- to 10-fold and desalted to 50 m*M* NaCl on a Centricon-100 (Amicon, Danvers, MA). At this point the sample contains only a single RNA by gel analysis but 20–25 proteins.

7. The sample is then split up and 200-μg fractions are loaded onto a 4-ml 15–30% (v/v) glycerol gradient in 20 m*M* Tris-HCl (pH 8.0), 50 m*M* NaCl, 5 m*M* EDTA, 0.1% (v/v) Tween 20, 1 m*M* DTT, 1 m*M* PMSF. The gradients are centrifuged for 6 hr at 60,000 rpm in an SW60Ti rotor at 4°. Pooled peak fractions containing RNase MRP activity (about 600 μl at 0.5 mg/ml from one preparation) can be pooled and frozen to −70°. The enzyme remains fully active for 4–5 weeks.

Substrate Labeling

Substrate RNA can be labeled internally by previously described methods[8]; on the 3′ end by using [^{32}P]pCp and T4 RNA ligase, or on the 5′ end by using [γ-^{32}P] ATP and polynucleotide kinase. When assaying yeast RNase MRP it is preferable to use a 3′-end-labeled substrate. The terminal 3′-phosphate blocks degradation by the numerous 3′ → 5′-exonucleases that are found in yeast (Fig. 1). This is less of a problem when using highly purified or mammalian enzymes. The plasmid p64/HS40-ori5/4.4, which generates a 145-nucleotide (nt) transcript, is routinely used as a template to generate the yeast substrate.[2]

1. The template, 2.5 μg of *Eco*RI-digested p64/HS40 ori5/4.4, is used in a 100-μl *in vitro* transcription reaction containing 40 m*M* Tris-HCl (pH 7.5), 6 m*M* $MgCl_2$, 2 m*M* spermidine, 5 m*M* NaCl, 10 m*M* DTT, a 500 μM concentration of each of the four nucleotides, 100 units of RNasin ribonuclease inhibitor, and 40 units of SP6 RNA polymerase. The reaction proceeds for 2 hr and the RNA transcript is separated from the unincorporated nucleotides by passing it over a Sephadex G-50 spin column.

2. The RNA is precipitated with ethanol and 3′-end labeled in a 10-μl reaction containing 50 m*M* HEPES (pH 7.5), 15 m*M* $MgCl_2$, 3 m*M* DTT, 30 μM ATP, 10% (v/v) dimethyl sulfoxide (DMSO), 50 μCi of [^{32}P]pCp, and 5 units of T4 RNA ligase. The reaction is incubated at 4° for 15 hr, stopped by the addition of 10 μl of formamide loading buffer [FLB; 80% (v/v) deionized formamide, 10 m*M*

[8] Promega Corporation, "Protocols and Applications Guide," 2nd Ed., Promega, Madison, Wisconsin, (1991).

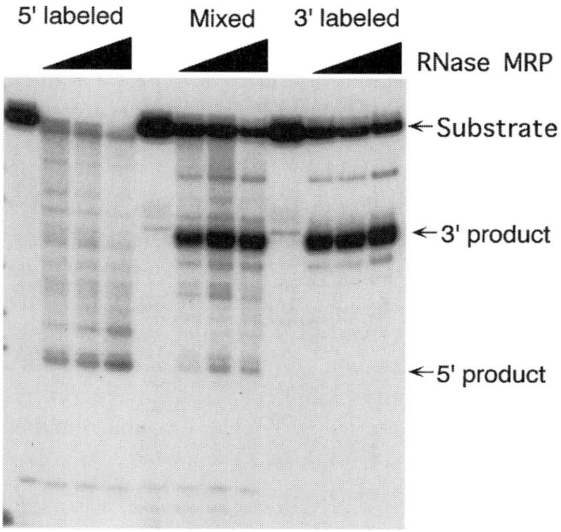

FIG. 1. Assay of RNase MRP activity on the ori5/4.4 substrate displays the difference between 5′- and 3′-labeled substrates. The yeast ori5/4.4 was ^{32}P labeled at either the 3′ or 5′ end and assayed with increasing amounts of yeast RNase MRP purified through a glycerol gradient. The prevalence of 3′ → 5′-exonucleases can be seen even in this highly purified preparation.

EDTA (pH 8.0), xylene cyanol FF (1 mg/ml), bromphenol blue (1 mg/ml)], heated to 95° for 5 min, and loaded into a single well of a 6% (w/v) acrylamide–7 *M* urea gel.

3. The labeled RNA is excised from the gel and eluted from the gel slice by incubation in 0.5 *M* ammonium acetate, 0.1% (w/v) sodium dodecyl sulfate (SDS), 1 m*M* EDTA for 12 hr at 37° without crushing the gel. The buffer is removed from the gel slice and precipitated with 2.5 volumes of ethanol, resuspended at 100,000 cpm/μl, and stored at −70° until use. Substrate is best used within 1 month.

RNA Cleavage Assay

1. The cleavage assay is carried out in a 25-μl reaction containing 20 m*M* HEPES (pH 7.5), 10 m*M* $MgCl_2$, 1 m*M* DTT, 50 m*M* KCl, bovine serum albumin (BSA, 50 μg/ml), 1 unit of RNasin ribonuclease inhibitor, and 10,000 cpm of substrate RNA at 37° for 5 to 30 min.

2. The reaction is stopped by the addition of 75 μl of 0.5 *M* sodium acetate (pH 5.3), and 5 μg of *Escherichia coli* tRNA. The sample is extracted with 25 : 24 : 1 phenol (pH 5.3)–chloroform–isoamyl alcohol and precipitated with ethanol. Reaction products are then analyzed on a 6% (w/v) acrylamide–7 *M* urea gel (see Fig.1).

In Vivo Analysis of Ribonuclease MRP Endoribonuclease Activity

RNase MRP functions in pre-rRNA processing in the nucleus. Cleavage of the rRNA precursor by RNase MRP is required to generate one of two species of 5.8S rRNA.[9] An alternative pathway, which is independent of RNase MRP, generates a longer form of 5.8S rRNA ($5.8S_L$) with seven extra nucleotides at the 5′ end. Thus depletion or mutation of RNase MRP components leads to the underaccumulation of the shorter 5.8S rRNA ($5.8S_S$) and buildup of the $5.8S_L$ rRNA.[9] The ratio of the two forms of 5.8S rRNA can be examined either by Northern analysis or by directly visualizing the 5.8S rRNA after ethidium bromide staining of total RNA. The latter is easier to perform, and requires no radioactive labeling.

Isolation of Total RNA from Yeast Cells

Visualization of 5.8S rRNA requires high-quality fresh total RNA. The method described here is a simple method that allows the simultaneous purification of RNA from up to 12 different samples in just over 1 hr.

1. Inoculate a 50-ml culture of YPD with an aliquot from a confluent overnight culture. The culture is grown with vigorous shaking at 30° to an OD_{600} of 1.0–1.5. We have found that a 1.25-ml inoculum in a 50-ml culture can be harvested in approximately 4 hr. Cells can also be grown in synthetic medium or shifted to various temperatures before harvesting.
2. Harvest the cells by centrifugation at 4000*g* for 5 min at 4°. Resuspend the cells in 1 ml of ice-cold AE buffer (50 m*M* sodium acetate, 10 m*M* EDTA, pH 5.3), transfer to a 1.5-ml microcentrifuge tube, and pellet by centrifugation at 16,000*g* for 10 sec in a microcentrifuge. The AE buffer is removed and the cells are resuspended by vortexing in the remaining liquid.
3. To the resuspended cells add 400 μl of AE and 40 μl of 10% (w/v) SDS. The tube contents are mixed and 400 μl of fresh phenol previously equilibrated with AE buffer (pH 5.3) is added. The tubes are mixed again, and placed in a 65° water bath for 5 min. The tubes are then removed from the 65° water bath and rapidly chilled in a dry ice–ethanol bath until phenol crystals appear (about 2 min). They are then placed back at 65° for an additional 5 min and rapidly chilled a second time. The tubes are then centrifuged at 16,000*g* for 10 min at room temperature to separate the aqueous and phenol phases.
4. The upper, aqueous phase is transferred to a clean tube and extracted with 400 μl of 25 : 24 : 1 phenol (pH 5.3)–chloroform–isoamyl alcohol at room temperature for 5 min. After centrifugation at 16,000*g* for 5 min at room temperature the aqueous phase is collected.
5. The RNA is precipitated by the addition of 40 μl of 3 *M* sodium acetate (pH 5.3) and 1 ml of ethanol and placed on dry ice for 10 min.

[9] M. E. Schmitt and D. A. Clayton, *Mol. Cell. Biol.* **13,** 7935 (1993).

6. Samples are centrifuged at 16,000g for 10 min at 4°. The liquid is removed and the RNA pellet is washed with cold 80% (v/v) ethanol. The RNA is dried in a vacuum centrifuge, resuspended in 50 μl of TE [10 m*M* Tris-HCl (pH 7.5), 1 m*M* EDTA], and stored at −70°. Typical yields are 50–100 μg of RNA. Throughout the preparation, normal precautions should be taken to avoid ribonuclease contamination. All nonvolatile solutions should be autoclaved before use.

Visualizing Ribosomal RNA

1. Approximately 5 μg of total RNA is mixed with an equal volume of formamide loading buffer [FLB; 80% (v/v) deionized formamide, 10 m*M* EDTA (pH 8.0), xylene cyanol FF (1 mg/ml), bromphenol blue (1 mg/ml)], and heated to 95° for 5 min. It is critical to maintain the sample volume under 10 μl. The RNA samples are immediately loaded onto a 6% (w/v) acrylamide–7 *M* urea polyacrylamide gel. The gel should be prepared fresh, and prerun in 1× TBE buffer until the gel is warm. The wells should be washed extensively with TBE to remove the urea before loading samples. Samples should be placed at the bottom of the wells.
2. The gel is run at 20 W to keep the gel warm and the RNA denatured. We typically run gels until the bromphenol blue reaches the bottom of the gel.
3. Immediately after electrophoresis, the gel is carefully transferred to a clean Pyrex dish with 200 ml of sterile doubly distilled H_2O. The gel is stained by the addition of 100 μl of ethidium bromide (5 mg/ml) and shaken gently for 15 min. The gel is destained by two 10-min washes in doubly distilled H_2O. The small rRNAs and tRNAs can be visualized with a UV transilluminator and photographed. In RNase MRP mutant cells, the ratio of the two forms of 5.8S rRNA is usually changed and a longer, 300-nt precursor can often also be observed. For documentation and quantitation we use a Bio-Rad FluorS MultiImager system.

Determination of RNA–Protein Interactions in Ribonuclease MRP by Coimmunoprecipitation

RNase MRP is a ribonucleoprotein complex that consists of a single RNA component and several protein components. In yeast, *S. cerevisiae,* nine protein components have been identified that associate with the MRP RNA *in vivo.* Using antibodies against protein subunits it is possible to precipitate the MRP RNA *in vitro.* This can provide an important tool for testing the association of various protein components with the RNA in various yeast mutants (see Fig. 2). Here we describe the coimmunoprecipitation method, using anti-Snm1p antibody as an example. Snm1p is a unique protein component of RNase MRP that specifically binds the MRP RNA.[10]

[10] M. E. Schmitt and D. A. Clayton, *Genes Dev.* **8,** 2617 (1994).

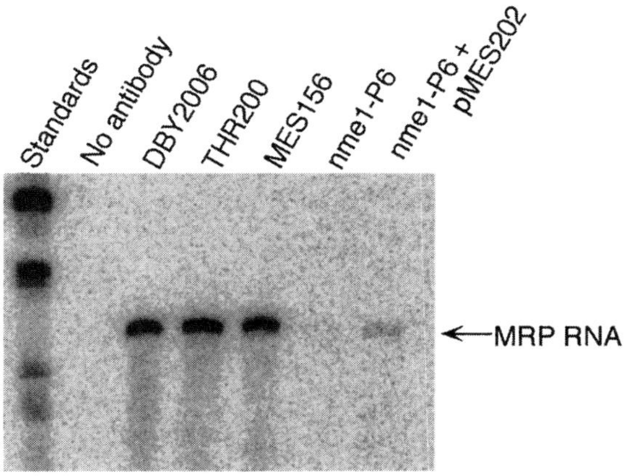

FIG. 2. Immunoprecipitation of MRP RNA, using anti-Snm1 antibodies. The mutation *nme1-P6* is a G122A change in the gene for the MRP RNA that confers temperature-sensitive growth. This mutation is defective in binding the Snm1 protein and can be suppressed by overexpressing the Snm1 protein (nme1-P6 + pMES202). DBY2006, wild-type strain with the *SNM1* gene on the chromosome; THR200, wild-type strain with the *SNM1* gene on a single-copy *CEN* plasmid; MES156, wild-type strain with the *SNM1* gene on a high-copy 2 μ plasmid.

Preparation of Yeast Whole-Cell Soluble Protein Extracts

1. Inoculate 100 ml of YPD [1% (w/v) yeast extract, 2% (w/v) peptone, 2% (w/v) dextrose] with an aliquot from a overnight culture. Grow with vigorous shaking at 30° to an OD_{600} of 1. Strains can also be grown in synthetic medium, or shifted to various temperatures to examine the effect of temperature-sensitive mutations.

2. Harvest cells by centrifugation at 4000*g* for 5 min at 4°. All procedures are performed at 4° unless otherwise noted. Resuspend in 1 ml of ice-cold sterile water and transfer to a 1.5-ml microcentrifuge tube. Wash the cells once in 1 ml of buffer B [20 m*M* Tris-HCl (pH 8.0), 150 m*M* KCl, 5 m*M* EDTA, 0.1% (v/v) Triton X-100, 10% (v/v) glycerol, 1 m*M* DTT, 1 m*M* PMSF]. Cells are recollected by centrifugation at 16,000*g* for 10 sec in a microcentrifuge.

3. The cells are resuspended in the remaining liquid and 400 μl of buffer B and 200 mg of glass beads (425–600 μm) are added. Samples are vigorously shaken for 15 min on a multitube platform Vortex Genie at 4°.

4. After cell breakage, the tubes are centrifuged at 16,000*g* for 2 min to pellet the glass beads and cell debris. The supernatant is transferred to a clean tube and centrifuged at 16,000*g* for an additional 15 min.

5. The supernatant (about 300 μl) is collected and stored at −70° until used. Protein concentration is determined by standard methodologies.

Immunoprecipitation of Ribonuclease MRP RNA by Anti-Snm1p Antibody

1. Swell 250 mg of protein A–Sepharose CL4B (Sigma, St. Louis, MO) in 2 ml of sterile water overnight. The swelled beads are washed three times with sterile water and equilibrated with buffer B.

2. All procedures are carried out in 1.5-ml microcentrifuge tubes. For each immunoprecipitation reaction incubate 50 μl of 200-mg/ml protein A–Sepharose CL4B beads with 3 μl of anti-Snm1p rabbit serum for 1 hr at 4°. One milliliter of buffer B is added to the mixture and the beads are pelleted by centrifugation at 4000*g* for 10 sec. The supernatant is removed and the beads are washed three more times with buffer B. The pelleted beads are resuspended in 80 μl of buffer B. Prebinding the antibody to the beads and washing them eliminate many of the nucleases present in serum that lead to degradation of the MRP RNA. Beads can be incubated with the antibody in batches, and split up for binding to different protein extracts.

3. For each sample to be analyzed, 80 μl of the protein A–Sepharose beads with bound antibody is incubated with protein extracts containing about 400 μg of protein for 2 hr at 4° with continuous gentle mixing.

4. The beads are then washed four times with 1 ml of buffer B to remove nonspecific binding. The beads from the last wash are brought up to 100 μl with buffer B and 100 μl of 5% (w/v) SDS–10% (v/v) 2-mercaptoethanol is added. The sample is mixed and heated to 95° for 5 min.

5. Two hundred microliters of 25 : 24 : 1 phenol (pH 5.3)–chloroform–isoamyl alcohol is added to each sample, and they are extracted at room temperature for 5 min. After centrifugation at 16,000*g* for 5 min, the upper aqueous phase is collected and 20 μl of 3 *M* sodium acetate, pH 5.3, and 500 μl of ethanol are added to precipitate the RNA.

6. The samples are placed on dry ice for 10 min and centrifuged at 16,000*g* for 15 min. They are then washed with 80% (v/v) ethanol and dried in a vacuum centrifuge.

7. The samples are resuspended in 3 μl of TE and 3 μl of formamide loading buffer (FLB), and heated to 95° for 5 min. Samples are immediately loaded onto a 6% (w/v) acrylamide–7 *M* urea polyacrylamide gel. Gel electrophoresis and Northern analysis are performed by standard techniques.

Acknowledgments

We thank M. Hale and B. A. Morisseau for comments and helpful discussions during the preparation of this manuscript. The authors' laboratory is supported by Grant NP-936 from the American Cancer Society and Grant 9850095T from the American Heart Association NYS Affiliate.

[13] *Escherichia coli* Ribonuclease III: Affinity Purification of Hexahistidine-Tagged Enzyme and Assays for Substrate Binding and Cleavage

By ASOKA K. AMARASINGHE, IRINA CALIN-JAGEMAN, AHMED HARMOUCH, WEIMEI SUN, and ALLEN W. NICHOLSON

Introduction

Ribonuclease III of *Escherichia coli* (EC 3.1.26.3) is a double-strand (ds)-specific, Mg^{2+}-dependent endoribonuclease. RNaseIII initially was detected as an activity in *E. coli* cell-free extracts that digests polymeric dsRNA to acid solubility,[1] and was shown subsequently to be involved in the maturation of the ribosomal RNA precursor, and bacteriophage T7 mRNA precursors. RNase III is now known to participate in the degradation as well as maturation of diverse cellular, phage, and plasmid RNAs (for reviews, see Refs. 2–4). Ongoing genome sequencing projects are revealing that RNase III is highly conserved in bacteria, with orthologs occurring (often in multiple copies) in plants, animals, fungi, and even a eukaryotic virus.[5] No RNase III ortholog has yet been noted in any sequenced archaeal genome. The reader is referred to the first article in this series for a bioinformatic analysis of the RNase III family.[5a]

The *E. coli* RNase III polypeptide (226 amino acids; molecular mass, 25.6 kDa) is encoded by the *rnc* gene, which is the first of three genes in the *rnc* (RNase III) operon, mapping at 55 min on the chromosome. Cells lacking RNase III are viable, but grow slowly and exhibit altered levels of specific proteins. RNase III-deficient strains also accumulate an ~5500-nucleotide (nt) species (30S RNA) that contains the rRNA sequences. As mentioned above, RNase III cleaves this transcript to provide the immediate precursors to the mature 16S, 23S, and 5S rRNAs. However, in the absence of RNase III, an alternative pathway provides functional rRNAs. Studies are indicating that pre-rRNA processing is a conserved function for RNase III family members.[6,7] *Escherichia coli* RNase III negatively regulates its expression by site-specific cleavage within the 5′ untranslated region of its mRNA, which

[1] H. D. Robertson, R. E. Webster, and N. D. Zinder, *J. Biol. Chem.* **243,** 82 (1968).
[2] J. J. Dunn, *in* "The Enzymes" (P. D. Boyer, ed.), p. 485. Academic Press, New York, 1982.
[3] D. Court, *in* "Control of Messenger RNA Stability" (J. G. Belasco and G. Brawerman, eds.), p. 71. Academic Press, New York, 1993.
[4] A. W. Nicholson, *FEMS Microbiol. Rev.* **23,** 371 (1999).
[5] I. S. Mian, *Nucleic Acids Res.* **25,** 3187 (1997).
[5a] L. Aravind and E. V. Koonin, *Methods Enzymol.* **341,** 3 (2001).
[6] S. Abou Elela, H. Igel, and M. Ares, *Cell* **85,** 115 (1996).
[7] H. Wu, H. Xu, L. J. Miraglia, and S. T. Crooke, *J. Biol. Chem.* **275,** 36957 (2000).

0076-6879/01 $35.00

destabilizes the transcript.[8,9] A full consideration of RNase III and its role in RNA maturation and decay, and gene regulation is beyond the scope of this article, and the reader is referred to other comprehensive reviews.[2–4,5a]

Escherichia coli RNase III is homodimeric in structure, and requires a divalent metal ion (preferably Mg^{2+}) for catalytic activity. Mn^{2+}, Ni^{2+}, and Co^{2+} can substitute for Mg^{2+}, whereas Ca^{2+}, Sr^{2+}, and Zn^{2+} are inactive. Zn^{2+} also inhibits the Mg^{2+}-dependent reaction.[10] The RNase III reaction employs a water molecule as nucleophile to cleave phosphodiesters, creating 5′-phosphate, 3′-hydroxyl product termini. Exhaustive cleavage of polymeric dsRNA provides duplex products averaging 12–15 bp in length, which corresponds to slightly greater than one turn of the A-helix (11 bp).[2,11] The dsRNA cleavage product termini exhibit two nucleotide, 3′ overhangs,[2–4,12] which places the scissile phosphodiesters on the same face of the helix, and across the minor groove.

Escherichia coli RNase III cleaves its cellular substrates in a site-specific manner. The cleavage sites are located within local secondary structure motifs, such as stem–loops, which have been termed processing signals.[12] The site of cleavage is determined by a combination of substrate sequence and structure. Substrates that exhibit strict double-helical structure undergo coordinate cleavage of both strands, while substrates with internal loops or bulges usually are cleaved at a single site, which is contained within or near the internal loop or bulge.[2–4,12] Specific Watson–Crick (WC) base pair sequences at defined positions relative to the cleavage site can inhibit cleavage, which have been termed "antideterminants,"[13] in analogy to the antideterminant sequences in tRNAs that control tRNA–protein transactions.[14] These substrate reactivity epitopes are not conserved among RNase III family members. For example, the *Saccharomyces cerevisiae* RNase III homolog, Rntlp, recognizes a specific hairpin tetraloop sequence, and cleaves within the stem at a defined distance from the tetraloop.[15,16]

The *in vivo* processing reactions of *E. coli* RNase III can be accurately reconstructed *in vitro* using small RNAs containing the critical reactivity epitopes, and using physiologically relevant ionic conditions (i.e., 1–10 mM Mg^{2+}, 150–250 mM salt).[17,18] However, lowering the salt concentration, or replacing Mg^{2+} by

[8] J. C. A. Bardwell, P. Regnier, A.-M. Chen, Y. Nakamura, M. Grunberg-Manago, and D. L. Court, *EMBO J.* **8,** 3401 (1989).

[9] J. Matsunaga, E. L. Simons, and R. W. Simons, *RNA* **2,** 1228 (1996).

[10] H. Li, B. S. Chelladurai, K. Zhang, and A. W. Nicholson, *Nucleic Acids Res.* **21,** 1919 (1993).

[11] H. D. Robertson and T. Hunter, *J. Biol. Chem.* **250,** 418 (1975).

[12] H. D. Robertson, *Cell* **30,** 669 (1982).

[13] K. Zhang and A. W. Nicholson, *Proc. Natl. Acad. Sci. U.S.A.* **94,** 13437 (1997).

[14] J. Rudinger, R. Hillenbrandt, M. Sprinzl, and R. Giegé, *EMBO J.* **15,** 650 (1996).

[15] G. Chanfreau, M. Buckle, and A. Jacquier, *Proc. Natl. Acad. Sci. U.S.A.* **97,** 3142 (2000).

[16] R. Nagel and M. Ares, Jr., *RNA* **6,** 1142 (2000).

[17] A. W. Nicholson, K. R. Niebling, P. L. McOsker, and H. D. Robertson, *Nucleic Acids Res.* **16,** 1577 (1988).

[18] B. S. Chelladurai, H. Li, and A. W. Nicholson, *Nucleic Acids Res.* **19,** 1759 (1991).

Mn^{2+}, promotes cleavage of additional ("secondary") sites, which are not normally recognized *in vivo*.[19,20]

The *E. coli* RNase III polypeptide contains two functional subdomains: one involved in substrate recognition, and the other important for catalysis. A conserved dsRNA-binding domain (dsRBD) is positioned near the C terminus and is required for productive binding of substrate. This domain is used by other dsRNA-binding proteins with diverse functions.[21] The N-terminal portion of the RNase III polypeptide contains the catalytic domain, which exhibits a number of highly conserved residues. One residue (Glu-117) has been shown to be important for phosphodiester hydrolysis.[22] The domain structure and conserved sequence elements indicate a common catalytic mechanism for RNase III family members, which has not yet been defined. The conserved structural organization of RNase III family members is discussed by Aravind and Koonin.[5a]

The existence of extensive biochemical and genetic data on *E. coli* RNase III function in cellular RNA metabolism, and the large set of characterized cellular, viral, and plasmid-encoded processing substrates, make this enzyme an excellent prototype for understanding the mechanism of action of other RNase III family members. Biochemical studies of RNase III and its specific mutants require (as with any enzyme) an efficient purification protocol. RNase III was first purified to near homogeneity from nonoverproducing *E. coli* cells.[19] Because RNase III is a low-abundance protein, large amounts of cells were required. The procedure also used dsRNA–agarose affinity chromatography, which exploited the ability of RNase III to bind dsRNA in the absence of Mg^{2+} (omission of divalent metal ion was necessary to avoid cleavage of the dsRNA affinity matrix). However, several *E. coli* RNase III mutants we are studying are defective in binding dsRNA, which necessitated a purification procedure that did not rely on dsRNA-binding ability. We describe here a protocol for the overproduction and affinity purification of *E. coli* RNase III carrying an N-terminal hexahistidine (His_6) sequence. We also describe the preparation of small processing substrates and their use in gel-based assays for RNase III binding and cleavage. The reader is referred to other protocols that describe the partial or complete purification of native *E. coli* RNase III from overproducing strains.[10,23–26]

Materials

Chemicals are molecular biology grade or reagent grade, and are generally obtained from Sigma (St. Louis, MO) or other major suppliers such as Fisher Scientific (Pittsburgh, PA) or VWR (San Francisco, CA). Restriction enzymes and

[19] J. J. Dunn, *J. Biol. Chem.* **251,** 3807 (1976).

[20] G. Gross and J. J. Dunn, *Nucleic Acids Res.* **15,** 431 (1987).

[21] I. Fierro-Monti and M. B. Mathews, *Trends Biochem. Sci.* **25,** 241 (2000).

[22] H. Li and A. W. Nicholson, *EMBO J.* **15,** 1421 (1996).

modifying enzymes are purchased from New England BioLabs (Beverly, MA). T7 RNA polymerase is purified in-house from an overproducing bacterial strain as described.[27] More recently we have used His_6-tagged T7 RNA polymerase, purified as described by Ni^{2+} affinity chromatography.[28] The pET plasmid vectors and *E. coli* DE3 strains are obtained from Novagen (Madison, WI). Calf intestinal alkaline phosphatase is obtained from Roche Molecular Biochemicals (Indianapolis, IN). Ultrapure ribonucleoside 5′-triphosphates are purchased from Amersham-Pharmacia Biotech (Piscataway, NJ) or from Roche Molecular Biochemicals. *Escherichia coli* bulk stripped tRNA (Sigma) is used as a carrier for radiolabeled RNA purification. It has previously been shown that tRNA has no major inhibitory effect on *E. coli* RNase III cleavage of substrate.[1] The tRNA is purified further by repeated phenol extraction. The extracted tRNA is ethanol precipitated, and the pellet is collected by centrifugation and washed with cold 70% (v/v) ethanol. The tRNA is briefly dried *in vacuo,* resuspended in water at a concentration of ~50 mg/ml, and stored in small aliquots at $-20°$. The A_{260}/A_{280} ratio should be ~2.0. Radiolabeled nucleotides are obtained from Dupont-NEN (Boston, MA). Isopropyl-β-D-thiogalactopyranoside (IPTG) stocks are prepared in sterile water (1 *M* concentration) and stored at $-20°$. For Polymerase chain reaction (PCR) cloning of the *rnc* gene and its mutants we use Vent DNA polymerase (New England BioLabs), as it has proofreading capability. To date we have encountered few instances of adventitious, PCR-induced mutations.

Overproduction and Purification of Hexahistidine-Tagged RNase III

The pET plasmid protein expression vectors[29] provide a convenient route for high-level overproduction of RNase III. We had purified previously native RNase III from an *E. coli* DE3 strain carrying plasmid pET-11(*rnc*).[10] This plasmid carries the *rnc* gene under the control of an IPTG-inducible T7 promoter and T7 translation initiation signal. To create pET-11 (*rnc*) the *rnc* gene is amplified by PCR, using a chromosomal DNA preparation as template, and then cloned into the *Nde*I and *Bam*HI sites of plasmid pET-11a. RNase III is overproduced at high levels in *E. coli* BL21(DE3),[30] with a substantial fraction of the protein

[23] S.-M. Chen, H. E. Takiff, A. M. Barber, G. C. Dubois, J. C. A. Bardwell, and D. L. Court, *J. Biol. Chem.* **265,** 2888 (1990).

[24] P. E. March and M. A. Gonzalez, *Nucleic Acids Res.* **18,** 3293 (1990).

[25] H. D. Robertson, *Methods Enzymol.* **181,** 189 (1990).

[26] N. Srivastava and R. A. K. Srivasatava, *Biochem. Mol. Biol. Int.* **39,** 171 (1996).

[27] J. Grodberg and J. J. Dunn, *J. Bacteriol.* **170,** 1245 (1988).

[28] B. He, M. Rong, D. Lyakhov, H. Gartenstein, G. Diaz, R. Castagna, W. T. McAllister, and R. K. Durbin, *Protein Expr. Purif.* **9,** 142 (1997).

[29] A. H. Rosenberg, B. N. Lade, D. Chui, S. Lin, J. J. Dunn, and F. W. Studier, *Gene* **56,** 125 (1987).

[30] F. W. Studier and B. A. Moffatt, *J. Mol. Biol.* **189,** 113 (1986).

TABLE I
STEADY-STATE KINETIC PARAMETERS OF PURIFIED His_6–RNase III AND RNase III[a]

Enzyme	k_{cat} (min^{-1})	K_m (μM)	k_{cat}/K_m (M^{-1}/min^{-1})
RNase III	7.7	0.26	3×10^7
His_6–RNase III	3.8	0.34	1.1×10^7

[a]The substrate used for both enzymes in enzymatic cleavage assays was internally ^{32}P labeled R1.1 RNA.[18] Reactions were performed at 37° in buffer that included 250 m*M* potassium glutamate and 10 m*M* $MgCl_2$ (see text). Data for RNase III are from Li *et al.*[10]

accumulating in the soluble portion of sonicated cell extracts. RNase III is purified in several steps, which include poly(I)–poly(C)–agarose affinity chromatography and gel filtration.[10]

The protocol summarized above would not be useful in purifying RNase III mutants that are defective in dsRNA binding. We therefore have applied an alternative approach, using the hexahistidine (His_6) sequence tag and immobilized metal (Ni^{2+} ion) resin.[31] This approach affords several advantages. First, the His_6 tag allows purification independent of the particular biochemical properties of an RNase III mutant. Second, the small size of the affinity tag minimizes any deleterious effect on RNase III structure and activity. Indeed, the N-terminal His_6 tag has only a minor effect on RNase III binding and cleavage of substrate (see below). Third, the His_6 tag can bind to Ni^{2+} resin in the presence of denaturants (e.g., urea), thereby allowing purification of RNase III mutants with reduced solubility. Fourth, the His_6 tag can be removed if necessary by thrombin cleavage. To provide RNase III with an N-terminal His_6 tag, the *rnc* gene is transferred from pET-11a (*rnc*) (see above) into the *Nde*I and *Bam*HI sites of plasmid pET-15b. Briefly summarized, His_6-tagged RNase III can be efficiently overproduced in an *E. coli* DE3 strain [e.g., BL21(DE3) or HMS174(DE3)], and accumulates in the soluble portion of sonicated cell extracts. Purified His_6–RNase III has a slightly lower catalytic efficiency (k_{cat}/K_m) than native RNase III (Table I), and exhibits the same cleavage site specificity as native enzyme (data not shown).

The overexpression and purification of His_6–RNase III mutants present an additional challenge, in that the activity of endogenous (chromosomal) RNase III must be suppressed. Such contamination, albeit small, could complicate the analysis of RNase III mutants with low activities. Although immobilized Ni^{2+} columns do not bind native RNase III, we employ an *E. coli* DE3 strain in which the activity of the chromosomally encoded RNase III is suppressed by a point mutation.

[31] J. Schmitt, H. Hess, and H. G. Stunnenberg, *Mol. Biol. Rep.* **18,** 223 (1993).

The *rnc105* mutation changes the highly conserved Gly-44 to aspartic acid in the catalytic domain and abolishes RNase III activity.[32,33] The bacterial strain BL21(DE3)*rnc105* has been prepared by Yamamoto and co-workers to overproduce a yeast RNase III ortholog.[34] To enhance the stability of the recombinant plasmid we have introduced the *recA*::Tn*9* allele into this strain by P1 transduction (the donor strain for the r*ecA*::Tn*9* allele is *E. coli* DB1318, provided by D. Court). BL21(DE3)*recA,rnc105* is slow growing, but allows overproduction of His_6–RNase III (see below). We have frequently observed that significant levels of His_6–RNase III are produced in the absence of IPTG induction. The reason for this is unclear, but may reflect the pleiotropic effects of the *rnc105* mutation on gene expression,[4] and the presence of a lambda (λ) lysogen, which uses RNase III in its gene regulatory circuitry.[4]

His_6–RNase III is prepared as follows. An overnight culture (in 5 ml of LB broth plus ampicillin at 100 μg/ml) (LB-Amp) is grown from a single colony of BL21(DE3)*rnc105,recA* cells freshly transformed with pET-15b(*rnc*). Although we have not noticed any significant instability of the pET-15b(*rnc*) plasmid in this strain, it is possible that plasmids carrying specific *rnc* gene mutants may be unstable. If necessary, the fraction of cells carrying plasmid can be determined by colony counts using LB-agar plates with or without ampicillin. An aliquot (1 ml) of the culture is added to 200–300 ml of LB-Amp at 37°. Alternatively, several fresh colonies can be collected from an LB-Amp agar plate and used as inoculant. Cultures are grown at 37° with vigorous aeration to an OD (600 nm) of 0.3–0.4. Immediately before IPTG addition an aliquot (1 ml) is removed for sodium dodecylsulfate–polyacrylamide gel electrophoresis (SDS–PAGE) analysis (see below). IPTG is added (final concentration, 0.5–1 m*M*), and incubation continued at 37° with vigorous aeration. Additional 1-ml aliquots are removed at 1, 2, and 3 hr after induction for SDS–PAGE analysis. After 3–4 hr, the culture is cooled on ice, and the cells collected by centrifugation at 4°. The cell pellet is stored at −20° prior to protein purification (see below).

RNase III overproduction is checked by SDS–PAGE. The collected aliquots (see above) are briefly centrifuged, and the pellets are fully resuspended in Laemmli gel loading buffer (~100 μl).[35] The samples are heated at 100° for 3–5 min, and aliquots are analyzed by electrophoresis in an 12% (w/v) polyacrylamide gel containing SDS.[35] Prestained protein size markers (Life Technologies, Bethesda, MD) are included in a side lane. The gel is stained with Coomassie Brilliant Blue R, and then destained and dried by standard methods. The image is captured by scanning or photography. The His_6–RNase III polypeptide electrophoreses as

[32] P. Kindler, T. U. Keil, and P. H. Hofschneider, *Mol. Gen. Genet.* **126,** 53 (1973).

[33] H. Nashimoto and H. Uchida, *Mol. Gen. Genet.* **201,** 25 (1985).

[34] Y. Iino, A. Sugimoto, and M. Yamamoto, *EMBO J.* **10,** 221 (1991).

[35] U. K. Laemmli, *Nature (London)* **227,** 680 (1970).

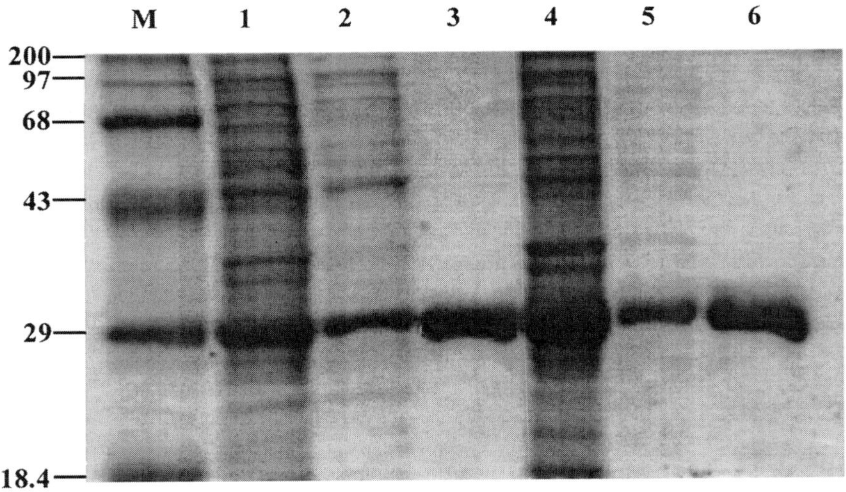

FIG. 1. Overproduction and purification of His_6-tagged RNase III. BL21(DE3)*rnc105recA* cells carrying pET-15b (*rnc*) were grown in LB broth containing ampicillin, and protein expression was induced with IPTG as described in text. Aliquots were taken and analyzed by SDS–PAGE [12% (w/v) polyacrylamide]. Proteins were visualized by Coomassie blue staining. Lane 1, total cell protein, 4 hr after IPTG addition. Lane 2, protein present in the soluble portion of sonicated cell lysates. Lane 3, protein in the soluble extract eluted from the Ni^{2+} affinity column. Lane 4, protein in the insoluble portion of sonicated cell lysates. Lane 5, protein present in the insoluble portion, solubilized with 6 *M* urea. Lane 6, protein present in the 6 *M* urea fraction, eluted from the Ni^{2+} affinity column. Lane M, prestained protein size markers (Life Technologies). The numbers on the left indicate the apparent molecular masses (kDa). The His_6–RNase III polypeptide comigrates with the 29-kDa marker.

a species of ~29 kDa molecular mass, and is the majority species in the cell by 4 hr after induction (Fig. 1, lane 1). As mentioned above, appreciable expression of the polypeptide is often obtained in the absence of IPTG.

Ni^{2+} Affinity Chromatography

Purification of His_6–RNase III is based on the protocol provided in the HisBind resin system manual (Novagen). We have incorporated a number of changes that have improved the yield and quality of enzyme. We have not found it necessary to include a protease inhibitor (e.g., phenylmethylsulfonyl fluoride) in the purification buffers, as long as the solutions are kept ice cold and affinity chromatography is performed as soon as possible after cell disruption. The following steps are performed at ~4°. The cell pellet is thawed and thoroughly resuspended in 30 ml of buffer A [500 m*M* NaCl, 20 m*M* Tris-HCl (pH 7.9)] containing 5 m*M* imidazole (Im). The cell suspension is transferred to a 30-ml Corex centrifuge tube, and subjected to repeated sonication bursts on ice, using an ultrasonic homogenizer (model

XL2007; 4-W power setting; Misonix, Farmingdale, NY). Sonication bursts are for 1 min, repeated four times with intermittent cooling on ice. Complete cell disruption is determined by visual examination, as indicated by the decrease in the initial viscosity to a final constant level. The sample is centrifuged at 7000 rpm for 20 min in a Sorvall SS34 rotor, and the supernatant is removed to a separate tube. If necessary, the centrifugation step is repeated to obtain complete clarification. His_6–RNase III is present in both the soluble and insoluble fractions of the sonicated cell mixture. Sufficient protein exists in the soluble fraction for purification. We have found that some RNase III mutants are obtained in largely insoluble form. If an RNase III mutant is being purified for the first time, aliquots are taken from the soluble and insoluble fractions for SDS–PAGE analysis to determine which fraction contains the protein. Purification of protein from the soluble fraction is described first, followed by the protocol for purification from the insoluble fraction.

The Ni^{2+} affinity column is prepared in a 10-ml glass pipette. The column can be operated in a cold room or in a cabinet refrigerator. Typically, 1–1.5 ml of HisBind resin is sufficient to purify protein from a 200- to 300-ml culture. A small plug of sterile glass wool is used as column support, and a short length of Tygon tubing is used to direct the eluent to the recipient tubes. An adjustable clamp is used to control the flow rate. The resin is washed with 10 column volumes of buffer A plus 5 m*M* Im, and then charged with a 50 m*M* $NiSO_4$ solution. The clarified protein solution (see above) is slowly applied to the column (approximately 1 hr of loading time for a 30-ml volume). The resin is washed with 10 column volumes of buffer A plus 5 m*M* Im, and then with 6 column volumes of buffer A plus 60 m*M* Im. The protein is eluted with three aliquots (1 ml each) of buffer consisting of 1 *M* NaCl, 400 m*M* Im, and 20 m*M* Tris-HCl (pH 7.9). The majority of His_6–RNase III elutes in the first three fractions, with the greatest amount in eluate fraction 2. The samples are combined, placed in dialysis tubing (SpectraP or membrane, 8000 MW cutoff; Spectrum, Laguna Hills, CA) and dialyzed against buffer (1 *M* NaCl, 400 m*M* Im, 60 m*M* Tris-HCl, pH 7.9) for ~2 hr, and then dialyzed for ~2 hr against the same buffer, but lacking Im. Dialysis is continued for 12–16 hr against buffer consisting of 1 *M* NaCl, 60 m*M* Tris-HCl (pH 7.9), 1 m*M* EDTA, 1 m*M* dithiothreitol (DTT). Purified enzyme is stored at -20° in 50% (v/v) glycerol, 0.5 *M* NaCl, 30 m*M* Tris-HCl (pH 7.9), 0.5 m*M* EDTA, 0.5 m*M* DTT at -20°. There is a negligible loss of activity over many months, and concentrations of ~1–2 mg/ml are stable. Attempts to store protein at higher concentrations led to some precipitation. Typically, ~2 mg of His_6–RNase III can be obtained from a 300-ml bacterial culture. The purity of His_6–RNase III is at least 90%, as estimated by gel electrophoresis (see Fig. 1, lane 3), and is free from contaminating nuclease activities.

We have found that proper execution of the dialysis procedure is critical for a successful purification. In a preliminary protocol we had originally used buffer A plus 1 *M* Im as elute buffer, followed by dialysis against buffer A without Im.

However, this caused protein precipitation. We suspected that the rapid drop in Im concentration was responsible for the precipitation. We therefore lowered the concentration of Im to 400 m*M* in the elute buffer and the first-step dialysis buffer. We also observed that inclusion of EDTA and DTT in the second dialysis step caused yellow–brown discoloration of the dialysate, and formation of similarly colored precipitates on the dialysis membrane with an accompanying loss of protein. We suspected that the combination of EDTA and DTT with Ni^{2+} ion (which coelutes with protein and is not efficiently removed by dialysis) promotes a chemical reaction that affects protein solubility. We therefore included 400 m*M* Im in the first-step dialysis buffer to help entrain the coeluting Ni^{2+}, and added EDTA and DTT only in the final dialysis step. Finally, the salt concentration influences RNase III solubility. We used 1 *M* NaCl in the dialysis buffer, because salt concentrations lower than ~500 m*M* caused progressively greater amount of protein precipitation, whereas concentrations greater than 1 *M* did not improve solubility. We did not observe any strong dependence of solubility on the type of metal ion (Na^+, K^+, NH_4^+).

As mentioned above, the purification of RNase III mutants presented additional challenges, because mutations can affect protein solubility. One approach to increase solubility is to overproduce the protein at lower temperatures (e.g., 25–30°). If this does not provide enough soluble protein in the sonicated cell lysate, His_6–RNase III and mutants can be purified from the insoluble fraction as follows. This protocol uses urea, rather than nonionic or ionic detergents, as we have found the latter reagents to be difficult to remove, and that at least for one detergent (Triton X-100), the secondary cleavage site activity of RNase III is enhanced (our unpublished observations, 2000). After sonication and centrifugation as described above, the supernatant is removed, and the inclusion body is washed several times with buffer A plus 5 m*M* Im. The pellet is then treated with buffer A plus 6 *M* urea on ice for 1–3 hr. The sample is centrifuged (10,000 rpm, 10–15 min) and the clarified supernatant is loaded on a Ni^{2+} column as described above. The column is washed with buffer A plus 5 m*M* Im, without urea. In this step the urea-solubilized protein undergoes renaturation on the Ni^{2+} column. The protein is eluted and dialyzed as described above. If necessary the solubility of the eluted protein (wild-type or mutant) can be maintained by including 10% (v/v) glycerol in the dialysis buffer. The purity of RNase III obtained by this route is also estimated to be >90% (see lane 6, Fig. 1).

RNase III Substrate Preparation

The original assays for *E. coli* RNase III took advantage of the ability of the enzyme to cleave polymeric dsRNA to acid solubility. The synthesis of dsRNA substrates can be accomplished by a number of routes, including transcription of a synthetic DNA copolymer [e.g., poly(dA-dT)] by *E. coli* RNA polymerase in the presence of radiolabeled ribonucleoside triphosphates.[1,11] The product is purified by CF11 cellulose chromatography, which separates dsRNA from ssRNA in ethanol-containing buffers.[36] With the development of phage RNA polymerase-based transcription systems, dsRNAs of virtually any sequence can be obtained by transcription of both strands of a specific DNA template, followed by annealing of the complementary RNAs. In one method, a plasmid carrying a defined sequence between convergent phage promoters (e.g., T7 and T3) is linearized in separate reactions, using a restriction site on either side of the target sequence. Alternatively, a DNA template with the same promoter configuration can be generated by PCR using primers containing the appropriate promoter sequence. The purified templates are transcribed in separate reactions *in vitro,* using the phage RNA polymerase. The complementary ssRNA sequences are purified and annealed, and the dsRNA is purified by gel electrophoresis. The dsRNA can be radiolabeled on either strand, or both strands. For examples of this protocol, see Refs. 37–39.

Although synthetic dsRNAs retain their value in detecting RNase III-like activities—especially in the absence of knowledge of the specific cellular substrate sequence requirements—they are not convenient for detailed enzymological studies. We are interested in how *E. coli* RNase III recognizes and cleaves its cellular and viral processing signals, which typically occur within much larger RNA sequences. These substrates typically are small (~40–60 nt) stem–loop structures. These substrates are readily generated by transcription of single-stranded oligodeoxynucleotides, using the method developed by Uhlenbeck and coworkers.[40,41] Several earlier studies from our laboratory used this technology to prepare a wide variety of *E. coli* RNase III substrates.[10,14,17,22] Figure 2 shows the structure of R1.1 RNA, an RNase III substrate we have used in a number of investigations. R1.1 RNA is based on the R1.1 processing signal, which is encoded in the phage T7 genetic early region between genes 1.0 and 1.1.[10,17,18] The 60-nt RNA is cleaved *in vitro* at a single site within the internal loop (indicated by the arrow in Fig. 2). The RNA can be enzymatically synthesized with T7 RNA polymerase and a 77-nt DNA template, annealed to an 18-nt "promoter oligonucleotide" (Fig. 2). The RNA synthesis and purification protocol is provided below.

General Precautions

Several steps are taken to avoid ribonuclease and metal ion contamination. Water is deionized and distilled. Gloves are worn, and sterile plasticware (tips,

[36] R. M. Franklin, *Proc. Natl. Acad. Sci. U.S.A.* **55,** 1504 (1966).

[37] L. Manche, S. R. Green, C. Schmedt, and M. B. Mathews, *Mol. Cell. Biol.* **12,** 5238 (1992).

[38] R. S. Tang and D. E. Draper, *Nucleic Acids Res.* **22,** 835 (1994).

[39] A. G. Polson and B. L. Bass, *EMBO J.* **13,** 5701 (1994).

[40] J. F. Milligan, D. R. Groebe, G. W. Witherell, and O. C. Uhlenbeck, *Nucleic Acids Res.* **15,** 8783 (1987).

[41] J. F. Milligan and O. C. Uhlenbeck, *Methods Enzymol.* **180,** 51 (1989).

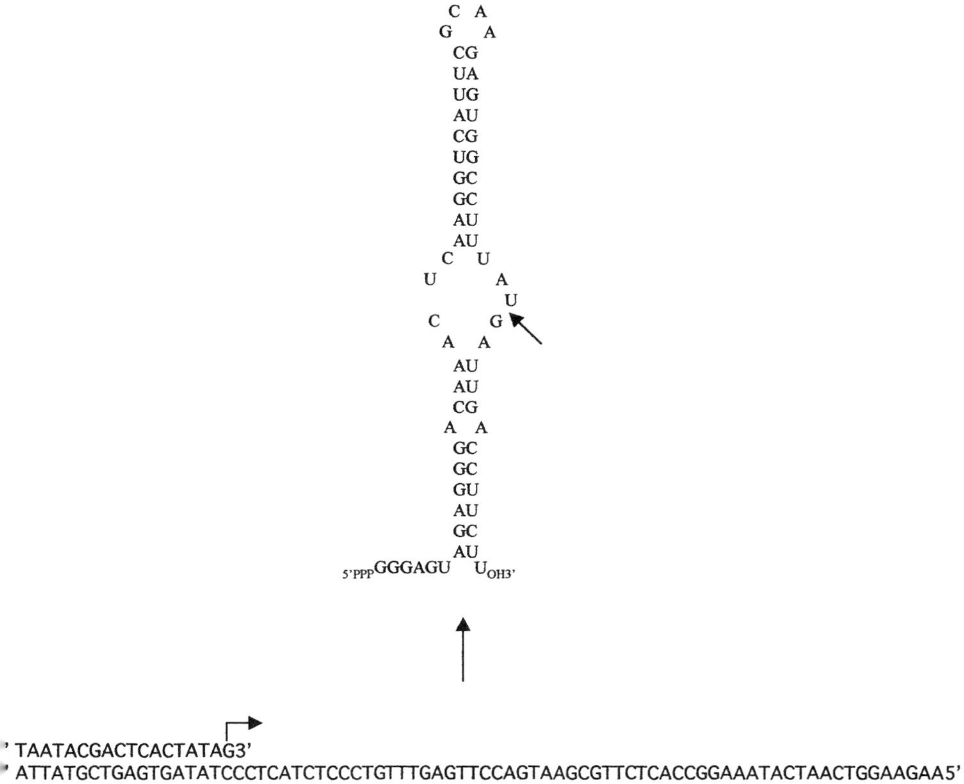

FIG. 2. Structure of R1.1 RNA, and the corresponding transcription template. The arrow indicates the site of RNase III cleavage in the internal loop. The 18-nt promoter oligonucleotide is shown annealed to the 77-nt transcription template. The bent arrow indicates the transcription start site.

tubes) is used. In this regard, radiolabeled RNAs generally can be recovered in higher yield by using siliconized microcentrifuge tubes. Liquid-handling devices (e.g., pipettors) are dedicated to RNA-containing samples. Glassware is thoroughly washed, rinsed thoroughly in deionized, distilled water, and baked before use. Glass plates for electrophoresis are treated with chromic acid, rinsed thoroughly in deionized, distilled water, and then dried (baking is optional). Using these procedures we have not found it necessary to include ribonuclease inhibitors (e.g., diethyl pyrocarbonate) in the buffer stocks and water.

Template Preparation

Synthetic oligodeoxynucleotides are obtained from a commercial source, and contain a T7 promoter sequence directly upstream of the sequence encoding the RNase III substrate. The 18-nt promoter oligonucleotide is also procured, which is complementary to the T7 promoter sequence.[40,41] Because of promoter sequence constraints the transcripts usually carry a specific 5′-end sequence (5′-pppGGGAGA...3′) corresponding to the T7 promoter +1 to +6 sequence. We have not found this sequence to affect substrate cleavage and binding activities (see below). There is some latitude allowed in the 5′ sequence, although changes from the canonical sequence (see above) can lower transcription yields.[40,41]

The oligodeoxynucleotides are purified by denaturing polyacrylamide gel electrophoresis. The crude, deprotected DNA (~50 nmol) is resuspended in TE buffer (~300 μl), combined with a one-third volume of dye solution [89 m*M* Tris base, 89 m*M* boric acid, 22 m*M* EDTA, 7 *M* urea, 20% (w/v) sucrose, 0.04% (w/v) each xylene cyanol and bromphenol blue], and electrophoresed (~22 V/cm) at room temperature in a 20% (w/v) polyacrylamide gel containing 7 *M* urea in TBE buffer. The gel region containing the full-length DNA is located by UV shadowing[42] and excised with a clean scalpel. The gel slice is crushed and incubated in 0.5 *M* ammonium acetate (pH 7.8) and 10 m*M* EDTA (~400 μl) for 12–16 hr at room temperature. The sample is centrifuged (14,000 rpm, 10 min, 4°), and the supernatant is carefully transferred to a fresh tube. One-tenth volume of 3 *M* sodium acetate (pH 5.2) is added, followed by 2.5 volumes of ethanol and incubation at −70° (if necessary, the supernatant volume can be reduced before ethanol addition by repeated *n*-butanol extraction). The sample is centrifuged at 4° or room temperature (14,000 rpm, 30 min), and the DNA pellet is briefly dried *in vacuo* and then resuspended in 50 μl of TE buffer [10 m*M* Tris-HCl (pH 8.0), 1 m*M* EDTA]. The amount of DNA is determined by the absorbance at 260 nm, and the purity is assessed by the A_{260}/A_{280} ratio, which should be >1.7. The DNAs are stored at −20°, but −80° is preferable for long-term storage.

RNA Synthesis

In a typical transcription reaction, DNA template (~1.6 μ*M*) and the promoter oligonucleotide (~1.8 μ*M*) are annealed in 100 μl of 10 m*M* Tris-HCl (pH 8.2) by heating for 5 min at 65°, followed by quick cooling on ice. One-tenth volume of the annealed template mix is added to a transcription reaction (typically a 50- to 100-μl volume) containing 40 m*M* Tris-HCl (pH 8.2), 20 m*M* $MgCl_2$, 10 m*M* spermidine, 0.01% (v/v) Triton X-100, polyethylene glycol (PEG) 8000 (80 mg/ml), 5 m*M* DTT, the four rNTPs (1 m*M* each), ~5–15 μCi of [α-^{32}P]UTP (or CTP) (3000 Ci/mmol), and ~400 units of T7 RNA polymerase. The reaction is incubated at 37° for 3–4 hr, and then stopped by adding one-third volume of dye mix containing 20 m*M* EDTA (see above), and then immediately loaded on a 15-cm, 20% (w/v) polyacrylamide gel containing 7 *M* urea in TBE buffer. The sample is

[42] S. M. Hassur and H. W. Whitlock, *Anal. Biochem.* **59,** 162 (1974).

electrophoresed at ~22 V/cm until the bromphenol blue runs off the bottom. The gel region containing the radiolabeled RNA is identified by autoradiography (nonradioactive RNAs are located by UV shadowing), excised, and soaked in 400 μl of extraction buffer (see above) containing ~1 μg of tRNA for 12–16 hr at room temperature. One-tenth volume of 3 *M* sodium acetate (pH 5.2) is added, followed by 2.5 volumes of ethanol and incubation at −70°. The sample is centrifuged for at least 30 min, and the RNA pellet is briefly dried and resuspended in TE buffer. For longer RNAs (>100 nt) efficient extraction can be achieved by soaking the gel slice instead in 400 μl of TE buffer (plus 1 μg of tRNA) and proceeding as described above.

For 5′-end radiolabeling, RNA (100 pmol) is dephosphorylated with ~1–2 units of calf intestine alkaline phosphatase in 50 m*M* Tris-HCl, 0.1 m*M* EDTA, pH 8.5 for 30 min at 37° (room temperature is also sufficient). The sample is phenol–chloroform extracted, and the RNA is recovered by ethanol precipitation as described above. The RNA is reacted in a small amount of TE buffer and stored at −20° until the next step. Approximately 25–50 pmol of the dephosphorylated RNA is treated with 2 units of T4 polynucleotide kinase and 5–10 μCi of [γ-^{32}P]ATP (3000 Ci/mmol) in the buffer supplied with the enzyme. The reaction is electrophoresed in a denaturing polyacrylamide gel, and the RNA is isolated as described above. The RNA is resuspended in a small amount of TE buffer and stored at −20° before further use.

Substrate Cleavage Assay

The original assay for RNase III used a synthetic substrate [e.g., [^{3}H]poly (A-U)], which was added to partially purified cell extracts, and the time-dependent decrease in trichloroacetic acid-precipitable radioactivity was determined.[1] A unit of RNase III activity was defined as the amount of enzyme that solubilizes 1 nmol of acid-precipitable (dsRNA) polymer phosphorus in 60 min at 35°.[1,2] With the availability of near-homogeneous enzyme and small, specific substrates, evaluation of enzyme activity can be made by determination of initial cleavage rates and the steady state kinetic parameters (see Table I). For enzymatic analyses in the steady state (e.g., determination of initial cleavage rates, and measurement of K_m and k_{cat} values), the RNase III concentration is typically ~10 n*M* (dimer form), and internally ^{32}P-labeled substrate is used at concentrations greater than ~50 n*M*. For initial velocity measurements the concentrations of enzyme and substrate are both lower than the K_m, such that changes in velocity would reflect changes in either K_m or k_{cat}.[43] Reactions are initiated by adding substrate, or enzyme, or divalent metal ion ($MgCl_2$). We have obtained the most consistent results by using Mg^{2+} addition as the initiating step. The addition of dye mix

[43] A. Fersht, "Enzyme Structure and Mechanism." W. H. Freeman, New York, 1985.

containing excess EDTA is an effective reaction quench. Polypropylene (0.65 ml) microcentrifuge tubes are convenient reaction vessels that provide reproducible results.

Immediately before the assay, an appropriate amount of ^{32}P-labeled RNA (either internally or 5′ labeled) is briefly heated (100°, 30 sec) in TE buffer, and then cooled on ice. This step removes intermolecular complexes that can form during storage at −20°. Aliquots are combined with RNase III in a buffer containing 30 m*M* Tris-HCl (pH 8), 250 m*M* potassium glutamate (or 160 m*M* NaCl), 5 m*M* spermidine (optional), tRNA (0.01 μg/μl), 0.1 m*M* EDTA, and 0.1 m*M* DTT. Samples are preincubated at 37° for 5 min, and $MgCl_2$ (prewarmed at 37°) is added to initiate cleavage (10 m*M* final concentration). Reactions are stopped by addition of EDTA–dye mix. Samples are loaded on a denaturing 15% (w/v) polyacrylamide gel containing TBE buffer and 7 *M* urea, and electrophoresed for 1–2 hr at 350 V. The top gel plate is removed, the excess buffer is removed, and the gel is wrapped in plastic. The cleavage reaction is followed by the rate of appearance of a specific cleavage product, and is quantitated by phosphorimaging (Storm 860 Phosphorimager system; Molecular Dynamics, Sunnyvale, CA) or by radioanalytic imaging (Ambis, San Diego, CA). To ensure linear kinetics the reaction times are kept short (0.5–3 min), such that only a small fraction of substrate is converted to product (typically <30%). For determination of steady state kinetic parameters, the substrate concentration is varied at a fixed enzyme concentration (usually ~10 n*M*), and saturation of initial velocity is determined by graphic analysis. The K_m and V_{max} values are determined by nonlinear least-squares curve fitting (Kaleidagraph; Synergy Software, Reading, PA), and the k_{cat} is determined from the V_{max}. For the k_{cat} determination it is necessary to have accurate determination of the protein concentration, and of the amount of [^{32}P]RNA recovered in the gel lanes.

Gel Mobility Shift Assay for Substrate Binding

The application of the gel mobility shift assay to monitor *E. coli* RNase III binding to substrate has been described.[22,44] Since then additional changes have been instituted. In the absence of divalent metal ion, some RNase III–substrate complexes cannot be directly observed in a nondenaturing polyacrylamide gel.[22] We have shown that Ca^{2+} can stabilize RNase III–substrate complexes during electrophoresis, while disallowing cleavage.[22] The gel shift assay also can be performed in the presence of Mg^{2+}, and using a catalytically inactive mutant of RNase III (e.g., the Glu117Lys or Glu117Ala mutants).[22] The gel shift protocol described below may not necessarily be applicable to the analysis of RNase III

[44] B. S. Chelladurai, H.-L. Li, K. Zhang, and A. W. Nicholson, *Biochemistry* **32,** 7549 (1993).

orthologs. The reader is referred to [14] and [15] in this volume[44a] for a description of gel shift conditions for two yeast RNase III homologs.

Immediately before the assay, 5′-^{32}P-labeled RNA is heated in TE buffer at 100° for 30 sec, and then placed on ice. This step removes intermolecular complexes that form during storage at −20°, which would otherwise be observed in a nondenaturing gel as slower moving species. RNA (~8000 cpm) is combined with RNase III in binding buffer [160 m*M* NaCl, 30 m*M* Tris-HCl (pH 8), 5–10 m*M* $CaCl_2$ (or $MgCl_2$ for catalytically inactive RNase III mutants), 0.1 m*M* EDTA, 0.1 m*M* DTT, 5% (v/v) glycerol, and tRNA at 0.01 μg/μl (250 m*M* potassium glutamate can be used in place of NaCl)]. Samples are incubated at 37° for 10 min, and then placed on ice for 20 min before loading on a 6% (w/v) polyacrylamide gel (acrylamide : bisacrylamide ratio, 80 : 1) containing 0.5× TBE buffer supplemented with 5–10 m*M* $CaCl_2$ (or $MgCl_2$). The binding reactions do not contain dyes, but the side lanes of the gel are loaded with aliquots of the bromphenol blue–xylene cyanol dye mix. The running buffer is also 0.5× TBE, supplemented with 5–10 m*M* $CaCl_2$ (or $MgCl_2$). The gel is preelectrophoresed (~7 V/cm) for 20 min, the samples are loaded, and the gel is electrophoresed (7.5 V/cm) at 4° for ~3–5 hr. After electrophoresis, the top plate is removed and the excess buffer is removed. The gel is covered in plastic wrap and the reactions are analyzed by phosphorimaging.

To determine the apparent dissociation constants (K'_D values) for the protein–RNA complex, protein titration experiments are performed in which increasing amounts of protein are added to a fixed amount of RNA. The amount of RNA in the complex is determined, and the fraction of substrate that is bound is plotted as a function of the reciprocal of the enzyme concentration. This yields a linear relation, with the slope equal to the K'_D.[45] In some instances the protein–RNA complex is unstable during gel electrophoresis, and instead of observing a specific complex, a "smear" of radioactivity occurs, which reflects dissociation of the complex during electrophoresis. In this case, quantitation can be carried out by measuring the amount of free (unbound) RNA, which is used to calculate the fraction of substrate bound.[46]

We have attempted to apply the nitrocellulose (NC) filter binding assay[47] to detect RNase III–substrate complexes. However, these complexes are not retained on an NC membrane under a wide variety of conditions (our unpublished observations, 2000). However, a protein–RNA complex involving a catalytically inactive RNase III mutant (see above) can be stably bound to NC in the presence of Mg^{2+}.

[44a] B. Lamontagne and S. Abou Elela, *Methods Enzymol.* **342,** [14] 2001 (this volume); and G. Rotondo and D. Frendewey, *Methods Enzymol.* **342,** [15] 2001 (this volume).

[45] J. Carey, V. Cameron, P. L. deHaseth, and O. C. Uhlenbeck, *Biochemistry* **22,** 2601 (1983).

[46] J. Carey, *Methods Enzymol.* **208,** 103 (1991).

[47] M. Yarus and P. Berg, *Anal. Biochem.* **35,** 450 (1970).

We are currently investigating the experimental parameters that influence the stability of RNase III–substrate complexes on NC membranes.

Summary

It is now evident that members of the RNase III family of nucleases have central roles in prokaryotic and eukaryotic RNA maturation and decay pathways. Ongoing research is uncovering new roles for RNase III homologs. For example, the phenomena of RNA interference (RNAi) and posttranscriptional gene silencing (PTGS) involve dsRNA processing,[48,49] carried out by an RNase III homolog. We anticipate an increased focus on the mechanism, regulation, and biological roles of RNase III orthologs. Although the differences in the physicochemical properties of RNase III orthologs, and distinct substrate reactivity epitopes and ionic requirements for optimal activity, may mean that the protocols describe here are not strictly transferrable, the affinity purification methodology, and substrate preparation and use should be generally applicable

Acknowledgments

The authors thank members of the laboratory (Dr. A. Pertzev, Dr. R. Nicholson, G. Li, and K. Fortin) and laboratory alumni (B. Chelladurai, Dr. H. Li, Dr. K. Zhang, S. Moturi, and B. Prince) for their advice about and interest in *E. coli* RNase III. We also appreciate a collaboration with Dr. Robert W. Simons and his group. We also thank Dr. M. Yamamoto for providing BL21(DE3)*rnc105*, Dr. Donald Court for *E. coli* strain DB1318 (*recA*::Tn*9*), and Steve Su and Bob Simons for advice about purification protocols. We thank Dr. R. Nicholson for a critique of the manuscript. This research is supported by research Grants GM56457 and GM56772 from the NIH.

[48] T. Tuschl, P. D. Zamore, R. Lehmann, D. P. Bartel, and P. A. Sharp, *Genes Dev.* **13,** 3191 (1999).
[49] E. Bernstein, A. A. Caudy, S. M. Hammond, and G. J. Hannon, *Nature* (*London*) **409,** 363 (2001).

[14] Purification and Characterization of *Saccharomyces cerevisiae* Rnt1p Nuclease

By BRUNO LAMONTAGNE and SHERIF ABOU ELELA

Introduction

Yeast Rnt1p is a member of the double-stranded RNA (dsRNA)-specific ribonuclease (RNase) III family identified by conserved dsRNA binding (dsRBD) and nuclease domains.[1,2] The protein consists of 471 amino acid residues with a molecular mass of 54.1 kDa. The canonical dsRBD motif is located at the C terminus (positions 372–440) with 25% identity to *Escherichia coli* RNase III and 31% identity to fission yeast pac1 (see [1] in this volume[2a]). However, unlike RNase III and pac1 dsRBDs, Rnt1p has a 32-amino acid extension at the C terminus. The Rnt1p 154-amino acid nuclease domain is similar in size to that of RNase III and pac1 and shares the same charged amino acid clusters. In addition, Rnt1p possesses the characteristic eukaryotic N-terminal domain that spans 191 amino acids. This domain has neither an apparent functional motif nor significant homology to the pac1 N-terminal domain.[2] Systematic deletions of Rnt1p domains indicate that the dsRBD is required and sufficient for RNA binding whereas the nuclease domain is required for RNA cleavage. On the other hand, the N-terminal domain plays an auxiliary role by ensuring protein stability and efficient RNA cleavage. Deletion of the N-terminal domain reduces the processing of the 25S rRNA 3′ end by about 30% *in vivo* and slows growth by 35–40%. Biochemical and genetic assays suggest that the N-terminal domain influences Rnt1p function by mediating both inter- and intramolecular interactions.[2]

Similar to the bacterial RNase III, Rnt1p is not essential but its deletion in yeast causes severe growth defects, temperature sensitivity, and budding anomalies.[1,3] The most visible function of Rnt1p is in pre-rRNA processing.[1] Inactivation of Rnt1p blocks the maturation of the 25S pre-rRNA 3′ end and slows the formation of the mature 5′ end.[1,4] In addition, Rnt1p initiates the 3′-end processing of all RNA polymerase II (Pol II)-transcribed small nuclear RNAs (snRNAs) including U1, U2, U4, and U5.[3,5–7] Deletion of *RNT1* blocks the 3′-end formation of U2 and the

[1] S. Abou Elela, H. Igel, and M. Ares, Jr., *Cell* **85,** 115 (1996).
[2] B. Lamontagne, A. Tremblay, and S. Abou Elela, *Mol. Cell. Biol.* **20,** 1104 (2000).
[2a] S. Nocke, T. N. Gonzalez, C. Sidrauski, M. Niwa, and P. Walter, *Methods Enzymol.* **342,** [1] 2001 (this volume).
[3] S. Abou Elela and M. Ares, Jr., *EMBO J.* **17,** 3738 (1998).
[4] J. Kufel, B. Dichtl, and D. Tollervey, *RNA* **5,** 909 (1999).
[5] R. L. Seipelt, B. Zheng, A. Asuru, and B. C. Rymond, *Nucleic Acids Res.* **27,** 587 (1999).

0076-6879/01 $35.00

large form of U5 (U5L). In contrast, the steady state level of mature U1, U4, and U5 small form (U5S) is unchanged in Δ*RNT1* cells because alternative processing pathways exist in the absence of Rnt1p.[5–7] Rnt1p also processes yeast polycistronic small nucleolar RNAs (snoRNAs) to separate them into individual RNAs.[8] Unlike the majority of snRNAs, snoRNA processing is almost completely blocked in the absence of Rnt1p. As in the case with U2 snRNA, essential snoRNAs such as U14 must remain partially functional without processing because *RNT1* deletion is not lethal.

Comparison of Rnt1p known substrates reveals the presence of a terminal tetraloop 13 to 16 base pairs from the cleavage site with the conserved sequence AGNN.[9] Deletion of this sequence abolishes cleavage, whereas deletions of sequences closer to the cleavage site do not affect the substrate recognition or cleavage.[3,9] In contrast, *E. coli* RNase III substrate recognition is independent of the tetraloop, and is influenced by antideterminant nucleotides near the cleavage site.[10] It is unclear how Rnt1p recognizes the tetraloop and selects the scissile bond. In any case, Rnt1p must use a different mechanism to cleave long intermolecular RNA duplexes without tetraloops.[1] Rnt1p functions as a 108-kDa homodimer, which is dependent on signals located at the N-terminal domain and the dsRBD. Biochemical and genetic assays suggest that the dimerization occurs in a parallel configuration.[2] This means that the N-terminal domain and the dsRBD self-dimerize. In addition, Rnt1p is capable of forming an intramolecular complex through an interaction between the N-terminal domain and the dsRBD. This intricate interaction pattern is required for efficient Rnt1p function *in vivo* and *in vitro*. Here we describe the basic methods used in our laboratory to purify and characterize Rnt1p activity *in vitro*.

Cloning and Expression of Hexahistidine-Tagged Rnt1p

The expression and purification of active Rnt1p is a critical step toward the biochemical analysis of dsRNA recognition and cleavage in yeast. For this reason the *RNT1* gene was expressed in bacteria and partially purified as a glutathione *S*-transferase (GST) fusion.[1] The GST-tagged protein is active and has been used to identify Rnt1p cleavage sites *in vitro*.[3,8] However, we have found that GST fosters an illegitimate dimerization pattern that may affect the enzyme native conformation. Moreover, removal of the GST domain through enzymatic cleavage is

[6] C. Allmang, J. Kufel, G. Chanfreau, P. Mitchell, E. Petfalski, and D. Tollervey, *EMBO J.* **18,** 5399 (1999).
[7] G. Chanfreau, S. A. Elela, M. Ares, Jr., and C. Guthrie, *Genes Dev.* **11,** 2741 (1997).
[8] G. Chanfreau, P. Legrain, and A. Jacquier, *J. Mol. Biol.* **284,** 975 (1998).
[9] G. Chanfreau, M. Buckle, and A. Jacquier, *Proc. Natl. Acad. Sci. U.S.A.* **97,** 3142 (2000).
[10] K. Zhang and A. W. Nicholson, *Proc. Natl. Acad. Sci. U.S.A.* **94,** 13437 (1997).

both costly and inefficient. To circumvent these problems we expressed Rnt1p as N-terminal hexahistidine (His_6)-tagged protein.

This enzyme is active and easy to produce in native form. The pQE31/RNT1 plasmid used for protein expression was produced by cloning a polymerase chain reaction (PCR)-amplified fragment of *RNT1* in the bacterial expression vector pQE31 (Qiagen, Mississauga, ON, Canada). The PCR fragment was inserted into the *Bam*HI–*Sal*I sites of pQE31 as described previously.[2] The expressed protein includes an N-terminal His_6 tag and terminates at the native Rnt1p stop codon. The protein can be expressed in most bacterial strains (e.g., JM101 or XL1-Blue) but the highest expression level and least protein degradation are obtained when expressed in the bacterial strain BL21. Normally, we purify Rnt1p from bacteria grown in 3 liters of LB medium containing ampicillin (100 mg/liter). The culture is incubated overnight at 30° with shaking. When the culture reaches an OD_{600} between 1.1 and 1.4, isopropyl-β-D-thiogalactopyranoside (IPTG) is added to a final concentration of 1 m*M* (to induce protein expression) and the culture is further incubated at 37° for 3 hr. At the end of the incubation period the cells are collected by centrifugation at 2500*g* for 12 min at 4°. The cell pellet corresponding to 1 liter of culture is washed with 10 ml of Ni^{2+} buffer [25% (v/v) glycerol, 1 *M* NaCl, 30 m*M* Tris, pH 8] and frozen at −20° until needed. Normally, a total of 10 g of cells is collected from 3 liters of culture. Protease inhibitors are added in the Ni^{2+} buffer from a 1000× cocktail containing 156 mg of benzamidine, 1 mg of aprotinin, 1 mg of leupeptin, 1 mg of pepstatin A, and 1 mg of antipain dissolved in 1 ml of dimethyl sulfoxide (DMSO; Sigma-Aldrich Canada, Oakville, ON, Canada). For cell lysis, each pellet corresponding to a 1-liter culture is resuspended in 5 ml of Ni^{2+} buffer and sonicated in a glass tube at 55% power for 10 min on ice, using a sonic dismembrator (Artek Systems, Farmingdale, NY). The cell debris is removed by centrifugation at 20,000*g* for 10 min at 4°. The resulting pellet is resuspended in 7.5 ml of Ni^{2+} buffer and resonicated for 10 min. The supernatants from both sonication steps are collected and recentrifuged for an additional 30 min as described above. The supernantant is then filtered through a 0.22-μm pore size filter before it is applied to a His-Trap metal chelating column (Pharmacia Biotech, Baie d'Urfé, QC, Canada).

Purification

Protein purifications are conducted with an AKTA Explorer fast protein liquid chromatography (FPLC) system (Pharmacia Biotech, Baie d'Urfé, QC, Canada). The protein is normally purified with two Ni^{2+}-NTA agarose columns followed by gel filtration on a Superdex 200 HR 10/30 column (Pharmacia Biotech). The first purification is conducted on a column with a 5-ml bed volume, preequilibrated with 10 m*M* imidazole in Ni^{2+} buffer. The sample volume varies between 30 and 50 ml and it is injected with a 50-ml Superloop (Pharmacia Biotech). The column is washed with 10 m*M* imidazole (10 column volumes), 20 m*M* imidazole

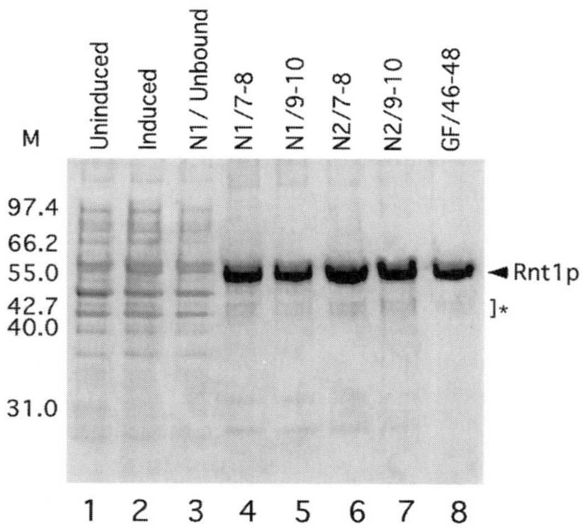

FIG. 1. Expression and purification of N-terminal His_6-tagged Rnt1p. Bacterial cells transformed with pQE31/RNT1 plasmid were grown overnight and the proteins were extracted before (lane 1) or after induction with IPTG (lane 2). Proteins extracted from induced cells were used to purify Rnt1p on two successive metal-chelating affinity columns (lanes 4 to 7) followed by a purification step on a gel-filtration column (lane 8). Equal amounts of protein from each purification step were fractionated by SDS–PAGE and stained with Coomassie Brilliant Blue R. N1/Unbound, proteins from the unbound fraction of the first metal-chelating affinity column; N1/7–8 and N1/9–10, proteins eluting in fractions 7 and 8 and in fractions 9 and 10 from the first column; N2/7–8 and N2/9–10, fraction numbers of proteins eluting from the second metal-chelating affinity column; GF/46–48, protein eluting in fractions 46 to 48 from the gel-filtration column. The position of Rnt1p is indicated by an arrowhead. The asterisk (*) indicates Rnt1p degradation product as judged by Western blot, using antibodies against the His_6 tag. The sizes of the protein molecular weight markers are indicated on the left-hand side.

(10 column volumes), and 60 m*M* imidazole (9 column volumes). At this stage most protein impurities have been removed except for the chaperonins (GroEL and GroES), which are removed by applying 5 column volumes of buffer containing 5 m*M* ATP, 50 m*M* KCl, 20 m*M* $MgCl_2$, and 50 m*M* Tris, pH 7.5.[11] The column is reequilibrated with 6 column volumes of 60 m*M* imidazole and washed with 2 column volumes of 100 m*M* imidazole. The elution is carried out with 3 column volumes of 150 m*M* imidazole. The protein is collected in 2 × 2.5 ml fractions corresponding to Rnt1p peaks (Fig. 1, lanes 4 and 5), which are pooled and diluted with buffer D (1 *M* NaCl, 30 m*M* Tris, pH 8) to 50 ml (15 m*M* final concentration of imidazole) and injected in a 1-ml Ni-NTA agarose column preequilibrated with 10 m*M* imidazole in buffer D. The column is washed with 15 column volumes of 60 m*M* imidazole and 2 column volumes of 100 m*M* imidazole in buffer D. The

[11] A. Thain, K. Gaston, O. Jenkins, and A. R. Clark, *Trends Genet.* **6,** 209 (1996).

elution is carried out with 6 column volumes of 250 m*M* imidazole in buffer D, and the protein is collected in 4 × 0.5 ml fractions. Pairs of fractions are pooled in dialysis bags (molecular mass limit of 14 kDa); one is dialyzed overnight against storage buffer [50% (v/v) glycerol, 0.5 *M* KCl, 30 m*M* Tris (pH 8), 0.1 m*M* dithiothreital (DTT), 0.1 m*M* EDTA] and the other against gel-filtration buffer [50 m*M* sodium phosphate (pH 7.5), 2 m*M* EDTA (pH 7.5), 0.1 m*M* DTT, and 0.5 *M* KCl] with at least three changes of buffer. The concentration of the protein fraction after dialysis against the storage buffer is normally 7 μg/μl, with a total of 1.5 mg of 90% pure protein (see Fig. 1, lanes 6 and 7).

The concentration of the protein fraction after dialysis against gel-filtration buffer is normally 0.5 μg/μl, again with a total of 1.5 mg of protein. A final step is performed with a Superdex 200 HR 10/30 gel filtration column (Pharmacia Biotech) precalibrated with gel-filtration buffer at 4°. The size of eluting protein fractions is determined with low (25–67 kDa) and high (162–669 kDa) molecular size markers (Pharmacia Biotech). Rnt1p is applied to the gel-filtration column in aliquots containing 250 μg of protein in gel-filtration buffer. Under these conditions, one major peak corresponding to Rnt1p tetramer migrates with an apparent molecular mass of 220 kDa. Application of higher amounts of proteins on the column causes protein oligomerization, while lower protein amounts allow the detection of Rnt1p dimers.[2] The major tetramer peak is collected in 250-μl fractions and dialyzed against storage buffer as described above.

After dialysis, the protein concentration ranged from 0.1 to 0.4 μg/μl with no visible contaminating protein (see Fig. 1, lane 8). The specific activity of the purified protein is about 56,500 units/mg and the enzyme unit is about 0.018 μg/μmol of substrate as judged by the RNA cleavage assay described below.

Substrate Production and RNA Cleavage Assay

There are several Rnt1p substrates available for *in vitro* cleavage assays, including the processing site near the 3′ ends of pre-rRNA, U2 snRNA, U5 snRNA, and several snoRNAs. In our hands the most efficient substrate is a T7 transcript of the rRNA 3′ end, which contains 152 nucleotides surrounding the 25S mature 3′ end and includes the 58-nucleotide stem–loop structure cleaved by Rnt1p (see Fig. 2B). The RNA used as a substrate in the enzymatic assays was generated by T7 RNA polymerase as described previously.[2] The RNA substrate was produced from a T7 promoter of the plasmid pRS316/3′ end digested with *Eco*RI. To construct this plasmid, an *Acc*I fragment of the rDNA was blunted and cloned in the pRS316 *Sma*I site.

The RNA cleavage activity of the purified protein fractions was tested *in vitro* with 0.056 μg from each fraction, which corresponds to 10 n*M* purified protein (Fig. 2A). The various protein fractions were incubated with 1 pmol of pre-rRNA

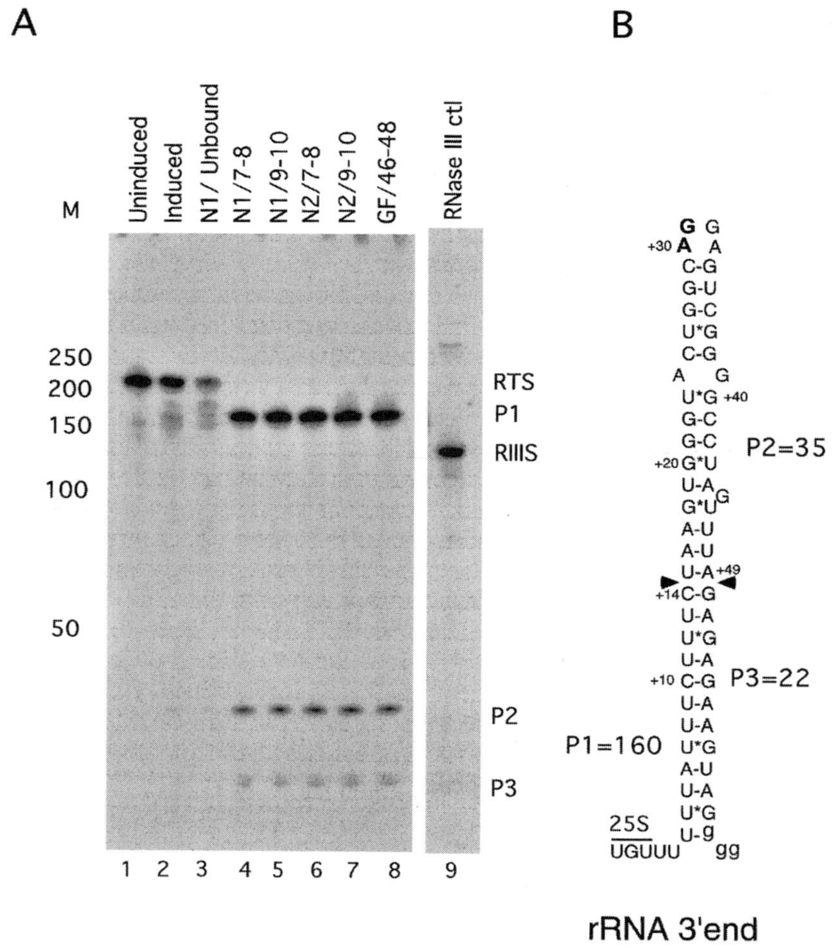

FIG. 2. *In vitro* cleavage assay of Rnt1p, using the 3′ end of 25S pre-rRNA as a model substrate. (A) The 220-nucleotide substrate was incubated with an equal amount of total protein from cells not expressing Rnt1p (lane 1), total protein from cells expressing Rnt1p (lane 2), protein in the unbound fraction of the first purification column (lane 3), protein eluting from fractions 7 and 8 (lane 4) and 9 and 10 (lane 5) from the first column, protein eluting from fractions 7 and 8 (lane 6) and 9 and 10 (lane 7) from the second column, and protein from fractions 46–48 of the gel-filtration column. In addition, protein from fractions 46–48 of the gel-filtration column were incubated with a 131-nucleotide RNA containing the T7 R1.1 processing site (RIIIS) recognized by the bacterial RNase III as control.[12] On the right-hand side, the position of the substrate and the different cleavage products are indicated as follow: RTS, full-length 220-nucleotide substrate; P1, 160-nucleotide 5′-end cleavage product; P2, the middle 35-nucleotide cleavage product; P3, the 22-nucleotide 3′-end cleavage product. The DNA molecular weight marker is indicated on the left-hand side. (B) Illustration of the 25S rRNA 3′-end model substrate. The *in vivo* cleavage sites are indicated by arrowheads. The numbers are relative to the 25S rRNA mature 3′ end. The vector sequence is indicated.

3′-end model substrate for 20 min at 30° in 10 μl of reaction buffer [30 mM Tris (pH 7.5), 150 mM KCl, 5 mM spermidine, 20 mM $MgCl_2$, 0.1 mM DTT, and 0.1 mM EDTA, pH 7.5]. In addition, the purified protein was also incubated under the same conditions with the *E. coli* RNase III T7 substrate[12] to detect any RNase III activity that may copurify with Rnt1p (Fig. 2A, lane 9). As seen in this typical experiment the uninduced cell lysate does not cleave Rnt1p substrate, while purified Rnt1p cannot cleave the RNase III substrate (Fig. 1A, lanes 1 and 9, respectively). A low amount of cleaved product is detected in the induced or unbound fractions, and full cleavage is obtained with the purified proteins (Fig. 2A, lanes 4 to 8).

The kinetic parameters for Rnt1p processing in the steady state are determined by measuring the initial cleavage rate of the 25S rRNA 3′-end processing signal as a function of its concentration. The cleavage reaction is performed as described above with an Rnt1p concentration of 10 nM (monomer) and RNA substrate concentrations that range from 0.05 to 7.05 μM. The cleavage rate is measured by quantifying the cleavage product separated on a 12% (w/v) denaturing gel, using the Molecular Analyst program (Bio-Rad, Hercules, CA). Rnt1p cleavage of the rRNA 3′-end processing signals follows substrate saturation kinetics with a K_m of 1.2 μM and k_{cat} of 5.5 min^{-1}. These values are similar to the previously reported kinetic parameters of bacterial RNase III cleavage.[13]

Mobility Shift Assay and In-the-Gel Cleavage Assay

Like the bacterial RNase III, Rnt1p can bind its RNA substrate without cleaving it, in the absence of divalent metal ions.[2] This useful feature allows analysis of RNA recognition and association independent of catalysis. Here we describe the mobility shift assay used in our laboratory to detect Rnt1p association with its substrate, and an in-the-gel cleavage assay used to measure the activity of different Rnt1p–RNA complexes. A typical gel mobility shift assay is conducted by incubating 2 fmol of internally radiolabeled RNA with increasing amount of Rnt1p ranging from 1 to 10 pmol (Fig. 3A). The binding reaction is carried out in 20 μl of binding buffer [20% (v/v) glycerol, 150 mM KCl, 30 mM Tris (pH 7.5), 5 mM spermidine, 0.1 mM DTT, and 0.1 mM EDTA] for 10 min on ice. The amount of KCl used reflects the physiological salt concentration in yeast cells. Half of the reaction is loaded on a 2-mm-thick 4% (w/v) nondenaturing polyacrylamide gel (1 : 80, bisacrylamide to acrylamide). The RNA–protein complexes are fractionated at 0.12 V/cm^2 for 30 min and then the voltage increased to 0.34 V/cm^2 for another 2 hr and 45 min. The resolution and stability of the RNA–protein complexes depend on the voltage and the migration distance. Higher voltage or

[12] H. L. Li, B. S. Chelladurai, K. Zhang, and A. W. Nicholson, *Nucleic Acids Res.* **21,** 1919 (1993).
[13] H. Li and A. W. Nicholson, *EMBO J.* **15,** 1421 (1996).

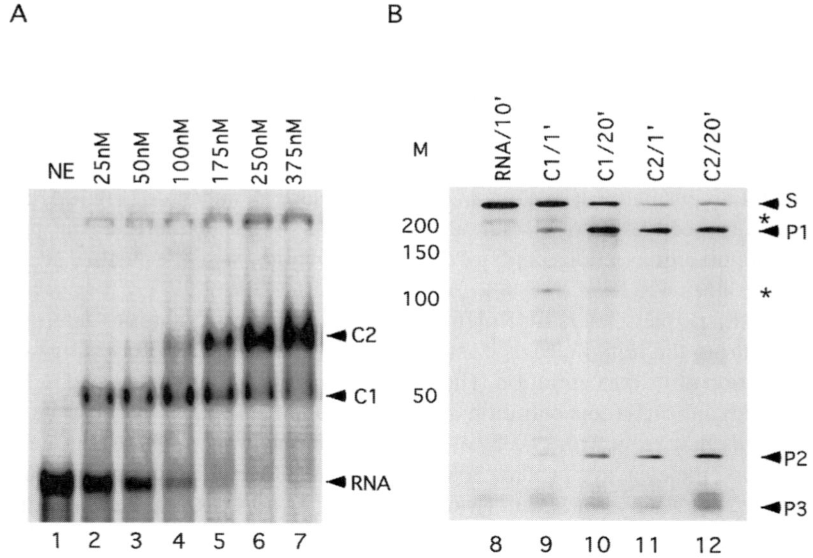

FIG. 3. Gel mobility shift assay, and in-the-gel cleavage assay of Rnt1p. Gel shift assay (A) was conducted by incubating the 25S pre-rRNA 3′-end substrate with increasing amount of Rnt1p in 150 mM KCl. The protein complexes were fractionated on a 4% (w/v) native polyacrylamide gel. The gel pieces corresponding to each complex were cut out and incubated in 150 mM KCl in the presence of 20 nM $MgCl_2$ to allow RNA cleavage (B). At the end of the incubation period (1 and 20 min) the gel pieces were removed and the RNA was extracted and loaded on a 12% (w/v) denaturing polyacrylamide gel. (A) Positions of the RNA, the first Rnt1p–RNA complex (C1), and the second Rnt1p–RNA complex (C2) are indicated on the right-hand side. The protein concentrations are indicated at the top. The band corresponding to the input RNA was used as control. (B) The asterisk (*) indicates Rnt1p-independent degradation product. The substrate and cleavage products are indicated on the right-hand side. The different complexes and the cleavage time are indicated at the top. The DNA molecular weight markers are indicated on the left-hand side.

longer migration may induce complex dissociation and lower band resolution. As seen in Fig. 3A, Rnt1p can form two complexes with the rRNA 3′-end substrate as a function of protein concentration with a K_d value of 43.5 nM. These two complexes represent either two different binding conformations, or reflect the binding of one or two protein molecules to the substrate (Fig. 3A).

The activity of the different Rnt1 protein complexes may be assayed by cleavage in the gel. In this method, we expect that active protein complexes that correctly position the RNA within the protein to spontaneously cleave the RNA on the addition of Mg^{2+}. On the other hands, incorrect complexes or complexes formed through the binding of a single protein molecule to the RNA will take longer time to cleave on the addition of Mg^{2+}. To perform in-the-gel cleavage, bands corresponding to different RNA–protein complexes (Fig. 3A, C1 and C2) are cut out of the gel and incubated in a cleavage buffer [30 m*M* Tris (pH 7.5), 5 m*M* spermidine, 0.1 m*M* DTT, 0.1 m*M* EDTA (pH 7.5), and 20 m*M* $MgCl_2$] at 30° for 1 or 20 min. At each time point, the gel pieces are removed, the RNA is eluted from the gel, and equal counts of the radioactive RNA are loaded on a 12% (w/v) denaturing polyacrylamide gel. As seen in Fig. 3B, the band corresponding to the RNA alone (lane 8) does not result in RNA cleavage. Complex 1 (lane 11) is inefficient and does not complete RNA cleavage even after 20 min. On the other hand, complex 2 readily cleaves the majority of the RNA substrate in 1 min (lane 12). Therefore, complex 2 is likely the main RNA–protein complex required for RNA cleavage by Rnt1p while complex 1 represents either a step toward the formation of complex 2 or malformed RNA–protein complex that may be reconfigured into active complex on addition of Mg^{2+}.

Summary

In this article, we have described methods used to purify Rnt1p and study its biochemical properties. Rnt1p can be easily purified from bacteria as N-terminal His_6-tagged protein and its activity may be monitored *in vitro*. Rnt1p cleaves the RNA by binding to a cleavage site followed by hydrolysis and product release. The kinetic parameters of Rnt1p are similar to those of other nucleases, including bacterial RNase III. The ability of Rnt1p to bind substrate without cleaving it in the absence of divalent metal ions provides a convenient means to study RNA recognition and binding independent of catalysis. The gel mobility shift and in-the-gel cleavage assays described here reveal the formation of two Rnt1p–RNA complexes with different cleavage activities, suggesting that the protein may bind the substrate in two different forms or through a two-step binding reaction.

Acknowledgments

We thank Allen Nicholson for providing a plasmid encoding the bacterial RNase III substrate (R1.1). We also thank Annie Tremblay for helpful comments on the manuscript. Research in our laboratory was supported by Grant MT-14305 from the Medical Research Council of Canada. The FPLC used for protein purification was purchased by a grant from Canada Foundation for Innovation. S.A. is a Chercheur-Boursier Junior II of the Fonds de la Recherche en Santé du Québec.

[15] Pac1 Ribonuclease of *Schizosaccharomyces pombe*

By GIUSEPPE ROTONDO and DAVID FRENDEWEY

Introduction

The Pac1 RNase of the fission yeast *Schizosaccharomyces pombe* is a double-strand-specific endoribonuclease (dsRNase) whose structure, biochemical properties, and biological functions place it in the RNase III family (see [13] in this volume[1]). The *S. pombe pac1*$^+$ gene was originally cloned by virtue of its ability to cause sterility when overexpressed,[1a,2] an intriguing property that has yet to be fully explained. The name *pac1*$^+$ derives from its ability to complement *pat1* mutants, which spontaneously initiate sexual development at the restrictive temperature.[3,4] We isolated the *pac1*$^+$ gene[5] as a multicopy suppressor of *snm1*, a temperature-sensitive (*ts*) mutant that maintains reduced steady state amounts of the spliceosomal small nuclear RNAs (snRNAs) and accumulates 3′-extended snRNA transcripts.[6] Subsequent work showed that the *snm1* mutation lies in the *pac1*$^+$ gene and established the requirement for the Pac1 RNase in the 3′ processing of the pre-U2 RNA and presumably other snRNA precursors.[7] Mutant *pac1* alleles, such as *snm1* (renamed *pac1-45*), also accumulate 3′-extended pre-rRNA,[7] and purified Pac1 RNase cleaves a pre-rRNA at *in vivo* RNA processing sites.[8,9] These results established a role for Pac1 in rRNA synthesis, a function shared by all members of the RNase III family (see [13] and [14] in this volume[1,9a]). This article describes the methods we and others have used to define the biological functions and biochemical properties of the Pac1 RNase.

Materials

Fission Yeast Strains

Schizosaccharomyces pombe strains used in the methods described below are listed in Table I. We have derived JP44 from isolate *ts*45 of our collection of

[1] A. K. Amarasinghe, I. Calin-Jageman, A. Harmouche, W. Sun, and A. W. Nicholson, *Methods Enzymol.* **342,** [13] 2001 (this volume).
[1a] H.-P. Xu, M. Riggs, L. Rodgers, and M. Wigler, *Nucleic Acids Res.* **18,** 5304 (1990).
[2] Y. Iino, A. Sugimoto, and M. Yamamoto, *EMBO J.* **10,** 221 (1991).
[3] Y. Iino and M. Yamamoto, *Mol. Gen. Genet.* **198,** 416 (1985).
[4] P. Nurse, *Mol. Gen. Genet.* **198,** 497 (1985).
[5] G. Rotondo and D. Frendewey, *Mol. Gen. Genet.* **247,** 698 (1995).
[6] J. Potashkin and D. Frendewey, *EMBO J.* **9,** 525 (1990).
[7] D. Zhou, D. Frendewey, and S. M. Lobo Ruppert, *RNA* **5,** 1083 (1999).
[8] G. Rotondo, J. Y. Huang, and D. Frendewey, *RNA* **3,** 1182 (1997).
[9] Y. F. Melekhovets, L. Good, S. Abou Elela, and R. N. Nazar, *J. Mol. Biol.* **239,** 170 (1994).

0076-6879/01 $35.00

TABLE I
FISSION YEAST STRAINS

Strain	Genotype
JP44	*h⁻ leu1-32 pac1-45*[a]
SP315	*h⁺ leu1-32 pat1-114*
SPB5	h^{90} *leu1-32 ura4-D18 ade6-M210*
SP631	h^{90}/h^{90} *leu1-32/leu1-32 ade6-704/ade6-704*

[a] We have renamed the original *snm1* allele[6] as *pac1-45*.

temperature-sensitive (*ts*) *S. pombe* strains.[6,10] M. McLeod (SUNY Downstate Medical Center, Brooklyn, NY) and D. Beach (Cold Spring Harbor Laboratory, Cold Spring Harbor, NY) kindly provided strains SP315, SPB5, and SP631 (Table I).

Nucleic Acids and Nucleotides

The following DNAs are the generous gifts of the listed providers: *S. pombe* U1, U2, and U3 genes (J. Wise, Case Western Reserve University, Cleveland, OH); *S. pombe* U4 and U6 genes (D. Tollervey, University of Edinburgh, Scotland); *S. pombe* U6 gene (Y. Ohshima, Kyushu University, Japan); *S. pombe* RNase P RNA subunit and $tRNA^{Ser}$-$tRNA^{Met}$ genes (D. Söll, Yale University, New Haven, CT); oligodeoxynucleotide templates for *Escherichia coli* RNase III substrates (A. Nicholson, Wayne State University, Detroit, MI); the *S. pombe* β-tubulin gene and an rDNA clone (M. Yanagida, Kyoto, Japan); and the pEM-7 template for HIV-1 TAR RNA (M. Mathews, UMDNJ-New Jersey Medical School, Newark, NJ). The following are purchased from the listed suppliers: DNA size markers (New England BioLabs, Beverly, MA); oligodeoxynucleotides (Oligos Etc., Wilsonville, OR); *E. coli* tRNA [(Roche, Indianapolis, IN), further purified by phenol–chloroform extraction and ethanol precipitation]; polynucleotides and nucleoside triphosphates (Amersham Pharmacia Biotech, Piscataway, NJ); pBluescript-SK(–) (Stratagene, La Jolla, CA); pRSET-A (Invitrogen, San Diego, CA); mono- and diphosphate nucleotide markers (Sigma, St. Louis, MO); and ^{32}P-labeled nucleoside triphosphates (New England Nuclear, Boston, MA).

Enzymes

Recombinant Pac1 RNase is purified as described below. *Escherichia coli* RNase III is the kind gift of A. Nicholson (Wayne State University). Other

[9a] B. Lamontagne and S. Abou Elela, *Methods Enzymol.* **342,** [14] 2001 (this volume).
[10] J. Potashkin, R. Li, and D. Frendewey, *EMBO J.* **8,** 551 (1989).

enzymes are listed with their suppliers: RNases A and T1, T3 and T7 RNA polymerases (Ambion, Austin, TX); RNase T2 and nuclease P1 (Calbiochem, La Jolla, CA); RNase-free DNase I (Promega, Madison, WI); shrimp alkaline phosphatase (United States Biochemical, Cleveland, OH); AmpliTaq DNA polymerase (Perkin-Elmer Cetus, Norwalk, CT); restriction endonucleases, T4 polynucleotide kinase, and CircumVent thermal cycle sequencing kit (New England BioLabs); SuperScript II RNase H^- reverse transcriptase (GIBCO-BRL, Gaithersburg, MD).

Other Materials

The following are listed with their suppliers: *E. coli* BL21(DE3)pLysS [(F^- *ompT hsdS*$_B$ ($r_B^- m_B^-$) *gal dcm* (DE3) pLysS(Cm^R)] and anti-T7-tag monoclonal antibody (Novagen, Madison, WI); RPA II RNase protection kit and phenol–chloroform mixtures (Ambion); Tri Reagent (Molecular Research Center, Cincinnati, OH); glycogen solution (Roche); 0.5-mm-diameter glass beads (B. Braun Melsungen, Melsungen, Switzerland); Centricon-10 and Spin-X filter units [Amicon (Danvers, MA) and Corning Costar (Corning, NY), respectively]; Duralon-UV nylon membranes (Stratagene); Bio-Spin 6 spin columns (Bio-Rad, Hercules, CA); GeneClean DNA purification kit (BIO 101, La Jolla, CA); RNasin RNase inhibitor and alkaline-phosphatase-conjugated goat anti-mouse and anti-rabbit IgG (Promega); Western Light kit (Tropix, Bedford, MA); Protein Gold reagent (Integrated Separation Systems, Natick, MA); Pro-Bond Xpress nickel affinity chromatography kit (InVitrogen); cellulose and polyethyleneimine cellulose thin-layer chromatography plates [Kodak (Rochester, NY) and Selecto Scientific (Norcross, Georgia)]; and siliconized glass wool (Sigma). Anti-Pac1 rabbit sera are raised against a denatured preparation of recombinant Pac1[11] by Cocalico Biologicals (Reamstown, PA). We prepare triethylammonium bicarbonate (pH 9.5) by bubbling CO_2 through triethylamine and store it in aliquots in amber vials at -20°. R. Simons (UCLA) has generously provided the anti-RNase III serum.

Methods

Assays for Pac1 RNase Function in Vivo

We have used three biological assays to test the function of *pac1*$^+$ gene constructs *in vivo*.[5] In these assays expression of *pac1*$^+$ on a multicopy plasmid in an appropriate genetic background either rescues a *ts* growth defect, which is scored as colony growth on agar plates, or blocks sexual development, which is scored by microscopic counting of zygotes, asci, or spores. We have used the plasmid vector pIRT31,[5] but similar vectors will work as well. The important features

[11] G. Rotondo and D. Frendewey, *Nucleic Acids Res.* **24,** 2377 (1996).

are an autonomous replication sequence and the *Saccharomyces cerevisiae LEU2* gene, which complements the *S. pombe leu1-32* mutation. All procedures employ standard *S. pombe* growth media and methods.[12,13]

Protocol 1: Rescue of Temperature-Sensitive Growth Impairment of pac1 Mutant Strain

1. Transform JP44 (Table I) with *pac1*$^+$ test constructs by the lithium acetate procedure or other suitable method. These procedures are described in the references cited above.
2. Spread an equal volume of the transformation mixes on two minimal medium agar plates and incubate at 23° for 16 hr.
3. Shift one of the plates to 37° and continue the incubation until colonies develop on the 23° plate.
4. Count the colonies on both plates.

A construct produces a complete positive score for function when approximately equal number of leu$^+$ transformants appear at 23 and 37°. A complete negative score results when no colonies appear at 37°, but the number at 23° is comparable to empty vector and wild-type *pac1*$^+$ controls. Because *pac1-45* is a "leaky" *ts* allele, the procedure outlined above provides a more stringent test of *pac1*$^+$ function than the alternative of allowing colonies to develop at the permissive temperature followed by replica plating to 37°. The assay is reliable as a general screen for modifications that affect *pac1*$^+$ function, but it has some inherent limitations. As the assay relies on multicopy suppression, some mutations that would be inactive as the sole genomic copy might score positive for function when overexpressed. Also, the test for function at 37° prevents the discovery of new *pac1 ts* alleles.

Protocol 2: Rescue of Temperature-Sensitive Lethality in pat1 Mutant Strain

1. Transform SP315 (Table I) with *pat1*$^+$ test constructs as in protocol 1.
2. Spread the transformation mixes on minimal medium agar plates and incubate at 23° until colonies appear (3 to 5 days).
3. Replicate the colonies onto two minimal medium plates.
4. Incubate one plate at 23° and the other at 37° until colonies develop.
5. Count the colonies on both plates.

The severity of the *pat1-114 ts* allele, which causes spontaneous, lethal meiosis in haploid strains[3,4] at the restrictive temperature, permits the less stringent replica

[12] S. Moreno, A. Klar, and P. Nurse, *Methods Enzymol.* **194,** 795 (1991).

[13] C. Alfa, P. Fantes, J. Hyams, M. McLeod, and E. Warbrick, "Experiments with Fission Yeast: A Laboratory Course Manual." Cold Spring Harbor Laboratory Press, Cold Spring Harbor, New York, 1993.

plating scheme. The criteria for positive and negative scores and the limitations are the same as for protocol 1.

Protocol 3: Induction of Sterility in Wild-Type Strain. The conjugation and sporulation assays in protocol 3 are based on those described by Neiman *et al.*[14]

3a: INHIBITION OF CONJUGATION

1. Transform the homothallic haploid strain SPB5 (Table I) with *pac1*$^+$ test constructs as in protocol 1.
2. Plate the transformants on minimal medium supplemented with adenine and uracil (75 mg/liter) and grow at 30°.
3. Pick several individual leu$^+$ transformants and grow to mid- to late exponential phase ($1\text{–}5 \times 10^7$ cells/ml) in liquid minimal medium supplemented with adenine and uracil at 30° with shaking aeration.
4. Harvest the cells by centrifugation and resuspend in liquid SSL-N,[15] a minimal medium without ammonia as a nitrogen source, to induce self-mating.
5. After 2 days of incubation at 23° without agitation, examine the cells microscopically and count the number of unmated cells, zygotes (fused mated pairs), asci (with four spores), and free spores.
6. Calculate the conjugation efficiency as the proportion of zygotes, asci, and free spores among the total number of cells counted.

3b: INHIBITION OF SPORULATION

1. Repeat the procedure outlined in protocol 3a, except transform the diploid strain SP631 (Table I). In diploid strains, nitrogen starvation induces meiosis and sporulation in azygotic asci directly without mating.
2. Count the number of unsporulated cells, asci, and free spores.
3. Calculate the sporulation efficiency as the proportion of asci and free spores.

To calculate the total number of cells, use the following evaluations: an unmated haploid cell (for conjugation, protocol 3a) or an unsporulated diploid cell (for sporulation, protocol 3b) = 1; a zygote or ascus = 2 for conjugation (protocol 3a) or 1 for sporulation (protocol 3b); a free spore = 0.5 for conjugation or 0.25 for sporulation of diploids.

Transformation with empty vector gives an approximately 5-fold greater efficiency of conjugation and sporulation than transformation with *pac1*$^+$, which does not completely block conjugation or sporulation, possibly because of random plasmid loss. A positive score for *pac1*$^+$ activity is obtained when a construct

[14] A. M. Neiman, B. J. Stevenson, H.-P. Xu, G. F. Sprague, Jr., I. Herskowitz, M. Wigler, and S. Marcus, *Mol. Biol. Cell* **4,** 107 (1993).

[15] R. Egel and M. Egel-Mitani, *Exp. Cell Res.* **88,** 127 (1974).

produces a sterile phenotype: conjugation and sporulation efficiencies much reduced compared with the vector-transformed control and approximately equal to that observed with wild-type *pac1*$^{+}$.

Analysis of Substrate RNAs in Vivo

We have investigated the participation of the Pac1 RNase in pre-rRNA and pre-snRNA processing and its indirect role in pre-mRNA splicing and tRNA 5′ processing by the procedures outlined below.

Protocol 4: Preparation of Total RNA from Fission Yeast. The RNA preparation method given here for small-scale cultures (10–20 ml) is simple, rapid, and produces high-quality RNA that can be used in any subsequent RNA analytical procedure. The method can be readily scaled up for larger cultures.

1. Grow *S. pombe* cultures to midlog phase (approximately 10^7 cells/ml) in a liquid rich medium or minimal supplemented medium with shaking aeration. For simple analysis of wild-type strains, grow cultures at 30°. For temperature shift experiments with *ts* strains, culture the cells to midlog phase at 23° and then shift a portion to 37° for various times.
2. Harvest the cells by centrifugation at 20° for 5 min at 2000*g*.
3. Immediately resuspend the cell pellet in 0.1 ml of Tri Reagent per 10^8 cells and transfer to a 1.7-ml polypropylene microcentrifuge tube.
4. Add glass beads until the volume of the suspension has approximately doubled but a meniscus remains above the beads. We bake the glass beads (see Materials), which require no prewashing, for 16 hr in glass bottles in a standard glass drying oven to destroy contaminating nucleases.
5. Disrupt the cells by three 1-min pulses of vigorous vortex mixing separated by 1-min rests on ice.
6. Add an additional 0.9 ml of Tri Reagent per 10^8 cells followed by brief mixing and incubation at room temperature for 5 min.
7. Extract the suspension with 0.2 ml of chloroform by vigorous vortex mixing for 15 sec followed by a 15-min incubation at room temperature.
8. Centrifuge at 16,000*g* for 15 min at 4° and remove the aqueous upper phase to a fresh tube.
9. Mix the aqueous phase with 0.25 ml each of 2-propanol and a solution of 1.2 *M* sodium citrate, 0.8 *M* NaCl and incubate for 10 min at room temperature.
10. Recover the RNA precipitate by centrifugation at 16,000*g* for 8 min at 4°. Wash the precipitate twice with 1 ml of 75% (v/v) ethanol at 4°, and allow the pellet to dry briefly at room temperature.
11. Dissolve the RNA in 0.1 ml of RNase-free water. Determine the concentration by absorbance at 260 nm.[16]

[16] J. Sambrook, E. F. Fritsch, and T. Maniatis, "Molecular Cloning: A Laboratory Manual," 2nd Ed. Cold Spring Harbor Laboratory Press, Cold Spring Harbor, New York, 1989.

Protocol 5: Northern Blot Analysis of Small Nuclear RNAs. For most experiments we analyze equal mass amounts of RNA from each sample. For comparative RNA analyses based on cell equivalents, we harvest equal numbers of cells (determined by counting with a hemocytometer) for each RNA preparation and standardize the samples for RNA recovery by adding 10^4 cpm of ^{3}H-labeled *S. pombe* total RNA to each cell pellet before RNA preparation.[6] We then load equal radioactive amounts of RNA in each well of the gel. We perform our hybridizations in a rotisserie oven (Bachofer, Reutlingen, Germany) in 100-ml glass test tubes sealed with silicone stoppers.

1. To 5 μl containing 10–20 μg of RNA add 5 μl of formamide loading solution [80% (v/v) formamide, 0.1% (w/v) xylene cyanol, 0.1% (w/v) bromphenol blue] and denature at 85° for 5 min.
2. Immediately load the RNA on an 8 or 10% (w/v) polyacrylamide (19 : 1, acrylamide : bisacrylamide)–7 *M* urea slab gel ($0.15 \times 20 \times 20$ cm) and fractionate by electrophoresis in 1× Tris–borate–EDTA buffer[16] (TBE) at approximately 40-W constant power.
3. Transfer the RNA to a nylon membrane by semidry electroblotting (Owl Separation Systems, Woburn, MA) in 1.2 m*M* Tris base, 0.6 m*M* sodium acetate, 0.03 m*M* EDTA (pH 7.8) for 1 hr at 0.4 A.
4. Photo-cross-link the RNA to the membrane by ultraviolet irradiation in a Stratalinker (Stratagene) or similar device according to the manufacturer instructions.
5. Wet the blot by placing it in a small puddle of prehybridization buffer [6× SSPE,[16] 0.1% (w/v) sodium dodecyl sulfate (SDS), *E. coli* tRNA (0.1 mg/ml)] on a sheet of plastic wrap. Roll the blot loosely around a 10-ml glass pipette, insert it into a hybridization tube containing 5 ml of prehybridization buffer, and unroll the blot onto the walls of the tube to press out any air bubbles between the membrane and the tube.
6. Close the tube and incubate with constant rotation for 1 to 2 hr at the hybridization temperature, which for oligodeoxynucleotide probes is usually 5–10° below the theoretical T_m; however, it is best to determine the optimum hybridization temperature empirically.
7. Remove the prehybridization buffer and replace with 5 ml of 6× SSPE, 0.1% (w/v) SDS containing 10^6 dpm of ^{32}P-labeled oligodeoxynucleotide probe and continue incubation for 16 hr. Oligodeoxynucleotides are 5′-end labeled with T4 polynucleotide kinase and [γ-^{32}P]ATP and purified by spin-column chromatography.[16]
8. Remove the blot from the tube and wash on a rotating table in 500 ml of 6× SSPE, 0.1% (w/v) SDS three times for 15 min at room temperature followed by a final wash at or below the hybridization temperature.
9. Dry the blot briefly on blotting paper, wrap it in plastic film, and autoradiograph at −70° with an intensifying screen.

10. Strip the probe from the blot by washing with several changes of 0.01× SSC,[16] 0.01% (w/v) SDS at 70° for 1–2 hr or by placing the blot in boiling distilled water for 1–2 min.

Protocol 6: RNase Protection Analysis of Pre-snRNA and Pre-rRNA 3′ Processing

6a: SYNTHESIS OF ANTISENSE RNA PROBES

1. Create templates for the synthesis of antisense RNA probes by subcloning gene sequences that include the 3′ flanking region into a plasmid, such as pBluescript, that carries promoters for phage T3 and T7 RNA polymerases separated by a multiple cloning site. Linearize the plasmids with a restriction enzyme that cuts within the snRNA or rRNA coding sequence and purify with GeneClean (BIO 101, La Jolla, CA). Alternatively, amplify the appropriate sequences from clones or genomic DNA by polymerase chain reaction (PCR) with primers that incorporate T3 and T7 phage promoter sequences. Gel purify the correct size PCR product.

2. Transcribe ^{32}P-labeled antisense RNAs in a 20-μl reaction containing approximately 1 μg of plasmid template or 40 ng of PCR template, 40 m*M* Tris-HCl (pH 7.9), 6 m*M* $MgCl_2$, 2 m*M* spermidine, 10 m*M* dithiothreitol (DTT) (include 50 m*M* NaCl for a T3 RNA polymerase reaction), 0.5 m*M* of each ribonucleotide triphosphate, 50 μCi of [α-^{32}P]UTP (800 Ci/mmol), 20 units of RNasin, and 20 units of either T7 or T3 RNA polymerase at 37° for 1 hr.

3. Digest the template DNA with 1 unit of RNase-free DNase I for 15 min at 37° and recover the [^{32}P]RNA by phenol–chloroform (5 : 1, pH 4.5) extraction and ethanol precipitation with 20 μg of RNase-free glycogen as carrier. Dissolve the precipitate in 5 μl of formamide loading solution (protocol 5) and heat denature at 85° for 5 min.

4. Purify the antisense RNA probe (loaded in three or four lanes) by electrophoresis on a 6% (w/v) polyacrylamide–urea gel (0.04 × 20 × 20 cm) in 1× TBE buffer. Retain the gel on one of the glass plates and cover with plastic wrap. Fix a piece of X-ray film to the gel with adhesive tape and autoradiograph for 30–60 sec. During the autoradiography, make multiple marks with a permanent marker from the film to the plastic-covered gel in an asymmetric pattern. Align the developed autoradiogram with the marks and fix it to the backing plate with tape. Using a radioactivity shield, excise the RNA bands with a scalpel or razor blade.

5. Place the gel pieces in a microcentrifuge tube containing 0.4 ml of 0.5 *M* ammonium acetate, 1 m*M* EDTA, 0.2% (w/v) SDS and elute by incubation at 37° for 12–16 hr or at 55° for 4 hr.

6. Recover the eluted RNA by ethanol precipitation with 20 μg of glycogen carrier and store in RNase-free water at −20°.

6b: RNase PROTECTION. We use the Ambion RPA II kit according to the manufacturer instructions.

1. Hybridize 20 μg of total RNA with 0.5–1 $\times$ 10^5 dpm of antisense RNA probe for 12–16 hr followed by digestion with 200 units of RNase T1.
2. Recover the digested RNAs by ethanol precipitation and dissolve in formamide loading solution (protocol 5).
3. Heat denature at 85° for 5 min and fractionate by electrophoresis on a 6% (w/v) polyacrylamide–urea gel in 1$\times$ TBE.
4. Visualize the RNase protection products by autoradiography at −70° with an intensifying screen.

Protocol 7: Primer Extension Analysis

1. To anneal the primer and template RNA, mix 20 μg of total RNA with 10^4–10^5 dpm of 5′-^{32}P-labeled oligodeoxynucleotide primer (protocol 5) in a 12-μl volume and heat at 85° for 10 min followed by a quick chill on ice.
2. Collect the tube contents by brief centrifugation and add 4 μl of 5$\times$ reverse transcription buffer [250 m*M* Tris-HCl (pH 8.3), 375 m*M* KCl, 15 m*M* $MgCl_2$], 2 μl of 0.1 *M* DTT, and 1 μl of a solution of all four deoxynucleoside triphosphates at 10 m*M* each. Mix and heat at 45° for 2 min.
3. Add 1 μl of SuperScript II reverse transcriptase and incubate at 45° for 50 min.
4. Stop the reaction by the addition of 1 μl of an RNase A/T1 mix (500 units of RNase A per milliliter; 20,000 units of RNase T1 per milliliter) followed by incubation at 37° for 15 min.
5. Recover the radioactive DNA by extraction with phenol–chloroform–isoamyl alcohol (25 : 24 : 1, pH 7.9) and ethanol precipitation. Dissolve the precipitate in 5 μl of formamide loading solution (protocol 5) and heat denature at 85° for 5 min.
6. Fractionate the cDNA products on an 8 or 10% (w/v) polyacrylamide gel (protocol 6) and visualize by autoradiography at −70° with an intensifying screen.

Assays for Pac1 RNase Function in Vitro

The experimental methods described below for the characterization of the Pac1 RNase employ a purified recombinant enzyme that is modified by attachment of polyhistidine and epitope tags to its amino terminus (see Fig. 1).

Protocol 8: Purification of N-Terminally Tagged Pac1 RNase

8a: CONSTRUCTION OF *pac1*$^+$ *Escherichia coli* EXPRESSION VECTOR. To facilitate purification of the Pac1 RNase, we introduced the coding sequence for a

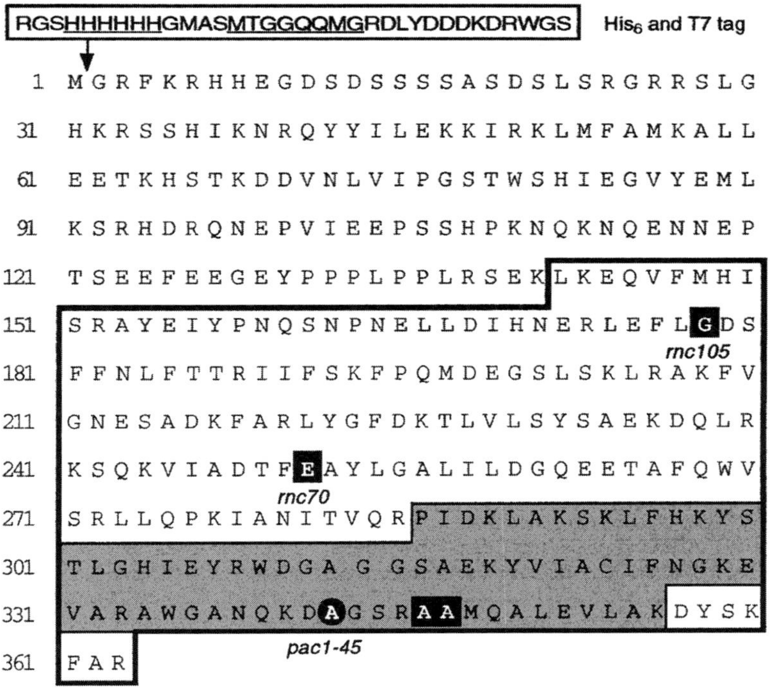

FIG. 1. Sequence of the *S. pombe* Pac1 protein. The 363-amino acid Pac1 protein (numbers at left indicate the amino acids at the start of each row) is similar (25% amino acid identity) to *E. coli* RNase III over its carboxyl-terminal two-thirds (enclosed by a black frame). Gly178 and Glu251 (in black boxes) correspond, respectively, to the positions of the inactivating *rnc105* and *rnc70* mutations in *E. coli* RNase III. The double-stranded RNA-binding domain (dsRBD) is shaded. Mutation of Ala346,347 (black box in dsRBD) to valine destroys Pac1 activity *in vivo*. The *pac1-45* Ala342Thr temperature-sensitive mutant allele is indicated by the black circle in the dsRBD. Shown at the top, the boxed sequence with attached arrow gives the position between the first and second amino acids of the 35-amino acid His_6 and T7 epitope (underlined) tag in the tPac1 protein.

35-amino acid tag between the codons for the initiating methionine and the second amino acid of Pac1 (Fig. 1) by amplification of the Pac1 coding sequence from the second to the last codons by PCR and insertion into the *Bam*HI and *Hin*dIII sites of pRSET-A. The tag includes six contiguous histidines (His_6), as a nickel affinity ligand, and part of the bacteriophage T7 gene 10 major capsid protein, as an epitope tag (T7-tag). Pac1 expression in this construct (pRSETpac) is under the control of an inducible T7 phage promoter.[17]

[17] F. W. Studier, A. H. Rosenberg, J. J. Dunn, and J. W. Dubendorff, *Methods Enzymol.* **185,** 60 (1990).

8b: EXPRESSION OF *tPac1* IN *Escherichia coli*. Protocol 8b is an adaptation of the method of Moore *et al.*[18] for the production of soluble recombinant proteins in *E. coli*.

1. Transform *E. coli* BL21(DE3)pLysS with pRSETpac. Grow a single transformant clone overnight at 30° in tryptone–phosphate broth [2% (w/v) Bacto-tryptone, 0.2% (w/v) Na_2HPO_4, 0.1% (w/v) KH_2PO_4, 0.8% (w/v) NaCl, 1.5% (w/v) yeast extract; 0.2% glucose (w/v)] containing ampicillin (250 μg/ml) and chloramphenicol (63 μg/ml).
2. The next day inoculate 250 ml of the same medium (ampicillin, 100 μg/ml; chloramphenicol, 25 μg/ml) with 3 ml of the overnight culture and grow with shaking aeration at 30° until the culture reaches an optical density at 600 nm of 0.6.
3. Add isopropylthio-β-D-galactoside (IPTG) to a final concentration of 0.4 m*M* and continue shaking for 3 hr.
4. Harvest the culture by centrifugation at 8000*g* for 10 min at 4°. Wash the cell pellet once with buffer A [Tris-HCl (pH 8.0), 0.1 m*M* EDTA, and 1 m*M* $MgCl_2$]; and store frozen at −70°.

8c: PREPARATION OF CLEARED CELL LYSATE

1. Resuspend the frozen cell pellet in 10 ml of nickel affinity chromatography binding buffer (see step 7 in the next section). Add egg white lysozyme to 100 μg/ml and incubate on ice for 15 min.
2. Sonicate with two 10-sec bursts at a medium intensity setting while holding the suspension on ice. Flash freeze the lysate in liquid nitrogen and quickly thaw at 37°. Repeat the freeze–thaw–sonication for three more cycles.
3. Add RNase A to 5 μg/ml and incubate on ice for 15 min.
4. Adjust the $MgCl_2$ concentration to 1 m*M*, add DNase I to 5 μg/ml, and incubate for 30 min at room temperature.
5. Remove insoluble debris by centrifugation at 3000*g* for 15 min at room temperature followed by passage through a 0.8-μm pore size cellulose acetate filter.

8d: NICKEL AFFINITY CHROMATOGRAPHY. These procedures are performed at room temperature.

1. Resuspend 1.8 ml of nickel affinity column resin by gentle inversion. Collect the resin in a 15-ml conical tube by a 2-min centrifugation at 800*g* in a swinging bucket rotor.
2. Wash the resin twice with 7 ml of sterile distilled water.

[18] J. T. Moore, A. Uppal, F. Maley, and G. F. Maley, *Protein Express. Purif.* **4,** 160 (1993).

3. Perform three additional washes with 7 ml of binding buffer (20 m*M* sodium phosphate, pH 7.8, 500 m*M* NaCl).

4. Bind tPac1 by resuspending the equilibrated resin in 5 ml of *E. coli* cell lysate followed by gentle rocking for 1 hr. Recover the resin by centrifugation and aspirate the supernatant. Repeat with a second 5 ml of lysate.

5. Wash the resin three times by resuspension and gentle rocking for 2 min with 4 ml of binding buffer. Collect the resin by centrifugation.

6. Wash the resin with 4 ml of washing buffer (20 m*M* sodium phosphate, pH 6.0, 500 m*M* NaCl). Repeat until the absorbance at 280 nm (A_{280}) of the resin supernatant is less than 0.01.

7. Elute the tPac1 protein by consecutive application of 5 ml of wash buffer containing imidazole at four increasing concentrations: 50, 200, 350, and 500 m*M*. Collect 1-ml fractions. Monitor the elution by A_{280} and Western blotting with an anti-T7-tag monoclonal antibody. In our purification, tPac1 elutes in two peaks at 350 and 500 m*M* imidazole.

8. We pool the tPac1 fractions, concentrate the protein by centrifugal filtration in a Centricon-10 to a volume of 0.5 ml, and then dilute the preparation to 1.5 ml with a storage buffer to a final composition of 500 m*M* NaCl, 20 m*M* sodium phosphate (pH 7.4), 67 m*M* imidazole, 1 m*M* DTT, 1 m*M* EDTA, and 30% (v/v) glycerol.

9. Partition into 0.1 ml aliquots and store at -20°.

Protocol 9: Preparation of RNA Substrates. We synthesize our substrate RNAs by *in vitro* transcription reactions (protocol 6) that incorporate ^{32}P-labeled nucleotides at relatively low specific radioactivity: all four unlabeled nucleoside triphosphates at 0.5 m*M* and the α-^{32}P-labeled nucleoside triphosphate (800 Ci/mmol) at 1 μCi/μl. 5′-End-labeled RNAs are produced in transcription reactions that incorporate [γ-^{32}P]GTP as the only labeled nucleotide. The method described below for the synthesis of generic dsRNA substrates of various lengths is based on the procedure of Manche *et al.*[19]

1. Digest pBS-SK(–) with *Pvu*II to release a fragment containing the opposing T3 and T7 promoters separated by the polylinker region.

2. Perform separate secondary digestions with *Apa*I, *Xho*I, *Eco*RI, *Bam*HI, and *Not*I, and purify the template DNAs by agarose gel electrophoresis.

3. Transcribe full-length T3 and T7 transcripts from the *Pvu*II fragment and T7 transcripts from the smaller templates, and purify the RNAs by DNase I digestion, phenol extraction, and ethanol precipitation.

4. Anneal the complementary ssRNAs by heating in 0.1 ml of 25 m*M* NaCl, 30 m*M* Tris-HCl (pH 7.6) at 90° for 5 min followed by slow cooling to room temperature.

[19] L. Manche, S. R. Green, C. Schmedt, and M. B. Mathews, *Mol. Cell. Biol.* **12,** 5238 (1992).

5. Remove the unpaired ends by digestion with a mixture of RNases A and T1 (protocol 7) and recover the RNA by phenol–chloroform extraction and ethanol precipitation.

6. Dissolve the dsRNA in formamide loading solution (protocol 5) and purify by electrophoresis on a 6% (w/v) polyacrylamide–urea gel (protocol 6). Do not heat-denature the dsRNAs before loading and run the gel at low voltage to prevent heat denaturation. This procedure produces sharper dsRNA bands than the use of native loading solutions and urea-free gels, and it allows heat-denatured single-stranded RNAs (ssRNAs) to be run on the same gel with dsRNAs.

7. Visualize the radioactive RNA by autoradiography, excise the bands from the gel, and elute (protocol 6).

8. Make single-stranded hairpin RNA substrates by transcription from either partially double-stranded synthetic DNA templates according to Milligan and Uhlenbeck[20] or from PCR-generated templates. Heat denature hairpin substrates before gel electrophoresis and then renature them after purification as described above for the annealing of dsRNA substrates.

9. Quantify the radioactive RNAs by liquid scintillation counting of an aliquot. Calculate their molar concentrations on the basis of the specific activity of the labeling nucleotide and the expected number of radioactive phosphates in the RNA.

Protocol 10: Standard Pac1 RNase Reaction Set-Up

1. Assemble a 10-μl reaction on ice in an RNase-free polypropylene microcentrifuge tube as follows.

2(*N*-Cyclohexylamino)ethanesulfonic acid (CHES)–OH (pH 8.5), 0.3 *M*	1 μl
DTT, 10 m*M*	1 μl
Poly(C), 10 μM	1 μl
Pac1 RNase (diluted 1 : 50; final concentration, 8 n*M*)	1 μl
Substrate RNA (final concentration, 40–100 n*M*)	1–5 μl
RNase-free distilled water	to 9 μl

2. Start the reaction by adding 1 μl of 50 m*M* $MgCl_2$.
3. Incubate at 30°.

Notes. For multiple reactions, make a proportional mix that contains the ingredients common to all the reactions, except $MgCl_2$, and deliver 9 μl to each tube. Dilute the Pac1 RNase immediately before use. Stop the reactions according to the procedures given below for the trichloroacetic acid (TCA) solubility and gel assays. Controls lack enzyme or $MgCl_2$ or are stopped immediately after the

[20] J. F. Milligan and O. C. Uhlenbeck, *Methods Enzymol.* **180,** 51 (1989).

addition of $MgCl_2$. As an estimate of initial rate for comparative activity assays, we stop the reactions after a 30-sec incubation.

Protocol 11: Trichloroacetic Acid Solubility Assay. The TCA solubility assay is fast and simple and can be used to rapidly assay multiple samples, but it detects only those products small enough to be soluble in TCA. It is therefore best used to assay degradative dsRNase activity.

1. Perform the Pac1 RNase reaction as in protocol 10.
2. Stop the reaction by adding 0.5 ml of ice-cold 5% (w/v) TCA.
3. Incubate for 15 min on ice.
4. Centrifuge a 0.5-ml aliquot at 16,000*g* for 4 min at room temperature in a Spin-X filter unit (Costar, Cambridge, MA).
5. Quantify the radioactivity of 0.4 ml of the filtrate by liquid scintillation counting in at least 10 volumes of scintillation fluid.

Protocol 12: Gel Electrophoresis Assay. Protocol 12 gives a better picture of the reaction than the TCA solubility assay and is preferred when analyzing site-specific cleavages of natural substrates.

1. Perform the Pac1 RNase reaction as in Protocol 10.
2. Stop the reaction by placing the tube in an ice–water bath and immediately add an equal volume of stop mix: 80% (v/v) formamide, 12 m*M* (ethylenedinitrilo) tetraacetic acid (EDTA), 0.1% (w/v) xylene cyanol, 0.1% (w/v) bromphenol blue.
3. Fractionate 5–10 μl by electrophoresis on a polyacrylamide–7 *M* urea gel. Load double-stranded substrates directly on the gel; denature single-stranded (hairpin) substrates at 85° for 5 min before loading.
4. Visualize the substrates and products by autoradiography at −70° with an intensifying screen. Estimate the sizes of the RNAs by comparison with ^{32}P-labeled DNA size markers. RNA runs slower than DNA of the same length.
5. To quantify substrate conversion or product formation, excise bands from the gel (protocol 6) and determine their radioactivity by liquid scintillation counting in 4 ml of scintillation fluid. Excise three "mock" bands from parts of the gel devoid of exposure and count separately. Take the average as background and subtract it from the values for the RNA bands.

Protocol 13: Identification of Cleavage Products and Mapping of Cleavage Sites. As an illustrative example, protocol 13 outlines the procedures that we use to define the Pac1 cleavages within the hairpin structure of the *S. pombe* pre-rRNA 3′ external transcribed spacer (3′-ETS)[8] (see Results and Fig. 4, later in this article).

1. Prepare a transcription template by PCR from rDNA with primers that incorporate phage promoter sequences and flank the hairpin structure in the 3′-ETS proposed to contain the *in vivo* processing sites.[9]

2. Transcribe an internally ^{32}P-labeled full-length substrate RNA from the template and gel purify.

3. Perform a 0.1-ml Pac1 RNase reaction under standard conditions (protocol 10). Sample the reaction at various times; analyze by gel electrophoresis, and purify the products (protocol 12).

4. With the 3′-ETS substrate (see Results and Fig. 4) we observe three products that appear at the same time, indicating two simultaneous cleavages.

5. To identify the 5′ cleavage product, perform a standard reaction with a 5′-labeled substrate. We observe a single radioactive product that is not converted to any other products and that comigrates with the smallest band produced by the internally labeled RNA. Its size relative to the DNA markers gives an initial estimate of the position of the 5′ cleavage.

6. Identify the 3′ product by primer extension (protocol 7) with the 3′ PCR primer that was used to make the substrate transcription template. (Include 5 μg of *E. coli* tRNA in the primer extension reactions.) When we perform primer extensions on the two unidentified products, only the largest cleavage product gives a primer extension product. The size of the cDNA is identical to the RNA cleavage product and indicates an approximate cut site.

7. By elimination, the middle size 3′-ETS product represents the central cleavage fragment. We confirm this assignment by primer extension with a primer complementary to sequences predicted to be near the 3′ end of the central fragment.

8. Map the cleavage sites by coelectrophoresis of the primer extension products from the 3′ and central cleavage fragments alongside DNA sequencing ladders of the 3′-ETS PCR template generated with the same 5′-end labeled primers used for the primer extensions.

Protocol 14: Determination of the Cleavage Chemistry. Cleavage chemistry refers to the position of phosphodiester bond cleavage and, therefore, the type of ends, 5′- or 3′-phosphate, left on the product RNAs. End group analysis also contributes to cleavage site mapping. This protocol outlines the methods we have used to show that Pac1 RNase cleavage produces 5′-phosphate and 3′-hydroxyl termini.[8,11]

1. On the basis of the primer extension mapping of the cleavage sites, label the substrate RNA with an [α-^{32}P]NTP that will produce labeled terminal nucleotides. For example, cleavage of the 3′-ETS RNA (see Fig. 4) can be represented as follows:

5′-GG...U//UC...AG//AG...UU-3′ (// indicates cleavage sites).

Labeling the substrate with [α-^{32}P]CTP or [α-^{32}P]UTP will label the 5′ terminal U of the middle fragment, while a substrate labeled with [α-^{32}P]GTP or [α-^{32}P]ATP will produce a labeled 5′-terminal A for the 3′ product.

2. Perform a 0.5-ml Pac1 RNase reaction under standard conditions (protocol 10) for 30 min and concentrate the reaction by ethanol precipitation.

3. Gel purify the products and coprecipitate each with 10 μg of *E. coli* tRNA. Dissolve in 10 μl of 10 m*M* Tris-HCl (pH 7.5), 1 m*M* EDTA.

4. To 4 μl of the products, add 1 μl of a mixture containing 5 units of RNase T1, 0.25 μg of RNase A, and 0.5 units of RNase T2 and digest at 37° for 2 hr. Add 1 μl of a mixture of the four 3′-monophosphate nucleotides (Np) and adenosine 3′,5′-bisphosphate (pAp) (~25 mg/ml each).

5. Spot all or a portion of the mixture 2 cm from the lower left corner of a plastic-backed cellulose thin-layer plate, 1 μl at a time, drying in between with a stream of air.

6. Separate the digestion products and markers by two-dimensional thin-layer chromatography[21] (2D-TLC): first dimension, isobutyric acid–0.5 *M* ammonium hydroxide (5 : 3, v/v), second dimension, 2-propanol–concentrated HCl–water (70 : 15 : 15, v/v/v). Perform the 2D-TLC in a sealed glass tank until the solvent fronts are about 1 cm from the top. Dry the plate thoroughly in a fume hood after each dimension.

7. Visualize the unlabeled nucleotide markers by ultraviolet absorbance (Gp is fluorescent) and mark their positions on the plate with a pencil. (*Note:* 3′,5′-bisphosphate nucleotides are no longer commercially available, but nucleotide spots can usually be unambiguously identified by reference to the excellent maps provided by Nishimura.[21])

8. Tape a piece of X-ray film to the TLC plate and make alignment marks (protocol 6). Visualize the labeled products by autoradiography with an intensifying screen. Align the autoradiogram with the chromatogram to identify the radioactive spots. The middle Pac1 RNase cleavage product will release a radioactive pUp from a U- or C-labeled substrate; the 3′ product will release a radioactive pAp from a G- or A-labeled substrate (see step 1 and Fig. 4).

9. Confirm the identity of the 5′-terminal pNp nucleotides by secondary analysis as follows. Align the autoradiogram with the TLC plate, place it on a light box, and mark the positions of the radioactive spots on the plastic backing. Stuff a wad of siliconized glass wool in an RNase-free P1000 tip and insert the pointed end of the tip in a hose attached to a vacuum source. Use the wide end of the tip to scrape the radioactive spot off the plate, at the same time "vacuuming" the pieces into the pipette tip. Elute the radioactive nucleotide by passing 200 μl of triethylammonium bicarbonate (TEAB, pH 9.5) through the tip into a centrifuge tube. Recycle the eluate back through the tip and repeat with a second 200 μl of fresh TEAB. Dry completely in a centrifugal evaporator; resuspend in 400 μl of water and dry down again. If TEAB odor persists, repeat the water wash.

[21] S. Nishimura, *in* "Transfer RNA: Structure, Properties, and Recognition" (P. Schimmel, D. Söll, and J. Abelson, eds.), p. 551. Academic Press, New York, 1979.

10. Dissolve the eluted nucleotides in 6 μl of 20 mM sodium acetate (pH 5.4). Transfer 3 μl to a fresh tube and add 0.5 μl of *E. coli* tRNA (20 mg/ml) and 0.5 μl of nuclease P1 (5 mg/ml in 20 mM sodium acetate, pH 5.4). Digest for 2 hr at 37°. *Note:* Nuclease P1 removes the 3′-phosphate from pNp.

11. Spot the P1 digest and one-half of the undigested sample on a cellulose TLC plate 2 cm from the bottom. Apply unlabeled 5′-monophosphate (pN) markers on adjacent spots. Separate in the second-dimension solvent (step 6).

12. Visualize the markers by UV absorbance and the radioactive products by autoradiography. A U-labeled middle fragment or an A-labeled 3′ fragment (step 1, Fig. 4) should release radioactive spots that migrate slower than the undigested pUp and pAp and comigrate with the pU and pA markers. A C-labeled middle fragment and a G-labeled 3′ fragment will release radioactive orthophosphate, which runs with the solvent front; the unlabeled pU and pA are not detected.

13. As a final confirmation, spot the remaining half of the undigested eluted pUp and pAp nucleotides on a polyethyleneimine (PEI) cellulose TLC plate. (Prepare the PEI plate by washing briefly in methanol; dry completely in a fume hood; mock chromatograph in water to the top of the plate, and dry again.) Apply 5′-diphosphate nucleotides (ppN) as markers on adjacent spots. Wash the spots by squirting water onto the plate from above the spots downward. Dry the plate and chromatograph in 1.75 M ammonium formate.[22] Wash the chromatogram twice for 10 min in 75% (v/v) ethanol and dry. Visualize the markers by UV absorbance and the radioactive products by autoradiography. The presumptive pUp spot from the middle fragment (step 1) should comigrate with the ppU marker and the presumptive pAp spot of the 3′ fragment should comigrate with the ppA marker.

Results

Structure–Function Analysis of pac1$^+$ Gene

The Pac1 protein sequence can be divided into a unique amino-terminal one-third, which is rich in polar amino acids, and a carboxyl-terminal two-thirds, which shares 25% amino acid identity with *E. coli* RNase III (boxed sequence in Fig. 1). The RNase III similarity domain can itself by subdivided into three blocks of sequence conservation[11]: two blocks surround the sites of *rnc105* and *rnc70* inactivating mutations in RNase III and the third block consists of the double-stranded RNA-binding domain (dsRBD, shaded in Fig. 1). The amino acids at the *rnc105* and *rnc70* sites (black boxes in Fig. 1) are absolutely conserved in all members of the RNase III family [see [13] in this volume[1]). Mutation of the corresponding residues in Pac1 (Gly178 to serine or aspartate and Glu251 to lysine) abolishes function *in vivo*[5] (protocols 1–3). Deletion of all or part of the dsRBD

[22] K. Randerath, R. C. Gupta, and E. Randerath, *Methods Enzymol.* **65,** 638 (1980).

also destroys *pac1*$^+$ activity.[5] The same is true for a double point mutation of two highly conserved alanines near the end of the dsRBD (Ala346,347 to valine, black boxes in Fig. 1). The larger valine side chains might disrupt the tight hydrophobic packing in this part of the dsRBD.[23,24] In the same vicinity, the *pac1-45* mutation (black circle, Fig. 1) changes a highly conserved alanine (Ala242) to threonine.[7] This mutation permits some activity; it is codominant with the wild-type allele. An alanine-to-valine change is found at the same position in the *E. coli* Rev3 mutant, which is an extragenic suppressor of a mutation in ribosomal protein S12.[25]

Role of Pac1 RNase in RNA Processing in Vivo

Strains carrying the *pac1-45* allele maintain a reduced steady state snRNA content and accumulate unprocessed, 3′-extended snRNA and rRNA precursors at the restrictive temperature.[6,7] The impairment in the synthesis of the RNase P RNA subunit in *pac1-45* mutants results in a secondary reduction in the efficiency of pre-tRNA 5′ processing. Likewise, the lowered spliceosomal snRNA content slightly reduces pre-mRNA splicing efficiency. The combined defects in RNA synthesis attenuate the growth potential of the *pac1-45* mutant. When shifted from 23 to 37°, a wild-type strain will approximately double its growth rate and will increase its steady state snRNA content, while *pac1-45* mutants continue to grow at a slow pace and fail to adjust their snRNA synthesis to the level required for rapid growth.[6,7]

General Properties of Pac1 RNase

To facilitate purification and detection of the Pac1 protein, we attached His_6 and epitope tags to the amino terminus. This tagged form of Pac1 (tPac1), when expressed in *S. pombe* on a multicopy plasmid under the control of the native *pac1*$^+$ promoter, retained full biological activity, and a protein of the expected size was detected by Western blotting with either an anti-T7 tag monoclonal antibody or an anti-Pac1 rabbit serum. To purify the tPac1 protein after expression in *E. coli* we used a method designed to promote the production of soluble protein.[18] With this method (protocol 8) approximately 50% of the tPac1 was expressed in a soluble form after a 3-hr induction (Fig. 2A, lane 4). At later times after induction the protein became insoluble and suffered extensive proteolytic breakdown. We purified the tPac1 by a single Ni-affinity chromatography step that produced essentially pure tPac1 (Fig. 2A, lane 5). The preparation contained a small proportion of apparent tPac1 breakdown products that reacted with the anti-Pac1 serum but

[23] A. Kharrat, M. J. Macias, T. J. Gibson, M. Nilges, and A. Pastore, *EMBO J.* **14,** 3572 (1995).
[24] M. Bycroft, S. Grünert, A. G. Murzin, M. Proctor, and D. St. Johnston, *EMBO J.* **14,** 3563 (1995).
[25] H. Nashimoto and H. Uchida, *Mol. Gen. Genet.* **201,** 25 (1985).

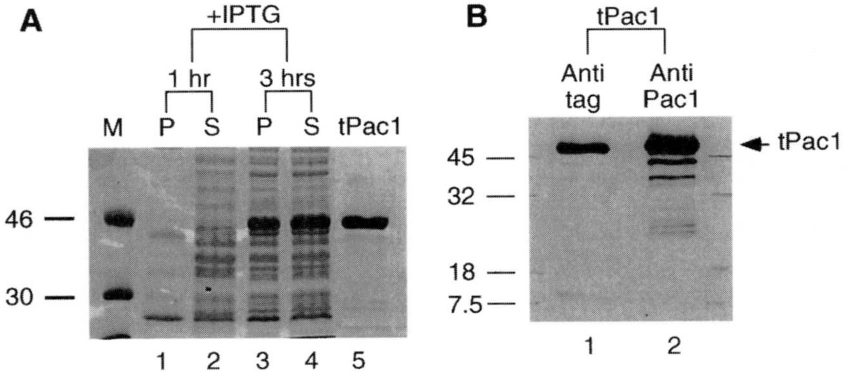

FIG. 2. Purification of His$_6$- and T7-epitope-tagged Pac1 RNase (tPac1). (A) Expression of tPac1 in *E. coli* BL21(DE3)pLysS was induced by IPTG for 1 hr (lanes 1 and 2) and 3 hr (lanes 3 and 4). Cell lysates were separated into insoluble pellet (P) and soluble supernatant (S) fractions. The tPac1 in the soluble fraction from the 3-hr IPTG induction was purified by Ni-affinity chromatography. Samples of the crude and purified protein were analyzed by SDS–polyacrylamide gel electrophoresis and visualized by Coomassie blue staining. Protein size standards were run in lane M; sizes (in kDa) are given on the left-hand side. (B) Purified tPac1 was analyzed by Western blotting and immunodetection with an anti-T7-epitope monoclonal antibody (lane 1) and with an anti-Pac1 rabbit serum (lane 2). The positions of protein size standards and their molecular masses (in kDa) are given on the left-hand side. The position of full-length tPac1 is indicated by the arrow on the right-hand side. Reprinted from G. Rotondo and D. Frendewey, *Nucleic Acids Res.* **24,** 2377 (1996) with permission of Oxford Univ. Press.

not with the anti-T7 tag antibody (Fig. 2B). The purified tPac1 protein was free of contaminating RNase III as judged by Western blotting with an anti-RNase III serum.[11]

The concentration of our tPac1 preparation was 180 μg/ml, which is equivalent to 4 μM on the basis of a theoretical molecular mass of 45.5 kDa. The enzyme has been stored at this concentration at -20° with negligible loss of activity for at least 5 years, but it loses activity on further dilution. By the unit definition used for *E. coli* RNase III[26]—the amount of enzyme that will solubilize 1 nmol of acid-precipitable polynucleotide phosphorus per hour—we estimate the specific activity of our Pac1 RNase preparation to be 5×10^5 units/mg protein, comparable to the most active RNase III preparation.[27] Expressed as micromole of substrate transformed per minute, the unit definition preferred by the International Union of Biochemistry, the tPac1-specific activity is approximately 0.2 unit/mg.

[26] J. J. Dunn, *in* "The Enzymes" (P. D. Boyer, ed.), Vol. 15, Part B, p. 485. Academic Press, New York, 1982.

[27] H.-L. Li, B. S. Chelladurai, K. Zhang, and A. W. Nicholson, *Nucleic Acids Res.* **21,** 1919 (1993).

Optimization of Pac1 RNase Assay Conditions

The Pac1 RNase exhibits poor activity under conditions established for *E. coli* RNase III.[27] Only 30% of a dsRNA substrate was rendered acid soluble (protocol 11) even when Pac1 was ~80-fold in excess of the substrate. Under RNase III conditions we detected no dsRNase activity at Pac1 concentrations less than 40 n*M*. After systematically altering the reaction components and conditions to achieve maximal Pac1 RNase activity, we arrived at the following standard reaction conditions: 30 m*M* CHES–OH (pH 8.5), 1 m*M* DTT, 5 m*M* $MgCl_2$, 1 μM poly(C). We routinely assay the Pac1 RNase at 30°; its activity is essentially identical at 37° and about 1.5 times lower at 20°. Under optimal conditions, dsRNase activity could be detected at Pac1 concentrations as low as 1 n*M*, and complete cleavage of a 20-fold molar excess of substrate was achieved in 10 min.

Ionic Requirements and Inhibitors

Monovalent cations inhibit the Pac1 RNase. Sodium chloride produces half-maximal inhibition at 75 m*M*, and similar results are obtained with KCl and NH_4Cl and with the glutamate salts of potassium and sodium. The Pac1 RNase requires magnesium for optimal activity; Ca^{2+}, Zn^{2+}, Ni^{2+}, and Co^{2+} cannot substitute for Mg^{2+}. Pac1 is active with Mn^{2+} as the divalent cation, but its specific activity is reduced and cleavage is more promiscuous with substrates for which the enzyme exhibits a cleavage site preference.[11] The requirement for a divalent cation is not relieved by polycations such as spermidine. The Pac1 RNase exhibits a broad activity optimum between pH 8 and 9 in several buffers. The enzyme is active at pH 7.5 in Tris-buffered reactions, but the activity is about 20% higher with Ches–OH buffer at pH 8.5. Pac1 has a single cysteine located in the dsRBD. Although the enzyme is not sensitive to *N*-ethylmaleimide (~60% inhibition at 10 m*M*), Pac1 is maximally active in 1 m*M* DTT. Vanadyl ribonucleoside complexes[28] are potent inhibitors of the Pac1 RNase (half-maximal inhibition at 2 μM). As expected for a dsRNase, Pac1 is also inhibited by ethidium bromide (half-maximal inhibition at 3 μM) and double-stranded ribopolynucleotides, but we also found that the enzyme was sensitive to single-stranded homopolymers of inosine, adenosine, and guanosine. Conversely, poly(C) stimulated Pac1 RNase activity at least 8-fold when present at 10 times molar excess over substrate.[11]

Substrate Preference and Cleavage Properties

The Pac1 RNase exhibits a general degradative activity with simple generic dsRNA substrates. At limiting enzyme concentration, Pac1 completely converts a 100-bp dsRNA into products of 10 to 40 nt in size (Fig. 3A, T3/T7 lanes, left).

[28] G. E. Lienhard, I. I. Secemski, K. A. Koehler, and R. N. Lindquist, *Cold Spring Harb. Symp. Quant. Biol.* **36,** 45 (1972).

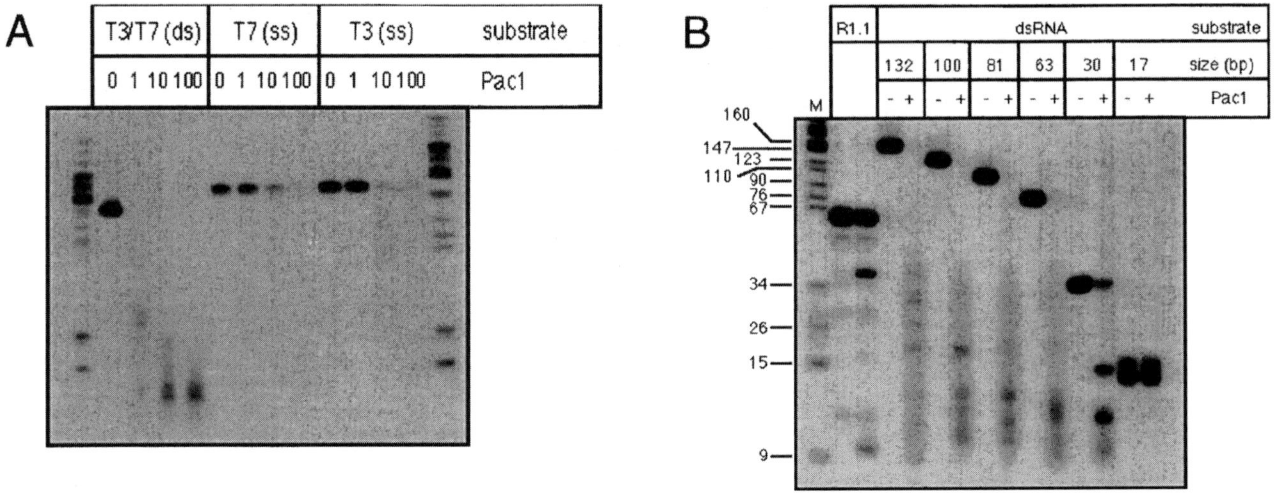

FIG. 3. Substrate cleavage properties of the Pac1 RNase. (A) Pac1 prefers dsRNA over ssRNA substrates. A ^{32}P-labeled 100-bp dsRNA [lanes labeled T3/T7 (ds)] or either of the single strands [lanes labeled T7 (ss) and T3 (ss)] (~6 n*M*) were incubated with Pac1 RNase at increasing concentrations (the numbers above the lanes indicate the relative Pac1 concentration; 1 = 2 n*M*) for 10 min at 30° in 30 m*M* Tris-HCl (pH 7.5), 25 m*M* KCl, 5 m*M* $MgCl_2$, 1 m*M* DTT. Samples of the reactions were analyzed by electrophoresis on a 10% (w/v) polyacrylamide–7 *M* urea gel followed by autoradiography. ^{32}P-labeled dsDNA size markers were run in the first and last lanes. The dsRNA reactions and the adjacent DNA markers were run without denaturing (native), while the ssRNA reactions and the adjacent DNA markers were heat-denatured before loading. Reprinted from G. Rotondo and D. Frendewey, *Nucleic Acids Res.* **24,** 2377 (1996) with permission of Oxford Univ. Press. (B) Determination of the minimal length helix for a Pac1 dsRNA substrate. ^{32}P-labeled dsRNAs of varying length (indicated in base pairs above the lanes) were incubated with (+) or without (–) Pac1 RNase at 2 n*M* and analyzed as described for (A). In each reaction, 15,000 cpm of RNA was used; therefore, concentrations varied from 45 p*M* for the longest dsRNA to 250 p*M* for the shortest. Reactions with the R1.1 hairpin RNA (27 n*M*; see Fig. 4) are shown in the second (minus Pac1) and third (plus Pac1) lanes. ^{32}P-labeled dsDNA size markers were run in the first lanes. Their sizes are given on the left. Reprinted from G. Rotondo, J. Y. Huang, and D. Frendewey, *RNA* **3,** 1182 (1997).

A complete digestion with excess enzyme further degrades the larger oligonucleotides into stable 10- to 20-nt products. The Pac1 RNase exhibits a stringent preference for dsRNA substrates; it cleaves the individual strands of the 100-bp dsRNA only when 3- to 30-fold in excess of RNA (Fig. 3A, T7 and T3 lanes, middle and right). The Pac1 RNase has a minimal helix length requirement: it efficiently degrades dsRNAs of 63–132 bp to 10- to 30-nt products but makes only one double-strand break near the middle of a 30-bp substrate (Fig. 3B). A 17-bp dsRNA is refractory to Pac1 cleavage, despite an 8-fold molar excess of enzyme over RNA. These results indicated a minimal helix requirement between 17 and 30 bp, a limit that we refined further with hairpin RNAs.

The R1.1 RNA, derived from an RNase III-processing site in the phage T7 polycistronic transcript,[29] is a 60-nt hairpin that consists of two 10-bp helices separated by an asymmetric bulge (Fig. 4). RNase III makes a single cut within the bulge of the R1.1 RNA.[29] Remarkably, the Pac1 RNase cleaves R1.1 at the same site. Thus, the *S. pombe* enzyme has conserved the ability to recognize the structural signature in R1.1 that directs a single-stranded cleavage at a defined site. When the bulge in R1.1 is closed by Watson–Crick pairs to create a 25-bp helix (R1.1[WC-L] RNA; Fig. 4), RNase III makes a double-strand cut, with one of the strands cleaved at the same site as in R1.1.[29] Pac1 also makes a double-strand break in the R1.1[WC-L] RNA, but its preferred cleavage sites are shifted a few base pairs toward the terminal end of the helix.

The Pac1 cleavages on the two strands of the R1.1[WC-L] helix are offset by two base pairs, a property characteristic of RNase III. Also, like RNase III, the Pac1 RNase breaks the phosphodiester bond to leave 5′-phosphate and 3′-hydroxyl termini on the products.[11] Pac1 cleaves helices that contain non-Watson–Crick base pairs, but it does not tolerate asymmetic bulges (the exception being R1.1 RNA). For example, the HIV-1 TAR RNA is completely resistant to Pac1 cleavage.[8] The TAR element is a hairpin similar in size to R1.1[WC-L] but containing three asymmetric bulges that distort the helix from linearity. Pac1 cleaves a TAR RNA derivative with the bulges removed, but at a slow rate.[8]

Cleavage of Natural Substrates

The *S. pombe* pre-rRNA 3′-ETS sequence can be modeled as a long hairpin that contains three *in vivo* processing sites[9] (indicated as A, B, and C in Fig. 4). *In vitro,* the Pac1 RNase makes a two-base pair staggered cut in a 3′-ETS RNA substrate at the B processing site and one nucleotide away from the C site (arrows in Fig. 4). Accurate Pac1 cleavage of the 3′-ETS *in vitro* and correct processing *in vivo* require maintenance of the helical structure at the cleavage sites and in the upper stem region.[30] The close agreement between *in vitro* cleavage and

[29] B. Chelladurai, H. Li, K. Zhang, and A. W. Nicholson, *Biochemistry* **32,** 7549 (1993).
[30] E. Ivakine, D. Frendewey, and R. Nazar, unpublished (1999).

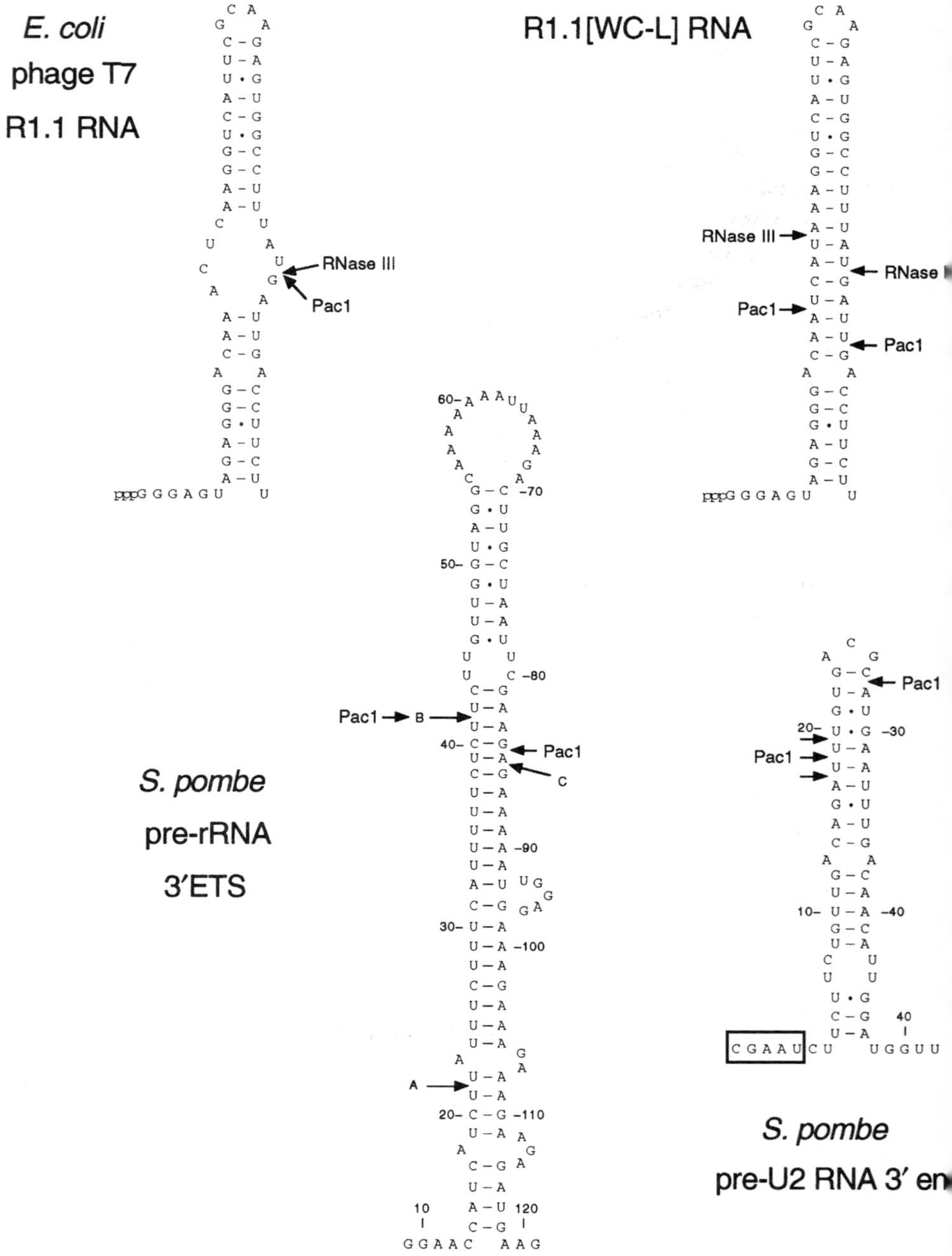
E. coli
phage T7
R1.1 RNA
RNase III
Pac1
R1.1[WC-L] RNA
RNase III
RNase
Pac1
Pac1
S. pombe
pre-rRNA
3′ETS
Pac1
B
Pac1
C
A
S. pombe
pre-U2 RNA 3′ en
Pac1
Pac1

the mapped processing sites strongly argues that Pac1 is directly responsible for 3′-ETS processing *in vivo,* at least at the B and C sites. Pac1 did not cleave at the A site *in vitro.* Either Pac1 is not the enzyme responsible for A site processing, or Pac1 requires an auxiliary factor, such as a small nucleolar ribonucleoprotein, to recognize the A site.

The sequences that flank the 3′ ends of the *S. pombe* spliceosomal snRNAs can be modeled as short hairpin structures. The proposed 3′-flanking hairpin for the pre-U2 RNA consists of a 21-bp helix containing a number of non-Watson–Crick base pairs and a three-nucleotide apical loop[7] (Fig. 4). This structure is supported by compensatory base pair mutations.[7] The Pac1 RNase cleaves a synthetic pre-U2 RNA substrate *in vitro* at four sites within the hairpin (Fig. 4). These cleavages are unusual in that they do not exhibit the characteristic two-base pair stagger we observed with other Pac1 cleavages, and one of the cuts is near the apical loop. Although the *in vitro* cleavages of the pre-U2 RNA have yet to be confirmed as *in vivo* processing sites, the ability of Pac1 to rescue pre-U2 RNA processing in an extract from a *pac1* mutant strain[7] is convincing evidence that the Pac1 RNase participates in the early events of pre-snRNA processing *in vivo.*

Concluding Remarks

The Pac1 RNase and its *Saccharomyces cerevisiae* ortholog Rnt1p share structural features, biochemical properties, and biological functions (see [14] in this volume[9a]). These yeast enzymes are the only well-characterized eukaryotic members of the RNase III family, but a search of the DNA databases suggests that RNase III-like proteins are ubiquitous in the higher eukaryotes. In fact, reports of RNase III-like activities from eukaryotic sources appeared more than 20 years ago.[31] It may be of evolutionary interest that obvious RNase III-like sequences are not represented in the sequenced Archaea genomes. Potential open reading frames for an unusual RNase III-like protein, whose structure is different from the Pac1 and Rnt1 proteins, appear to be present in *S. pombe, Arabidopsis thaliana, Caenorhabditis elegans,* and human genomic DNA.[11,32] The sequences

FIG. 4. Pac1 RNase hairpin substrates. The proposed secondary structures for four hairpin RNAs that are efficient substrates for the Pac1 RNase *in vitro.* The R1.1 RNA (upper left) is a 60-nt RNA derived from an *E. coli* RNase III processing site in a phage T7 polycistronic mRNA.[26,29] The R1.1[WC-L] RNA (*top right*) is a modified form of R1.1 in which the central bulge is closed with Watson–Crick base pairs.[29] The large hairpin at bottom left is derived from the *S. pombe* pre-rRNA 3′ETS.[9] The small hairpin at bottom right is part of the *S. pombe* pre-U2 snRNA 3′ transcribed extension.[7] The numbering of the *S. pombe* RNAs indicates the distance beyond the ends of the mature rRNA and U2 RNA (the last 5 nucleotides of the mature U2 RNA are boxed, lower right). Arrows labeled A, B, and C on the 3′ETS RNA refer to *in vivo* processing sites.[9] The positions of RNase III or Pac1 RNase cleavages *in vitro* are given by arrows.

of these putative proteins predict enzymes having both dsRNase and RNA helicase activities. Their RNA helicase similarities are contained in large amino-terminal regions and all share the sequence DECH, rather than the DEAD or DEAH elements commonly found in RNA helicases. The RNase III similarity lies at the carboxyl terminus in these helicase–RNases, but it differs from those found in bacterial RNase III, Pac 1, and Rnt1. The helicase–RNase proteins have a duplication of part of the RNaseIII-like sequence, and all contain large insertions between the blocks of homologous sequence surrounding the *rnc105* and *rnc70* sites[11] (Fig. 1). The *A. thaliana* protein also has a second dsRBD at its carboxyl terminus.[32] None of these putative proteins has been biochemcially characterized, so it remains to be proved that they have both dsRNase and RNA helicase activities. But disruption of the gene (known as *CAF*) for the *A. thaliana* protein causes defects in flower development.[32] The function of the *S. pombe* member of this family is not known, but it cannot compensate for a deletion of the *pac1*$^+$ gene, which is lethal.[2] This novel class of helicase–RNase proteins deserves further investigation.

Another intriguing phenomenon that warrants more detailed analysis is Pac1-induced sterility. Pac1 overexpression has been proposed to cause sterility by preventing the accumulation of the *mei2*$^+$ mRNA,[2,33] but evidence of the direct action of the Pac1 RNase in the destruction of RNAs required for sexual development has yet to be reported. Pac1 might not play a role in normal sexual differentiation; we have not observed any problems with mating and sporulation in *pac1* mutant strains. Alternatively, Pac1-induced sterility could be related to the function of the putative *S. pombe* helicase–RNase protein. If this protein participated in the control of sexual development, then its function might be abrogated by ectopic expression of the related Pac1 RNase. The requirement of the Caf protein for sexual development (flowering) in *A. thaliana,*[32] and the analogous role of RNase III in the control of conjugation in *E. coli,*[34] suggest the participation of dsRNases in sexual development from bacteria to higher eukaryotes.

The developments in the phenomenon of dsRNA-targeted gene inactivation, or RNAi,[35] have fueled speculation as to the role of dsRNase activities in this process. It would be a great surprise if an RNase III-like enzyme were not involved in the RNAi mechanism. The putative helicase–RNase proteins offer particularly attractive candidates, as both RNA unwinding and double-strand cleavage are central features of the current hypotheses of the RNAi mechanism. One of the biological functions proposed for RNAi is antiviral defense, especially in plants. In this regard, it is interesting that transgenic expression of the *S. pombe* Pac1 RNase

[31] K. Ohtsuki, Y. Groner, and J. Hurwitz, *J. Biol. Chem.* **252,** 483 (1977).

[32] S. E. Jacobsen, M. P. Running, and E. M. Meyerowitz, *Development* **126,** 5231 (1999).

[33] Y. Watanabe, Y. Iino, K. Furuhata, C. Shimoda, and M. Yamamoto, *EMBO J.* **7,** 761 (1988).

[34] G. Koraimann, C. Schroller, H. Graus, D. Angerer, K. Teferle, and G. Högenauer, *Mol. Microbiol.* **9,** 717 (1993).

[35] A. Fire, *Trends Genet.* **15,** 358 (1999).

in plants affords some protection against viral infection.[36] It might be feasible to engineer RNase III-like enzymes with high affinity for viral dsRNA elements. Such enzymes could have therapeutic potential in both plants and animals. Future research is bound to reveal novel uses and new functions for this ancient class of enzyme.

Acknowledgments

Our work on the *S. pombe* Pac1 RNase was initiated by R. Li, who isolated our first collection of mutants, which included *ts*45 carrying the *pac1-45* allele. The work of J. Potashkin on the description of the *snm1* mutant was instrumental in defining the role for the Pac1 RNase in snRNA synthesis. Over the years we have enjoyed the invaluable assistance of S. Serrano, M. Gillespie, R. Massey, D. Rush, and J. Niemniec. We are grateful to numerous colleagues who shared materials crucial to the success of our research on Pac1. Their contributions are acknowledged in the Materials section. We thank P. Cowin for patient assistance in the preparation of the manuscript. This work was supported by NIH Grant GM38242 to D.F. and by the Long Island Biological Association, the American Cancer Society, New York University Medical Center, the Center for AIDS Research at NYU, a Whitehead Fellowship for Junior Faculty (D.F.), and an Irma T. Hirschl Career Scientist Award (D.F.).

[36] Y. Watanabe, T. Ogawa, H. Takahashi, I. Ishida, Y. Takeuchi, M. Yamamoto, and Y. Okada, *FEBS Lett.* **372,** 165 (1995).

[16] *Dictyostelium* Double-Stranded Ribonuclease

By JINDRICH NOVOTNY, SONJA DIEGEL,
HEIKE SCHIRMACHER, AXEL MÖHRLE, MARTIN HILDEBRANDT,
JÜRGEN OBERSTRASS, and WOLFGANG NELLEN

Introduction

Double-strand-specific ribonucleases (dsRNases) have been shown to play an important role in the processing of ribosomal RNAs and they are implicated in the degradation of dsRNA resulting from the interaction of antisense transcripts with mRNA targets.[1–4] In bacteria, RNase III is the double-stranded endoribonuclease responsible for the degradation of some antisense–sense RNA duplexes.[5] The yeast

[1] R. A. Young and J. A. Steitz, *Proc. Natl. Acad. Sci. U.S.A.* **75,** 3593 (1978).
[2] J. J. Dunn and F. W. Studier, *Proc. Natl. Acad. Sci. U.S.A.* **70,** 3296 (1973).
[3] P. Blomberg, E. G. H. Wagner, and K. Nordström, *EMBO J.* **9,** 2331 (1990).
[4] W. Nellen and C. P. Lichtenstein, *Trends Biochem. Sci.* **18,** 419 (1993).
[5] K. Gerdes, A. Nielsen, P. Thorsted, and E. G. Wagner, *J. Mol. Biol.* **226,** 637 (1992).

0076-6879/01 $35.00

genes *RNT1* from *Saccharomyces cerevisiae* and *PAC1* from *Schizosaccharomyces pombe,* which encode proteins with homology to *Escherichia coli* RNase III, have been cloned.[6–8] Both enzymes show dsRNase activity and play a role in ribosomal RNA processing.[9] In higher eukaryotes, RNase III homologous sequences have been identified for *Arabidopsis thaliana, Caenorhabiditis elegans,* and others. They are part of large (approximately 190-kDa) peptides that also contain a DExH/DEAD-box type RNA helicase domain and whose exact function is so far unknown.[10] dsRNase activities have been documented in several organisms.[11–14]

In *Dictyostelium discoideum,* endogenous antisense RNA as well as antisense transgenes regulate gene expression by reducing mRNA stability.[15,16] When both orientations of a single locus, or sense and antisense transcripts encoded in different loci, are transcribed, neither mRNA nor antisense RNA is accumulated. Similarly, when a sense transcript is allowed to accumulate in the cytosol and antisense transcription is then induced, the preexisting RNA is rapidly lost. Interestingly, this posttranscriptional gene-silencing mechanism appears to be regulated: under certain experimental conditions, in late developmental stages of *Dictyostelium* and in at least one specific mutant, antisense-mediated RNA degradation does not occur.

As in most other organisms, sense–antisense hybrids cannot be detected by conventional hybridization experiments in *Dictyostelium.* A dsRNase may therefore cause the degradation of the postulated double-stranded RNA and might also be a target for the regulation of antisense mechanisms. By using an assay to identify nucleases that specifically hydrolyze double-stranded RNA, a predominantly cytosolic dsRNase was identified in *Dictyostelium* extracts. The enzyme had no effect on single-stranded RNA or DNA but had some activity similar to RNase H. *Dictyostelium* dsRNase (DdsRNase) is nonsequence specific and digests homopolymer duplexes as well as random sequence dsRNA targets generated from various *Dictyostelium* gene sequences. dsRNA digestion depends on divalent metal ions, with a vast preference for Mg^{2+}. The minimum target size appears to be 25 bp and the digestion products are predominantly distinct 25-bp or 25-nucleotide (nt) fragments. The target size of the dsRNase provides an explanation of why cellular

[6] S. Abou Alela, H. Igel, and M. Ares, Jr., *Cell* **85,** 115 (1996).
[7] H. P. Xu, M. Riggs, L. Rodgers, and M. Wigler, *Nucleic Acids Res.* **18,** 5304 (1990).
[8] Y. Iino, A. Sugimoto, and M. Yamamoto, *EMBO J.* **10,** 221 (1991).
[9] G. Rotondo, J. Y. Huang, and D. Frendewey, *RNA* **3,** 1182 (1997).
[10] S. E. Jacobsen, M. P. Running, and E. M. Meyerowitz, *Development* **126,** 5231 (1999).
[11] H. Wu, A. R. MacLeod, W. F. Lima, and S. T. Crooke, *J. Biol. Chem.* **273,** 2532 (1998).
[12] J. M. Meegan and P. I. Marcus, *Science* **244,** 1089 (1989).
[13] J. Rech, G. Cathala, and P. Jeanteur, *J. Biol. Chem.* **255,** 6700 (1980).
[14] J. Rech, G. Cathala, and P. Jeanteur, *Nucleic Acids Res.* **3,** 2055 (1976).
[15] M. Hildebrandt and W. Nellen, *Cell* **69,** 197 (1992).
[16] T. E. Crowley, W. Nellen, R. H. Gomer, and R. A. Firtel, *Cell* **43,** 633 (1985).

RNAs with complex secondary structures (which usually contain fewer than 25 consecutive paired bases) are not digested. DdsRNase has been purified and elutes from a gel-filtration column as a large, approximately 450-kDa multicomponent complex. The enzyme is thus distinct from the RNase III family and other previously described dsRNase activities by structure and biochemical activity. In this article we describe a purification for DdsRNase and present a partial biochemical characterization of the enzyme.

Materials and Methods

Generation of Substrates

Substrates for testing dsRNase activity are generated either by *in vitro* transcription of GEM vectors (Promega, Madison, WI), using SP6 and T7 RNA polymerase, or by using synthetic polynucleotides. Specifically, the following templates are used: the 259-bp *Pvu*ll–*Dra*l gene fragment EB-4-11 of the *Dictyostelium* PSV-A (EB-4) gene 1 is used as a standard blunt-ended double-stranded RNA substrate.[17] For the long strand of the loop substrate, the 703-bp *Pvu*ll fragment EB-4-1 is used, and for the short strand, the internal 61-bp *Dra*I fragment is deleted.[17] Polymerase chain reaction (PCR) products from these constructs are also employed for generating *in vitro* transcripts. dsRNA substrates made from other *Dictyostelium* genes (e.g., actin) are also used (data not shown).

For the R1.1 substrate, which contains the processing signal for *E. coli* RNase III upstream of the 1.1 gene of phage T7, a PCR fragment that allows T7 *in vitro* transcription of the 126-nt target RNA is synthesized.[18,19] Primers T7 (5′-TAATACGACTCACTATAG-3′) and R1.1 reverse (5′-TATTAACCGGAAG-AAGGTC-3′) are used for PCR amplification the template pET-4, containing the T7 promoter–R1.1 fusion (kindly provided by C. Conrad and G. Klug, University of Giessen, Giessen, Germany).

In vitro transcriptions are performed in a 50-μl volume, containing T7 buffer [40 m*M* Tris-HCl (pH 7.9), 6 m*M* $MgCl_2$, 2 m*M* spermidine, 10 m*M* dithiothreitol (DTT)], 1 μg of the template, 60 U of RNasin (MBI-Fermentas, St. Leon-Rot, Germany), rNTPs (0.5 m*M* each), and 60 U of T7 RNA polymerase (MBI-Fermentas). In labeling reactions, rUTP is replaced by 20 μCi of [α-^{32}P]rUTP (approximately 100 Ci/mmol).

Unincorporated nucleotides are removed on a Sephadex G-50 spin column, RNA is than precipitated with 100% ethanol, washed with 70% (v/v) ethanol, and dried in a Speed Vac machine (Savant, Holbrook, NY). Complementary transcripts

[17] M. Hildebrandt, U. Saur, and W. Nellen, *Dev. Genet.* **12,** 163 (1991).

[18] D. C. Schweisguth, B. S. Chelladurai, A. W. Nicholson, and P. B. Moore, *Nucleic Acids Res.* **22,** 604 (1994).

[19] B. S. Chelladurai, H. Li, and A. W. Nicholson, *Nucleic Acids Res.* **19,** 1759 (1991).

are combined in a buffer containing 89 mM Tris (pH 8.3), 89 mM boric acid, 10 mM $MgCl_2$, 265 mM sodium chloride, 26.5 mM sodium citrate, and 1.9 mM EDTA and denatured for 5 min at 95° in a heating block and then slowly cooled to room temperature, with an overnight hybridization. Samples are treated with 134 U of DNase I (Gibco-BRL/Life Technologies, Rockville, MD) for 10 min at 37°. For the blunt substrate, both strands are labeled, and unpaired RNA and overhanging strands are removed by treatment with 10 ng of RNase A for 5 min at 37°. For the loop substrate only the longer strand is labeled and the RNase A digestion is omitted. Samples are extracted with phenol and separated on a Sephadex G-50 spin column. RNA substrates are then precipitated with ethanol, washed with 70% (v/v) ethanol, dried, and resuspended in doubly distilled water.

The synthetic homopolynucleotides poly(rI), poly(rC), poly(rA), and poly(dT) (Pharmacia, Piscataway, NJ) are dissolved in water. To obtain a heterogeneous size distribution ("ladder") poly(rI) is hydrolyzed with 100 mM sodium carbonate (pH 10.2) at 70°. Aliquots are taken at 10-min intervals and assayed by denaturing gel electrophoresis, and the gel is stained with reduced silver.[20,21] Appropriate fractions with a size distribution from 1 nt to approximately 80 nt are pooled and used for hybridization. Because of sufficient size heterogeneity it is not necessary to hydrolyze poly(rA). Twenty micrograms are used for end labeling in a volume of 50 μl containing T4 polynucleotide kinase buffer [50 mM Tris-HCl (pH 7.6), 10 mM $MgCl_2$, 5 mM DTT, 0.1 mM spermidine, 0.1 mM EDTA], 10 U of T4 polynucleotide kinase (MBI-Fermentas), and 20 μCi of [γ-^{32}P]ATP. The reaction is carried out for 30 min at 37°, and samples are extracted twice with phenol, once with phenol–chloroform, precipitated with ethanol, and resuspended in doubly distilled water.

To obtain homopolymeric double-stranded RNA and RNA–DNA hybrids, labeled hydrolyzed poly(rI) is hybridized to nonlabeled poly(rC), and labeled poly(rA) is hybridized to nonlabeled poly(dT). Strands are combined, denatured at 95° in a heating block, and slowly cooled (overnight) to room temperature.

Assay Conditions

Cytosolic extracts or purified fractions are tested for dsRNase activity in a 50- or 20-μl reaction, containing 1 ng of labeled dsRNA. Standard assay conditions for DdsRNase are as follows: 50 mM Tris-HCl (pH 8.0), 25 mM KCl, 5 mM $MgCl_2$, 2 mM DTT, tRNA (250 μg/ml), and 15% (v/v) glycerol. Reactions are performed as specified for 15 min to overnight at room temperature. The RNase A inhibitor RNasin (MBI-Fermentas) is added where indicated. Reactions are stopped by (1) heating to 95° followed by phenol extraction and precipitation with ethanol or (2)

[20] F. Sanger, S. Nicklen, and A. R. Coulson, *Proc. Natl. Acad. Sci. U.S.A.* **74,** 5463 (1977).
[21] J. Schumacher, H. L. Sänger, and D. Riesner, *EMBO J.* **2,** 1549 (1983).

precipitation of the entire reaction with ethanol, or (3) an aliquot is directly mixed with sample buffer (see below) and loaded on a gel. Samples from (1) and (2) are washed with 75% (v/v) ethanol, dried, resuspended in sample buffer [deionized formamide, 9 m*M* Tris–borate, 0.2 m*M* EDTA, 0.25% (w/v) bromphenol blue, 0.25% (w/v) xylene cyanol], heated at 95° for 5 min, cooled on ice, and separated on sequencing-type acrylamide–urea gels.[20] Gels are dried on a vacuum gel dryer and exposed either to X-ray film or on an imaging plate for analysis in a Fuji X Bas 1500 bioimaging analyzer (Raytest, Straubenhardt, Germany). The same procedure is used for experiments with RNase H and RNase III and for other nucleic acid samples (DNA and DNA–RNA hybrids).

Enzyme Purification

Dictyostelium discoideum strain Ax2 cells (10^{10} or more) are collected by centrifugation from axenic growth medium, washed with ice-cold 17 m*M* phosphate buffer (pH 6.0), and resuspended in 80 ml of lysis buffer containing 50 m*M* morpholineethanesulfonic acid (MES), 37.5 m*M* Tris (pH 8.0), 2 m*M* DTT, and phenylmethylsulfonyl fluoride (PMSF, 0.1 ng/ml; Sigma, St. Louis, MO). The suspension is sonicated for 3 to 5 min in a Hielscher (Berlin, Germany) sonicator with maximum intensity and cell lysis is monitored in the microscope. Glycerol is added to a final concentration of 15% (v/v). The suspension is centifuged at 10,000*g* for 30 min at 4° in a fixed-angle rotor C 0650 (Avanti; Beckman, Fullerton, CA) to remove cell debris. The supernatant is then centrifuged at 100,000*g* for 1 hr at 4° in a Ti 55.2 fixed-angle ultracentrifuge rotor (Beckman) to remove remnants of membranes, cell organelles, and large particles. All procedures are carried out at 4°. The resulting supernatant (cytosolic or crude extract) contains approximately 10 mg of protein per milliliter. Aliquots are tested for enzymatic activity and used for further fractionation.

Chromatography on DEAE (DE-52)-Cellulose. To prepare the DE-52 anion-exchange column matrix 30 g of diethylaminoethyl cellulose (Whatman, Clifton, NJ) is resuspended in 200 ml of 250 m*M* MES, 375 m*M* Tris (pH 8.0). The supernatant is decanted and the matrix is first equilibrated with 250 ml of buffer A [37.5 m*M* MES, 50 m*M* Tris (pH 8.0), 5 m*M* $MgCl_2$, 1 m*M* DTT, 5% (v/v) glycerol], then with 500 ml of buffer B [37.5 m*M* MES, 50 m*M* Tris (pH 8.0), 5 m*M* $MgCl_2$, 1 *M* NaCl, 1 m*M* DTT, 5% (v/v) glycerol], and finally washed with 250 ml of buffer A. Buffers are removed by filtration through a glass filter funnel. The slurry is then incubated with 30 ml of crude extract at 4° for 1–2 hr with gentle stirring. The suspension is poured into a glass column (Econo column 100 ml; Bio-Rad, Hercules, CA), and the column is packed by gravity flow. The column bed (50 ml) is first washed with 200 ml of buffer A, and then 200 ml of 30% (v/v) buffer B in buffer A (300 m*M* NaCl final concentration) is applied at a flow rate of 0.5 ml/min. Fractions of 3 ml are collected, with the activity eluting at 250 m*M* NaCl. Active fractions are pooled and aliquots are stored at −80°; the majority of the eluate is applied to the phosphocellulose column.

Phosphocellulose Chromatography. Phosphocellulose usually serves as a cation-exchange matrix. However, nucleic acid-binding proteins may bind strongly to the phosphate groups, with the resin instead functioning as an affinity matrix.

One gram of cellulose phosphate (Whatman) is resuspended in 50 ml of 0.5 *M* NaOH and rinsed over a glass filter with water until the pH of the suspension is below pH 11. The phosphocellulose is then resuspended in 50 ml 0.5 *M* HCl and again rinsed with water until the pH of the suspension is above pH 3.0. The material is then equilibrated with 0.5 *M* phosphate buffer (pH 7.0) and chilled to 4° for column preparation. A glass column (Econo column, 100 ml; Bio-Rad) is packed and the final bed volume (30 ml) is washed with the following buffers.

Binding buffer (150 ml): 250 m*M* MES, 375 m*M* Tris (pH 6.5), 5 m*M* $MgCl_2$, 1 m*M* DTT, 5% (v/v) glycerol
Buffer 1 (150 ml): 250 m*M* MES, 375 m*M* Tris (pH 7.5), 5 m*M* $MgCl_2$, 1 m*M* DTT, 5% (v/v) glycerol
Buffer 2 (300 ml): 250 m*M* MES, 375 m*M* Tris (pH 7.5), 100 m*M* Na_2HPO_4, 100 m*M* NaH_2PO_4, 5 m*M* $MgCl_2$, 1 m*M* DTT, 5% (v/v) glycerol

The material is finally preequilibrated with 150 ml of binding buffer. The pooled active DE-52 fractions (up to 30 ml, corresponding to approximately 20 mg of protein) are mixed with an equal volume of binding buffer and loaded on the column. The column is first washed with 60 ml (2 column volumes) of binding buffer and then with the same amount of buffer 1. The activity is then eluted with a steep gradient up to 90% (90 m*M* phosphate) of buffer 2. The column is run at 0.5 ml/min and the enzyme elutes at approximately 70 m*M* phosphate. Fractions of 2.5 ml are collected and tested for activity. Active fractions are pooled and stored at −80°, or directly prepared for chromatography on a Q5-Sepharose column.

Chromatography on Q5-Sepharose. Q-Sepharose (Bioscale, 5 ml; Bio-Rad) is an anion-exchange column and is used as the last step of purification. Pooled active fractions from the phosphocellulose column (approximately 7 ml) are concentrated and desalted by ultrafiltration in a Centricon-30 unit (Amicon, Danvers, MA) and equilibrated in 7 ml of buffer A [37.5 m*M* MES, 50 m*M* Tris (pH 8.0), 5 m*M* $MgCl_2$, 1 m*M* DTT, 5% (v/v) glycerol] and loaded on the column. After loading, the column is first washed with 20 ml of buffer A, followed by a shallow gradient of buffer A and 50% buffer B [37.5 m*M* MES, 50 m*M* Tris (pH 8.0), 5 m*M* $MgCl_2$, 1 *M* NaCl, 1 m*M* DTT, 5% (v/v) glycerol], reaching 500 m*M* NaCl after 50 ml is applied. Thirty fractions of 1 ml are collected, with the main peak of activity eluting in fractions 4 and 5 at approximately 150 m*M* NaCl. The column is run at 1 ml/min. Active enzyme fractions are frozen in small aliquots at −80° and are stable for several months. After thawing, activity can be preserved for about 2 days by storing tubes on ice in a cold room.

Gel Filtration. Gel filtration is performed on a Superose 6 PC3.2/30 gel-filtration column (2.4 ml; Pharmacia Biotech), using the Smart system (Pharmacia Biotech). The column is equilibrated with buffer A [37.5 m*M* MES, 50 m*M* Tris (pH 8.0), 5 m*M* $MgCl_2$, 1 m*M* DTT, 5% (v/v) glycerol].

Calibration runs are done with a high molecular mass protein marker mixture containing 5 μg of thyroglobulin (669 kDa), 5 μg of ferritin (440 kDa), and 10 μg of catalase (232 kDa) in 50 μl of buffer A. For DdsRNase separation, 50 μl of crude extract is loaded. Protein is eluted with buffer A and fractions of 50 μl are collected. The peak of activity appears at 1.4 ml, corresponding to approximately 450 kDa. In runs on a Superose 12 column (Pharmacia), the activity elutes in the exclusion volume. However, in all gel-filtration experiments, the recovered activity is strongly reduced to less than 30% of the load. Losses are even more pronounced when more purified fractions are applied. We assume that the large complex dissociates during the run, resulting in a substantial loss of activity.

Other Purification Attempts

Because of high losses of activity, a HiLoadSuperdex 200 (Pharmacia) gel-filtration column cannot be used for preparative purposes. Further purification of DdsRNase on the Superose S (Pharmacia) cation-exchange matrix, on a phenyl Superose (Pharmacia) hydrophobic interaction column, and on a poly(rI) · poly(rC) affinity matrix have failed. The activity is either lost (Superose S, phenyl Superose) or does not bind to poly(rI) · poly(rC) in the absence of magnesium ions. On spermine agarose and NADP agarose the activity is lost and on most dye columns (Sigma RDL-6 kit) the activity fails to bind. From heparin Sepharose, no activity can be eluted; this is due to the inhibitory effect of heparin, which can not be separated from the enzyme after chromatography. Surprisingly, DdsRNase can not be further purified by sucrose or glycerol gradient centrifugation. In contrast to the molecular weight estimated by gel filtration, the majority of the activity is recovered from the top of the gradient.

RNA Secondary Structure Prediction

For the prediction of RNA secondary structures, the MFold 3.0 algorithm developed by Zuker *et al.* is used.[22]

[22] M. Zuker, D. H. Mathews, and D. H. Turner, *in* "RNA Biochemistry and Biotechnology" (J. Barcizsewski and B. F. C. Clark, eds.), p. 11. NATO ASI Series. Kluwer Academic Publishers, Dordrecht, The Netherlands, 1999.

TABLE I
PURIFICATION OF *Dictyostelium* DOUBLE-STRAND-SPECIFIC RNase[a]

Fraction	Volume (ml)	Total protein (mg)	Total activity (units)[b]	Specific activity (units/mg)	Purification factor (fold)
Crude extract (100,000 *g* ultracentrifugation)	50	650	50,000	76	—
DE-52 column	40	28	40,000	1,428	18
Phosphocellulose	15	3.15	15,000	4,762	62
Q5-Sepharose	3	0.150	4,500	30,000	394

[a] The enzymatic activity was determined by quantitation of radioactivity in substrate and product bands after gel electrophoresis, using the evaluation program of the Fuji X Bas 1500 bioimaging analyzer (Raytest). Units of enzyme were determined in 50-μl reactions, under standard conditions, containing 7000 cpm and 1 nmol of dsRNA (310-bp dsRNA fragment, label quantitated in a Wallac liquid scintillation counter, amount in micrograms quantitated by UV_{260} absorbance). Protein quantitations were done by the Amido black assay. The purification factor of 394 is probably an underestimate because of a loss of active enzyme (compare protein pattern in Fig. 1).

[b] One unit is the amount of enzyme required to digest 100 fmol of dsRNA in 1 hr at 22°.

Gel Electrophoresis of Nucleic Acid Samples

^{32}P-labeled nucleic acid samples (RNA and DNA) are run on 8 or 12% (w/v) denaturing, sequencing type polyacrylamide–urea gels.[20]

Sodium Dodecyl Sulfate Gel Electrophoresis of Protein Samples

Protein samples are supplemented with 2× Laemmli buffer,[23] heat denatured, and then separated on a 4 to 17% (w/v) sodium dodecyl sulfate (SDS)-polyacrylamide gradient gel. Gels are stained with Coomassie Brilliant Blue R250.

Purification of *Dictyostelium Double-Strand-Specific Ribonuclease*

The purification described above is summarized in Table I. With a recovery of approximately 10% of the activity, a purification of 400-fold is achieved. We assume that this is an underestimate due to loss of activity because the analysis by SDS gel electrophoresis (see below) suggests a much higher purification.

Figure 1 shows the proteins recovered after the different chromatographies. In the final step (Q-Sepharose), six major bands (34, 54, 62, 77, 82, and 88 kDa) are detected, with an additional weak band seen in the range of 94 kDa. If it is assumed

[23] U. K. Laemmli, *Nature* (*London*) **227,** 680 (1970).

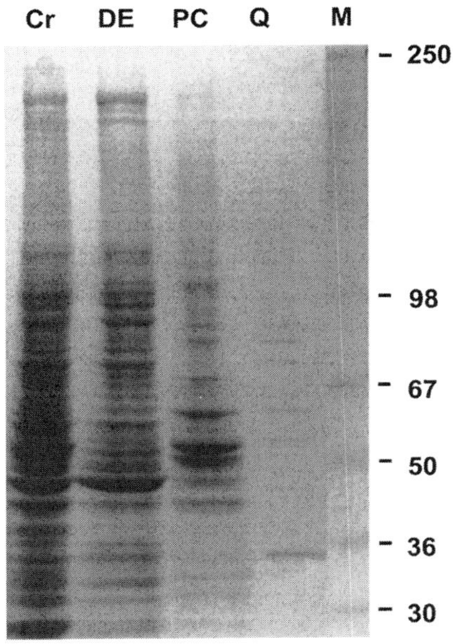

FIG. 1. SDS–polyacrylamide gel analysis of DdsRNase purification fractions. Aliquots of active fractions were taken at the various stages of the purification procedure and analyzed on a 4–17% (w/v) gradient SDS–polyacrylamide gel. The gel was stained with Coomassie Brilliant Blue R250. Cr, Crude cytosolic fraction (100,000*g* supernatant); DE, pooled active fractions from a DE-52 column; PC, pooled active fractions from a phosphocellulose column; Q, pooled active fractions from a Q5-Sepharose column; M, protein markers (size in kDa).

that all these polypeptides are part of the DdsRNase complex and that they all occur only once in the complex, simply adding the molecular masses of the bands results in approximately 490 kDa; this correlates surprisingly well with the size of the complex as determined by gel filtration. In fact, an approximately 450-kDa band is also seen by native gel electrophosesis of Sepharose-Q fractions.[24]

Because the purification scheme and the properties of DdsRNase are similar to those of bacterial RNase III, we tested antibodies directed against RNase III and the yeast homolog Pac1 [generously provided by R. Simons (University of California, Los Angeles, CA) and D. Frendewey (New York University, New York, NY) respectively] on our purified preparations, but no cross-reactivity could be detected. Although some of these antibodies appear to be rather species specific, the results suggest that DdsRNase does not contain an RNase III-like component.

[24] J. Novotny, M. Guttenberger, and W. Nellen, in preparation.

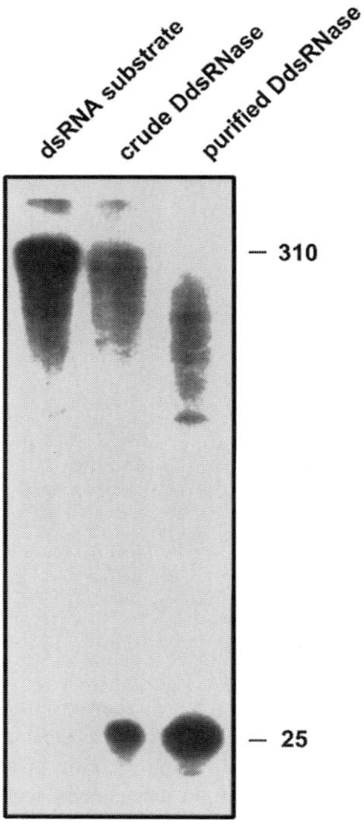

FIG. 2. *Dictyostelium* dsRNase (DdsRNase) processing of dsRNA. The substrate is a 310-bp blunt-ended dsRNA. With a crude DdsRNase preparation a reduction of the substrate and an accumulation of a characteristic 25-nt product are observed. A one-step purified preparation (Q-Sepharose) shows increased activity with a smear of partially degraded substrate and a further accumulation of the product. All samples were heat denatured before gel loading.

Characterization of Dictyostelium Double-Strand-Specific Ribonuclease

In the standard assay for dsRNase activity, two complementary *in vitro*-transcribed and hybridized RNAs were used. To remove residual single strands and single-stranded overhangs, the hybrids were trimmed with RNase A. With crude cytosolic extracts or with more purified fractions, double-helical RNA was digested and distinct degradation products of approximately 25 nucleotides were recovered (Fig. 2). Because we used denaturing gel electrophoresis it is not clear so far whether these products are single or double stranded. To determine the minimal target size, a ladder of synthetic homopolymeric poly(rI) · poly(rC) was subjected to DdsRNase digestion. All molecules larger than 25 bp were degraded, indicating that more than 25 bp of target was required for DdsRNase (Fig. 3). The

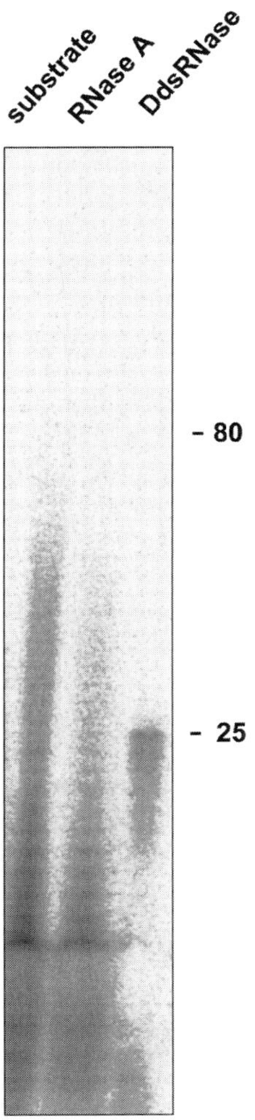

FIG. 3. DdsRNase activity on synthetic dsRNA homopolymers. Denaturing polyacrylamide–urea gel showing DdsRNase activity (Q-Sepharose purified) on an end-labeled poly(rI) · poly(rC) substrate ladder. The substrate control lane shows the size distribution of the fragments: after treatment with RNase A, the size distribution is essentially the same. After treatment with DdsRNase, all fragments larger than 25 bp are digested and an accumulation of radioactivity is observed in the 25-nt position. The reduction of fragments in the low molecular weight range may be due to other activities still contained in the purified DdsRNase fraction. Samples were heat denatured before gel loading.

simultaneous reduction of low molecular weight dsRNA species could be due to other activities still contaminating the enzyme preparation.

For further characterization, the following properties of the activity were defined: the enzyme was sensitive to heat (10 min, 60°), proteinase K, and vanadyl ribonucleoside complex but resistant to RNasin (MBI-Fermentas). RNase A and RNase T1 did not interfere with the enzyme, indicating that it did not contain an exposed RNA component. Heparin inhibited activity, probably by competition with the substrate. Sensitivity to the alkylating agent *N*-ethylmaleimide (NEM) suggested the presence of histidine and/or serine groups in the active center of the enzyme whereas iodoacetamide (IAA) had no significant effect on the activity (data not shown).

EDTA caused reversible inhibition of the enzyme. Mg^{2+} most efficiently restored activity but could be replaced by Co^{2+} and Mn^{2+}; however, with these ions the activity was significantly reduced. Other divalent metal ions such as Zn^{2+}, Ca^{2+}, Fe^{2+}, Ni^{2+}, and Cu^{2+} had no effect in the range of 0.5 to 2 m*M* (Fig. 4). Titration for the optimal Mg^{2+} concentration showed an increase in activity up to 5 m*M*; and at 10 m*M* a slight decrease could be observed (not shown). Therefore, a Mg^{2+} concentration of 5 m*M* was used for the standard assays. The pH optimum was at approximately pH 8; this pH was used for the standard assay.

To determine the specificity of the enzyme, activity was also assayed on dsDNA, ssRNA, and RNA–DNA hybrids. As shown in Fig. 5, no degradation of ssRNA or dsDNA was observed. In contrast, DNA–RNA hybrids [poly(rA) · poly (dT)] were efficiently digested (Fig. 6) and generated products up to 5 nt, similar to *E. coli* RNase H. However, the dsRNA-specific products of 25 nt were not observed. We cannot yet determine whether this activity is an intrinsic property of DdsRNase or whether it is based on a loosely associated or copurified protein.

To further characterize DdsRNase, we used different substrates that contained single- and double-stranded RNA regions either as 5′ overhangs, 3′ overhangs, or internal single-stranded loops. Figure 7 shows the loop substrate as an example: whereas the double-stranded regions were digested by DdsRNase, the single-stranded loop remained intact. It should be noted that the size of the loop is 61 nt whereas the recovered fragment is approximately 80 nt. This suggests that DdsRNase does not cut exactly at the dsRNA–ssRNA junction but rather inside the dsRNA region. A similar apparent "extension" of the single-stranded region was observed when substrates with 5′ or 3′ single-stranded overhangs were used (data not shown).

The substrate specificity of DdsRNase was then compared with the activity of recombinant RNase III enzymes from *E. coli* and *Rhodobacter capsulatus* (kindly provided by C. Conrad and G. Klug, University of Giessen, Germany)

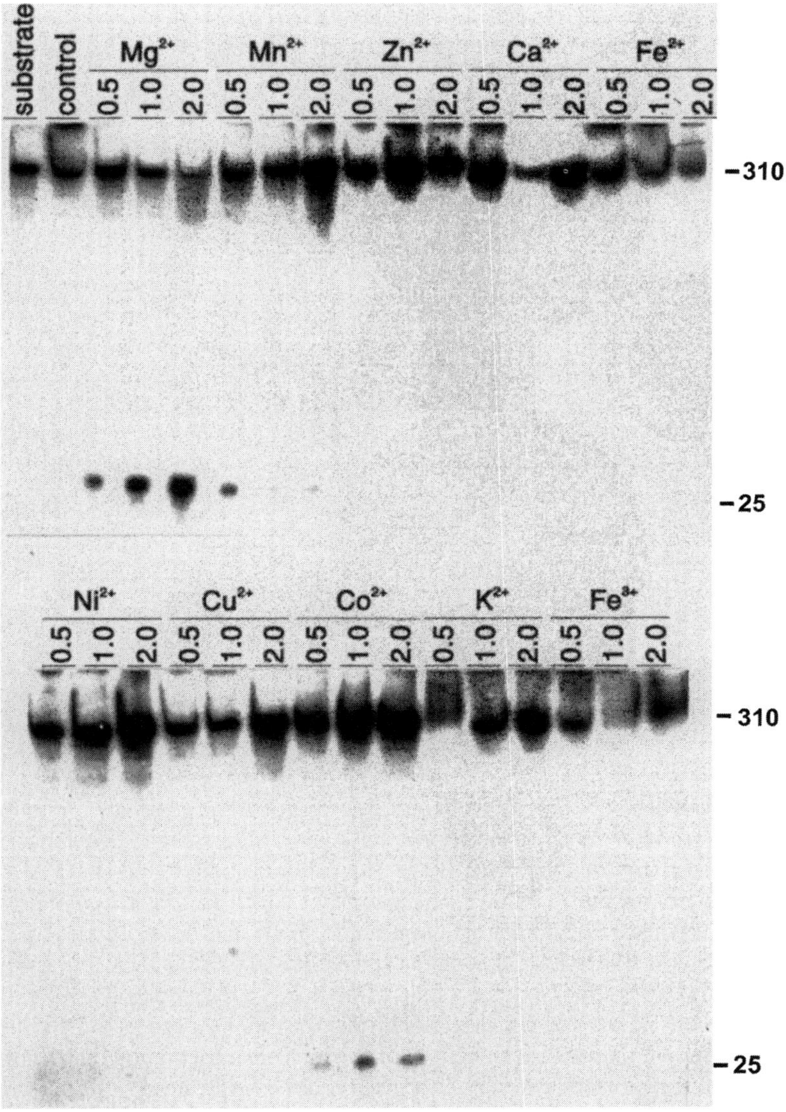

FIG. 4. Examination of divalent metal ion requirement for DdsRNase activity. A cytosolic extract was prepared without Mg^{2+}; endogenous divalent cations were chelated with 0.2 m*M* EDTA. The control reaction shows no activity on the 310-bp substrate. Divalent cations were added at 0.5, 1.0, and 2.0 m*M* concentrations as indicated. Reactions were done for 20 min at 20°. Samples were heat denatured before gel loading.

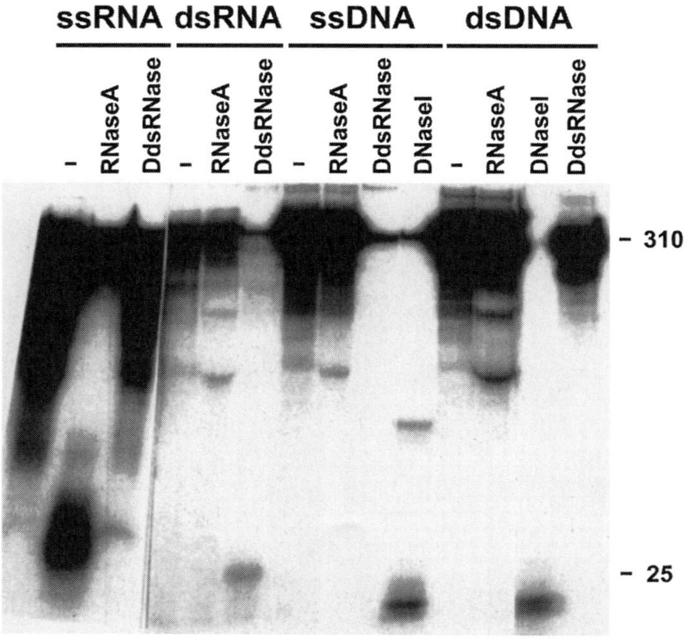

FIG. 5. Specificity of DdsRNase. ssRNA, ssDNA, and dsDNA were used as targets for crude DdsRNase. Control reactions were carried out with RNase A and DNase I, respectively. A dsRNA sample is shown for comparison. Samples were heat denatured before gel loading.

on a 310-bp dsRNA.[25] Figure 8A shows the specific 25 nt cleavage products of the DdsRNase while both RNaseIII enzymes generate a smear. To compare the cleavage specificity on an RNaseIII-specific substrate, the T7–R1.1 RNA (provided by C. Conrad and G. Klug), which contains an RNase III processing signal (Fig. 8C), was used as a target.[18,19] Figure 8B shows that R1.1 was nonspecifically degraded to a low molecular weight smear by a crude DdsRNase preparation. Most likely this was due to other RNases abundantly present in this preparation. In contrast, the DdsRNase preparation purified over three columns did not digest this substrate. This was in agreement with the previous observation that at least a double-helical region of more than 25 bp was required for cleavage. RNase III from *R. capsulatus* digested the substrate as expected. Note that the size of the major cleavage product (25 nt) is only by coincidence the same as the DdsRNase cleavage product.

[25] C. Conrad, R. Rauhut, and G. Klug, *Nucleic Acids Res.* **26,** 4446 (1998).

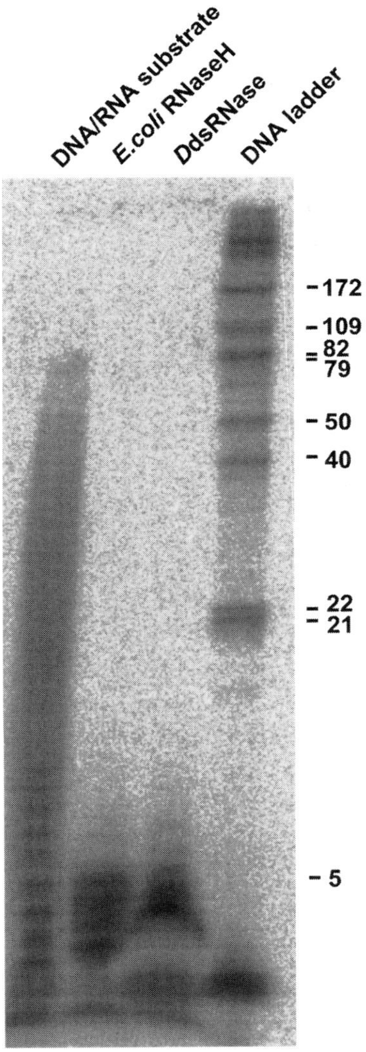

FIG. 6. RNase H-like activity in purified DdsRNase. A homopolymeric poly(rA) · poly(dT) ladder was used as a target for DdsRNase. The RNA component was end labeled with [γ-^{32}P]ATP. The size distribution of the substrate is shown in the first lane. Digestion with 1 unit of RNase H from *E. coli* (MBI-Fermentas) results in complete degradation of the substrate to 1- to 5-nt products. A similar result is observed with DdsRNase (purified over three columns); the dsRNA-specific product of 25 nt is not found. The molecular weight marker (DNA ladder) is a *Sau*3A digestion of pGEM 3Z (Promega) ^{32}P labeled by filling in overhangs with the Klenow fragment of DNA polymerase I. All samples were heat denatured before gel loading.

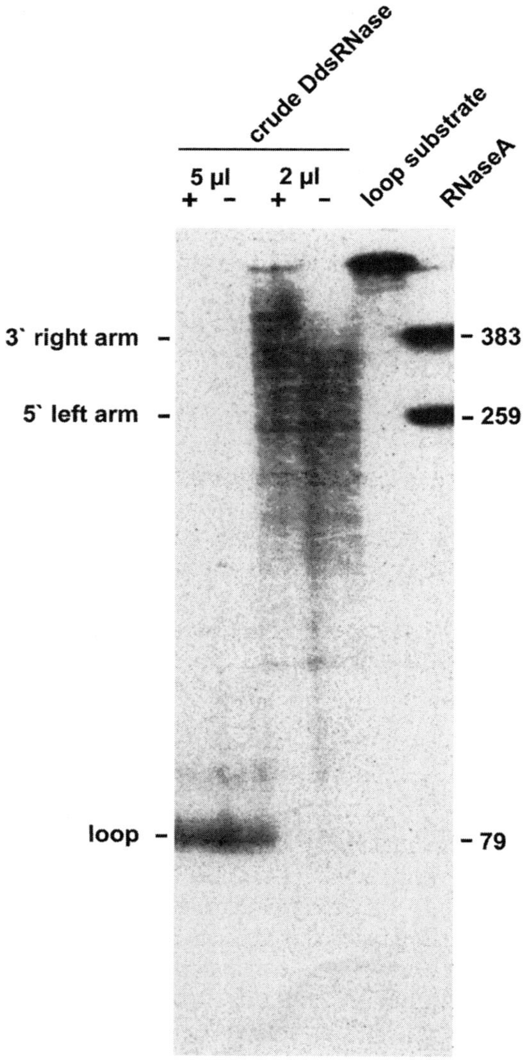

FIG. 7. DdsRNase assay on a "mixed substrate." A 703-bp dsRNA molecule containing an internal single-stranded loop of 61 nt was used as a substrate. Digestion with RNase A results in cleavage of the loop and release of the double-stranded arms. Treatment with 2 and 5 μl of the crude DdsRNase (1 μl is an equivalent of 10^4 cells) generates the specific 25-nt products (not visible) and a fragment of 79 nt representing the single-stranded region of the substrate; intermediary products are visible. Reactions were performed in the absence (–) or presence (+) of 60 U of RNasin (MBI-Fermentas). The same results are obtained with the purified DdsRNase. All samples were heat denatured before gel loading.

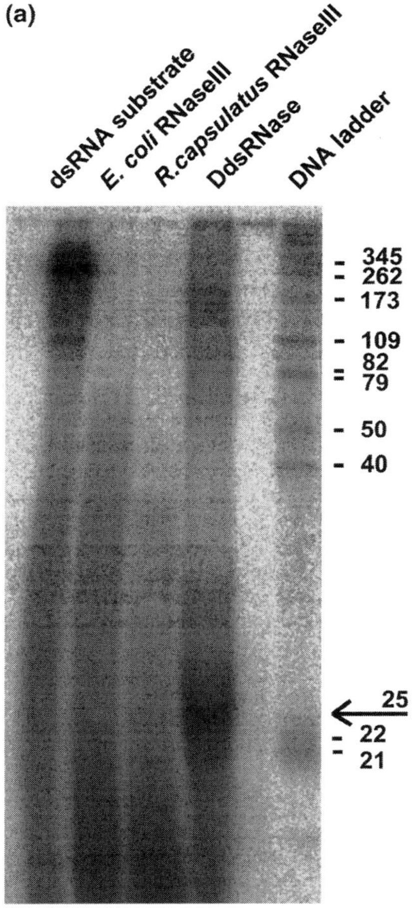

FIG. 8. Comparison of DdsRNase and RNase III activity. (A) With the standard 310-bp dsRNA substrate, recombinant *E. coli* RNase III (0.2 μg/μl) and recombinant *R. capsulatus* RNase III (0.3 μg/μl) generate a smear whereas purified DdsRNase produces the typical 25-nt degradation product. (B) With the specific *in vitro*-synthesized RNase III substrate T7–R1.1 (126 nt) incubated under standard conditions with *R. capsulatus* RNase III, specific cleavage products including the expected 25-nt fragment are found. In contrast, a crude DdsRNase preparation unspecifically degrades the substrate to a low molecular weight smear whereas the purified DdsRNase preparation (DE-52, phosphocellulose, Q5-Sepharose) displays no significant activity on this substrate. Size markers are the same as in Fig. 6. (C) Secondary structure of the T7–R1.1 transcript computed by the Mfold 3.0 algorithm [M. Zuker, D. H. Mathews, and D. H. Turner, *in* "RNA Biochemistry and Biotechnology" (J. Barcizsewski and B. F. C. Clark, eds.), p. 11, NATO ASI Series. Kluwer Academic Publishers, Dordrecht, The Netherlands, 1999]; the RNase III processing site is indicated by an arrow.

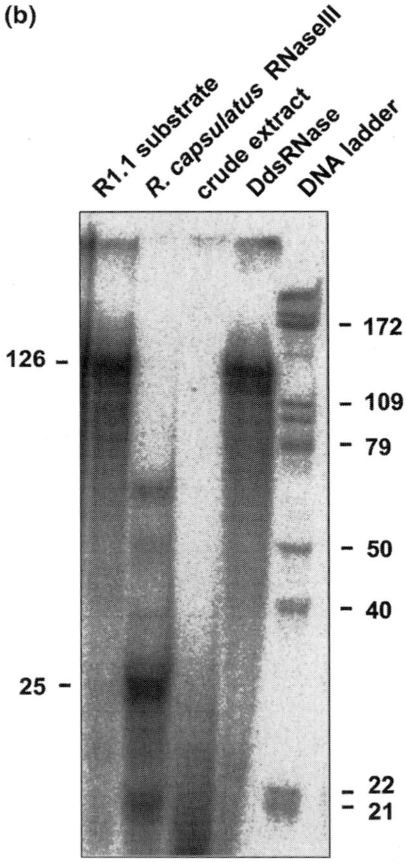

FIG. 8b

Conclusions

DdsRNase is so far a unique enzymatic activity that specifically degrades dsRNA but has no effect on ssRNA or dsDNA. In its biochemical properties it is distinct from other identified dsRNases. In particular, substrate requirements and degradation products are different from RNase III-related enzymes, which also seem not to share antigenic determinants with DdsRNase. In contrast to most other nucleases, the enzymatic activity appears to reside in a large multicomponent complex consisting of at least six different polypeptides. This is reminiscent of prokaryotic degradosomes, which contain several different RNases and additional enzymes for processing and degradation of RNAs.[26] However, isolated components of degradosomes are mostly active nucleases while it appears as if the loss of

(c)

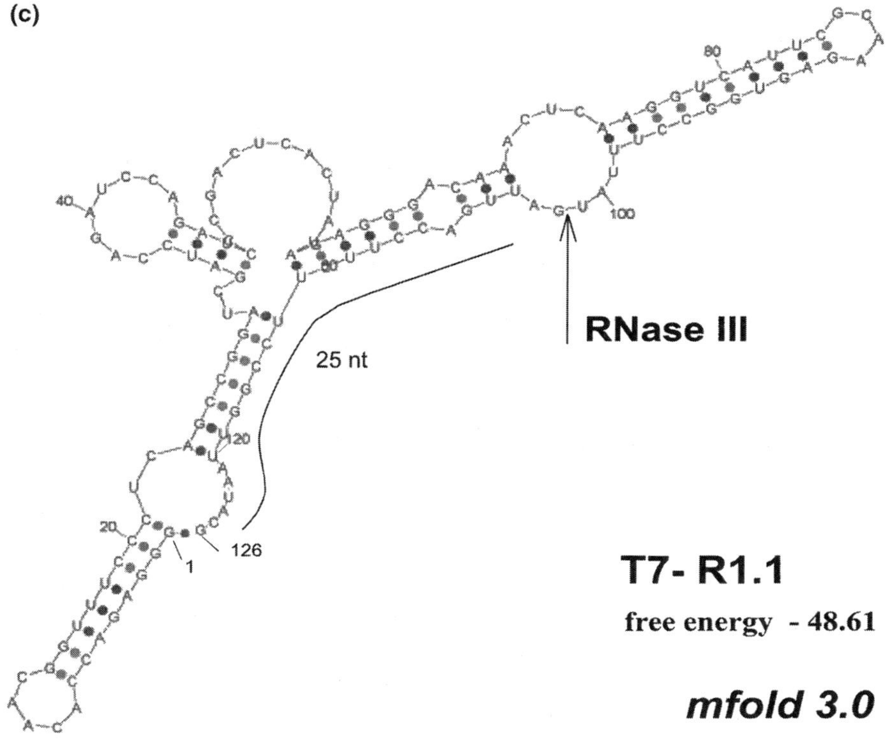

FIG. 8c

subunits causes loss of function of DdsRNase. This may explain the high loss of activity during gel filtration, where no lower molecular weight fractions displayed any activity.

We originally searched for a dsRNase activity to explain the specific destabilization of highly expressed mRNA when a corresponding antisense RNA is expressed at similar high levels. DdsRNase may actually serve this function. In addition, the unusual degradation products of 25 nt (or 25 bp), are of the same size as the small sense and antisense RNAs that were implied to play a role in cosuppression (posttranscriptional gene silencing, PTGS) in plants.[27] It may be assumed that a plant homolog to DdsRNase could thus degrade dsRNA and at the same time generate the mobile signal that can spread in a PTGS plant.

Interestingly, RNAs of 25 nt have also been suggested to specify RNase activity in the posttranscriptional RNA interference (RNAi)-mediated gene silencing in

[26] G. A. Coburn, X. Miao, D. J. Briant, and G. A. Mackie, *Genes Dev.* **13,** 2594 (1999).
[27] A. J. Hamilton and D. C. Baulcombe, *Science* **286,** 950 (1999).

Drosophila.[28] A genetic link between cosuppression and RNA interference has been shown for *C. elegans* and for *Neurospora crassa*.[29,30]

Acknowledgments

Petra Kroeger and Beate Schnell are acknowledged for contributions that are not part of this article, but that added to the general understanding of DdsRNase activity. We thank Christian Conrad and Gabriele Klug (University of Giessen) for providing recombinant RNase III and the pET-4 R1.1 plasmid, David Frendewey (New York University) for providing the Pac1 antisera, and Robert Simons (University of California, Los Angeles) for providing the RNase III antisera. Part of this work was carried out at the Department of Cell Biology, Max-Planck-Institute for Biochemistry (Martinsried, Germany). This work was supported by a grant from the Deutsche Forschungsgemeinschaft to W.N.

[28] S. M. Hammond, E. Bernstein, D. Beach, and G. J. Hannon, *Nature (London)* **404,** 293 (2000).
[29] R. F. Ketting and R. H. A. Plasterk, *Nature (London)* **404,** 296 (2000).
[30] C. Catalanotto, G. Azzalin, G. Macino, and C. Cogoni, *Nature (London)* **404,** 245 (2000).

[17] Double-Stranded RNA Nuclease Associated with Rye Germ Ribosomes

By MARIA A. SIWECKA

Introduction

Ribonucleolytic activities are known to be associated with plant ribosomes, but the biological significance of this activity *in vivo* has not yet been clarified.[1] It is known that *Drosophila* ribosomal protein S3 contains DNase activity.[2] We have demonstrated that both ribonucleolytic and deoxyribonucleolytic activities are associated with rye germ ribosomes.[3,4] The two enzymatic activities are strongly bound to ribosomes and their subunits, but part of the activity can be released by washing the ribosomes with 0.5 *M* ammonium chloride treatment.[4] It was then noted that the material extracted from ribosomes by ammonium chloride treatment is capable of degrading a double-stranded RNA from a virus of *Penicillium chrysogenum* as well as the double-stranded poly(A) · poly(U) complex.[5] This material,

[1] H. Bielka, *in* "Eucaryotic Ribosome" (H. Bielka, ed.), p. 98. Academic Verlag, Berlin, 1982.
[2] D. M. Wilson III, W. A. Deutsch, and M. R. Kelley, *J. Biol. Chem.* **269,** 25359 (1994).
[3] M. A. Siwecka, T. Gołaszewski, and J. W. Szarkowski, *Bull. Acad. Polon. Sci. Sér. Sci. Biol.* **25,** 15 (1977).
[4] M. A. Siwecka, M. Rytel, and J. W. Szarkowski, *Acta Biochim. Polon.* **26,** 97 (1979).
[5] M. A. Siwecka, M. Rytel, and J. W. Szarkowski, *Phytochemistry* **21,** 273 (1982).

0076-6879/01 $35.00

therefore, appears to contain a double-stranded RNA hydrolyzing enzyme. It is worth pointing out, however, that a type I nuclease, isolated from an ammonium chloride wash of rye germ ribosomes and purified to a homogeneity, showed no activity toward double-stranded RNA.[6]

On the basis of the previous observations, attempts were made to isolate the enzyme(s) degrading double-stranded RNA from the wash of rye germ ribosomes,[7] and the ionic strength and pH conditions optimal for activity were established.[8]

The isolation and partial purification of a nuclease from tobacco anthers that degrades double-stranded RNA was reported by Matoušek *et al.*[9] These authors also detected such an activity in hop tissues infected with hop latent viroid.[10]

The nucleolytic enzymes of higher plants have been grouped into four enzyme types: RNase I, RNase II, nuclease I, and exonuclease I. Nuclease type I enzymes are endonucleases that degrade both single-stranded DNA and RNA, but not double-stranded RNA (for review, see Wilson[11]). Nuclease II from rye germ ribosomes and the nuclease from tobacco anthers cannot be classified into any of the four classic types of enzymes, and therefore represent higher plant nucleases of a new type.

Described here is a two-step procedure to isolate and purify a novel nuclease able to degrade double-stranded RNA, and some properties of this enzyme (action on single- and double-stranded RNAs, optimum pH, effects of ionic strength and cations on enzyme activity, and hydrolysis of polynucleotides). Data on the ability of the nuclease to cleave *Lupinus luteus* 5S rRNA is also presented. The enzyme is called rye germ ribosomal nuclease II, to distinguish it from nuclease I, the enzyme previously purified from rye germ ribosomal washes.[6]

Materials and Methods

Plants

Commercial rye (*Secale cereale* L.) germs are used as the source of ribosomes. Starch contamination is eliminated by sieving. Meshes of 1.2 and 0.6 mm are used. The germs remaining on the 0.6-mm sieve are taken for the experiments.

[6] M. A. Siwecka, M. Rytel, and J. W. Szarkowski, *Acta Biochim. Polon.* **36,** 45 (1989).

[7] M. A. Siwecka, M. Rytel, and J. W. Szarkowski, *in* "Metabolism and Enzymology of Nucleic Acids, including Gene Manipulations" (J. Zelinka and J. Balan, eds.), p. 245. Plenum Press, New York, 1988.

[8] M. A. Siwecka, M. Rytel, and J. W. Szarkowski, *Acta Physiol. Plant.* **5,** 105 (1983).

[9] J. Matoušek, L. Trnena, R. Oberhauser, C. P. Lichtenstein, and W. Nellen, *Biol. Chem. Hoppe Seyler* **375,** 261 (1994).

[10] J. Matoušek, L. Trnena, L. Svoboda, P. Oriniakova, and C. P. Lichtenstein, *Biol. Chem. Hoppe Seyler* **376,** 715 (1995).

[11] C. M. Wilson, *in* "Isozymes: Current Topics in Biological and Medical Research," Vol. 6, p. 33. A. R. Liss, New York, 1982.

Isolation of Cytoplasmic Ribosomes

All operations are carried out at 4°. Rye germs (300 g) are homogenized in a mortar and then in an ice-cooled mixer in 10 m*M* Tris-HCl, pH 8.0, containing 0.25 *M* sucrose, 10 m*M* $MgCl_2$, and 10 m*M* 2-mercaptoethanol. Protein inhibitors are added to the buffer: aprotinin (0.2 μg/ml), leupeptin (0.2 μg/ml), and benzamidine up to a concentration of 0.1 m*M*. The homogenate is filtered through two layers of nylon cloth and centrifuged at 12,000*g* for 15 min. Triton X-100 or Brij 35 is added to the supernatant to a 0.5% (v/v) final concentration, and the supernatant is centrifuged at 40,000*g* for 30 min. The sediment is discarded, and the supernatant is recentrifuged at 160,000*g* for 2 hr. The resulting crude ribosomal pellet is resuspended in 10 m*M* Tris-HCl, pH 8.0, containing 10 m*M* KCl and 10 m*M* $MgCl_2$. The ribosome suspensions (2 ml) are applied on top of two-layer discontinuous sucrose gradients that contain 1 ml of 1.5 *M* sucrose and 1 ml of 1 *M* sucrose. The sucrose is dissolved in the same buffer as the ribosomes. The samples are centrifuged for 4 hr at 140,000*g* and the pellet is collected.

Release of Proteins with Nucleolytic Activity from Ribosomes

The ribosomal pellet is washed twice with 30 m*M* Tris-HCl, pH 8.0, containing 0.6 *M* KCl, 7 m*M* $MgCl_2$, 10 m*M* 2-mercaptoethanol, and 10% (v/v) glycerol. The suspension is kept at 4° for 18 hr with constant stirring and then centrifuged at 160,000*g* for 2 hr at 4°. The supernatants are used for further purification of the enzyme (washes 1 and 2).

Enzyme Purification

All purification steps are carried out at 4°. The washes (6 ml, protein at 1.2 mg/ml) are applied to a Sephadex G-100 (60 × 1.5 cm) column and eluted with 10 m*M* Tris-HCl, pH 8.0, containing 10 m*M* 2-mercaptoethanol and 10% (v/v) glycerol. Fractions active toward poly(I) · poly(C) (see below) are collected and concentrated either in dialysis bags in Ficoll 400 or by centrifugation with a Centricon 10 (Amicon, Danvers, MA). For further purification an octyl-Sepharose column (3.0 × 2.6 cm) or a phenyl-Sepharose column (3.5 × 2.3 cm) is used. Before its application to the column, the sample is dialyzed against 10 m*M* sodium phosphate buffer, pH 6.8, containing 10 m*M* 2-mercaptoethanol, and then $(NH_4)_2SO_4$ is added to 25% saturation. Proteins are eluted from the column with 10 m*M* Tris-HCl, pH 6.8, containing $(NH_4)_2SO_4$ to 25% saturation and 10 m*M* 2-mercaptoethanol. The fractions showing activity toward poly(I) · poly(C) are pooled and the $(NH_4)_2SO_4$ is removed by dialysis. The fractions are concentrated and glycerol is added to 20–50% (v/v) concentration. This material is stored at −70° until use.

Determination of Nucleolytic Activity

Nucleolytic activity toward poly(I) · poly(C), RNA, or DNA is determined by a modified version of the method of Anfinsen *et al.*[12] The reaction mixture contains, in a total volume of 0.5 ml of buffer A, B or C, 0.1 ml of substrate [poly(I) · poly(C), or RNA or DNA, 0.5 mg/ml], and from 0.01 to 0.10 ml of the enzyme solution protein (from 1 to 0.1 mg/ml).

Buffer A [for activity on poly(I) · poly(C)]: 10 m*M* Tris-HCl, pH 8.5
Buffer B (for activity on RNA): 10 m*M* Tris-HCl, pH 7.8
Buffer C (for activity toward DNA): 50 m*M* citric acid–sodium phosphate, pH 5.0

The samples are incubated at 37° for 15 min. The reaction is stopped by addition of 20 m*M* lanthanum acetate (1 ml) in 12% (w/v) perchloric acid, the sample is cooled to 0°, and the precipitate is eliminated by centrifugation at 6000*g* for 15 min. Absorbance of the supernatant is measured spectrophotometrically at 260 nm. One unit (1 U) of nucleolytic activity is defined as the amount of enzyme causing an absorbance increase ($\Delta A_{1\ \text{cm},\ 260}$) of 0.1. Specific activity is expressed as units per milligram of protein.

Protein Determination

Protein content is determined by the method of Bradford[13] or by a modified version of the procedure of Lowry, using the Bio-Rad (Hercules, CA) assay.[14] Bovine serum albumin (BSA) is used as standard.

Sodium Dodecyl Sulfate–Polyacrylamide Gel Electrophoresis

Denaturing polyacrylamide gel electrophoresis of protein is performed as described by Laemmli.[15] Gels are stained with 0.025% (w/v) Coomassie Brilliant Blue or by silver staining, using a silver kit (Bio-Rad).

Molecular Mass Determination

Molecular masses are determined electrophoretically in 10% (w/v) polyacrylamide gels containing 0.1% (w/v) sodium dodecyl sulfate SDS. An electrophoresis calibration kit (Pharmacia Fine Chemicals, Piscataway, NJ) is used for determination of low molecular weight proteins.

[12] C. B. Anfinsen, R. R Redfield, W. L. Choate, J. Page, and W. R. Carrol, *J. Biol. Chem.* **207,** 201 (1954).
[13] M. M. Bradford, *Anal. Biochem.* **72,** 248 (1976).
[14] G. L. Petersen, *Anal. Biochem.* **100,** 201 (1979).
[15] U. K. Laemmli, *Nature (London)* **227,** 681 (1970).

Isolation of Lupinus luteus 5S rRNA

Lupin 5S rRNA is isolated as described by Barciszewska *et al.*[16] and purified by Sephadex G-75 column chromatography, followed by high-performance liquid chromatography (HPLC) on a TSK DEAE 5PW gel.[17]

Labeling of 5S rRNA

The lupin 5S rRNA is dephosphorylated by bacterial alkaline phosphatase, and then 5′-end labeled with [γ-^{32}P]ATP and T_4 polynucleotide kinase. Labeled lupin 5S rRNA is purified on denaturing 12.5% (w/v) polyacrylamide gels, located by autoradiography, excised, and eluted from the gel with 0.3 *M* potassium acetate, pH 5.1, containing 1 m*M* EDTA and 0.1% (w/v) SDS. The eluted 5S rRNA is precipitated with ethanol, dissolved in water, and stored at −20°.[18]

Digestion of Lupinus luteus 5S rRNA

Digestion of 5′-end-labeled lupin 5S rRNA with nuclease II is performed in 10 m*M* Tris-HCl (pH 8.5)–5 m*M* dithiothreitol (DTT) for 30 min at 37°. The reaction is stopped by addition of 8 *M* urea containing 20 m*M* EDTA and directly loaded on a 15% (w/v) denaturing polyacrylamide gel.[19]

Analysis of Reaction Products

To assign cleavage sites in lupin 5S rRNA, the products of nuclease II digestion are compared with the products of alkaline degradation and limited T1 nuclease digestion. Partial T1 nuclease digestion of 5S rRNA is performed under denaturing conditions (50 m*M* sodium citrate, pH 4.5, with 0.1 unit of enzyme) for 10 min at 55°. A 5S rRNA ladder is created by incubation in formamide containing 0.5 m*M* $MgCl_2$ at 100° for 15 min.[20]

Electrophoresis

Electrophoresis is performed with denaturing polyacrylamide gels [15% (w/v) acrylamide, 0.75% (w/v) bisacrylamide, 7 *M* urea in 50 m*M* Tris–borate (pH 8.3) and 1 m*M* EDTA]. The gel size is 40 × 30 × 0.04 cm. Gels are run at 1500 V for 2–5 hr, followed by autoradiography at −80°.

[16] M. Barciszewska, T. Mashova, L. L. Kisselev, and J. Barciszewski, *FEBS Lett.* **192,** 289 (1985).
[17] J. Ciesiołka and W. J. Krzyżosiak, *Biochem. Mol. Biol. Int.* **39,** 319 (1996).
[18] D. Michałowski, J. Wrzesiński, J. Ciesiołka, and W. J. Krzyżosiak, *Biochimie* **78,** 131 (1996).
[19] J. Wrzesiński, D. Michałowski, J. Ciesiołka, and W. J. Krzyżosiak, *FEBS Lett.* **374,** 62 (1995).
[20] M. Napierała and W. J. Krzyżosiak, *J. Biol. Chem.* **272,** 31079 (1997).

TABLE I
PURIFICATION OF RYE GERM RIBOSOMAL NUCLEASE II BY SEPHADEX G-100 AND OCTYL-SEPHAROSE COLUMN CHROMATOGRAPHY[a]

Purification step	Total protein (mg)	Total activity (U)	Specific activity toward poly(I) · poly(C) (U/mg protein)	Purification factor (fold)	Yield (%)
Ribosomes	405.0	11,218.5	27.7	1.0	100.0
KCl wash 1	60.5	3,309.4	54.7	2.0	29.5
KCl wash 2	14.0	658.0	47.0	1.7	5.9
Sephadex G-100	17.0	2,981.8[b]	175.4	6.3	26.5
Octyl-Sepharose	0.4	378.0	945.0	34.1	3.4

[a] Data are average values obtained from 30 independent enzyme purification experiments.
[b] KCl wash 1 plus wash 2.

Purification of Nuclease

The nuclease is released from ribosomes by two washes with appropriate buffer containing 0.6 *M* KCl. The results of the two-step purification procedure are shown in Tables I and II. The elution profiles obtained from the octyl-Sepharose and phenyl-Sepharose columns are shown in Fig. 1. The enzyme activity elutes as a single peak with 10 m*M* sodium phosphate buffer, pH 6.8, 25% saturated with ammonium sulfate. The adsorbed proteins, eluting with a gradient of decreasing ammonium sulfate concentration, show no activity toward poly(I) · poly(C) (not shown). The specific activity of the recovered enzyme from the octyl-Sepharose column is about 34-fold higher toward poly(I) · (C) than the corresponding activity from ribosomes, and the yield of this procedure is 3.4% (Table I).

TABLE II
PURIFICATION OF RYE GERM RIBOSOMAL NUCLEASE II BY SEPHADEX G-100 AND PHENYL-SEPHAROSE COLUMN CHROMATOGRAPHY[a]

Purification step	Total protein (mg)	Total activity (U)	Specific activity toward poly(I) · poly(C) (U/mg protein)	Purification factor (fold)	Yield (%)
Ribosomes	440.0	12,320.0	28.0	1.0	100.0
KCl washes 1 and 2	56.0	3,679.2	65.7	2.3	29.9
Sephadex G-100	21.0	2,730.0	130.0	4.6	22.9
Phenyl-Sepharose	0.3	248.7	829.0	29.6	2.0

[a] Data are average values obtained from 30 independent enzyme purification experiments.

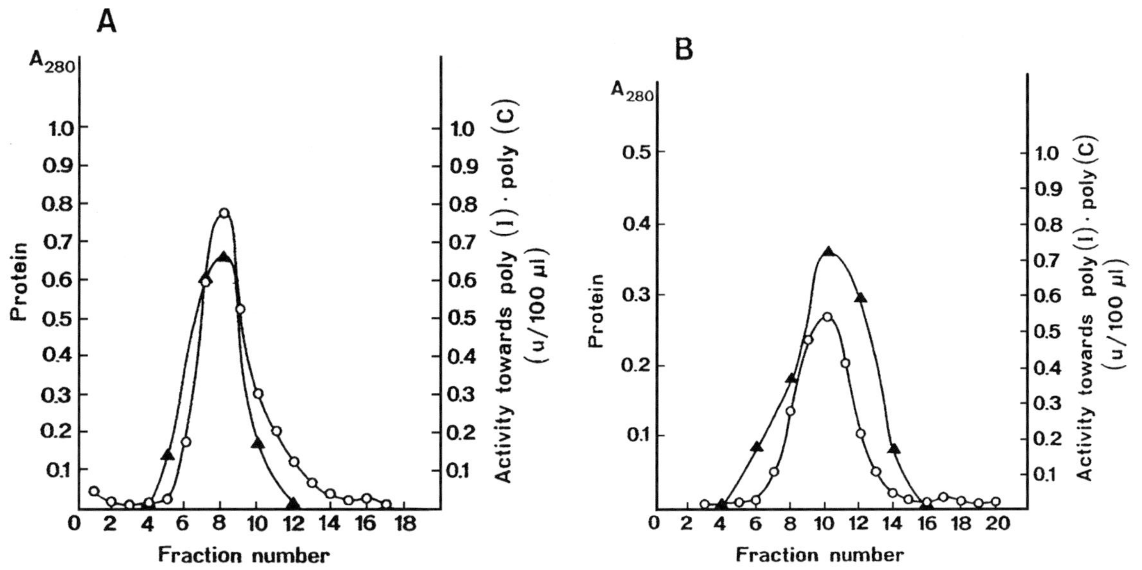

FIG. 1. Purification of ribosomal nuclease II onan octyl-Sepharose column (A) and a phenyl-Sepharose column (B). For elution, 10m*M* sodium phosphate buffer, pH 6.8, containing 10m*M* 2-mercaptoethanol and $(NH_4)_2SO_4$ to 25% saturation was used. (○) Protein concentration, (▲) activity toward poly(I) · (C). [Reprinted from M. A. Siwecka, *Acta Biochim. Polon.* **44,** 61 (1997), with permission.]

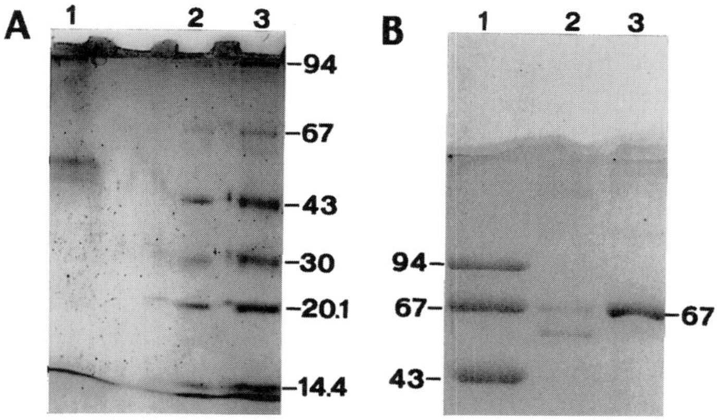

FIG. 2. Polyacrylamide gel electrophoretic analysis of rye germ nuclease II. (A) A 10% (w/v) polyacrylamide gel stained with Coomassie Brilliant Blue. Lane 1, proteins from the phenyl-Sepharose column. Lanes 2 and 3, protein standards: phosphorylase *b* (94 kDa), bovine serum albumin (67 kDa), ovalbumin (43 kDa), carbonic anhydrase (30 kDa), soybean trypsin inhibitor (20.1 kDa), and α-lactalbumin (14.4 kDa). (B) A 10% (w/v) Polyacrylamide gel that has been silver stained. Lane 1, protein standards: phosphorylase *b* (94 kDa), bovine serum albumin (67 kDa), and ovalbumin (43 kDa). Lane 2, proteins from the octyl-Sepharose column. Lane 3, bovine serum albumin (67 kDa). [Reprinted from M. A. Siwecka, *Acta Biochim. Polon.* **44,** 61 (1997), with permission.]

After phenyl-Sepharose column chromatography (Table II), an ~30-fold increase in enzyme specific activity toward poly(I) · poly(C) is achieved. This value is similar to that obtained in the previous procedure, while the yield of active material in the latter procedure is somewhat lower (2%).

The preparations of purified nuclease obtained after either procedure are analyzed by 10% (w/v) polyacrylamide gel electrophoresis in the presence of SDS. Both preparations show two bands (Fig. 2). The apparent molecular masses of the two bands, 62 and 57 kDa, are determined by comparison with protein standards.

Properties of Enzyme Preparation

Activators and Inhibitors

The activity toward poly(I) · poly(C) of washes 1 and 2, or of the enzyme preparations obtained after Sephadex G-100 filtration, is not affected by the addition of Mg^{2+} (0.5 m*M*) or Na^+ or K^+ (50 m*M*) ions. Mn^{2+} (5 m*M*) and 2-mercaptoethanol (10 m*M*) stimulate the activity at all purification steps (to 125 and 175% of initial activity, respectively).

pH Optimum

As found earlier for the ribosomal washes,[8] the activity toward poly(I) · poly(C) of the enzyme preparations obtained after Sephadex G-100 filtration and phenyl- or octyl-Sepharose column chromatography attains a maximum at pH 8.5. The pH optimum for activity toward RNA is observed at pH 7.8, and the pH optimum for activity toward denatured DNA is at pH 5.0.

Stability

The preparations of rye germ ribosomal nuclease obtained by either procedure are unstable. The ribosomal washes are most stable. After storage of the ribosomal washes for 2 months at −20° a 22% loss of activity toward poly(I) · poly(C) is observed. The enzyme preparations recovered from Sephadex G-100 kept under identical conditions lose half of their activity. Solutions of the purified enzyme kept at 4° for 1 week lose all activity.

Addition of glycerol to a concentration of 20–50% (v/v) stabilizes the purified enzyme. A similar effect is observed with 30% (w/v) trehalose. When glycerol or trehalose is added to purified enzyme after one thawing and freezing cycle it retains almost 90% of its initial activity.

Substrate Specificity

The relative activities of rye germ ribosomal nuclease II toward double- and single-stranded RNA and double- and single-stranded DNA are shown in Table III. The highest activity is observed toward dsRNA from *Penicillium chrysogenum* virus, but the enzyme also degrades highly polymerized RNA from wheat germ, and native and denatured DNA from calf thymus.

Nuclease II completely converts the closed, double-stranded, supercoiled form of phage ΦX174 DNA into the open (circular) form, and then to the linear form

TABLE III
DEGRADATION OF NUCLEIC ACID SUBSTRATES BY RYE GERM RIBOSOMAL NUCLEASE II

Substrate	Specific activity (U/mg protein)	Relative rate of hydrolysis (%)
Double-stranded RNA from virus of *Penicillium chrysogenum*	1116	100
Single-stranded RNA from wheat germ	891	80
Native DNA from calf thymus	785	70
Denatured DNA from calf thymus	640	57

TABLE IV
DEGRADATION OF HOMOPOLYNUCLEOTIDES AND DOUBLE-STRANDED COMPLEXES BY RYE GERM RIBOSOMAL NUCLEASE II

Substrate	Specific activity (U/mg protein)	Relative rate of hydrolysis (%)
Poly(I) · poly(C)	1045	100
Poly(A) · poly(U)	805	77
Poly(U)	794	76
Poly(A)	783	75
Poly(C)	616	59
Poly(I)	606	58
Poly(G)	125	12
Poly(dI) · poly(dC)	75	7

(M. A. Siwecka, unpublished data, 2000). This indicates that the enzyme acts endonucleolytically on double-stranded DNA.

Nuclease II shows different rates of activity toward various polynucleotides (Table IV). The highest activity is observed toward double-stranded complexes: poly(I) · poly(C) and poly(A) · poly(U), and then toward the single-stranded homopolyribonucleotides poly(U) and poly(A). The lowest activity is observed toward poly(G) and double-stranded complex poly(dI) · poly(dC).

Poly(I) · poly(C) is used as substrate to determine enzyme activity toward dsRNA at all steps of the purification procedure.

Nuclease II-Induced Cleavages in Lupin 5S rRNA

To gain more information regarding the structure and sequence specificity of nuclease II, experiments are performed with a substrate of well-defined structure, lupin 5S rRNA. Nuclease II cleaves lupin 5S rRNA at C10, U12, C15, U21, C26, C39, C49, and U53 (strong cleavage intensity). Moderate cleavage intensities are observed at C18 and C36, and weak cleavage intensities at U73, U76, and U79 (Figs. 3 and 4).

Nuclease S1 and RNase V1 cleavages induced in lupin 5S rRNA have been previously analyzed by Joachimiak *et al.*[21] They have observed strong digestion by S1 nuclease of loops A and B, and moderate digestion of loops C and D. Strong digestion by nuclease II is observed in loops A, B, and C (Fig. 4). Sites of strong

[21] A. Joachimiak, R. Basavappa, M. Z. Barciszewska, M. Nalaskowska, and J. Barciszewski, *Acta Biochim. Polon.* **37,** 359 (1990).

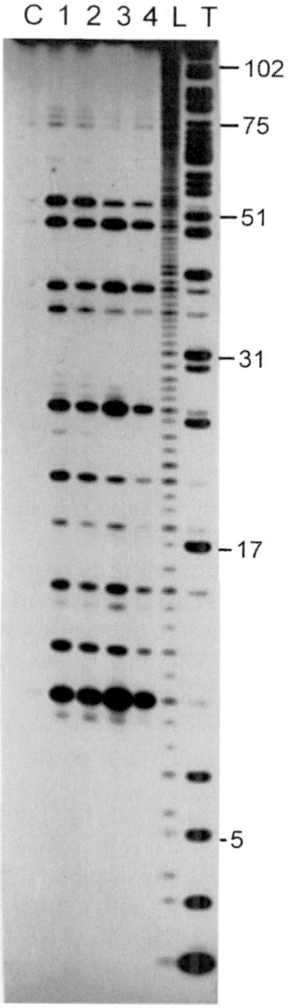

FIG. 3. Hydrolysis of 5′-^{32}P-labeled lupin 5S rRNA with rye germ nuclease II. Lane C, reaction control; lane L, formamide ladder; lane T, limited cleavage by RNase T1; lanes 1–4, 0.5, 1.0, 2.5, and 5.0 units of rye germ nuclease II [10 m*M* Tris-HCl (pH 8.5), 5 m*M* DTT, 30 min, 37°].

nuclease V1 digestion are observed in stems I, II, III, IV, and V, but also in loop B (C20, U21, and A22), at bulge C49, and in loop E (A71 and U73).[21]

It should be pointed out that some of the nuclease II and nuclease V1 nicks occur at the same sites: C18, U21, C49, and U73. It has also been found that all nuclease II cleavages occur at CpA or UpA phosphodiester bonds at the 5′ end

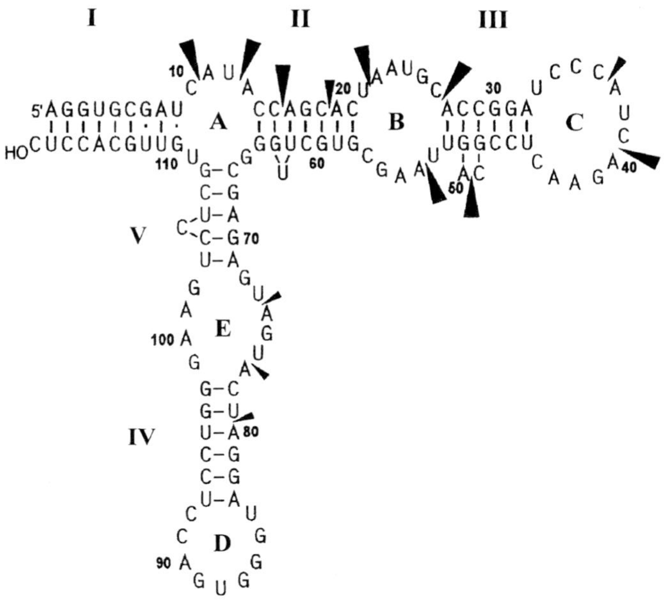

FIG. 4. Secondary structure model of *Lupinus luteus* 5S rRNA and specific cleavages by rye germ nuclease II. Arrowheads represent cleavage sites, with the size of the arrowhead indicating the relative level of cleavage.

of the substrate. The enzyme recognizes these CpA and UpA bonds if they are in double-stranded stems (C15, C18, and U79), in the bulge (C49), or in single-stranded loops (C10, U12, U21, C26, C36, C39, U53, U73, and U76).

Thus, it appears from this study that nuclease II is a base-specific enzyme that recognizes CpA and UpA bonds, both in base-paired and non-base-paired regions of RNA.

Conclusions

A novel nuclease associated with ribosomes and able to degrade double-stranded RNA was isolated from rye germ. Two methods were developed to purify of the enzyme; the first, described previously, uses ion-exchange chromatography on CM-cellulose and chromatofocusing on PBE 94 and PBE 118 gels. The second, presented here, uses affinity chromatography with octyl- or phenyl-Sepharose. The two methods yielded similar results: the nuclease preparations showed a specific activity toward poly(I) · poly(C) about 30-fold higher than the nuclease present in the ribosomes. The previous five-step method yielded an unstable product.[7] The newly introduced addition of glycerol stabilizes enzyme activity toward poly(I) · poly(C).

In addition, the recovery of protein was 5-fold higher than in the previous method. The final products obtained by the two methods (data from the first method not shown) gave, on SDS electrophoresis, two bands with molecular masses of 62 and 57 kDa. After elution from native and SDS gels the proteins were active toward the poly(I) · poly(C) substrate (not shown).

Gel electrophoretic analysis of tobacco anther proteins revealed four sugar-nonspecific nucleases active against DNA and RNA and three faster migrating RNases that were all able to digest dsRNA.[9] Bovine seminal ribonuclease, which degrades both single- and double-stranded RNAs, is isolated as a dimer, and dimerization is important for dsRNA cleavage.[22] The relation between the level of activity (toward ssRNA and dsRNA) of rye germ ribosomal nuclease II and the structure of the enzyme needs further investigation.

In addition to its activity toward dsRNA and ssRNA, nuclease II cleaves native and denatured DNA. These results indicate that it is a sugar-nonspecific nuclease.[11] However, the enzyme differs from the other plant nucleases, classified by Wilson as nucleases of type I,[11] by its ability to cleave dsRNA. Thus it appears that rye germ ribosomal nuclease II and the nucleases from tobacco anther[9] are examples of a new type of plant nuclease.

Like other plant nucleases rye germ nuclease II is most active toward poly(U) and poly(A).[11] Nuclease I, previously isolated from an ammonium chloride wash of rye germ ribosomes,[4] is most active toward poly(C).

More detailed studies with *Lupinus luteus* 5S rRNA confirmed the results concerning base specificity of rye germ nuclease II. It was established that nuclease II is base specific and recognizes and cleaves UpA and CpA bonds, not only in base-paired stems but also in internal loops.

Any mechanistic similarities between dsRNA cleavage by nuclease II and that of other dsRNA-enzymes such as RNase III are still uncertain. According to Sorrentino and Libonati[23] the efficiency of enzymatic cleavage of dsRNA by pancreatic-type RNases appears to be a function of the basicity of the enzyme. More precisely, enzyme activity can be related to the number of specifically located basic amino acid residues in the RNase molecule. The cleavage event could be the consequence of the preferential binding of the RNase—which is dependent on the number and location of the positive charges in the enzyme—to short sequences of the double-stranded RNA transiently made single stranded by the spontaneous thermal fluctuations of the RNA secondary structure.

The metabolic role of plant nucleases associated with ribosomes has so far not been elucidated.[1,4,8] Perhaps these enzymes have regulatory functions in mRNA

[22] G. D'Alessio, A. D. Di Donato, A. Parente, and R. Piccoli, *Trends Biochem. Sci.* **16,** 104 (1991).
[23] S. Sorrentino and M. Libonati, *FEBS Lett.* **404,** 1 (1997).

decay or in controlling the level of antisense RNA–mRNA transcripts.[24] It has been suggested that ribosomes carry a nuclease capable of degrading mRNA, and that ribosome movement causes changes in mRNA conformation that unmask nuclease target sites.[24] Indeed, in preliminary experiments it was found that rye germ ribosomal nuclease II degrades repeat sequences of certain mRNAs.

It was suggested[25] that posttranscriptional gene silencing (or cosuppression) in plants requires the participation of dsRNA nuclease activity. Gallie[25] proposed three potential mechanisms to explain cosuppression. The first is an aberrant RNA model, in which transgene RNA from a complex multicopy T-DNA may form extended regions of double-stranded RNA, which may be degraded by active dsRNA nucleases or may be recognized by dsRNA-activated kinases, which in turn activate dsRNA nucleases. The second mechanism is an RNA threshold model. High levels of transgene RNA from simple multicopy T-DNA loci may form localized regions of dsRNA that may be degraded by dsRNA nucleases. Alternatively, the high level of RNA may serve as a template for endogenous RNA-dependent RNA polymerase to produce antisense RNA that forms dsRNA with the template and is a target for dsRNA nucleases. The third model proposed for the mechanism of cosuppression is transgene-mediated viral resistance: replication of sense viral RNA by an RNA-dependent RNA polymerase would result in production of antisense RNA, followed by synthesis of sense genomic and subgenomic RNAs. Expression of viral transgene RNA in combination with replicating antisense viral RNA may cause activation of dsRNA nucleases and repression of further viral replication.

It is conceivable that nuclease II associated with rye germ ribosomes may be involved in posttranscriptional silencing.

Acknowledgments

The author thanks Dr. M. Napierała for providing lupin 5S rRNA and performing the experiment with nuclease II (Fig. 3), and is indebted to Professor A.-L. Haenni and Professor M. Libonati for critically reading the manuscript. Some of these investigations were supported by the State Committee for Scientific Research (KBN), Grant 6P2030400.

[24] G. Brawerman, *in* "Control of Messenger RNA Stability" (J. Belasco and G. Brawerman, eds.), p. 157. Academic Press, London, 1993.

[25] D. R. Gallie, *Curr. Opin. Plant Biol.* **1,** 166 (1998).

[18] Yeast mRNA Decapping Enzyme

By TRAVIS DUNCKLEY and ROY PARKER

Introduction

An important step in gene regulation occurs at the level of mRNA stability (for reviews, see Refs.1–3). In the yeast *Saccharomyces cerevisiae* mRNAs are degraded through two general pathways. The predominant decay pathway occurs via shortening of the poly(A) tail, followed by removal of the 5′ cap structure by the Dcp1p decapping enzyme.[4] Decapping of the mRNA exposes the body of the message to Xrn1p-mediated 5′→3′-exonucleolytic degradation.[5–8] The second general decay pathway occurs by poly(A) shortening followed by 3′→5′-exonucleolytic degradation of the mRNA.[8,9]

Decapping is a key step in the major pathway of 5′→3′ mRNA decay in yeast because removal of the cap structure is a prerequisite to 5′→3′-exoribonucleolytic degradation of the mRNA.[4–6] Different mRNAs also have message-specific rates of decapping, suggesting that the process of removing the 5′ cap structure is controlled.[7,8] Moreover, the process of mRNA surveillance, wherein aberrant mRNAs are recognized, occurs by rapid deadenylation-independent decapping.[10] A yeast decapping enzyme, encoded by the *DCP1* gene, has been identified and shown to be required for all decapping in yeast.[4,11] In this article we briefly review the process of mRNA decapping including the properties of Dcp1p and the proteins that modulate decapping rates *in vivo* (for a more comprehensive review of decapping see Tucker and Parker[12]). In addition, we describe techniques for analyzing mRNA decapping *in vitro*.

Dcp1p: The Enzyme

Purification of Dcp1p and subsequent *in vitro* characterization of the enzyme have yielded several important observations.[11,13] First, the Dcp1 protein

[1] J. Ross, *Microbiol. Rev.* **59,** 423 (1995).
[2] G. Caponigro and R. Parker, *Microbiol. Rev.* **60,** 233 (1996).
[3] A. Jacobson and S. W. Peltz, *Annu. Rev. Biochem.* **65,** 693 (1996).
[4] C. Beelman, A. Stevens, G. Caponigro, T. LaGrandeur, L. Hatfield, D. Fortner, and R. Parker, *Nature (London)* **283,** 642 (1996).
[5] C. J. Decker and R. Parker, *Genes Dev.* **7,** 1632 (1993).
[6] C. L. Hsu and A. Stevens, *Mol. Cell. Biol.* **13,** 4826 (1993).
[7] D. Muhlrad, C. Decker, and R. Parker, *Genes Dev.* **8,** 855 (1994).
[8] D. Muhlrad, C. Decker, and R. Parker, *Mol. Cell. Biol.* **15,** 2145 (1995).
[9] J. S. Anderson and R. Parker, *EMBO J.* **17,** 1497 (1998).
[10] D. Muhlrad and R. Parker, *Nature (London)* **370,** 578 (1994).

0076-6879/01 $35.00

```
                                              23                                               71
                                              |                                                |
DCP1 (NC_001147.1)                            NFNVIGRYDPKIKQLLFHTPHASLYKWDFKKDEWNKLEYQGVLAIYLRD
Candida albicans (54/66)                      NFNVISRYDPAIKQLLCNTSHCVLYNFNDETEEWVKSDFQGTLALYVRD
Arabidopsis (AAC17938.1) (22/51)              NLTVLQRIDPYIEEILITAAHVTFYEFNIELSQWSRKDVEGSLFVVKRS
Drosophila (AAF47089.1) (28/56)               NLAAIKKIDPYAKEIVDSSSHVAFYTFNSSQNEWEKTDVEGAFFIYHRN
Homo Sapiens (CAB77023.1)  (26/51)            SLAALKQHDPYITSIADLTGQVALYTFCPKANQWEKTDIEGTLFVYRRS
```

FIG. 1. Proteins related to Dcp1p. Shown is an alignment with Dcp1p and several related proteins in the database. Only the most highly conserved region of the protein is shown. Percentage identity and similarity are indicated in parentheses. Boxed residues denote completely conserved residues that are known to be critical for Dcp1p function.[14] Lighter shaded boxes indicate similarity and darker boxes indicate identity. Only amino acids conserved between Dcp1p and at least two other homologs are shaded. The *Candida albicans* sequence was obtained from the Candida database (*http://alces.med.umn.edu/Candida.html*).

is sufficient for the decapping of mRNAs *in vitro.* This indicates that Dcp1p is the yeast mRNA decapping enzyme. Second, Dcp1p cleaves cap structures to the products m^7GDP and the full-length mRNA containing a 5′-phosphate. Third, Dcp1p preferentially cleaves substrates containing 7-methylguanine over those containing an unmethylated cap. This demonstrates that the 7-methyl modification on the cap structure contributes to the specificity of Dcp1p. Fourth, Dcp1p shows enhanced decapping activity on longer mRNA substrates. *In vitro,* Dcp1p does not efficiently decap messenger RNAs shorter than 25 nucleotides in length, indicating that Dcp1p likely recognizes a portion of the mRNA in addition to the 7-methyl modification. Consistent with Dcp1p requiring recognition of the mRNA body, the *in vitro* activity of Dcp1p is inhibited by the addition of uncapped mRNA but not by the addition of cap analog, m^7GpppG.[11] Fifth, the DCP1 protein requires divalent cation in the form of either Mg^{2+} or Mn^{2+}, with a preference for magnesium.[11] Sixth, purified Dcp1p for model substrates shows a K_m of approximately 15 n*M*.[11,13] Last, Dcp1p is known to be a phosphoprotein when purified from yeast, although the precise modification and its significance for enzymatic activity are currently unknown. Nevertheless, a requirement for posttranslational modification may explain why Dcp1p expressed in *Escherichia coli* is enzymatically inactive.

Because decapping of mRNAs is likely a conserved process it is anticipated that there will be homologs of Dcp1p in other eukaryotes. Examination of the available databases identifies a family of proteins that are similar to Dcp1p (Fig. 1). Importantly for Dcp1p, mutational analyses have demonstrated that altering the residues conserved among this family of proteins has yielded the strongest loss of function alleles in Dcp1p (Tharun and Parker[14]; boxed residues in Fig. 1). This

[11] T. E. LaGrandeur and R. Parker, *EMBO J.* **17,** 1487 (1998).

[12] M. Tucker and R. Parker, *Annu. Rev. Biochem.* **69,** 571 (2000).

[13] A. Stevens, *Biochem. Biophys. Res. Commun.* **96,** 1150 (1980).

[14] S. Tharun and R. Parker, *Genetics* **151,** 1273 (1999).

suggests that, although the overall similarity is not high, these proteins may be functional homologs of Dcp1p.

Additional Decapping Factors

In addition to Dcp1p, several additional proteins have been shown to affect the process of mRNA decapping (Table I[15–26]). Among these is the *DCP2*-encoded protein. The *DCP2*-encoded protein was identified as a high-copy suppressor of a conditional *dcp1* mutant.[15] Subsequent analysis of the *dcp2*Δ mutant demonstrated that Dcp2p is absolutely required for decapping of both normal and aberrant mRNAs. Dcp2p physically associates with the Dcp1p decapping enzyme and it appears that the interaction between Dcp1p and Dcp2p is required for the activation of Dcp1p (Dunckley and Parker[15]; and our unpublished observations, 2000). The specific mechanism whereby Dcp2p activates Dcp1p is unclear. However, because Dcp2p possesses a functional MutT motif, the Dcp2p-mediated activation of Dcp1p likely requires cleavage of an as yet unknown pyrophosphate bond (for a review of the MutT motif see Bessman *et al.*,[27] and Koonin[28]).

The Lsm (like Sm) proteins represent another set of proteins that influence the mRNA decapping rate. There are seven Lsm proteins (Lsm1 to Lsm7) that affect mRNA decapping.[16–18] The Lsm proteins are related to the Sm proteins, which are components of small nuclear RNAs (snRNAs) and appear to form a seven-membered protein complex that binds to mRNA to promote mRNA decapping.[17,18] In addition, the Lsm protein complex involved in mRNA decay stably interacts with the *PAT1/MRT1* gene product, which is an additional decapping factor.[17,18,29] However, neither the Lsm complex nor Pat1/Mrt1p is required for decapping through the mRNA surveillance pathway. Thus, the pat1/Mrt1–Lsm complex functions in

[15] T. Dunckley and R. Parker, *EMBO J.* **18,** 5411 (1999).
[16] R. Boeck, B. Lapeyre, C. E. Brown, and A. B. Sachs, *Mol. Cell. Biol.* **18,** 5062 (1998).
[17] S. Tharun, W. He, A. E. Mayes, P. Lennertz, J. D. Beggs, and R. Parker, *Nature (London)* **404,** 515 (2000).
[18] E. Bouveret, G. Rigaut, A. Shevchenko, M. Wilm, and B. Serapin, *EMBO J.* **19,** 1661 (2000).
[19] S. W. Peltz, A. H. Brown, and A. Jacobson, *Genes Dev.* **7,** 1737 (1993).
[20] Y. Cui, K. W. Hagan, S. Zhang, and S. W. Peltz, *Genes Dev.* **9,** 423 (1995).
[21] F. He and A. Jacobson, *Genes Dev.* **9,** 437 (1995).
[22] G. Caponigro and R. Parker, *Genes Dev.* **9,** 2421 (1995).
[23] D. C. Schwartz and R. Parker, *Mol. Cell. Biol.* **19,** 5247 (1999).
[24] D. C. Schwartz and R. Parker, submitted (2001).
[25] S. Zhang, C. J. Williams, K. Hagan, and S. W. Peltz, *Mol. Cell. Biol.* **19,** 7568 (1999).
[26] D. Zuk, J. P. Belk, and A. Jacobson, *Genetics* **153,** 35 (1999).
[27] M. J. Bessman, D. N. Frick, and S. F. O'Handley, *J. Biol. Chem.* **271,** 25059 (1996).
[28] E. V. Koonin, *Nucleic Acids Res.* **21,** 4847 (1993).
[29] L. Hatfield, C. A. Beelman, A. Stevens, and R. Parker, *Mol. Cell. Biol.* **16,** 5830 (1996).

TABLE I
PROTEINS THAT AFFECT mRNA DECAPPING

Gene	Protein function	Refs.
DCP2	Activates Dcp1p	*a*
LSM1–LSM7	Protein complex that binds mRNA to stimulate decapping	*b*
PAT1/MRT1	Interacts with Lsm complex to stimulate decapping	*c*
UPF1, UPF2, UPF3	Activates mRNA surveillance	*d*
PAB1	Inhibits decapping	*e*
eIF-4E	Binding to cap inhibits mRNA decapping rate	*f*
VPS16	Mutation inhibits decapping activity	*g*
GRC5, SLA2, MRT4, THS1	Affects decapping rates	*h*

[a] Dunckley and Parker (1999).[15]
[b] Boeck *et al.* (1998),[16] Tharun *et al.* (2000),[17] and Bouveret *et al.* (2000).[18]
[c] Tharun *et al.* (2000)[17] and Bouveret *et al.* (2000).[18]
[d] Peltz *et al.* (1993),[19] Cui *et al.* (1995),[20] and He and Jacobson (1995).[21]
[e] Caponigro and Parker (1995).[22]
[f] Schwartz and Parker (1999)[23] and Schwartz and Parker (2000).[24]
[g] Zang *et al.* (1999).[25]
[h] Zuk *et al.* (1999).[26]

the deadenylation-dependent decapping pathway to bind to mRNAs to promote decapping by a yet to be determined mechanism.

A different set of three proteins functions exclusively to activate decapping through the mRNA surveillance pathway (reviewed in Jacobson and Peltz[3]). These proteins, referred to as Upf1, Upf2, and Upf3, interact with each other as well as with translation termination factors.[30] This has led to the idea that the Upf protein complex binds to the translation termination machinery and signals a rapid, deadenylation-independent decapping event if translation termination is premature.

Translation and Decapping

In addition to the activators of decapping discussed above, translation initiation factors are in some cases inhibitors of decapping. For example, the poly(A) tail and the associated poly(A) binding protein inhibit decapping.[22] Similarly, several lines of evidence suggest that there is a competition between the cap binding protein eIF-4E and Dcp1 for the cap structure. For example, eIF-4E inhibits decapping in a purified system.[24] Moreover, *in vivo* mutations in eIF-4E increase the rates of

[30] K. Czaplinski, M. J. Ruiz-Echevarria, S. V. Paushkin, X. Han, Y. Weng, H. A. Perlick, H. C. Dietz, M. D. Ter-Avanesyan, and S. W. Peltz, *Genes Dev.* **12,** 1665 (1998).

decapping and can suppress partial loss of function alleles in Dcp1p.[23,24] These observations imply that more highly translated mRNAs are stable, at least in part, because they are bound to initiation factors more frequently than are poorly translated mRNAs, thereby slowing the decapping rate for highly translated mRNAs. An important goal of future work is to understand the relationship between the decapping activator proteins and the inhibition of decapping imposed by the translation initiation machinery.

Methods

A useful tool for the analysis of Dcp 1 has been the ability to assay the decapping reaction *in vitro*. This has allowed an analysis of the effects of various proteins on the activity of Dcp1p and has enabled the analysis of the effects of various *dcp1* mutants on enzymatic activity. A simple *in vitro* decapping assay utilizing synthetic, cap-labeled substrates and examining the reaction products by thin-layer chromatography has been developed by modifications of the methods of Stevens[31] and Zhang *et al.*[32] This assay can be performed with purified Dcp1p, and also proteins that affect decapping such as Dcp2p[15] or eIF-4E,[24] thereby assessing their effects on Dcp1p activity. In addition, such an assay can be applied to crude extracts, thereby allowing the analysis of decapping in other organisms,[32] or biochemical mutant screens.[25] Below we describe the *in vitro* decapping assay from the initial purification of Dcp1p to the analysis of the reaction products. Methods for each of these procedures have been published previously.[4,11,14,31,32] However, we present them here, with modifications to the Dcp1p purification protocol, as a convenient reference.

Purification of Dcp1p

Dcp1p has been purified from yeast by using a number of different epitope tags. The best yields have been obtained from overexpressing, from the glyceraldehyde phosphate dehydrogenase (GPD) promoter, the Dcp1p protein containing a hexahistidine (His_6) epitope at its N terminus. The following purification procedure should yield between 1 and 5 mg of highly purified Dcp1p.

1. Grow 2 liters of a yeast strain expressing the Dcp1p fusion protein from the GPD promoter to an OD_{600} of 1.0 to 1.5. Pellet the cells and wash once with 30 ml of cold (4°) doubly distilled H_2O. Resuspend the washed cell pellet in 6 ml of lysis buffer [50 m*M* $NaPO_4$ (pH 8.0), 300 m*M* NaCl, 2 m*M* 2-mercaptoethanol, 1 tablet

[31] A. Stevens, *Mol. Cell. Biol.* **8,** 2005 (1988).

[32] S. Zhang, C. J. Williams, M. Wormington, A. Stevens, and S. W. Peltz, *Methods* **17,** 46 (1999).

of EDTA-free protease inhibitor cocktail from Roche/Boehringer Mannheim (Indianapolis, IN)].

2. Add 4.5 ml of acid-treated glass beads. Lyse the cells by vortexing for 30 sec at 4° and placing the tube on ice for 30 sec. Repeat this for 10 cycles of vortexing and cooling.

3. Centrifuge the lysate at 4° for 20 min at 15000 rpm. Add the supernatant fraction to 1 ml of Ni^{2+}–nitrilotriacetic acid (NTA) agarose that has been equilibrated in lysis buffer. Incubate the slurry at 4° for 1 hr with gentle agitation. Load the slurry onto a 15-ml column. Reapply the flow-through to the column and then wash sequentially with 15 ml of buffer 1 (lysis buffer without protease inhibitors), 30 ml of buffer 2 [50 m*M* $NaPO_4$ (pH 6.0), 1 *M* NaCl, 5 m*M* 2-mercaptoethanol], and 10 ml of buffer 3 [50 m*M* $NaPO_4$ (pH 7.0), 25 m*M* NaCl, 5 m*M* 2-mercaptoethanol, 10% (v/v) glycerol]. For the final two washes use 10 ml of buffer 3 containing 50 m*M* imidazole and then 3 ml of buffer 3 containing 75 m*M* imidazole. The low concentrations of imidazole in the last two washes help to increase the purity of the preparation by eluting proteins that may bind nonspecifically to Ni^{2+}.

4. Elute the His_6–Dcp1p from the column by washing with 4 ml of buffer 3 containing 250 m*M* imidazole. Finally, because the imidazole that is required for the elution of hexahistidine fusion proteins can interfere with the *in vitro* decapping assay described below, the purified Dcp1p should be dialyzed against 4 liters of 5 m*M* Tris-HCl (pH 7.6), 10 m*M* NaCl, 200 m*M* dithiothreitol (DTT). This sample may then be concentrated to 500 μl. For storage in aliquots at $-80°$ nonidet P-40 (NP-40) and glycerol should then be added to a final concentration of 0.1 and 20% (v/v), respectively.

Preparation of Substrate RNA

Any mRNA may be *in vitro* transcribed and used for a decapping assay because the decapping product is the same for any mRNA.[11]

1. Transcribe the mRNA by standard *in vitro* transcription methods. To ensure that all of the mRNA is of uniform length, the RNA should be gel purified in a 6% (w/v) denaturing polyacrylamide gel. By UV shadowing, the full-length mRNA can be observed. The full-length mRNA is excised from the gel and eluted by any standard technique. The eluted mRNA should be phenol–chloroform extracted and ethanol precipitated.

2. Cap the mRNA *in vitro,* using [α-^{32}P]GTP, a methyl donor for formation of the 7-methyl modification on the cap, and vaccinia virus guanylyltransferase (Gibco-BRL, Gaithersburg, MD). Specifically, the capping reaction may be performed as follows: combine 25 pmol of mRNA, 45 pmol of [α-^{32}P]GTP, 4.5 μl of 10× capping buffer [500 m*M* Tris-HCl (pH 7.7), 20 m*M* $MgCl_2$, 10 m*M* DTT, 60 m*M* KCl], 3 μl of 10 m*M* *S*-adenosylmethionine, 1 μl (40 units) of rRNasin, 14 μl

of doubly distilled H_2O, and 6 μl (20 units) of guanylyltransferase and incubate the reaction for 2 hr at 37°. Add 150 μl of RNase-free TE and remove the unincorporated GTP by purification over a Sephadex G-50 column. To ensure that the capped RNA is not degraded during purification, we often spin the RNA directly into phenol–chloroform preequilibrated in TE and quickly extract the samples. To increase the yield of capped mRNA, wash the column once with 100 μl of TE and collect the elution as described previously in phenol–chloroform. The purity of the substrate can by analyzed by polyethyleneimine (PEI)–cellulose thin-layer chromatography (TLC) (described below) or polyacrylamide gel electrophoresis.

In Vitro Decapping Assay

The *in vitro* decapping assay requires three components: a decapping protein, a cap-labeled mRNA substrate, and decapping buffer. This assay monitors the release of the decapping product, m^7GDP, from the cap-labeled mRNA substrate. The reaction is performed as follows: combine 1.5 μl of 10× decapping buffer [500 m*M* HEPES (pH 7.0), 10 m*M* $MgCl_2$, 10 m*M* DTT, 0.5% (v/v) NP-40], 1 μl of purified Dcplp (100 ng), 1 μl of capped mRNA (30,000 cpm), and 11.5 μl of doubly distilled H_2O. Incubate at 30° for 20 min. Stop the reaction by adding 1 μl of 500 m*M* EDTA. To ensure a strong signal, spot at least 10,000 cpm onto a PEI–cellulose TLC plate that has been prerun for several minutes (see below). To analyze decapping over a time course (Fig. 2B), remove aliquots of the reaction at

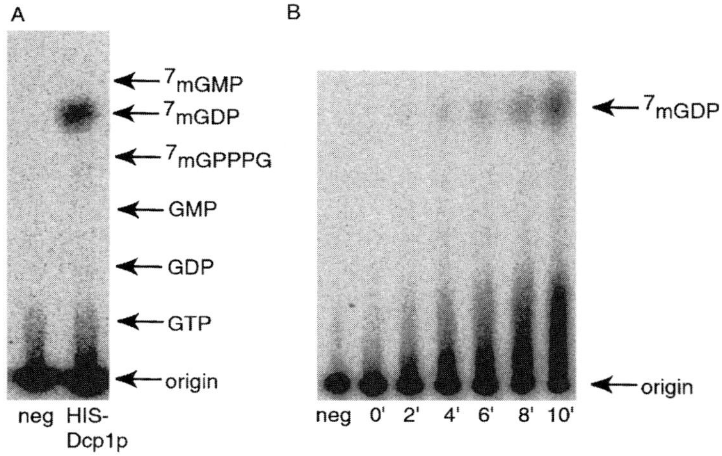

FIG. 2. *In vitro* decapping assays. (A) A representative decapping assay using purified His_6–Dcp1p and the decapping protocol described herein is shown. The His_6–Dcp1p sample shows the amount of m^7GDP released after a 20-min reaction. Indicated to the side of the TLC plate for comparison are the relative mobilities of other guanine nucleotide derivatives. (B) A representative decapping time course is shown. Protein purification and decapping assays were performed as described in Methods.

the desired times and rapidly stop the reaction by adding EDTA and placing the aliquots on ice.

Thin-Layer Chromatography

Thin-layer chromatography (TLC) is a relatively rapid and straightforward technique for separating small molecules on the basis of their relative hydrophobicity. To use this technique to analyze the products of a decapping reaction, the PEI–cellulose TLC plates must be prerun in 450 m*M* ammonium sulfate for 5 min. After the prerunning, the individual decapping reactions are spotted near the bottom of the TLC plate slightly above the running buffer (450 m*M* ammonium sulfate). The plates are then incubated in the buffer for 2 hr. With PEI–cellulose TLC, more positively charged molecules migrate more rapidly up the TLC plate. This allows the separation of the decapping product from the full-length mRNA, which remains at the origin (Fig. 2A). The relative positions where various other nucleotides run in this type of TLC plate is shown in Fig. 2A.

This protocol can be adapted in several ways, depending on the objective of the experiment. For example, to rapidly screen yeast strains for decapping defects, the assay can be performed with 20 to 50 μg of a crude yeast cell extract.[32] This assay also is applicable to studying the decapping activity of *Xenopus laevis* oocyte extracts.[32] That cell types other than yeast possess mRNA decapping activity highlights the point that mRNA decapping is a conserved process.

[19] RNA Lariat Debranching Enzyme

By SIEW LOON OOI, CHARLES DANN III, KIEBANG NAM, DANIEL J. LEAHY, MASAD J. DAMHA, and JEF D. BOEKE

Introduction

Nucleic acids are usually linked by 3′–5′ phosphodiester bonds. Interestingly, in both prokaryotes and eukaryotes, there exist low levels of nucleic acids containing 2′–5′ phosphodiester bonds. This linkage results in unusual branched nucleic acids that can form either a fork or lariat structure (Fig. 1A and B). Examples of branched RNAs include intron lariats in eukaryotes,[1] Y-like transplicing

[1] H. Domdey, B. Apostol, R. J. Lin, A. Newman, E. Brody, and J. Abelson, *Cell* **39,** 611 (1984).

0076-6879/01 $35.00

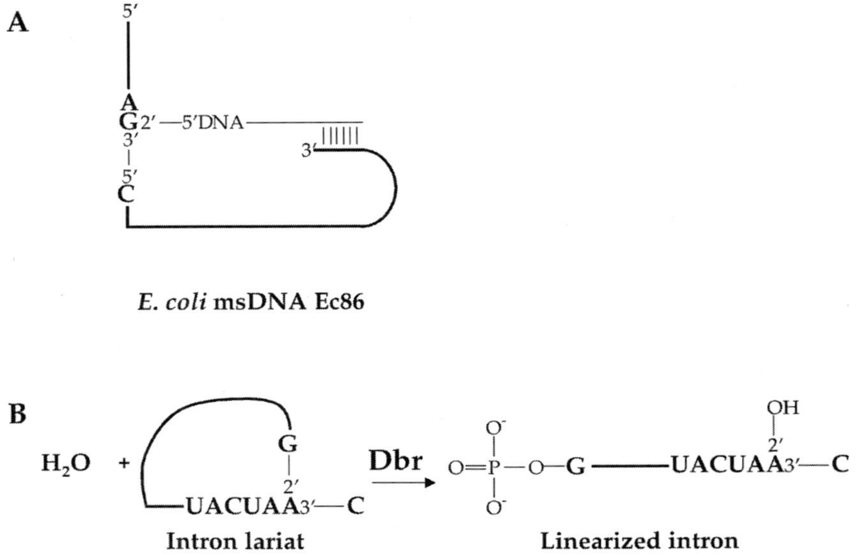

FIG. 1. (A) Fork structure of *E. coli* branched msDNA Ec86. The msDNA is linked to the G residue of msRNA via a 2′–5′ phosphodiester bond. (B and C) Debranching reaction. The 2′–5′ phosphodiester bond at the branch point A residue of intron lariat is hydrolyzed by lariat debranching enzyme to produce linear RNA with a 2′-hydroxyl group at the branch point A. The arrow indicates the bond hydrolyzed by debranching enzyme. The 3′ functional group of the branch point ribose is represented by a C residue; however, the *in vivo* structure of the 3′ functional group is not known. RNA residues are represented by boldface letters and lines.

intermediates in trypanosomes,[2,3] and group II intron lariats in eukaryotic organelles and prokaryotes.[4] Multicopy single-stranded DNAs (msDNAs)[5–7] form a set of RNA molecules with a DNA branch found in certain prokaryotes. In msDNAs, the DNA is linked to the RNA via a 2′–5′ phosphodiester bond, creating a branched fork structure (Fig. 1A).

Branched nucleic acids were first discovered by Wallace and Edmonds in the population of nuclear polyadenylated RNAs.[8] These branched nucleic acids are mostly derived from intron lariats, one of the intermediate structures in pre-mRNA

[2] R. E. Sutton and J. C. Boothroyd, *Cell* **47,** 527 (1986).
[3] W. J. Murphy, K. P. Watkins, and N. Agabian, *Cell* **47,** 517 (1986).
[4] F. Michel and J. L. Ferat, *Annu. Rev. Biochem.* **64,** 435 (1995).
[5] T. Furuichi, A. Dhundale, M. Inouye, and S. Inouye, *Cell* **48,** 47 (1987).
[6] T. Furuichi, S. Inouye, and M. Inouye, *Cell* **48,** 55 (1987).
[7] D. Lim and W. K. Maas, *Cell* **56,** 891 (1989).
[8] J. C. Wallace and M. Edmonds, *Proc. Natl. Acad. Sci. U.S.A.* **80,** 950 (1983).

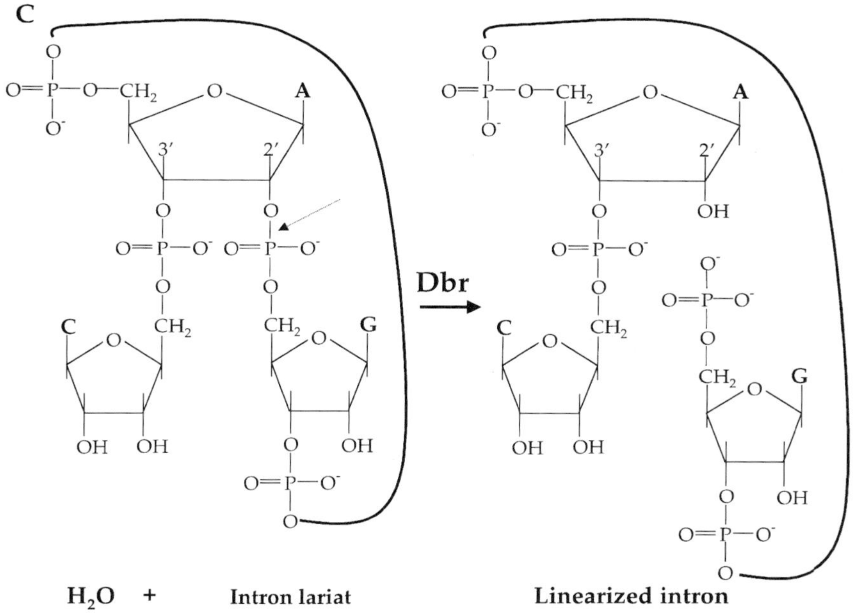

FIG. 1. (*Continued*)

splicing. Pre-mRNA splicing, a eukaryote-specific process, generates two products: the spliced exons and intron lariats. The 2′–5′ phosphodiester bonds of intron lariats are then hydrolyzed by the RNA lariat debranching enzyme (Dbr).[9–11] Dbr is thus a 2′→5′-phosphodiesterase. It specifically hydrolyzes the 2′–5′ phosphodiester linkage of RNA intron lariats between the G residues of the 5′ splice site and the A residues of the branch point (Fig. 1B and C). This cleavage converts the intron lariat into a linearized intron and is the rate-limiting step in intron degradation pathway. Linearized introns can then be rapidly degraded by cellular exonucleases.

Genetics and Biology

The gene that encodes RNA lariat debranching enzyme (Dbr) is highly conserved, and has been cloned from the yeast *Saccharomyces cerevisiae*,[11] *Schizosaccharomyces pombe*,[12] *Caenorhabditis elegans*,[12] and human[13] (Fig. 2). Dbr also

[9] B. Ruskin and M. R. Green, *Science* **229,** 135 (1985).

[10] J. Arenas and J. Hurwitz, *J. Biol. Chem.* **262,** 4274 (1987).

[11] K. B. Chapman and J. D. Boeke, *Cell* **65,** 483 (1991).

[12] K. Nam, G. Lee, J. Trambley, S. E. Devine, and J. D. Boeke, *Mol. Cell. Biol.* **17,** 809 (1997).

[13] J. W. Kim, H.-C. Kim, G.-M. Kim, J.-M. Yang, J. D. Boeke, and K. Nam, *Nucleic Acids Res.* **28,** 3666 (2000).

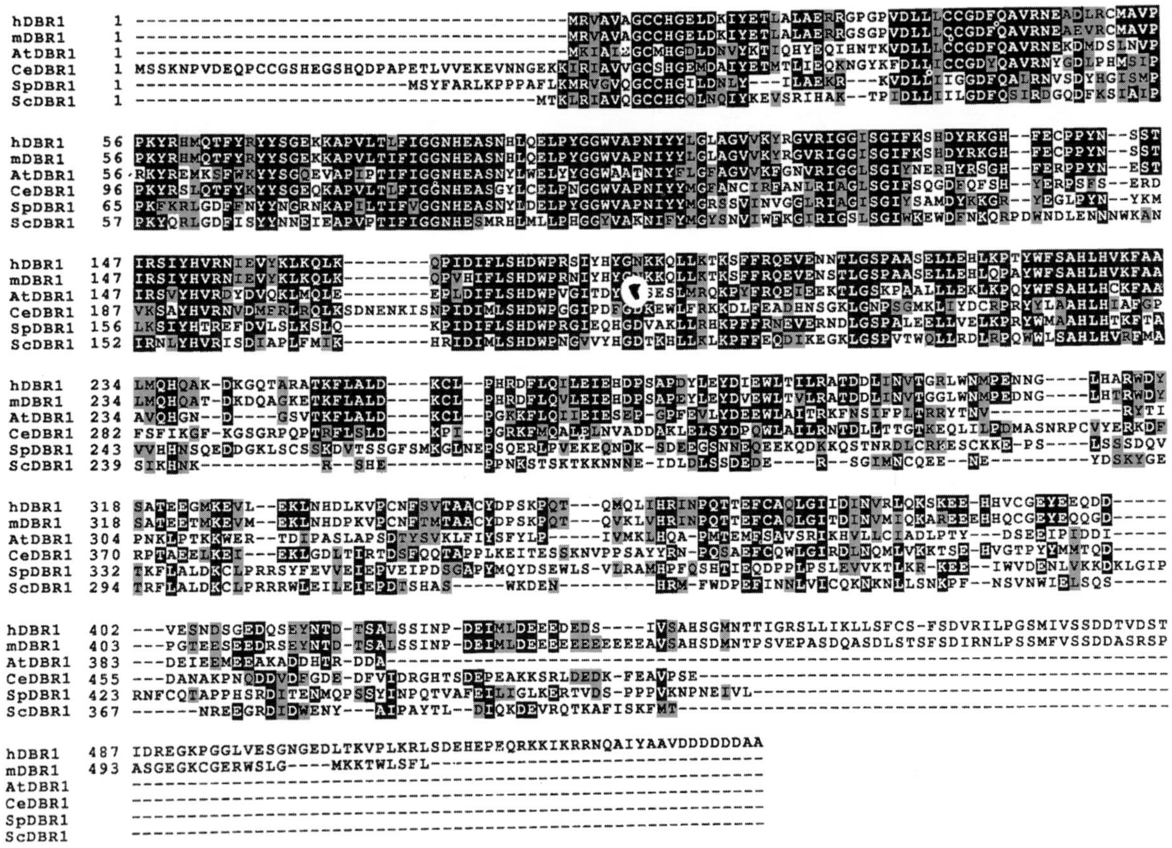

FIG. 2. Multiple amino acid sequence alignment of *S. cerevisiae* (ScDBR1), *S. pombe* (SpDBR1), *C. elegans* (CeDBR1), *Arabidopsis thaliana* (AtDBR1), mouse (mDBR1), and human (hDBR1) RNA lariat debranching enzyme. The PILEUP program from the Genetics Computer Group (GCG) was used to align the amino acid sequences. Amino acid identities are white letters in black boxes; similarities are shaded. [Adapted with permission from Fig. 1 of Kim *et al.*[13]]

shares the GD/GNH signature with other phosphoesterases, suggesting a common evolutionary origin.[14] The original mutant allele *dbr1-1* was isolated in *S. cerevisiae* on the basis of its reduced Ty1 transposition frequency.[11] *Saccharomyces cerevisiae dbr1* mutants are viable, have modest growth defects, and accumulate intron lariats missing the linear sequences on the 3′ side of the branch point.[11] *Saccharomyces cerevisiae dbr1* mutants have mild pseudohyphal growth defects[15] and intronic small nucleolar RNAs (snoRNAs) accumulate in the lariat form.[16,17] Pre-mRNA splicing is not affected, as expression of many mature mRNAs analyzed does not differ from that of wild-type cells. *Schizosaccharomyces pombe dbr1* mutants also accumulate introns.[12] However, in contrast to *S. cerevisiae dbr1* mutants, which have modest growth defects, *S. pombe dbr1* mutants have a severe growth phenotype. The phenotypes of *dbr1* mutants are intriguing, but it is not clear whether the enzyme plays a direct role in the processes affected.

Chemistry and Substrate Requirement

The RNA lariat debranching enzyme (Dbr) hydrolyzes only the 2′–5′ phosphodiester bonds, yielding a 2′-hydroxyl group at the branch attachment site and a 5′-phosphate end(Fig. 1B and C).[10] Dbr leaves all 3′–5′ phosphodiester bonds intact, distinguishing it from the snake venom phosphodiesterase, which hydrolyzes both 2′–5′ and 3′–5′ phosphodiester bonds.

Dbr is not capable of hydrolyzing oligonucleotides consisting of adenosine residues joined by 2′–5′ phosphodiester bonds that entirely lack 3′–5′ phosphodiester bonds.[9] An H[9] or OH[18,19] group at the 3′ position will not support cleavage, while a phosphate group[18,19] at the 3′ position supports cleavage extremely inefficiently. This suggests that Dbr needs a 3′–5′ phosphodiester bond at the branch point for its activity. Dbr cannot remove a 2′ phosphomonoester, distinguishing it from other phosphatases.[10]

Branched DNA[18] and RNA molecules with DNA branches[20] could also serve as substrates for Dbr. With respect to nucleotide sequence requirement, at the branch position, an A residue is debranched more efficiently than C.[20] At the 2′ position, an A residue is debranched more efficiently than both C and T pyrimidine residues.[20] This suggests that a purine is preferred at the 2′ position. Substrates

[14] E. V. Koonin, *Protein Sci.* **3,** 356 (1994).

[15] H. U. Mosch and G. R. Fink, *Genetics* **145,** 671 (1997).

[16] S. L. Ooi, D. A. Samarsky, M. J. Fournier, and J. D. Boeke, *RNA* **4,** 1096 (1998).

[17] E. Petfalski, T. Dandekar, Y. Henry, and D. Tollervey, *Mol. Cell. Biol.* **18,** 1181 (1998).

[18] S. L. Ooi, A. Liscio, J. D. Boeke, and M. J. Damha, unpublished results (1999).

[19] A. Liscio, M.Sc. Thesis. McGill University, Montreal, Canada, 2000.

[20] K. Nam, R. H. Hudson, K. B. Chapman, K. Ganeshan, H. J. Damha, and J. D. Boeke, *J. Biol. Chem.* **269,** 20613 (1994).

with purine residues at both the 2′ and 3′ positions are debranched significantly better than those with pyrimidine residues at these positions.[20–22]

Properties of 2′–5′ Phosphodiester Bonds

The 2′–5′ phosphodiester bond makes the adjacent 3′–5′ bond resistant to cleavage by most nucleases such as RNase A, RNase T2, and nuclease P1. The 2′–5′ phosphodiester bond also efficiently blocks reverse transcription from the 3′ end of the branch point.[9,23] However, it is possible to perform reverse transcription with a primer that spans the branch point. This has been done to make a cDNA library of introns from the *S. cerevisiae dbr1* mutant.[24]

RNA Lariat Debranching Enzyme as Tool to Probe RNA Structure

Dbr has been used (1) to aid the mapping of the positions of RNA branches generated during *in vitro* pre-mRNA splicing,[25,26] (2) to verify the Y structure RNA intermediate generated during *trans*-splicing in trypanosomes,[2,3] (3) to help deduce the structure of multicopy single-stranded DNAs in the bacterium *Stigmatella aurantiaca*[5] and *Escherichia coli*,[7] (4) to determine the nucleotide sequence requirements for lariat debranching enzyme,[20] and (5) to verify that an *in vitro*-isolated ribozyme has 2′–5′ branch formation activity.[27]

Sources of RNA Lariat Debranching Activity

RNA lariat debranching activity can be prepared from HeLa, *S. cerevisiae*, and *E. coli* cell extracts. Preparation of human HeLa cell nuclear extract containing RNA lariat debranching activity has been reviewed by Ruskin and Green.[28] A further purification of the debranching enzyme from HeLa cells of about 700-fold has been reported.[10]

Here, we describe the preparation of yeast and *E. coli* extracts containing lariat debranching activity and the purification of yeast Dbr (yDbr) protein from *E. coli* extracts expressing yDbr. yDbr from these sources have been used to debranch total RNA from *S. cerevisiae dbr1* mutants, lariats derived from *in vitro*-spliced RNA, group II intron lariats, msDNA substrates, and *in vitro*-synthesized branched RNAs and DNAs.

[21] R. H. E. Hudson, Ph.D. Thesis. University of Toronto, Toronto, Canada, 1995.
[22] R. S. Braich, Ph.D. Thesis. McGill University, Montreal, Canada, 2000.
[23] A. R. Krainer, T. Maniatis, B. Ruskin, and M. R. Green, *Cell* **36,** 993 (1984).
[24] M. Spingola, L. Grate, D. Haussler, and M. Ares, Jr., *RNA* **5,** 221 (1999).
[25] R. Reed and T. Maniatis, *Cell* **41,** 95 (1985).
[26] B. Ruskin, J. M. Greene, and M. R. Green, *Cell* **41,** 833 (1985).
[27] T. Tuschl, P. A. Sharp, and D. P. Bartel, *EMBO J.* **17,** 2637 (1998).
[28] B. Ruskin and M. R. Green, *Methods Enzymol.* **181,** 180 (1990).

For most applications, purified enzyme will be the most suitable source, as the extracts contain many contaminating nucleases. However, certain assays are not affected by nuclease contamination. For these assays, extracts may suffice.

Preparation of Yeast Extracts Containing RNA Lariat Debranching Activity

Yeast extracts are prepared from a *dbr1* null mutant (KC99) containing plasmid pKC55.[11] Plasmid pKC55 overexpresses yeast *DBR1* under the control of the *GAL* promoter. The backbone of pKC55 is pCGS109, a 2μm–URA3 marked yeast plasmid. The *GAL* promoter can be induced by growing yeast cells on medium containing 2% (w/v) galactose.

1. Yeast strain KC99 containing pKC55 is grown for 8 hr in 200 ml of SC–Ura medium (synthetic complete medium without uracil) containing 1% (w/v) raffinose as carbon source. This step allows the yeast cells to utilize all residual glucose so that the *GAL* promoter can be rapidly and efficiently induced. Raffinose is used as a carbon source because it neither inhibits nor induces the *GAL* promoter, whereas glucose represses the *GAL* promoter.
2. Galactose is then added to a final concentration of 2% (w/v), and the culture is induced for 18 hr. It is not necessary to remove the raffinose.
3. The cell pellets are collected and resuspended in 10 ml of buffer 1 [1 *M* sorbitol, 50 m*M* Tris-HCl (pH 8.0), 10 m*M* $MgCl_2$, and 3 m*M* dithiothreital (DTT)] in glass tubes.
4. Glass beads are then added to the cell suspension to the meniscus and cells are lysed by rigorous vortexing for 15 min.
5. A 0.5-ml volume of buffer 2 [0.3 *M* HEPES (pH 7.8), 1.4 *M* KCl, 30 m*M* $MgCl_2$] is added to the cell lysate.
6. The cell lysate is then centrifuged at 41000 rpm in a Beckman (Fullerton, CA) Ti 70.1 rotor for 1 hr at 4°.
7. The pellet is discarded and the supernatant can be used directly.

Preparation of Escherichia coli Extracts Containing RNA Lariat Debranching Activity

Escherichia coli extracts are prepared from a strain containing pKC65 expressing yeast *DBR1* driven by the T7 promoter.[20] The backbone for pKC65 is pET11a. The parental strain is *E. coli* BL21(DE3), a lysogen of a λ derivative containing the T7 RNA polymerase gene driven by the *lacUV5* promoter. The transcription of *DBR1* is therefore indirectly controlled by the *lacUV5* promoter, which in turn can be induced by the addition of isopropyl-1-thio-β-D-galactopyranoside (IPTG).

Escherichia coli S100 cytoplasmic extract containing lariat debranching activity can be prepared by following steps 1 and 2 of the yDbr purification protocol. In this case, the amount of bacteria inoculated and all of steps 1 and 2 can be scaled down 100-fold. Bacterial culture (160 ml) can be induced in 1 m*M* IPTG and it is not necessary to use a fermentor. This should provide about 17 mg of total protein, sufficient for about 100 reactions.

Purification of Hexahistidine-Tagged Yeast RNA Lariat Debranching Enzyme

Histidine-tagged yDbr protein can also be purified from *E. coli* strain KNe136 expressing N-terminal hexahistidine (His_6)-tagged yeast *DBR1* driven by the T7 promoter. The backbone for KNe136 is pET15b. The parental strain is *E. coli* BL21(DE3). The transcription of *DBR1* is indirectly controlled by the *lacUV5* promoter, which in turn can be induced by the addition of IPTG.

Preparation of Bacterial Hexahistidine Yeast RNA Lariat Debranching Enzyme Pellet

1. Bacterial strain KNe136 is grown at 37° overnight in 100 ml of 2× YT (yeast extract–Bacto-Tryptone) containing ampicillin (100 μg/ml).
2. The next day, the entire culture is inoculated into 1 liter of 2× YT containing ampicillin (100 μg/ml) until the culture reaches an OD_{600} of 0.8. This takes about 1.5 to 2 hr.
3. IPTG is added to the culture to a final concentration of 0.2 m*M* to activate *DBR1* transcription.
4. After 4 hr of induction, *E. coli* cells are spun down at 4000*g* for 10 min at 4°, and the supernatant is discarded.
5. Cells are resuspended in 50 ml of 1× phosphate-buffered saline (PBS), pH 7.0, containing 0.2% (w/v) NaN_3, 0.5 m*M* phenylmethyl sulfoxyl fluoride (PMSF), and leupeptin (1 μg/ml).
6. Triton X-100 is added to 1% (v/v) immediately before sonication.
7. Cells are sonicated four times for 25 sec, with 45 sec between each sonication.
8. Cells are spun at 4000*g* for 10 min at 4°. The supernatant is used for subsequent purification.

Nickel–Nitrilotriacetate Agarose Purification of Hexahistidine Yeast RNA Lariat Debranching Enzyme

1. Five milliliters of nickel–nitrilotriacetate (Ni–NTA) agarose resin (Qiagen, Valencia, CA) is preequilibrated with 8 column volumes of 1× PBS, pH 7.0, in a gravity flow column (Bio-Rad, Hercules, CA).
2. The cell lysate is applied to the column, and the sample is allowed to load by gravity.
3. The column is washed with 8 column volumes of 1× PBS, pH 7.0, containing 20 m*M* imidazole.
4. Elution is performed with 4 column volumes of 1× PBS, 300 m*M* imidazole (pH 7.0), and eluates are collected in 1.5-ml fractions.

5. Fractions are checked by sodium dodecyl sulfate–polyacrylamide gel electrophoresis (SDS–PAGE). His_6–Dbr migrates at approximately 53 kDa.

Purification of Yeast RNA Lariat Debranching Protein from Escherichia coli Extract Expressing Native Yeast Dbr

Purification Strategy

Yeast Dbr can be purified from the bacteria containing pKC65 used to prepare *E. coli* extracts containing debranching activity.[20] The *DBR1* gene is induced with IPTG. The bacterial pellet is then prepared from the induced culture, the cells are lysed, and particulates are removed by centrifugation.

The whole cell extract containing yDbr protein is then subjected to column chromatography. Because yDbr is a nucleic acid-binding enzyme with a predicted p*I* of 4.1 at pH 7.0, and is likely a positively charged enzyme, the extract is first run through a DEAE-cellulose anion-exchange column to remove contaminating negatively charged proteins. About 90% of debranching activity is retained in the flow-through fraction.

In the second step, the nucleic acid-binding property of Dbr is exploited. The flow-through fraction from step 1 is subjected to chromatography with phosphocellulose, a negatively charged ion-exchange resin with special affinity for nucleic acid-binding enzyme. Most of the debranching activity can be detected in the 1.0 *M* KCl eluate.

Removal of remaining proteins is further achieved by using Superose 6 and heparin–agarose columns to separate proteins on the basis of size and binding properties. The form of the enzyme isolated by this method is highly active.

STEP 1: PREPARATION OF BACTERIAL PELLET

1. Bacteria containing pKC65 are grown at 37° overnight in 50 ml of Luria broth (LB) containing ampicillin (100 μg/ml).
2. The next day, the entire 50 ml of culture is inoculated into 16 liters of LB containing ampicillin (100 μg/ml).
3. The culture in incubated in a fermentor at 37° until the culture reaches an OD_{600} of 0.6.
4. IPTG is added to the culture to a final concentration of 1 m*M* to activate *DBR1* transcription.
5. Cells are harvested 3 to 4 hr after IPTG induction.
6. The pellet (~15 g) is frozen in liquid nitrogen.

STEP 2: PREPARATION OF S100 EXTRACT

1. One 15-g *E. coli* cell pellet (equivalent to 15 liters of liquid culture) is thawed at room temperature and resuspended in 60 ml of buffer A [20 m*M* HEPES-KOH (pH 7.6), 1 m*M* DTT, and 10% (v/v) glycerol] containing 40 m*M* KCl.

2. The cell membrane is disrupted with 106-μm glass beads (Sigma, St. Lewis, MO), using a Bead-Beater (Biospec Products, Bartlesville, OK) with a 350-ml vessel at 4°.

3. The crude extract is centrifuged at 10,000*g* for 10 min at 4° to remove unbroken cells and glass beads.

4. The supernatant is centrifuged in a new tube for 60 min at 100,000*g* at 4° to prepare the S100 cytoplasmic extract.

5. One 15-g pellet should yield about 1.7 g of protein with 1470 units of activity per milligram of protein. It may be difficult to accurately estimate the specific activity in crude lysates because of contaminating nucleases.

All further enzyme purification steps are carried out at 0–4°.

STEP 3: DEAE-CELLULOSE CHROMATOGRAPHY

1. A 250-ml DEAE-cellulose (DE-51; Whatman, Clifton, NJ) column (34 × 2.5 cm) is equilibrated with 5 column volumes of buffer A containing 40 m*M* KCl.

2. The entire 125 ml of S100 is loaded onto the DEAE-cellulose column. About 90% of debranching activity should be detected in the flow-through.

STEP 4: PHOSPHOCELLULOSE CHROMATOGRAPHY

1. A 150-ml phosphocellulose (P-11; Whatman) column is equilibrated with 5 column volumes of buffer A containing 40 m*M* KCl.

2. The entire 125 ml of DEAE-cellulose column flow-through in buffer A containing 40 m*M* KCl is directly loaded onto the phosphocellulose column.

3. The column is washed with 5 column volumes of buffer A containing 250 m*M* KCl.

4. The yDbr protein is step eluted with 1 to 2 column volumes of buffer A containing 0.5 *M* KCl followed by 1 to 2 column volumes of buffer A containing 1.0 *M* KCl step gradient in buffer A. Most of the debranching activity is detected in the 1.0 *M* KCl eluate. Some debranching activity may be detected in the 0.5 *M* KCl eluate. However, this fraction is contaminated with a nuclease activity and should be discarded.

5. The 1.0 *M* KCl fractions are pooled and diluted to 125 m*M* KCl by adding buffer A. After dilution, the volume of phosphocellulose fraction is about 150 ml.

6. The 150-ml diluted phosphocellulose fraction is concentrated to a total protein concentration of 5 mg/ml with a Diaflow concentrator equipped with a YM10 membrane (Amicon, Danvers, MA). This step should yield about 4 ml of protein at 5 mg/ml, or 20 mg of protein in total.

STEP 5: SUPEROSE 6 CHROMATOGRAPHY

1. A 25-ml Superose 6 HR 10/30 fast protein liquid chromatography column (Pharmacia Biotech, Piscataway, NJ) is equilibrated with two column volumes of CVs buffer A containing 125 m*M* KCl.
2. All (about 4 ml) of the concentrated eluate is applied to the column by loading 200 μl at each injection to get the highest resolution. The entire eluate requires about 20 injections.
3. yDbr is eluted in 1 column volume of buffer A containing 125 m*M* KCl.
4. Eluate is collected in 0.4-ml fractions.
5. Each fraction is assayed for debranching activity. Fractions with debranching activity should be pooled.

STEP 6: HEPARIN–AGAROSE CHROMATOGRAPHY

1. A 5-ml heparin–agarose type I (Sigma) column is equilibrated with 5 column volumes of buffer A containing 125 m*M* KCl.
2. The pooled fractions from the Sepharose 6 column are loaded onto the heparin–agarose column.
3. The column is washed with 5 column volume of buffer A containing 125 m*M* KCl.
4. Elution is performed with 1 column volume of buffer A containing 250 m*M* KCl.
5. yDbr activity should be detected in the 125 m*M* and 250 m*M* KCl. More yDbr protein can be eluted with 1 column volume of buffer A containing 500 m*M* KCl. However, the 250 *M* KCl fractions are the most pure fractions. Thus, we usually discard other fractions and use only the 250 m*M* KCl fractions.
6. The 250 m*M* KCl fractions are pooled and dialyzed against buffer A containing 125 m*M* KCl and 50% (v/v) glycerol. This yields about 360 μg of enzyme.
7. Purified yDbr is then aliquoted and stored at -80° or in liquid nitrogen. Purified yDbr can be stored at -20° for 6 months with less than 2-fold loss of activity. The enzyme appears to be sensitive to repeated freezing and thawing.

Comments on Purification. This protocol provides yeast Dbr purified at least 280-fold relative to the S100 cytoplasmic extract. One 15-g bacterial pellet is used for each purification, and this should yield about 360 μg of purified yDbr. The bacterial inoculation and extract preparation procedure in steps 1 and 2 can be scaled up 10-fold to prepare about 150 g of bacterial pellet, sufficient for 10 purifications. The theoretical molecular mass of yDbr is 47,741 Da. However, the form of yDbr we purified has an electrophoretic mobility suggesting a molecular mass of 42,000 Da. It is not clear why the mobility of the purified yDbr differs from the predicted mobility, but the purified form may be a proteolytic fragment. (Refer to Nam *et al.*[20] for a silver stain gel of fractions from S100, DEAE flow-through, phosphocellulose, Superose 6, and heparin–agarose.)

Biochemical Requirements for Enzymatic Debranching of Purified Yeast RNA Lariat Debranching Enzyme

Parameter	Optimum range
pH	7–8
$MnCl_2$	0.1 m*M*
$MgCl_2$	0.25 m*M*
$CaCl_2$	0.5 m*M*
Temperature	30–37°

Temperature. Debranching activity (>50%) can be detected between 16 and 42°. The optimum reaction temperature is 30–37°. Debranching activity is totally inactivated at 55°.

pH. Debranching activity (>50%) can be detected between pH 6 and 8.5. The optimum reaction pH is between pH 7 and 8. Debranching activity is totally inactivated at pH 9.0.

Divalent cations. In contrast to partially purified Dbr from HeLa cells, purified yDbr does not require any added divalent cation for activity. However, low concentrations of divalent cations (0.1–0.5 m*M*) enhance yDbr activity. The optimum concentrations of $MnCl_2$, $MgCl_2$, and $CaCl_2$ are 0.1, 0.25, and 0.5 m*M*, respectively. EDTA will inhibit yDbr only at a concentration of 25 m*M* or above. Significant debranching activity (~70%) can still be detected in the presence of 10 m*M* EDTA. This may be because Dbr contains two tightly bound metal ions, as do some distantly related phosphoesterases.[14,29–31]

High concentrations of KCl (>500 m*M*) inhibit more than 90% of the debranching activity. RNasin (1.25 units/μl) or yeast tRNA (0.25 μg/μl) also inhibit debranching activity by more than 90%.

Unit Definition of Purified Yeast RNA Lariat Debranching Enzyme

One unit is the amount of yDbr that hydrolyzes 1.0 fmol of msDNA Ec86 completely under standard conditions. ^{32}P-labeled msDNA Ec86, prepared as described below, is incubated at 30° for 30 min in a 20-μl reaction mixture containing 20 m*M* HEPES (pH 7.6), 125 m*M* KCl, 0.5 m*M* $MgCl_2$, 1.0 m*M* DTT, and 10% (v/v) glycerol.

[29] M. P. Egloff, P. T. Cohen, P. Reinemer, and D. Barford, *J. Mol. Biol.* **254,** 942 (1995).

[30] J. Goldberg, H. B. Huang, Y. G. Kwon, P. Greengard, A. C. Nairn, and J. Kuriyan, *Nature (London)* **376,** 745 (1995).

[31] T. Klabunde, N. Strater, R. Frohlich, H. Witzel, and B. Krebs, *J. Mol. Biol.* **259,** 737 (1996).

Preparation of Multicopy Single-Stranded DNA Ec86 Substrate

The msDNA Ec86 substrate is prepared from *E. coli* strain containing pDB808 expressing msDNA Ec86.[7] The parental strain is *E. coli* DH5α. Total nucleic acid is prepared from the cell pellet, followed by RNase A treatment to degrade RNAs including the RNA portion of msDNA. RNase A treatment results in intact msDNA containing a triribonucleotide linked to the 5′ end of the DNA portion of the msDNA via a 2′–5′ phosphodiester bond.[7] The triribonucleotide residues are AG(2′-DNA)C, in which the middle G residue is branched. The DNA portion of msDNA is 86 nucleotides long. The RNase A-treated total nucleic acid samples are then resolved on a polyacrylamide gel to isolate msDNAs. The isolated msDNAs are then 5′-end labeled, loaded on a preparative polyacrylamide gel, purified, eluted, and used as substrates.

1. The bacterial strain containing pDB808 is grown at 37° in 50 ml of LB containing ampicillin (100 μg/ml) to an OD_{600} of 0.6.
2. Cells are collected and resuspended in 2.5 ml of ice-cold TE [10 m*M* Tris-HCl, 1 m*M* EDTA (pH 7.6)].
3. To isolate total nucleic acid, the cell pellet is gently mixed with 5% (w/v) SDS to a final concentration of 0.5% (v/v).
4. The suspension is boiled for 90 sec to disrupt the cells and immediately cooled on ice.
5. One-quarter volume of ice-cold 3 *M* potassium acetate (pH 5.5) is added, gently mixed, and incubated on ice for 10 min.
6. The mixture is then centrifuged at maximum speed for 10 min at 4° in a microfuge.
7. The supernatant is collected and transferred to a new tube.
8. The supernatant is phenol–chloroform extracted, ethanol precipitated, and resuspended in TE buffer. This is designated as the total nucleic acid sample.
9. To prepare the msDNA substrate, the total nucleic acid sample is treated with RNase A (50 μg/ml) for 30 min at 37°.
10. The RNase A-treated nucleic acid sample is then resolved on an 8% (w/v) nondenaturing polyacrylamide gel to isolate msDNAs.
11. msDNAs are visualized by ethidium bromide staining. msDNAs migrate as two species of 76 and 67 DNA nucleotides, respectively. Either the bottom band or both bands can be eluted. When only the bottom band is eluted, when it is run on a gel it still migrates as two species after elution.
12. msDNAs are eluted from the gel by cutting the gel slice into small pieces, and soaking them in elution buffer [30 m*M* Tris-HCl (pH 7.5), 300 m*M* NaCl, and 3 m*M* EDTA] overnight at 4°.
13. The msDNAs are then ethanol precipitated, and resuspended in TE buffer.

14. Depending on the number of reactions, the required amount of eluted msDNAs (50 fmol/reaction) is then labeled with [γ-^{32}P]ATP, using T4 kinase.

15. Labeled msDNAs are then resolved in an 8% (w/v) polyacrylamide–8 *M* urea gel, and purified from the gel by cutting the gel slice into small pieces, and soaking them in elution buffer [30 m*M* Tris-HCl (pH 7.5), 300 m*M* NaCl, and 3 m*M* EDTA] overnight at 4°.

16. The next day, the eluate is phenol–chloroform extracted, ethanol precipitated, and resuspended in TE buffer. The amount of labeled gel purified msDNA can then be quantitated by scintillation counting, by assuming 100% labeling and determining the yield of disintegrations per minute (dpm). The substrate is now ready to be used.

17. Labeled msDNA (50 fmol) is used for each debranching assay under the conditions described below for purified lariat debranching enzyme.

18. The reaction is then resolved on a 20% (w/v) polyacrylamide–8M urea gel. Untreated end-labeled substrate is loaded on the same gel as control.

Substrates for Debranching Assay

Alternative substrates for the debranching assay include total RNA, *in vitro*-transcribed pre-mRNA that has been subjected to *in vitro* splicing reaction, gel-purified branched RNAs, and *in vitro*-synthesized branched nucleic acids. When preparing a substrate, it is important to keep the substrate requirements of Dbr in mind. Substrates can be labeled before the debranching reaction. *In vitro*-transcribed RNA can be labeled by incorporating [α-^{32}P]NTP during *in vitro* transcription. Synthetic branched nucleic acids can be 5′-end labeled with [γ-^{32}P]ATP, using T4 kinase, or 3′-end labeled with RNA ligase and [5′-^{32}P]Cp. It is important to note that T4 kinase also has 3′-phosphatase activity. If this is a problem, T4 kinase without 3′-phosphatase activity is available from Roche Molecular Research (Indianapolis, IN). Substrates such as total RNA can be used unlabeled, and after the debranching assay, the substrates can be resolved on gels and visualized by Northern blotting.

In Vitro Synthesis of Branched Nucleic Acids

It is also possible to chemically synthesize branched DNA, branched RNA, and branched DNA–RNA chimeras via automated solid-phase methods. Initial methods utilized a 2′,3′-bifunctional phosphoramidite reagent[32] combined with RNA synthesis[33] to generate the branch point.[34–36] The limitation is that in the products the sequences of the 2′ and 3′ branches are identical. However, using new

[32] M. J. Damha and K. K. Ogilvie, *J. Org. Chem.* **53,** 3710 (1988).

[33] M. J. Damha and K. K. Ogilvie, *Methods Mol. Biol.* **20,** 81 (1993).

methods,[37–39] branched oligonucleotides such as A(2′-GUAUGU)3′-CAAGUU,[38] where the sequences at the 2′ and 3′ positions are different, could be made. This technology greatly complements and helps with characterization of branched structure.

In addition, msDNA analogs[40] containing RNA linked to the DNA via 2′–5′ phosphodiester bonds and Y-shaped branched DNAs[41] such as TACTAA(2′-GTATGT)3′-CAAGTT could be made. In these methods, one chain is first grown in the conventional 3′-to-5′ direction, and the branched nucleotide 2′-OH is protected with a *tert*-butyldimethylsilyl (TBDMSi) group. After acetylating ("capping") the 5′ end of the sequence, the TBDMSi group is cleaved with fluoride ions. The 2′-linked sequence is then synthesized in the "reverse" 5′-to-3′ direction with commercially available 5′-phosphoramidite reagents (ChemGenes, Waltham, MA).

It is important to note that the above described chemical synthesis, particularly those developed in our laboratories (M.J.D.), are carried out with conventional DNA/RNA synthesizers, commercially available solid supports, and nucleoside phosphoramidite reagents (ChemGenes). Detailed protocols for the assembly, deprotection, and purification of the synthetic branched RNA substrates are provided.[32–36,38,40,41]

RNA Lariat Debranching Assay

The lariat or fork structure of a branched nucleic acid causes it to migrate more slowly than a linear nucleic acid of the same size on denaturing polyacrylamide gels. After enzymatic debranching, whether a substrate has been debranched can be deduced by a shift in its electrophoretic mobility from that of a slower migrating species to one that migrates at the same rate as its linear counterpart.

Reaction Conditions

Reaction conditions using different sources of Dbr are summarized in the tabulations below.

[34] M. J. Damha and S. V. Zabarylo, *Tetrahedron Lett.* **30,** 6295 (1989).
[35] M. J. Damha, K. Ganeshan, R. H. Hudson, and S. V. Zabarylo, *Nucleic Acids Res.* **20,** 6565 (1992).
[36] R. H. E. Hudson, K. Ganeshan, and M. J. Damha, *in* "Carbohydrate Modifications in Antisense Research" (P. D. Cook, and Y. S. Sanghvi, eds.), Vol. 580, pp. 133–152. American Chemical Society, Washington, D.C., 1994.
[37] C. Sund, P. Agback, L. H. Koole, A. Sandstrom, and J. Chattopadhyaya, *Tetrahedron* **48,** 695 (1992).
[38] K. Ganeshan, K. Nam, R. H. E. Hudson, T. Tadey, R. Braich, W. C. Purdey, J. D. Boeke, and M. J. Damha, *Nucleosides Nucleotides* **14,** 1009 (1995).
[39] M. Groti, R. Eritja, and B. Sproat, *Tetrahedron* **53,** 11317 (1997).
[40] M. J. Damha and R. S. Braich, *Tetrahedron Lett.* **39,** 3907 (1998).
[41] R. S. Braich and M. J. Damha, *Bioconjug. Chem.* **8,** 370 (1997).

With Yeast Extracts Containing Lariat Debranching Activity

Reagent	Total amount
Yeast extract	18 μg
yeast extract debranching buffer	20 m*M* HEPES (pH 7.0), 3 m*M* $MgCl_2$, bovine serum albumin (BSA; 0.25 mg/ml, final concentration)
Substrate	1 fmol–5 pmol substrates
tRNA	1.25 μg
Water	Fill up to a final volume of 40 μl

With Escherichia coli Cytoplasmic Extract Containing Lariat Debranching Activity

Reagent	Total amount
E. coli S100 extract	~140 μg in ~10 μl
E. coli extract debranching buffer	20 m*M* HEPES (pH 7.6), 40 m*M* KCl, 3m*M* $MgCl_2$, 1 m*M* DTT, 10% (v/v) glycerol (final concentration)
Substrate	1 fmol–5 pmol substrates
Water	Fill up to a final volume of 30 μl

With Hexahistidine-Tagged or Purified Yeast Lariat Debranching Enzyme

Reagent	Total amount
Debranching buffer	20 m*M* HEPES (pH 7.6), 125 m*M* KCl, 0.5 m*M* $MgCl_2$, 1 m*M* DTT, 10% (v/v) glycerol (final concentration)
Substrate	1 fmol–5 pmol substrates
Purified yDbr	50–100 units
or	
His_6–Dbr	50–100 ng
Water	Fill up to a final volume of 15 μl

The amount of substrates used can range from femtomoles to picomoles. About 5 μg of total RNA can also be used as substrate. Although the compositions of different debranching buffers differ slightly, it is likely that debranching buffer for purified yDbr would work for all sources of debranching activity.

1. The debranching assay is performed at 30° for 30–60 min.
2. If extracts are used as sources of debranching activity, the sample should be phenol–chloroform extracted, ethanol precipitated, and resuspended in formamide loading buffer [95% (v/v) formamide, 20 m*M* EDTA, 0.05% (w/v) bromphenol blue, and 0.05% (w/v) xylene cyanol]. If purified or histidine-tagged yDbr is used, then the debranching reaction is stopped directly by the addition of an equal volume of formamide loading buffer without the extraction and precipitation steps.
3. The sample is then resolved on a denaturing polyacrylamide gel.

Section II

Processing and Degradative Exoribonucleases

A. $5' \rightarrow 3'$-Exoribonucleases
Articles 20 through 24

B. $3' \rightarrow 5'$-Exoribonucleases
Articles 25 and 26

[20] 5′-Exoribonuclease 1: Xrn1

By AUDREY STEVENS

Introduction

The detection, purification, and partial characterization of 5′-exoribonuclease 1 (Xrn1) from *Saccharomyces cerevisiae* as a 160-kDa RNase was reported in 1980[1] and 1985.[2] The enzyme is a processive exonuclease hydrolyzing RNA from the 5′ end with the production of 5′-mononucleotides. The *XRN1* gene was cloned in several laboratories, and the results showed that it encodes a 175-kDa protein. The gene is not essential; however, its disruption leads to disparate phenotypes and slow growth.[3] The isolation of the gene in five laboratories by unrelated approaches (see reviews by Kearsey and Kipling[4] and by Liu and Gilbert[5]) is likely due to the disparate phenotypes found. Other gene designations that resulted from the multiple gene isolations include *SEP1*,[6] *DST2*,[7] *KEM1*,[8] and *RAR5*.[9] The enzyme shows a 5′-exonuclease activity with either RNA or DNA, a DNA strand exchange activity that has been studied in some detail,[10,11] and a G4-DNA cleavage activity.[5] That the protein is a cytoplasmic protein has been carefully analyzed.[12]

The first evidence of the involvement of Xrn1 in RNA metabolism was the finding that yeast cells with a disrupted gene accumulate an internal transcribed spacer fragment of pre-rRNA.[13] It was then found that short-lived mRNAs lacking both the poly(A) tail and the cap structure also accumulate.[14] The findings suggested that Xrn1 is involved in cytoplasmic RNA metabolism and mRNA is hydrolyzed after its deadenylation and decapping. Decker and Parker[15] uniquely showed that when a poly(G) tract is inserted into *MFA2* mRNA, RNA fragments consisting of

[1] A. Stevens, *J. Biol. Chem.* **255,** 3080 (1980).
[2] A. Stevens and M. K. Maupin, *Arch. Biochem. Biophys.* **252,** 339 (1985).
[3] F. W. Larimer, C. L. Hsu, M. K. Maupin, and A. Stevens, *Gene (Amst.)* **120,** 51 (1992).
[4] S. Kearsey and D. Kipling, *Trends Cell Biol.* **1,** 110 (1991).
[5] Z. Liu and W. Gilbert, *Cell* **77,** 1083 (1994).
[6] D. Tishkoff, A. W. Johnson, and R. D. Kolodner, *Mol. Cell. Biol.* **11,** 2593 (1991).
[7] C. C. Dykstra, K. Kitada, A. B. Clark, R. K. Hamatake, and A. Sugino, *Mol. Cell. Biol.* **11,** 2583 (1991).
[8] J. Kim, P. O. Ljungdahl, and G. R. Fink, *Genetics* **126,** 799 (1990).
[9] D. Kipling, C. Tambini, and S. E. Kearsey, *Nucleic Acids Res.* **19,** 1385 (1991).
[10] A. W. Johnson and R. D. Kolodner, *J. Biol. Chem.* **266,** 14046 (1991).
[11] A. W. Johnson and R. D. Kolodner, *J. Biol. Chem.* **269,** 3637 (1994).
[12] W. D. Heyer, A. W. Johnson, U. Reinhart, and R. D. Kolodner, *Mol. Cell. Biol.* **15,** 2728 (1995).
[13] A. Stevens, C. L. Hsu, K. Isham, and F. W. Larimer, *J. Bacteriol.* **173,** 7024 (1991).
[14] C. L. Hsu and A. Stevens, *Mol. Cell. Biol.* **13,** 4826 (1993).
[15] C. J. Decker and R. Parker, *Genes Dev.* **7,** 1632 (1993).

0076-6879/01 $35.00

the G tract and downstream sequences accumulate during the mRNA turnover process because of an apparent block of 5′-exoribonuclease action. In yeast cells with a disrupted *XRN1* gene, the fragments are not found.[16] The overall role of Xrn1 in the mRNA turnover pathway has been reviewed by Caponigro and Parker.[17] Further findings of its role in RNA processing are described by Kressler *et al.*[18]

This chapter describes the purification of Xrn1 from yeast cells containing the gene in a *GAL10* expression vector. Assay methods for analysis of 5′-exoribonuclease activity as well as the substrate specificity of Xrn1 are also described.

Purification of Xrn1

Johnson and Kolodner[10] constructed a *GAL10* plasmid (pRDK249) bearing the coding region of the *XRN1* gene and transformed it into the haploid protease-deficient yeast strain BJ5464. BJ5464/RDK249 cells are grown in medium containing the nonfermentable carbon sources lactose and glycerol, and collected 6.5 hr after the addition of galactose. At that time, Xrn1 accounts for approximately 6% of the soluble protein of extracts of the cells.

The protein is purified from yeast cells containing the expression plasmid by a procedure, described below, that was also used for its purification from *S. cerevisiae* BJ926 in the same laboratory.[6,10] The procedure for purification from *S. cerevisiae* S288C, used by Stevens[1] and Stevens and Maupin,[2] is an alternative procedure utilizing different ion-exchange columns.

The chemicals used for the procedure are from the following sources: PBE 94 and Sephacryl S-200 are from Amersham Pharmacia Biotech (Piscataway, NJ). Single-stranded DNA cellulose is prepared by the method of Alberts and Herrick[19] and contains DNA at about 0.5 mg/ml. Benzamidine hydrochloride, leupeptin, pepstatin A, phenylmethylsulfonyl fluoride (PMSF), and glass beads (0.45 mm) are from Sigma (St. Louis, MO).

The purification steps are as follows.

1. Fraction 1: Preparation of crude extract: Cells (20 g) are thawed on ice and suspended in an equal volume of buffer A [20 m*M* Tris-HCl (pH 7.5), 10% (v/v) glycerol, 10 m*M* 2-mercaptoethanol, 1 m*M* EDTA, and 1 m*M* PMSF]. The buffer also contains 150 m*M* NaCl, 10 m*M* $NaHSO_3$, 2 m*M* benzamidine, 4 μM pepstatin A, and 2 μM leupeptin. All procedures are carried out at 0–4°. Cells are broken with a Bead-Beater (Biospec, Bartlesville, OK) equipped with a 50-ml chamber that is partially filled with glass beads. For breakage, five 30-sec pulses are used with 2-min intervals of cooling between the pulses. The cell lysate is

[16] D. Muhlrad, C. J. Decker, and R. Parker, *Genes Dev.* **8,** 855 (1994).

[17] G. Caponigro and R. Parker, *Microb. Rev.* **60,** 233 (1996).

[18] D. Kressler, P. Linder, and J. De La Cruz, *Mol. Cell. Biol.* **12,** 7897 (1999).

[19] B. Alberts and G. Herrick, *Methods Enzymol.* **21,** 198 (1971).

poured off and the beads are washed with 20 ml of the same buffer. The lysate combined with the wash is centrifuged in a Beckman (Fullerton, CA) 45 Ti rotor for 30 min at 40,000 rpm, and the supernatant is collected.

2. Fraction II: DE52 column chromatography: A 1.8 $cm^2 \times 14$ cm DE52 column is used and equilibrated with buffer A containing 150 m*M* NaCl, 2 m*M* benzamidine, 2 μM leupeptin, and 2 μM pepstatin A. After application of the Spinco (Palo Alto, CA) supernatant at a rate of 25 ml/hr, the column is washed with the same buffer. Absorbance of the eluate is measured at 280 nm and the flow-through fraction is collected.

3. Fraction III: DNA–cellulose chromatography: A 1.8 $cm^2 \times 11$ cm column of single-stranded DNA–cellulose is prepared and equilibrated with buffer A containing 150 m*M* NaCl, 2 m*M* benzamidine, 2 μM leupeptin, and 2 μM pepstatin A. After addition of 1 m*M* PMSF to fraction II, it is applied to the column at a rate of 17 ml/hr. After washing the column with 2.5 column volumes of the same buffer, the protein is eluted with a 150-ml linear gradient from 150 to 750 m*M* NaCl in buffer A supplemented with 2 m*M* benzamidine, 2 μM pepstatin A, and 2 μM leupeptin in the same buffer. Xrn1 elutes at about 250 m*M* NaCl, and the active fractions are combined.

4. Fraction IV: PBE 94 column chromatography: Fraction III is diluted 1:1 with buffer A containing 2 m*M* benzamidine, 2 μM pepstatin A, and 2 μM leupeptin. A 0.38 $cm^2 \times 9.8$ cm column of PBE 94 is prepared and equibrated with buffer A containing 125 m*M* NaCl, 2 μM pepstatin A, and 2 μM leupeptin. Fraction III is applied to the column at a rate of 11 ml/hr, and the column is washed with 5 ml of the same buffer. For the elution of the enzyme, a linear gradient of 30 ml from 125 to 600 m*M* NaCl in the same buffer is used. Fractions containing Xrn1 are pooled and then concentrated to 0.6 ml with a Centricon microconcentrator (Amicon, Danvers, MA) with a 100,000 molecular weight cutoff.

5. Fraction V: Sephacryl S-200 column chromatography: A 2 $cm^2 \times 68$ cm column of Sephacryl S-200 is prepared and equilibrated with buffer B [20 m*M* Tris-HCl (pH 7.5), 10% (v/v) glycerol, 10 m*M* 2-mercaptoethanol, 0.1 m*M* EDTA, and 0.1 m*M* PMSF] containing 250 m*M* NaCl, 0.2 m*M* PMSF, 1 μM leupeptin, and 1 μM pepstatin A. Fraction IV is applied at a rate of 10 ml/hr, and the column is washed with the same buffer. Fractions containing Xrn1 are combined and dialyzed against buffer B containing 100 m*M* NaCl and 20% (w/v) polyethylene glycol 8000 (Kodak, Rochester, NY). The fraction is then dialyzed against buffer B containing 100 m*M* NaCl and 60% (v/v) glycerol. The final protein concentration is about 2.2 mg/ml, and the enzyme is stored at $-20°$. The enzyme has been found to be stable for as long as 3 months.

This purification procedure is summarized in Table I. Single-stranded T7 DNA was used as the assay substrate. An electrophoretic analysis of Xrn1, using sodium dodecyl sulfate (SDS)–polyacrylamide gels, showed only a single 160-kDa protein band with the purified enzyme.[10]

TABLE I
PURIFICATION OF Xrn1

	Fraction	Total protein (mg)	Total activity (units)	Specific activity (units/mg)
I.	Crude extract	800	36,000	45
II.	DE52	800	27,200	34
III.	DNA–cellulose	9	4,680	520
IV.	PBE 94	5.3	4,611	870
V.	Sephacryl S-200	4.2	4,074	970

Assay Methods

Assay of Enzyme Activity

Uniformly labeled RNA is used to assay Xrn1. 5′- or 3′-terminally labeled RNA can be used for specific purposes such as assaying the direction of hydrolysis. 18S ribosomal RNA of yeast and 16S ribosomal RNA of *Escherichia coli* are excellent substrates because they contain a 5′-phosphate end group that is required for best hydrolysis. The labeled ribosomal RNAs can be prepared by labeling cultures of yeast cells or *E. coli* cells with ^{32}Pi or [^{3}H]adenine.[13] Labeled poly(A) is also an excellent substrate but requires a source of polynucleotide phosphorylase for its preparation.[1] Transcription plasmids with phage polymerase promoters and specific genes can also be used to prepare labeled RNA substrates. With the synthetic RNAs, the terminal 5′-triphosphate end group needs to be converted to a 5′-phosphate end group by use of tobacco acid pyrophosphatase, according to the directions of the manufacturer (Epicentre Technologies, Madison, WI).[20]

The reaction mixtures for the assay contain (in 50–150 μl): 0.5 to 2 nmol of labeled substrate, 30 mM Tris-HCl buffer (pH 8.0), 2 mM $MgCl_2$, 50 mM NH_4Cl, 0.5 mM dithiothreitol, and 1 μg of acetylated albumin (BRL, Gaithersburg, MD) per microliter. The mixtures are incubated at 37° for 10 min and the reactions are stopped by the addition of an equal volume of 7% (w/v) perchloric acid and 50 μg of acetylated albumin. After 10 min in ice, the mixtures are centrifuged for 5 min at high speed (10,000 to 13,000 rpm) in a microcentrifuge. Radioactivity is determined with 50% of the supernatant solution. A reaction mixture lacking enzyme is used as a control. A unit of Xrn1 activity was first defined as the amount of enzyme required to release 1 nmol of labeled mononucleotide (per 50 μl of reaction mixture) in a 30-min period of incubation.[1] However, Johnson and Kolodner[10] used a 20-min period of incubation. The units also depend on the substrate used for assay (see substrate specificity below).

[20] A. Stevens and T. L. Poole, *J. Biol. Chem.* **270,** 16063 (1995).

TABLE II
SUBSTRATE SPECIFICITY OF Xrn1

Substrate	Relative activity[a]
Comparison A	
Poly(A)	100
18S yeast rRNA	77
T7 single-stranded DNA (*Hae*III digested)	5
T7 native DNA (*Hae*III digested)	<1
Comparison B	
Poly(A)	100
TAT RNA with 5′-phosphate terminus	148
TAT RNA with 5′-cap structure	0.8
TAT RNA with 5′-triphosphate terminus	2.8

Abbreviation: TAT, Tyrosine aminotransferase.

[a] The reaction mixtures (50 μl) were as described under Assay of Enzyme Activity, using in each case an amount of Xrn1 that gave measurable hydrolysis. In comparison A about 1.5–1.85 nmol of the ^{3}H-labeled polynucleotides was used and in comparison B, about 0.32 nmol was used. The TAT RNAs were prepared as described in Stevens and Poole.[20]

As described previously,[20] both the specific RNA used and the type of 5′-end group can influence the rate of 5′-exonuclease hydrolysis. The substrate specificity of Xrn1 is shown in Table II. The relative activities of ribosomal RNA and T7 single- and double-stranded DNAs as compared with poly(A) are shown in Table II (comparison A). DNA is hydrolyzed poorly as compared with poly(A). Values with plasmid-prepared synthetic RNAs are usually in the range of 30–200% that of poly(A). In comparison B (Table II), poly(A) is compared with a synthetic RNA, tyrosine aminotransferase RNA (TAT RNA), with different 5′-end groups. TAT RNA with a 5′-phosphate end group is hydrolyzed better than poly(A) (1.48-fold). TAT RNA with a 5′-cap structure or a 5′-triphosphate end group is hydrolyzed at less than 1% and at about 2%, respectively, of the RNA with the 5′-phosphate end group. Thus, it is important to convert either a cap structure or a triphosphate end group to a 5′-phosphate end group with tobacco acid pyrophosphatase.

Assay of Direction of Xrn1 Hydrolysis

Assay of the direction of hydrolysis (5′→3′) by the enzyme can be done by measuring the rate of release of label, using as a substrate uniformly labeled RNA bearing a second label at the 5′ end.[1] The usual assay conditions as described above are used. Reactions can be carried out at 0° and with a large amount of enyzme to saturate as many ends as possible. 5′-Exoribonucleolytic hydrolysis results in a

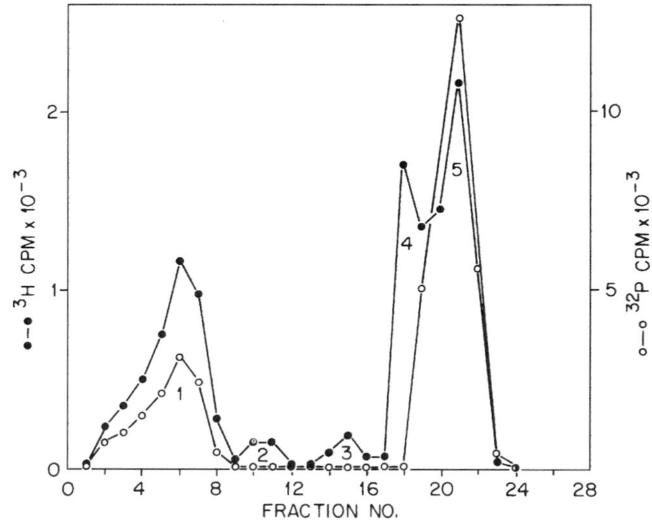

FIG. 1. Paper chromatography of the products of the hydrolysis of [^{3}H] poly(A)$_5$, labeled at the 5′ terminus with ^{32}P. [^{3}H] Poly(A)$_5$ (3.6 nmol) labeled at the 5′ end with ^{32}P was incubated at 37° in a 50-μl reaction mixture with 19 units of Xrn1 as described under Assay of Enzyme Activity. After 60 min, paper chromatography and determination of label in the products were carried out as described.[1]

much faster release of the 5′ label. Results with Xrn1 are described in Stevens.[1] Evidence of a 5′ → 3′ direction of hydrolysis can also be obtained by determining label in the products of hydrolysis of oligonucleotides with both uniform and 5′-^{32}P label. Figure 1 shows the products of Xrnl hydrolysis of [^{3}H] poly(A)$_5$ containing 5′-terminal ^{32}P label. Both ^{3}H and ^{32}P label are found in poly(A)$_5$ (peak 1, Fig. 1) and 5′-AMP (peak 5, Fig. 1). ^{3}H label is found in poly(A)$_4$ (peak 2, Fig. 1), poly(A)$_3$ (peak 3, Fig. 1), and poly(A)$_2$ (peak 4, Fig. 1), but no ^{32}P label is found, as expected for a 5′→3′-exoribonuclease.

Assay of Stalling of Enzyme by RNA Secondary Structure

A processive exoribonuclease reaction is one in which shortened substrate molecules are not usual intermediates in the hydrolysis reaction with a substrate lacking strong secondary structure. Xrn1 was determined to be processive by molecular sieve chromatography showing that at any step of poly(A) hydrolysis, only poly(A) of the original size and the product 5′-AMP are found.[1] Denaturing gel electrophoresis can also be used to measure the processivity of Xrn1 with different RNA substrates. The enzyme is stalled by strong secondary structure and can be used for the determination of the presence and site of such structure.

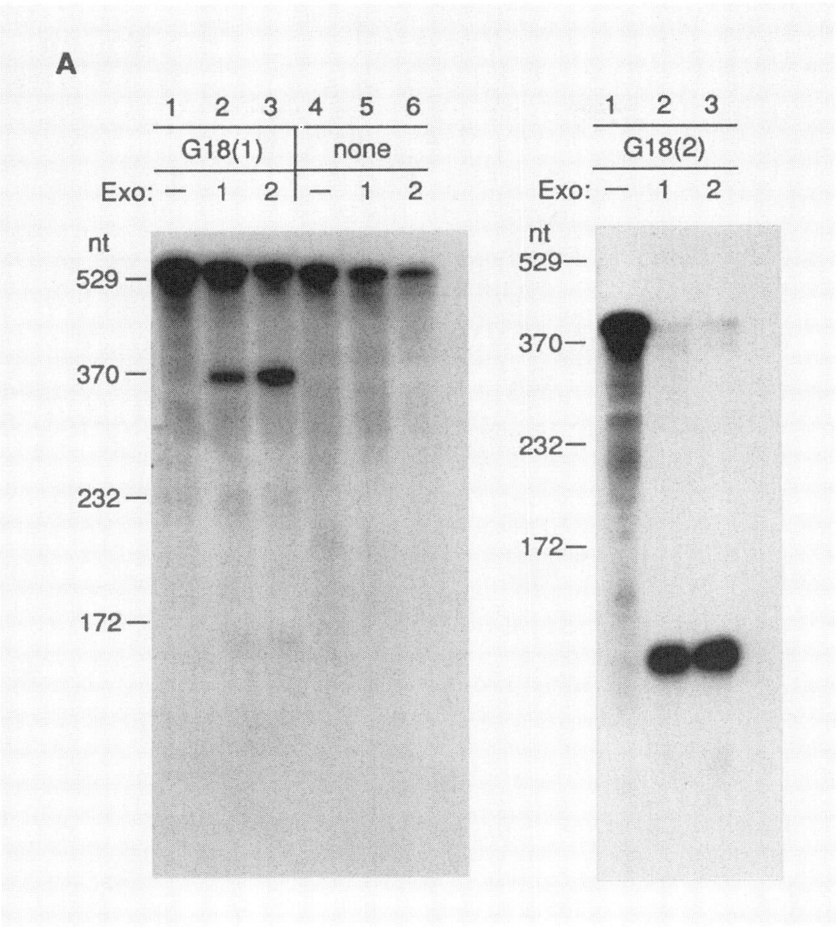

FIG. 2. Analysis of Xrn1 (exo-1) and Rat1 (exo-2) hydrolysis products from oligo$(G)_{18}$-containing RNAs. [Oligo$(G)_{18}$-containing transcription plasmids were a kind gift of D. Muhlrad and R. Parker, University of Arizona, Tucson, AZ.] (A) $[^{32}P]$RNA containing an oligo$(G)_{18}$ tract about 370 nucleotides from the 3′ end [G18(1)] and RNA containing an oligo(G) tract about 130 nucleotides from the 3′ end [G18(2)] were used and compared with RNA containing no tract (none). (B) RNA containing G_{18}, G_{16}, and G_9 tracts were compared as substrates with both Xrn1 and Rat1. For both (A) and (B), Xrn1 (4.4 units) and Rat1 (52 units) were used. Digestion products were analyzed on 6% (w/v) polyacrylamide gels (1.5 mm) containing 6 *M* urea.

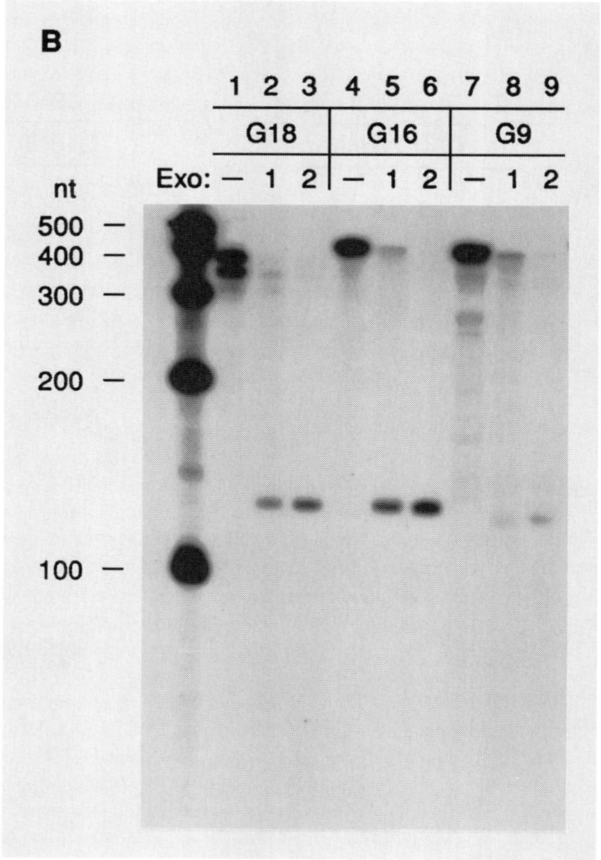

FIG. 2. (*Continued*)

Both Vreken and Raue[22] and Parker *et al.*[15,16] showed that when an oligo$(G)_{18}$ tract is inserted into a yeast gene there is a strong stall in the turnover of the mRNA in yeast, causing the accumulation of mRNA fragments containing the G_{18} tract and downstream sequences. *In vitro,* too, the best example of the stalling of Xrn1 by structure is found with RNAs that contain tracts of G residues.[21] Both Xrn1 (exo-1) and Rat1 (exo-2) have been examined in the same experiments. Results with the two enzymes using two oligo$(G)_{18}$-containing *MFA2* mRNA constructs of Muhlrad *et al.*[16] and a control construct are shown in Fig. 2A. $(G)_{18}(1)$ RNA contains the $(G)_{18}$ tract about 370 nucleotider from the 3′end of the RNA and a

[21] T. L. Poole and A. Stevens, *Biochem. Biophys. Res. Commun.* **235,** 799 (1997).

fragment of such a length accumulates highly on incubation with both enzymes. RNA containing the G_{18} tract about 130 nucleotides from the 3′ end yielded a shorter fragment of approximately 130 nucleotides. RNA containing no G_{18} tract showed no intermediates in the hydrolysis. The results show that RNA fragments accumulate that are shortened from the 5′ end to the site of the secondary structure in the RNA substrate. Figure 2B shows a comparison of the stalling found with RNAs with G_{18}, G_{16}, and G_9 tracts inserted at the same position. Xrn1 is highly stalled by both the G_{18} and G_{16} tracts and less, but detectably, by a G_9 tract. Stem structures with $\Delta G = -32$ and -80 kcal/mol have also been examined in experiments similar to those used for the G tracts.[21] Fragments caused by stalling of Xrn1 can be detected by gel electrophoresis. The stall fragments in these cases are about 5–15% of the substrate RNA.

XRN1 Homologs

The mouse *XRN1* (*mXRN1*) gene has been cloned, and the enzyme has been purified from *S. cerevisiae* after expression of the mouse gene in a *GAL10* plasmid as described above for the yeast *XRN1* gene.[23] The amino acid sequence of the protein is homologous to the yeast gene. Characterization of the enzyme is not yet complete. Shobuike *et al.*[24] also described a mouse Xrn1.

Xrn1 homologs from *Schizosaccharomyces pombe*[25] and *Drosophila melanogaster*[26] have also been described. The *Drosophila* exoribonuclease is developmentally regulated.

[22] P. Vreken and H. A. Raue, *Mol. Cell. Biol.* **12,** 2986 (1992).

[23] V. I. Bashkirov, H. Scherthan, J. A. Solinger, J. M. Buerstedde, and W. D. Heyer, *J. Cell Biol.* **136,** 761 (1997).

[24] T. Shobuike, S. Sugano, T. Yamashita, and H. Ikeds, *Nucleic Acids Res.* **23,** 357 (1995).

[25] P. Szankasi and G. R. Smith, *Curr. Genet.* **30,** 284 (1996).

[26] D. D. Till, B. Linz, J. E. Seago, S. J. Elgar, P. E. Marujo, M. L. Elias, C. M. Arraiano, J. A. McClellan, J. E. McCarthy, and S. F. Newbury, *Mech. Dev.* **79,** 51 (1998).

[21] Rat1p Nuclease

By ARLEN W. JOHNSON

Introduction

Rat1p is an essential 5′-exoribonuclease from yeast that is localized to the nucleus. It is homologous to the cytoplasmic enzyme Xrn1p[1] and these two enzymes are functionally interchangeable when directed to the appropriate cellular compartment.[2] Orthologs of Rat1p are found throughout eukaryotes and some of these are described elsewhere in this volume.[3,4] Prokaryotic enzymes of similar function have not been identified. However, the 5′-exonuclease domain of these enzymes is related to the 5′-exonuclease domain of T4 RNase H, *Escherichia coli* DNA polymerase I, and human flap endonuclease.[5] These enzymes are primarily involved in DNA-processing reactions, removing short, displaced 5′ ends or RNA primers. Thus, the Rat1p and Xrn1p-like 5′-exoribonucleases may have evolved from structure-specific DNA-processing enzymes. This line of thinking suggests that these enzymes evolved primarily as RNA-processing factors and that their role in RNA turnover was ancillary.

Rat1p was first purified on the basis of its exoribonuclease activity. Two polypeptides (~120 and 45 kDa) were found to cochromatograph with the exonuclease activity. N-terminal sequencing of the larger polypeptide identified it as the protein product of the *RAT1* gene, found previously from genetic approaches for genes involved in mRNA nuclear export[6] and RNA polymerase I transcription.[7] The 45-kDa polypeptide corresponded to a previously uncharacterized gene that we have named *RAI1*.[8] By genetic and biochemical criteria Rat1p and Rai1p form a complex. Deletion of *RAI1* leads to a slow growth phenotype and is lethal in combination with a *rat1-1* temperature-sensitive mutation. The slow growth defect is corrected by increased gene dosage of *RAT1* or by directing Xrn1p to the nucleus. These results suggest that Rai1p is required *in vivo* for full Rat1p activity. This is supported by *in vitro* results showing that Rai1p stabilizes and enhances

[1] A. Stevens, *Methods Enzymol.* **342,** [20] 2001 (this volume).
[2] A. W. Johnson, *Mol. Cell. Biol.* **17,** 6122 (1997).
[3] I. V. Chernukhin, J. E. Seago, and S. Newbury, *Methods Enzymol.* **342,** [24] 2001 (this volume).
[4] J. P. Kastenmayer, M. A. Johnson, and P. J. Green, *Methods Enzymol.* **342,** [22] 2001 (this volume).
[5] J. A. Solinger, D. Pascolini, and W. D. Heyer, *Mol. Cell. Biol.* **19,** 5930 (1999).
[6] D. C. Amberg, L. A. Goldstein, and C. N. Cole, *Genes Dev.* **6,** 1173 (1992).
[7] G. Di Segni, B. L. McConaughy, R. A. Shapiro, T. L. Aldrich, and B. D. Hall, *Mol. Cell. Biol.* **13,** 3424 (1993).
[8] Y. Xue, X. Bai, I. Lee, G. Kallstrom, J. Ho, J. Brown, A. Stevens, and A. W. Johnson, *Mol. Cell. Biol.* **20,** 4006 (2000).

0076-6879/01 $35.00

the exonuclease activity of Rat1p. Thus, Rat1p and Rai1p exist *in vivo* as a complex. This complex is referred to as exoribonuclease 2 or Xrn2 and the individual polypeptides as Rat1p and Rai1p.

The nuclear localization of Rat1p depends on a bipartite nuclear localization signal (NLS). The protein is found throughout the nucleoplasm and is not restricted to the nucleolus, although the known substrates for Xrn2 are required for ribosomal RNA processing (see below). Mutations in this NLS lead to mislocalization of the protein to the cytoplasm and complementation of an *xrn1* mutant. Because mislocalization of Rat1p is required for complementation of an *xrn1* mutation, Rat1p function is normally restricted to the nucleus. Rat1p localization is not significantly altered in an *rai1* deletion mutant, indicating that Rai1p is not required for normal Rat1p nuclear accumulation. Thus, the apparent reduction of Rat1p activity in an *rai1* mutant is not due to mislocalization or Rat1p. We have not been able to ascribe a catalytic or RNA-binding function to the Rai1 polypeptide.

Cellular Function of Xrn2

The function of Xrn2 is best defined in ribosome biogenesis where it is required for processing 5.8S rRNA and small nucleolar RNAs (snoRNAs). Two forms of 5.8S rRNA ($5.8S_L$ and $5.8S_S$) arise from two different processing pathways.[9] In a wild-type cell $5.8S_S$ accounts for approximately 90% of the total 5.8S rRNA, indicating that the pathway generating $5.8S_S$ is the major pathway in the cell. The 5′ end of $5.8S_S$ is formed from exonucleolytic processing by Xrn2. Not surprisingly, there is a reduced level of $5.8S_S$ in an *rat1-1* temperature-sensitive mutant and in a *rai1* deletion mutant. In addition, an *rai1* deletion mutant displays defects in processing the 3′ end of 5.8S rRNA, resulting in the accumulation of 7S rRNA. This may be due to physical coupling between Xrn2 and the exosome, responsible for the 3′-processing reactions, or it could result indirectly from a defect in snoRNA processing that in turn affects 3′ processing of 5.8S rRNA.

The second class of known substrates of Xrn2 is snoRNAs. These small RNAs serve as guide RNAs for posttranscriptional modifications, methylations, and pseudouridylations of rRNAs. SnoRNAs are processed from a variety of precursor molecules including introns and polycistronic messages. The *RAT1* gene has been shown to be required for normal 5′ processing of multiple intronic and exonic snoRNAs.[10–12]

[9] Y. Henry, H. Wood, J. P. Morrissey, E. Petfalski, S. Kearsey, and D. Tollervey, *EMBO J.* **13,** 2452 (1994).

[10] E. Petfalski, T. Dandekar, Y. Henry, and D. Tollervey, *Mol. Cell. Biol.* **18,** 1181 (1998).

[11] L. H. Qu, A. Henras, Y. J. Lu, H. Zhou, W. X. Zhou, Y. Q. Zhu, J. Zho, Y. Henry, F. M. Caizergues, and J. P. Bachellerie, *Mol. Cell. Biol.* **19,** 1144 (1999).

[12] T. Villa, F. Ceradini, C. Presutti, and I. Bozzoni, *Mol. Cell. Biol.* **18,** 3376 (1998).

It is likely that Xrn2 has roles in addition to processing 5.8S and snoRNAs. Its activity in 5′ processing of 5.8S and snoRNAs suggests that Xrn2 could act in 5′ processing of other small RNA species whose 5′ end results from an endonuclease cleavage to yield a 5′-monophosphate. In addition to its role in processing, the activity of Xrn2 in trimming the 5′ ends of intron-encoded snoRNAs suggests that Xrn2 may also act in the general degradation of debranched introns. Although *rat1* mutants alone do not display defects in intron degradation, this might simply be due to redundancy with other degradation factors. Another potential substrate of Xrn2 activity is the fragment of RNA polymerase II transcripts downstream of the cleavage site for polyadenylation. Currently, the fate of this RNA species is not known, but degradation by a 5′-exoribonuclease is possible. Last, the identification of mutations in *RAT1* that affect poly (A) mRNA export[6] and transcriptional activation of RNA polymerase I[7] suggests the existence of additional but as yet unidentified substrates.

Protein Purification

Xrn2 was first purified from a ribosomal salt wash.[13] In retrospect the presence of Xrn2 in this fraction probably did not reflect its *in vivo* function and may have resulted from the cosedimentation of Xrn2-containing nuclear material with ribosomes. Here, we provide a means of rapidly purifying recombinant Xrn2 from yeast. To facilitate the purification of the Xrn2 complex, we have made galactose-inducible vectors for the co-overexpression of Rat1p and a glutathione *S*-transferase (GST)–Rai1p fusion protein. GST–Rai1p is functional and forms a complex with Rat1p *in vivo*[8] and the GST moiety provides a convenient means of purifying the complex on glutathione–Sepharose beads. We have not been able to reconstitute the Rat1p/GST–Rai1p complex *in vitro* from the proteins purified separately from yeast. In addition, it is our experience that expression of Rat1p is mildly toxic to *E. coli*. Thus, at present, purification of the proteins as a complex from yeast is the most efficient strategy.

Strains, Plasmids, and Media

The protease-deficient strain BJ5464 (*MATα ura3-52 trp1 leu2-delta1 his3-delta 200 pep4::HIS3 prb1-delta1.6R can1 GAL*) (ATCC 208288) is used for the expression and purification of Xrn2 and GST–Rai1p. The free Rat1p subunit is purified from the protease-deficient *rai1* deletion strain AJY942 (*MATα ura 3-52 trp1 leu2Δ1 his3Δ200 pep4::HIS3 prb1Δ1 rai1::LEU2*) containing pAJ508. Plasmids are pAJ245 (2μm *GAL10::GST::RAI1 URA3*) and pAJ508 (2μm *GAL10::RAT1*

[13] A. Stevens and T. L. Poole, *J. Biol. Chem.* **270,** 16063 (1995).

TRP1). Plasmids are introduced into yeast by a standard transformation protocol.[14] A synthetic complete (SC) dropout medium is used to select for plasmids during cell growth. SC medium contains yeast nitrogen base (6.7 g/liter) without amino acids and with ammonium sulfate; adenine, arginine, histidine, methionine, tryptophan, and uracil (60 mg/liter); lysine and tyrosine (90 mg/liter); phenylalanine (150 mg/liter); and leucine (180 mg/liter). For routine propagation of strains 2% (w/v) glucose is used for the carbon source. However, because glucose represses galactose induction, raffinose is used in place of glucose for cultures that will subsequently be induced with galactose. For the preparation of plates, agar is added to 2% (w/v). Dropout medium is prepared as described for SC medium except that specific amino acid or base supplements are omitted as required. Yeast strains can be stored indefinitely in 15% (v/v) glycerol at temperatures below −60°.

Purification of Exoribonuclease 2

Cell Growth. Four days before beginning cultures, a frozen stock of strain BJ5464 containing pAJ245 and pAJ508 is streaked for single colonies on an SC Ura− Trp− (glucose) plate and incubated at 30°. A 2-ml starter culture in SC Ura− Trp− is inoculated with several fresh colonies and grown for 16–20 hr with shaking at 30° in SC Ura− Trp− (glucose). The culture is diluted into 50 ml of SC Ura− Trp− medium containing raffinose in a 500-ml flask. The density should be $\sim 1 \times 10^7$ cells/ml. After 8–10 hr of growth at 30° with shaking the entire 50-ml culture is added to 1 liter of prewarmed SC Ura− Trp− (raffinose) in a 2-liter flask. The culture is grown at 30° with shaking for 16–20 hr to a density of $1\text{–}1.5 \times 10^7$ cells/ml. Galactose is then added to 1% (w/v) and the cells are grown for an additional 7 hr. Cells are harvested by centrifugation at 5000 rpm for 5 min at 4° in a Beckman (Fullerton, CA) JLA10.5 rotor. The cells are washed once by resuspending the pellet in phosphate-buffered saline (PBS buffer: 140 m*M* NaCl, 2.7 m*M* KCl, 10 m*M* Na_2HPO_4, 1.8 m*M* KH_2PO_4, pH 7.4) and recentrifugation as described above. The cell pelletes are stored at −80°. The yield is approximately 2 g of wet cell paste per liter of culture.

Preparation of Crude Extracts. All procedures are performed at 0–4°. The cell pellet is thawed on ice and resuspended in 2 volumes of PBS buffer supplemented with 1 m*M* EDTA, 0.5 m*M* dithiothreitol (DTT), leupeptin (0.5 μg/ml), pepstatin A (0.7 μg/ml), and 0.5 m*M* phenylmethylsulfonyl fluoride (PMSF). (Prepare a stock of 200 m*M* PMSF in 2-propanol and add to buffers immediately before use.) The resuspended cells are transferred to a thick-walled glass tube such as a Corex 30-ml round-bottom centrifuge tube. An equal volume of acid-washed glass beads (0.4 mm; Sigma, St. Louis, MO) is added to the cell suspension and

[14] D. Gietz, A. St. John, R. A. Woods, and R. H. Schiestl, *Nucleic Acids Res.* **20,** 1425 (1992).

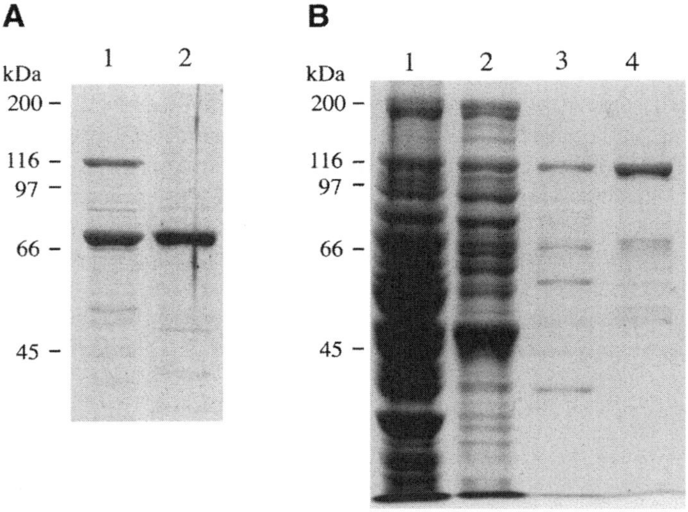

FIG. 1. Purification of proteins for *in vitro* analysis. Proteins were separated by SDS–PAGE and visualized by staining with Coomassie blue. (A) GST–Rai1p complexed with Rat1p (lane 1) and GST–Rai1p (lane 2) were overexpressed and purified as described. (B) Rat1p was purified from AJY942 (*rai1*Δ) containing pAJ508 as described. Lane 1, crude extract; lane 2, PEI fraction; lane 3, ssDNA–cellulose fraction; lane 4, Source 15Q fraction. The positions of molecular weight markers are indicated. [Adapted from Fig. 8 of Xue *et al., Mol. Cell. Biol.* **20,** 4006 (2000).]

cells are disrupted by vortexing six times for 0.5 min with 1-min rests on ice in between vortexings. For larger cell masses (≥20 g), a Bead-Beater (Biospec, Bartlesville, OK) is useful. The cell lysate is recovered with a chilled pipette and the beads are washed with a small volume of the same buffer. The cell lysate and wash are combined and centrifuged at 20,000*g* for 30 min, and the supernatant is recovered as the crude extract.

Glutathione–Sepharose 4B Chromatography. Glutathione–Sepharose 4B (Pharmacia, Piscataway, NJ) is added to the crude extract [0.25 ml of a 50% (v/v) slurry per 15 mg of total protein] and mixed by gentle rocking for 1 hr at 4°. The sample is then transferred to a small disposable fritted column such as a Poly-Prep chromatography column (Bio-Rad, Hercules, CA). After washing the column with 10 volumes of PBS buffer supplemented with 1 m*M* EDTA, 0.5 m*M* DTT, leupeptin (0.5 μg/ml), pepstatin A (0.7 μg/ml), and 0.5 m*M* PMSF, proteins are eluted with 1 ml of glutathione elution buffer: 50 m*M* Tris-HCl, 10m*M* reduced glutathione, and 140m*M* KCl. Eluted proteins are monitored by sodium dodecyl sulfate–polyacrylamide gel electrophoresis (SDS–PAGE) and fractions containing Xrn2 are stored in aliquots at −80°. Xrn2 run on SDS–PAGE as two polypeptides of approximately 116 and 45 kDa (see Fig. 1A, lane 1). Alternatively, if the protein preparation is to be used regularly, it can be dialyzed against PBS buffer supplemented with 1m*M* EDTA, 0.5 m*M* DTT, leupeptin (0.5μg/ml), pepstatin A (0.7 μg/ml), and 0.5 m*M* PMSF and 50% (w/v) glycerol and stored at -20°. This avoids freezing the sample.

Notes about Purification. This procedure yields active Xrn2 containing an excess of Rai1p. The excess Rai1p appears to be due to aggregation or multimerization of Rai1p and is not easily separated from the stoichiometric Xrn2 complex (A. W. Johnson, unpublished observation, 2000). However, because Rai1p itself does not appear to bind RNA and does not display any hydrolytic activities, the excess Rai1p in this preparation does not appear to adversely affect Xrn2 activity.[8] Nevertheless, we have constructed a low-copy galactose-inducible GST–*RAI1* expression vector (pAJ293) that yields lower levels of GST–Rai1p and, when coexpressed in yeast with pAJ508, allows the purification of Xrn2 containing closer to stoichiometric amounts of the two subunits. We have not yet characterized the activity of Xrn2 prepared in this manner.

Purification of Rat1p Subunit of Xrn2

If desired, Rat1p and GST–Rai1p can be purified separately from yeast.

Cell Growth. Strain BJ5464 containing plasmid pAJ508 is grown and induced as described above for the preparation of Xrn2, with the exception that growth medium contains uracil. The cell pellet is washed with 20m*M* Tris-HCl (pH 7.6), 10% (w/v) glycerol, 20 m*M* KCl, 1 m*M* EDTA, 0.5 m*M* DTT, and 0.5 m*M* PMSF and recentrifuged, and the cell pellet is stored at -80° until use.

Preparation of Crude Extract. All purification steps are carried out at 0–4°. Rat1p purification is monitored by SDS–PAGE. The cell pellet is thawed on ice and resuspended in 2 volumes of extraction buffer [buffer A: 20 m*M* Tris-HCl (pH 7.6) 10% (w/v) glycerol, 1m*M* EDTA, and 0.5m*M* PMSF supplemented with 20 m*M* KCl, 0.5 m*M* DTT, leupeptin (0.5 μg/ml), and pepstain A (0.7 μg/ml)]. The cell suspension is disrupted as described above. The cell lysate is then clarified by centrifugation at 20,000*g* for 2.5 hr, and the supernatant is recovered (fraction I).

Polyethyleneimine Precipitation. Fraction I is diluted with extraction buffer to give a protein concentration of 20 mg/ml. Polyethyleneimine [PEI, 5% (w/v)] is slowly added to the diluted fraction I with stirring to give a final PEI concentration of 0.075% (w/v). After 30 min of stirring, the suspension is centrifuged for 15 min at 20,000*g*. The supernatant is discarded and the pellet is washed once with extraction buffer. Proteins are eluted by extensively mixing the pellet with 20 ml of buffer A supplemented with 200 m*M* KCl, 0.5m*M* DTT, leupeptin (0.5 μg/ml), and pepstatin A (0.7 μg/ml). The sample is clarified by centrifugation and the supernatant is recovered (fraction II).

Single-Stranded DNA–Cellulose Column Chromatography. Prepare a single-stranded DNA (ssDNA)–cellulose (Amersham, Arlington Heights, IL) column, using 1 ml of resin per 35 mg of protein in fraction II. The column is equilibrated with buffer A supplemented with 200 m*M* KCl, 0.5 m*M* DTT, leupeptin (0.5 μg/ml), and pepstatin A (0.7 μg/ml). Fraction II is loaded onto the column at 10 ml/cm^2/hr. The column is washed with 10 bed volumes of the same buffer. Proteins are eluted with a linear KCl gradient (200 to 500 m*M* and total volume of 15 bed volumes) in buffer A supplemented with 0.5 m*M* DTT, leupeptin (0.5 μg/ml), and pepstatin A (0.7 μg/ml). Fractions are collected and fractions spanning the elution peak of Rat1p, which elutes at ~280 m*M* KCl, are pooled (fraction III).

Source 15Q FPLC. A Source 15Q column (1 ml of resin per 35 mg of fraction II protein) is equilibrated with buffer B [20 m*M* Tris-HCl (pH 8.5), 10% (w/v) glycerol, 1m*M* EDTA, and 0.5 m*M* PMSF] supplemented with 70 m*M* KCl, 0.5 m*M* DTT, leupeptin (0.5 μg/ml), and pepstatin A (0.7 μg/ml). Fraction III is diluted with buffer B supplemented with 0.5 m*M* DTT, leupeptin (0.5 μg/ml), and pepstatin A (0.7μg/ml) to a conductivity corresponding to 70 m*M* KCl in buffer B, and then loaded onto the column at 1 ml/min. After washing the column with 3 column volumes of the same buffer, proteins are eluted with a linear KCl gradient (70 to 500 m*M*, total volume is 10 bed volumes) in buffer B supplemented with 0.5 m*M* DTT, leupeptin (0.5 μg/ml), and pepstatin A (0.7 μg/ml). Fractions are collected and Rat1p-containing fractions are pooled. Fractions can be saved in small aliquots at $-80°$. Figure 1B illustrates a typical purification of free Rat1p.

Notes on Purification of Free Rat1p. Rai1p is required for maximum Rat1p activity *in vivo*. This is reflected in the reduced stability of the exonuclease activity of free Rat1p compared with complexed Rat1p *in vitro*.[8] In addition, free Rat1p is highly susceptible to proteolysis during the initial steps of purification. In our hands, the use of ssDNA–cellulose chromatorgraphy as the second step is essential for high yields of intact Rat1p. We also found that the behaviour of free Rat1p during PEI precipitation was unusual. Rat1p is effeciently precipitated at low (0.075%, w/v) PEI concentrations but at slightly higher concentrations (>0.5%) it is not precipitated. This is probably explained by coprecipitation of Rat1p with RNAs or ribonucleoproteins (RNPs) at low PEI concentration. At higher concentrations, Rat1p may be displaced from the RNAs and consequently is not coprecipitated with nucleic acids.

Purification of GST–Rai1p Subunit of Xrn2

GST–Rai1p is purified from yeast overexpressing the fusion protein alone, according to the procedure given above for purification of Xrn2 (see Fig. 1A, lane 2). The amount of fusion protein in the extracts is in sufficient abundance over Rat1p that Rat1p contamination of the GST–Rai1p preparation is minimal (<0.5%). GST–Rai1p purified from *E. coli* or from yeast behaves as a large complex, probably because of aggregation and/or multimerization of the protein. This may account for our inability to reconstitute Xrn2 from the purified subunits.

Assaying Xrn2

Preparation of Substrate

Because Xrn2 is most active toward RNAs containing a 5′-monophosphate and is inhibited by the presence of 5′-cap structures and 5′-triphosphates, it is often desirable to remove such modifications. An RNA transcribed *in vitro* and free of unincorporated nucleotides is treated with tobacco acid pyrophosphatase (TAP) to hydrolyze the cap or triphosphate to yield a 5′-monophosphate. The TAP reaction (100 μl) contains 50 m*M* sodium acetate (pH 5.0), 1 m*M* EDTA, 0.1% (v/v) 2-mercaptoethanol, 0.01% (v/v) Triton X-100, 40 pmol of transcript, and 10 units of TAP (Epicentre, Madison, WI). This amount of TAP was found to be optimal for the activation of *in vitro* transcripts for degradation by Xrn1p, which also requires a 5′-phosphate for efficient degradation. Higher amounts of TAP, either commercially prepared or prepared by us, led to lower levels of activation of *in vitro* transcripts for degradation by Xrn1p, probably because of hydrolysis of the 5′-phosphate at saturating TAP concentration. After incubation at 37° for 15 min, the reaction mixture is extracted with phenol–chloroform (pH 4.5) and ethanol precipitated, and the RNA is resuspended in H_2O.

In Vitro Exoribonuclease Assays

The assay is based on the release of acid-soluble radioactivity from ^{32}P-labeled *in vitro*-transcribed RNA treated with TAP. The reaction (30 μl) contains 0.05 pmol of TAP-treated ^{32}P-labeled transcript, 20 m*M* Tris-HCl (pH 8.5), 75 m*M* NaCl, 5 m*M* $MgCl_2$, bovine serum albumin (BSA, 0.1 mg/ml), 0.5 m*M* DTT, and purified protein as required. After incubation at 30° for 10 min, the reactions are stopped on ice by the addition of 10μ of stop buffer [ssDNA (1 mg/ml) and 20 m*M* EDTA]. Sixty microliters of 0.6 *M* trichloroacetic acid (TCA) is then added and reactions are incubated on ice for 10 min and centrifuged for 10 min at 15,000*g*, and the radioactivity in the supernatant is measured by liquid scintillation counting.

Enzymatic Activities of Xrn2

Xrn2 is similar in enzymatic properties to Xrn1p.[1] It is a processive 5′-exoribonuclease that releases nucleoside 5′-monophosphates.[13] Under standard reaction conditions the rate of reaction is approximately 940 nucleotides/min on mRNA. It also degrades DNA, although at a rate 5–8% that of its activity on RNA.[15] As mentioned already, the enzyme requires a 5′-phosphate for maximum activity. The presence of a 5′-cap structure or 5′-triphosphate on an RNA inhibits the enzyme activity by 98 and 86%, respectively.[13] The exonuclease activity of Xrn2 is also

[15] T. L. Poole and A. Stevens, *Biochem. Biophys. Res. Commun.* **235,** 799 (1997).

inhibited by adenosine 3′,5′-bisphosphate (pAp).[16] pAp is a by-product of sulfur assimilation and accumulates in cells in which the phosphatase Hal2p is inhibited. Although the mechanism of pAp inhibition of Xrn2 has not been characterized it is likely competitive because pAp mimicks the 5′ end of RNAs that are Rat1p substrates. Whether there is a biological reason for Xrn2 to be sensitive to pAp or whether it is the unavoidable consequence of structural requirements in the active site of the enzyme remains an interesting question.

Xrn2 is also inhibited by the presence of strong secondary structures within RNAs. Oligo(G) tracts of 16 to 18 nucleotides in length effectively block degradation by Xrn2.[15] RNAs containing such oligo(G) tracts are degraded from the 5′ end up to the 5′ side of the oligo(G) tract; the remaining 3′ fragment is resistant to further digestion. Oligo(G) tracts of nine nucleotides lead to a modest accumulation of the 3′ fragment, indicating that they are slowly degraded by Xrn2. Stem–loop structures are less efficient than oligo(G) tracts at inhibiting degradation by Xrn2 and their effectiveness at inhibiting degradation depends on their position within an RNA. Stem–loop structures located between 9 and 20 nucleotides from the 5′ end of an RNA are more effective than when positioned further downstream,[15] suggesting that stem loops close to a 5′ end inhibit degradation by reducing binding or initiation of hydrolysis. The effects of proteins bound to RNA on the degradation by Xrn2 are largely unexplored. Xrn2 is not inhibited by the presence of poly(A)-binding protein bound to poly(A) RNA, but the effects of other protein RNA complexes on Xrn2 activity have not been reported. Considering that Xrn2 acts as a specific processing factor on snoRNAs and 5.8S rRNA, it is likely that specific RNA or RNA–protein structures limit its activity during processing reactions. The *in vitro* reconstitution of such 5′-processing reactions will provide insight into the mechanism that limits the exonuclease reaction for specific processing events.

Acknowledgments

This work was supported by NIH Grant GM53655 to A. Johnson. I thank J. Brown and G. Kallstrom for commenting on the manuscript.

[16] B. Dichtl, A. Stevens, and D. Tollervey, *EMBO J.* **16,** 7184 (1997).

[22] Analysis of XRN Orthologs by Complementation of Yeast Mutants and Localization of XRN–GFP Fusion Proteins

By JAMES P. KASTENMAYER, MARK A. JOHNSON, and PAMELA J. GREEN

Introduction

The XRN family of 5′→3′-exoribonucleases was first described in *Saccharomyces cerevisiae,* in which it consists of two related proteins, Xrn1p and Xrn2p/Rat1p. These enzymes are similar in sequence and enzymatic activity but differ in their intracellular locations and cellular functions. Xrn1p is an abundant cytoplasmic enzyme and plays a major role in the degradation of mRNAs in the cytoplasm, as well as trimming the 5′ ends of rRNAs and degrading rRNA processing intermediates (Refs. 1–4 and [20] in this volume[4a]). In contrast to Xrn1p, Xrn2p/Rat1p is targeted to the yeast nucleus, exists in a complex with Rai1p,[5] and functions in the processing of rRNA and small nucleolar RNAs (snoRNAs) (Refs. 6–8 and [21] in this volume[8a]). In most other eukaryotes for which sequence is available, the XRN family also consists of a single member of the Xrn1p-like class and a single member of the Xrn2p/Rat1p-like class. However, the role of these XRN-like enzymes in mRNA degradation or RNA processing is unknown. Because certain aspects of mRNA degradation appear to differ between yeast and multicellular eukaryotes, and because this difference could be due to mechanisms involving XRN enzymes, an investigation of these XRNs in warranted.[9] Further, higher plants apparently lack XRN1 orthologs.[9] These observations raise the possibility that XRN enzymes from multicellular eukaryotes may differ in enzymatic activity or cellular function from their yeast counterparts.

Insight into the possible cellular function of XRN orthologs can be gained by examining their exoribonuclease activities and intracellular locations. The

[1] W.-D. Heyer, A. W. Johnson, U. Reinhart, and R. D. Kolodner, *Mol. Cell. Biol.* **15,** 2728 (1995).
[2] C. L. Hsu and A. Stevens, *Mol. Cell. Biol.* **13,** 4826 (1993).
[3] Y. Henry, H. Wood, J. P. Morrissey, E. Petfalski, S. Kearsey, and D. Tollervey, *EMBO J.* **13,** 2452 (1994).
[4] A. Stevens, C. L. Hsu, K. R. Isham, and F. W. Larimer, *J. Bacteriol.* **173,** 7024 (1991).
[4a] A. Stevens, *Methods Enzymol.* **342,** [20] 2001 (this volume).
[5] Y. Xue, X. X. Bai, I. Lee, G. Kallstrom, J. Ho, J. Brown, A. Stevens, and A. W. Johnson, *Mol. Cell. Biol.* **20,** 4006 (2000).
[6] A. W. Johnson, *Mol. Cell. Biol.* **17,** 6122 (1997).
[7] E. Petfalski, T. Dandekar, Y. Henry, and D. Tollervey, *Mol. Cell. Biol.* **18,** 1181 (1998).
[8] T. Villa, F. Ceradin, C. Presutti, and I. Bozzoni, *Mol. Cell. Biol.* **18,** 3376 (2000).
[8a] A. W. Johnson, *Methods Enzymol.* **342,** [21] 2001 (this volume).
[9] J. P. Kastenmayer and P. J. Green, *Proc. Natl. Acad. Sci. U.S.A.* **97,** 13985 (2000).

0076-6879/01 $35.00

exoribonuclease activity of recombinant XRN enzymes has been examined on RNA substrates *in vitro,* an approach used to study the mouse Xrn1p ortholog mXRN1p.[10] Intracellular location has been examined by immunocytochemistry with anti-XRN antibodies,[10] and localization of XRN–green fluorescent protein (GFP) fusion proteins.[6,9] Although these approaches can yield detailed information about the potential cellular function of XRN enzymes, a few simple experiments performed before such a detailed analysis can give rapid insight. These experiments have the advantage that they are easy to perform, and their results can enhance subsequent studies of the XRN enzymes both *in vitro* and in their native contexts.

Analysis of XRN Activity on Poly(G)-Containing mRNAs

By making use of RNA substrates with labeled 5′ or 3′ ends, Stevens and co-workers demonstrated that both Xrn1p and Xrn2p/Rat1p function as RNases that preferentially degrade RNAs from the 5′ end *in vitro*.[11] A modification of these studies was their use of RNA substrates that contained secondary structures known to block RNA degradation *in vivo*. By introducing stem–loops or poly(G) tracts into substrate RNAs, it was shown that sequences 3′ of the stable structure are not effectively degraded by Xrn1p or Xrn2p/Rat1p, and that these structures inhibited the XRN's progression through the RNA.[12] Therefore, this approach confirms that the XRNs are active as exoribonucleases, demonstrates that they are blocked by poly(G) tracts, and shows that they degrade RNA from the 5′ end.

Experiments have shown that the expression of genes in yeast that contain poly(G) tracts of 18 guanosines results in the accumulation of mRNA degradation intermediates that begin at the poly(G) tract and end at the poly(A) tail.[16] That such poly(G)-stabilized mRNA decay intermediates do not arise when the *XRN1* gene is deleted led to the hypothesis that Xrn1p degrades mRNAs from the 5′ end and that its activity is inhibited by poly(G) tracts.[16] The demonstration that Xrn1p action is indeed blocked by poly(G) tracts *in vitro* is strong support for this hypothesis. Therefore, the accumulation of poly(G)-stabilized mRNA decay intermediates, when poly(G)-containing genes are expressed in yeast, is a demonstration of the enzymatic function of Xrn1p. As discussed below, the accumulation of poly(G)-stabilized mRNA decay intermediates in the absence of Xrn1p function

[10] V. I. Bashkirov, H. Scherthan, J. A. Solinger, J.-M. Buerstedde, and W.-D. Heyer, *J. Cell Biol.* **136,** 761 (1997).

[11] T. L. Poole and A. Stevens, *Biochem. Biophys. Res. Commun.* **235,** 799 (1997).

[12] A. Stevens, *Biochem. Biophys. Res. Commun.* **86,** 1126 (1979).

[13] J. F. Gera and E. J. Baker, *Mol. Cell. Biol.* **18,** 1498 (1998).

[14] J. E. Russell, J. Morales, and S. A. Liebhaber, *Prog. Nucleic Acid Res. Mol. Biol.* **57,** 249 (1997).

[15] M. L. Sullivan and P. J. Green, *RNA* **2,** 308 (1996).

[16] D. Muhlrad, C. J. Decker, and R. Parker, *Genes Dev.* **8,** 855 (1994).

can serve as a rapid method to analyze the activity of XRN orthologs from other organisms.

Given the utility of the poly(G) tract approach to understanding mRNA decay in yeast, the application of a similar approach to other eukaryotes may also lead to insight into mRNA degradation. The accumulation of poly(G)-stabilized mRNA decay intermediates similar to those observed in yeast could indicate that XRN-mediated degradation occurs. Such intermediates have been observed in *Chlamydomonas reinhardtii,*[13] which may indicate that degradation by an XRN-like enzyme occurs; however, in several plant and mammalian systems, expression of poly(G) tract-containing genes does not result in such intermediates.[9] The analysis of poly(G)-containing mRNAs in other eukaryotes may indicate whether the absence of poly(G)-stabilized mRNA decay intermediates is the result of an mRNA degradation mechanism common to multicellular eukaryotes, or is particular to specific groups of eukaryotes. Such knowledge should aid in determining the extent to which the mechanism of mRNA decay in multicellular eukaryotes differs from the major mRNA decay pathway in yeast.

Analysis of Reporter RNAs Containing Poly(G) Tracts in *Arabidopsis thaliana*

This section describes an analysis of poly(G)-containing transcripts produced in stably transformed tobacco cells and transgenic *Arabidopsis thaliana* plants, and provides an example of controls and experimental considerations that can enhance such an analysis. Testing a number of reporter RNAs with different structures and degradation kinetics is advantageous, as the accumulation of poly(G)-stabilized mRNA decay intermediates could depend on these factors. In addition, because mRNA degradation can vary in different cell types,[14] or between cell culture and transgenic organisms,[15] examining the expression of poly(G)-containing reporter genes in several systems has the greatest potential for identifying mRNA decay intermediates. Most importantly, it is necessary to demonstrate that the gel system used to analyze the poly(G)-containing reporter RNAs is able to detect possible intermediates that could arise not only from $5' \rightarrow 3'$ degradation, but also from $3' \rightarrow 5'$ degradation, or both simultaneously.

We monitored the accumulation of poly(G)-containing RNAs in both tobacco suspension cells as well as in transgenic *Arabidopsis* plants. Similar results were obtained with both of these systems. *Arabidopsis* is easy to transform and these experiments describe the analysis of transgenic plants expressing stable copies of poly(G)-containing transgenes. Similar analyses should be applicable to transient expression in systems where stable transformation is not feasible. Poly(G) tracts of 25 guanosines were inserted into the 3′ untranslated regions (UTRs) of reporter constructs that had previously been used to test mRNA sequences for their function as instability determinants in plant cells. These constructs provided the potential

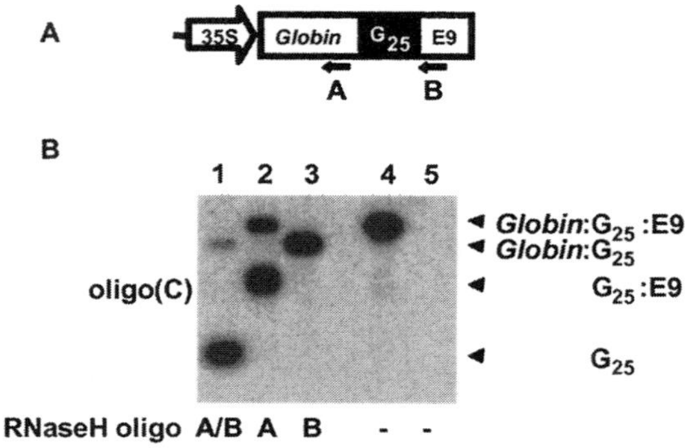

FIG. 1. Analysis of poly(G)-containing mRNAs in transgenic *Arabidopsis*. (A) Schematic diagram of 35S–globin–G_{25}–E9 transgene. Oligonucleotides used for RNase H cleavage in part B are indicated by arrows. (B) Northern blot analysis of total RNA from transgenic plants expressing 35S–globin–G_{25}–E9. Lanes 1–4 contain 20 μg of total RNA from plants expressing the transgene shown in (A). For lanes 1–3, the RNA was incubated with RNase H and the oligonucleotides indicated at the bottom of the panel before electrophoresis. For lane 4, the RNA was not treated with RNase H. Lane 5 contains 20 μg of total RNA from plants expressing a 35S–globin–E9 transgene, which lacks a poly(G) tract. Radiolabeled oligo(C) probe was used in the hybridization. 35S, cauliflower mosaic virus promoter; globin, mammalian β-globin-coding region; E9, rbcs small subunit polyadenylation signal from pea.

to test whether insertion of a poly(G) tract into a stable mRNA could lead to the accumulation of poly(G) intermediates, or whether rapid mRNA turnover mediated by a dimer of the DST element,[17] an AUUUA repeat,[18] or the *SAUR-AC1* 3′ UTR[19] would be required for poly(G)-stabilized intermediate accumulation.

RNA was extracted from transgenic plants expressing these poly(G)-containing genes, and analyzed by Northern blot. To demonstrate that all potential poly(G)-stabilized mRNA decay intermediates could be readily detected by the gel system employed, RNA fragments of expected size were produced as controls. These RNA fragments were not synthesized, but instead were generated by directed cleavage of the reporter RNAs isolated from the transgenic plants. This was accomplished by incubating the isolated RNA with RNase H and DNA oligonucleotides complimentary to the poly(G)–reporter RNA. RNase H cleaves the RNA strand of an RNA : DNA duplex, so that RNA fragments of specific length can be generated. Figure 1A shows the position of the DNA oligonucleotides used to generate fragments of the globin–G_{25} reporter, which approximate potential mRNA decay intermediates expected to arise as a result of degradation from either end, or both

[17] T. C. Newman, M. Ohme-Takagi, C. B. Taylor, and P. J. Green, *Plant Cell* **5,** 701 (1993).

[18] M. Ohme-Takagi, C. B. Taylor, T. C. Newman, and P. J. Green, *Proc. Natl. Acad. Sci. U.S.A.* **90,** 11811 (1993).

[19] P. Gil and P. J. Green, *EMBO J.* **15,** 1678 (1996).

simultaneously. These RNA fragments were analyzed by Northern blot using high-percentage agarose gels, and compared with total RNA not treated with RNase H. Equal amounts of RNase H-treated and untreated RNA were analyzed, so that the amounts of the RNA fragments would be in the range expected for bona fide intermediates of mRNA degradation. As can be seen in Fig. 1B, reporter RNA fragments approximating degradation from 5′→3′ decay (lane 2), from 3′→5′ decay (lane 3), or from simultaneous 5′→3′ and 3′→5′ decay (lane 1), were detected in the gel system. In contrast, only the full-length globin transcript was observed when the RNA sample was analyzed without prior RNase H treatment (lane 4). Similar results were obtained in the analysis of reporter RNAs bearing the DST element (DST × 2), the AUUUA repeat, or the *SAUR-AC1* 3′ UTR. This experiment indicated that although the gel system was adequate to detect poly(G)-stabilized intermediates, these intermediates did not accumulate *in vivo*. The absence of poly(G)-stabilized mRNA decay intermediates in plant and mammalian cells is unlikely to be due solely to the activity of XRN family members, and may represent a difference in mRNA degradation mechanisms between yeast and multicellular eukaryotes.[9]

RNase H Cleavage to Generate RNA Size Controls

DNA oligonucleotides complementary to the 3′ end of the globin-coding sequence (oligonucleotide A in Fig. 1A) or to the 5′ end of the E9 3′ UTR (oligonucleotide B in Fig. 1A) were used to cleave the reporter transcripts at specific sites: oligonucleotide A, 5′-TAATAGAAATTGGACAGCAA-3′; oligonucleotide B, 5′-CCCAATGCCATAATACTCG-3′. Oligonucleotide design for RNase H cleavage follows the same guidelines as for sequencing primers; however, it is often necessary to test several primers as cleavage efficiencies can vary over a wide range (e.g., a portion of full-length globin-G_{25} remains intact despite cleavage with oligonucleotide A; Fig. 1B, lane 2). Twenty micrograms of total RNA is incubated with 2 μg of oligonucleotide at 65° for 30 min in a water bath. To anneal the oligonucleotides to their cognate mRNAs, the water bath is slowly cooled to at least 30°. The reactions are then placed at room temperature for 5 min. RNase H digestions are incubated at 37° for 1 hr in 50 μl of RNase H reaction mix containing 4 m*M* Tris-HCl (pH 8.0), 10 m*M* $MgCl_2$, 20 m*M* KCl, 1 m*M* dithiothreitol (DTT), and 2 units of RNase H (GIBCO-BRL, Gaithersburg, MD). The reactions are ethanol precipitated and analyzed by Northern blot.

Northern Blot Analysis

Standard Northern blotting procedures are altered in order to resolve potentially small poly(G)-stabilized mRNA decay intermediates. Gels containing 2.2% (w/v) NuSieve 3 : 1 agarose (FMC, Rockland, ME)–0.2% (v/v) formaldehyde can be used to effectively resolve RNA species. These gels are blotted by standard methods and probed with a radiolabeled oligonucleotide C probe. We have also used polyacrylamide gels [6% (w/v) acrylamide, 7 *M* urea] to achieve enhanced resolution. These gels were run and transferred to blots as described.[13]

Use of Yeast *xrn1*Δ and Poly(G)-Containing mRNAs to Analyze XRN Enzyme Activity

Both Xrn1p and Xrn2p/Rat1p of yeast, as well as the three Xrn2p/Rat1p orthologs from *A. thaliana,* when expressed in yeast, are blocked by poly(G) tracts.[9] It is likely that blockage by poly(G) tracts is an inherent property of XRN enzymes. This property is experimentally advantageous, as it allows exoribonuclease activity to be rapidly determined through complementation of a specific yeast *xrn1* mutant strain as described below.

A simple test of the potential exoribonuclease activity of an XRN enzyme is to express it in an *xrn1*Δ yeast strain and analyze the degradation of poly(G)-containing mRNAs. If an XRN protein is active as an exoribonuclease, and degrades transcripts from the 5′ end, then it will likely generate poly(G)-stabilized mRNA decay intermediates similar to that of Xrn1p. Studies of mouse mXRN1[10] and *Drosophila* Pacman (Ref. 10 and [24] in this volume[21]) have made use of *xrn1* complementation. Expression of mXRN1 or Pacman in an *xrn1* mutant was shown to reduce the abundance of endogenous mRNAs that normally accumulate to high levels in the absence of Xrn1p. It was further shown that this reduction in abundance when Pacman was expressed in the *xrn1* mutant was due to increased mRNA degradation rates. An advantage of the analysis of poly(G)-containing mRNAs is that this approach does not require complex mRNA half-life determinations, as the accumulation of an intermediate is monitored, rather than a change in the stability of mRNA in the transformed strain relative to the mutant. In addition, it enables the direction of catalysis to be determined and allows the effect of poly(G) tracts on the progression of these enzymes to be addressed. This experiment can yield results for both nuclear-targeted as well as cytoplasmic exoribonucleases, and therefore can be used to study Xrn1p orthologs as well as Xrn2p/Rat1p orthologs.[9]

Complementation of the *xrn1*Δ mutant was used to study the three XRN-like enzymes of *A. thaliana*. The AtXRNs were expressed from a 2μ plasmid (p1954 in Fig. 3A) in an *xrn1*Δ strain that also expresses two poly(G)-containing reporter genes, *PGK1* and *MFA2*. An analysis of AtXRN2 is shown in Fig. 2. Expression of AtXRN2 in the *xrn1*Δ strain leads to a decrease in the accumulation of the full-length *PGK1* reporter RNA, and the formation of a poly(G)-stabilized *PGK1* mRNA decay intermediate similar to wild type. This indicates that AtXRN2 is

[20] D. C. Amberg, A. L. Goldstein, and C. N. Cole, *Genes Dev.* **6,** 1173 (1992).

[21] I. V. Chernukhin, J. E. Seago, and S. F. Newbury, *Methods Enzymol.* **342,** [24] 2001 (this volume).

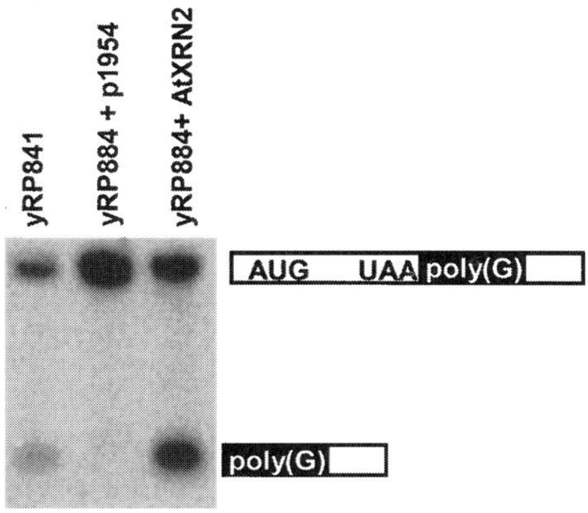

FIG. 2. Analysis of the enzymatic activity of AtXRN2 on poly(G)-containing mRNAs when expressed in yeast lacking Xrn1p (*xrn1*Δ). AtXRN2 was inserted in p1954 (see Fig. 3) and expressed in an *xrn1*Δ strain. The *xrn1*Δ strain expresses two poly(G)-containing RNAs, *PGK1* and *MFA2*, each of which has an 18-guanosine tract in the 3′ UTR. The accumulation of the poly(G) reporter mRNA *PGK1*, and its corresponding poly(G)-stabilized mRNA decay intermediate were analyzed by Northern blot. Twenty micrograms of total RNA was analyzed, and the blot was hybridized with a radiolabeled oligonucleotide complementary to the poly(G)-containing PGK1 mRNA. The structure of the poly(G) reporter, and of the poly(G)-stabilized intermediate generated by Xrn1p or AtXRN2, are shown at right.

active as a 5′→3′-exoribonuclease, and is blocked by poly(G) tracts when expressed in yeast. Similar results were obtained with all three AtXRNs.[9]

*xrn1*Δ Complementation

Yeast strains (wild type) yRP841 (*MAT*α, *trp1-*Δ*1, ura3-52, leu2-3,112, lys2-201, cup*:: *LEU2pm*), and (*xrn1*Δ) yRP884 (*MAT***a,** *trp1-*Δ*1, ura3-52, leu2-3,112, lys2-201, cup*::*LEU2pm, XRN1*::*URA3*) are described.[22] Both of these strains harbor chromosomal genes encoding *PGK1* and *MFA2* reporter RNAs with poly(G) tracts of 18 guanosines in their 3′ UTRs. These genes are under the control of the *GAL1* upstream activating sequence, requiring growth in galactose to induce expression. Expression of high levels of heterologous proteins can be accomplished with the yeast expression vector pG1.[23] We have modified the polylinker of this vector to include an additional unique restriction site (*Not*I; Fig. 3A, top) and generated a vector that can be used to generate GFP fusions for protein localization studies in yeast (Fig. 3A, bottom).

The yRP841 and yRP884 strains can be easily transformed with the vectors described in Fig. 3, using standard methods. We recommend the following

[22] G. Caponigro and R. Parker, *Genes Dev.* **9,** 2421 (1995).

[23] M. Schena, D. Picard, and K. R. Yamamoto, *Methods Enzymol.* **194,** 389 (1991).

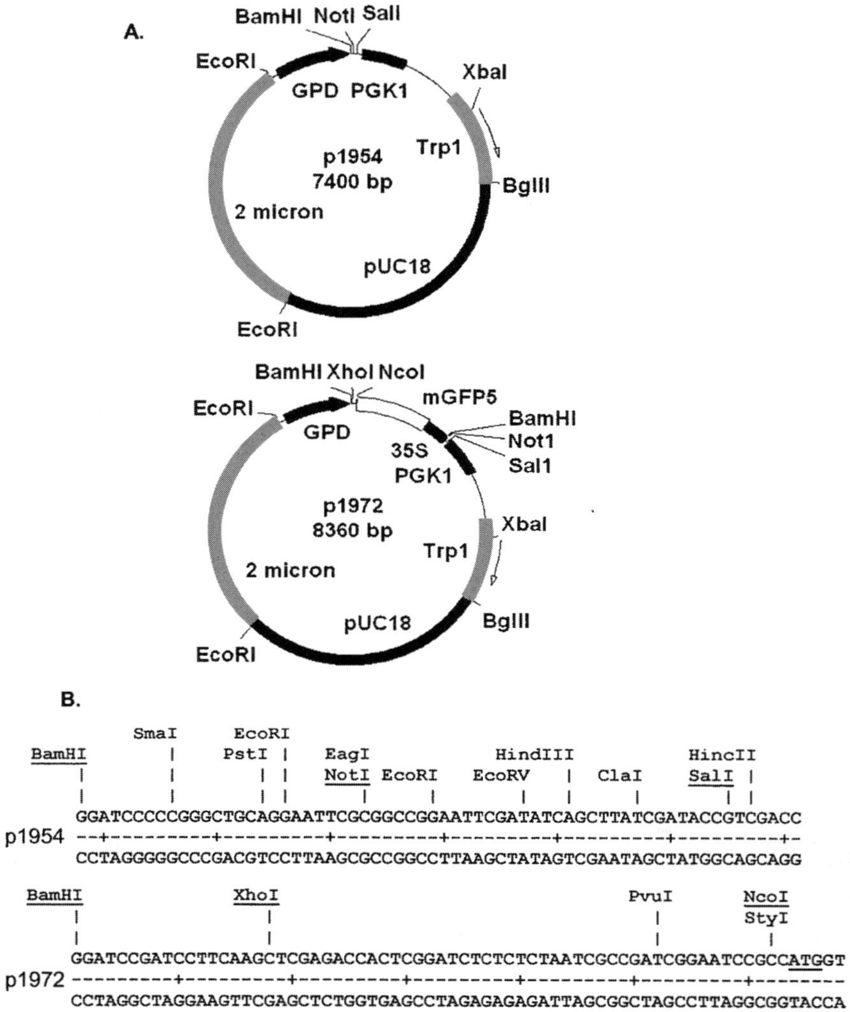

FIG. 3. Derivatives of pG1 vector[25] for expression of proteins in yeast. (A) p1954 is a multicopy plasmid (2μ) with the GPD promoter and *PGK1* terminator. For expression in yeast, sequences can be inserted between the promoter and terminator, using the unique *Bam*HI, *Sal*I, or *Not*I site. p1972 is similar to p1954 but contains the mGFP5 derivative of GFP.[29] Fusion of GFP to the C terminus of a protein is accomplished by insertion of an open reading frame into the *Nco*I site. This vector could also be used to express proteins (not as a GFP fusion), using the *Xho*I site not available in p1954. (B) Polylinker sequence of p1954 and sequence surrounding the *Nco*I site of p1972. Unique enzyme sites are underlined. The ATG of GFP is also underlined. p1972 also contains the cauliflower mosaic virus 35S terminator, 35S.

procedure, which includes a small-volume subculturing step, to obtain good growth of the transformed strains and high levels of reporter gene expression. A single colony of each transformant is used to inoculate 1 ml of SD medium containing 2% (w/v) glucose, which is grown for 2 days at 28° in a shaker. Twenty microliters of each overnight culture is used to inoculate 1 ml of SD containing 2% (w/v) galactose [to induce expression of poly(G) reporters], which is grown for an additional 1 to 2 days. The cells grown in SD + galactose are used to inoculate 30 ml of SD + galactose to an OD_{600} of 0.05. It is possible to inoculate the 30-ml SD + galactose culture directly with the initial SD + glucose overnight culture; however, this can result in a significant lag in growth and poor growth. The 30-ml culture is grown to an OD_{600} of 0.3–0.4 (usually takes 2 days) and is harvested by centrifugation in 50-ml conical tubes. Total RNA is isolated by the method described in Parker *et al.*,[24] and analyzed using standard Northern blotting techniques. The probes used for analysis of the *PGK1* and *MFA2* reporter RNAs are oligonucleotides PGK1 (5′-AATTGATCTATCGAGGAATTCC-3′) and MFA2 (5′-ATATTGATTAGATCAGGAATTCC-3′) as described in Caponigro and Parker.[22] The oligonucleotides are 5′-end labeled as follows: 300 ng of oligonucleotide and 400 μCi of [γ-P^{32}]ATP (ICN, Costa Mesa, CA) are incubated with 10 units of T4 polynucleotide kinase (Roche, Nutley, NJ) and kinase buffer- (supplied by the manufacturer) in a 20-μl total volume at 37° for 1 hr. The labeled oligonucleotide can be isolated from free nucleotide by gel-filtration chromatography, using a NucTrap column (Strategene, La Jolla, CA).

Determining Intracellular Localization of XRN Family Members

The cellular function of XRN family members is dependent on intracellular location. As mentioned above, yeast Xrn1p is a cytoplasmic mRNase, whereas Xrn2p/Rat1p is nuclear. However, at least in *A. thaliana,* an Xrn2p/Rat1p ortholog is cytoplasmic.[9] This indicates that the intracellular location of an XRN family member may not be reliably predicted on the basis of its similarity to either Xrn1p or Xrn2p/Rat1p. In addition, although AtXRN2 is targeted to the nucleus, high levels of expression in yeast led to the accumulation of a poly(G) intermediate, similar to the one produced by cytoplasmic Xrn1p (Kastenmayer and Green[9]; and Fig. 2), most likely due to some accumulation of AtXRN2 in the cytoplasm. Therefore, although *xrn1*Δ complementation can demonstrate the exoribonuclease activity of an XRN family member, it apparently does not distinguish between cytoplasmic and nuclear enzymes. To fully address the possible cellular function of XRN enzymes, an investigation of their intracellular location in a homologous system is required. Before in-depth localization studies are performed, insight into

[24] R. Parker, D. Herrick, S. W. Peltz, and A. Jacobson, *Methods Enzymol.* **194,** 415 (1990).

intracellular localization can be gained by an additional yeast complementation experiment using the *xrn2/rat1* mutant *rat1-1*^{ts}.

If expression in the *xrn1*Δ strain indicates that an XRN enzyme is active as an exoribonuclease, complementation of an *xrn2/rat1* mutant can then be tested. Xrn2p/Rat1p is encoded by an essential gene, and cells harboring the *rat1-1*^{ts} mutation rapidly arrest their growth at the nonpermissive temperature.[20] Rescuing the *rat1-1*^{ts} mutation appears to require an active exoribonuclease in the nucleus.[6] This indicates that complementation of *rat1-1*^{ts} by an XRN protein can serve as an assay for exoribonuclease activity and nuclear targeting.

Complementation studies of *rat1-1*^{ts} were employed to gain insight into the intracellular locations of the AtXRNs. The results of these experiments indicated that complementation of *rat1-1*^{ts} by heterologous proteins is dependent on the same intracellular location requirements as for the endogenous yeast proteins: AtXRNs targeted to the nucleus complement *rat1-1*^{ts} whereas a cytoplasmic AtXRN did not.[9] In addition, as described below, the localization of the AtXRNs in plant cells was consistent with complementation of *rat1-1*^{ts} serving as an indicator of intracellular location. Therefore, *rat1-1*^{ts} complementation may serve as an indication of the intracellular location of XRN proteins not only when expressed in yeast, but perhaps also in their native contexts.

Complementation of rat1-1^{ts}

The yeast strains employed are FY86 (*MAT*α, *ura3-52, his3*Δ*200, leu2*Δ*1*) and DAt1-1 (*MAT*α, *ura3-52, leu2*Δ*1, trp1*Δ*63, rat1-1*).[20] This particular *rat1-1*^{ts} strain allows for the use of the same expression vectors employed in *xrn1*Δ complementation, as transformants can be selected for growth in the absence of tryptophan. After transformation, single colonies are used to inoculate 1 ml of SD medium. After growth overnight, the cultures are diluted to equal OD_{600} with SD medium and the cultures are streaked on plates. The best results have been obtained by streaking out 3 μl of a dilution to an OD_{600} of approximately 1. This is because no growth of the *rat1-1*^{ts} cells is observed at the nonpermissive temperature, whereas with greater volumes or higher OD_{600} some growth can occur, which can complicate the analysis. Three microliters of the diluted cells is pipetted onto two SD plates and spread with a sterile loop. One plate is incubated at the permissive temperature (26°) and the other at the nonpermissive temperature (37°). Incubation of these plates for 3 days will result in abundant growth of the wild-type and complemented *rat1-1*^{ts} strains. The *rat1-1*^{ts} strain transformed with the yeast *XRN2/RAT1* gene is used as a positive control. It should be noted that complementation of the *rat1-1*^{ts} strain by heterologous proteins may not result in growth at 37° equivalent to *rat1-1*^{ts} complemented with *RAT1*.[9] Therefore, if it appears that growth does not occur in 3 days, plates should be incubated for a longer period to determine whether cells are growing, albeit slowly. The *rat1-1*^{ts} strain transformed with vector only (p1954) will not grow even when incubated for 7 days. In addition, the growth of the transformed *rat1-1*^{ts} strain at the permissive temperature could also be informative. Overexpression of Xrn1p in wild-type cells has been reported to adversely affect growth rate,[25] and it is possible that overexpression of other XRN proteins could affect growth of the *rat1-1*^{ts} strain. If so, this may be evident at the permissive temperature. Expression of the AtXRNs, using the vectors in Fig. 3, had no apparent effect on the growth of the *rat1-1*^{ts} strain at the nonpermissive temperature, indicating that the level of expression obtained with these vectors does not appear to inhibit growth. A final note with respect to *rat1-1*^{ts} complementation relates to the solid SD medium used for growing the transformed strains. Although some protocols indicate that SD medium can be autoclaved, we have found that the complemented *rat1-1*^{ts} grows poorly on autoclaved medium. The SD medium should be filter sterilized.

Localization of XRN–GFP Fusion Proteins in Yeast

A more direct approach to protein localization is the analysis of fusions of the XRNs to the green fluorescent protein (GFP), a technique successfully used to study Xrn2p/Rat1p of yeast, and the AtXRNs.[6,9] Analysis of XRN–GFP fusions expressed in yeast can be used to confirm the results of *rat1-1*^{ts} complementation. We have constructed a derivative of the pG1 vector that allows for the expression in yeast of proteins with GFP fused to the C terminus (p 1972; Fig. 3). Insertion of an open reading in the unique *Nco*I site is used to generate the XRN–GFP fusion. Before localization studies are performed in any system, it should be demonstrated that the XRN–GFP fusions retain exoribonucleolytic activity and that *rat1-1*^{ts} complementation is not affected by the fusion. This is easily accomplished by testing whether the XRN–GFP fusion retains the ability to generate a poly(G)-stabilized mRNA decay intermediate when expressed in the *xrn1*Δ strain and does not differ in ability to complement *rat1-1*^{ts} relative to the XRN protein without the GFP fusion. In the case of the AtXRNs, fusion of GFP did not affect *rat1-1*^{ts} complementation and AtXRN4–GFP exhibited the same activity as AtXRN4 in the *xrn1*Δ strain (Fig. 4). Similarly, fusion of GFP to the C terminus of the yeast XRN proteins does not appear to adversely affect their function.[6] If the XRN–GFP protein retains exoribonucleolytic activity and is not altered in *rat1-1*^{ts} complementation, its localization can then be determined by flourescence microscopy. Analysis of the AtXRN–GFP fusions revealed that they were localized in yeast in a manner expected on the basis of the results of complementation of *rat1-1*^{ts} (Ref. 9, and J. P. Kastenmayer and P. J. Green, unpublished data, 2000).

[25] V. I. Bashkirov, J. A. Solinger, and W. D. Heyer, *Chromosoma* **104,** 215 (1995).

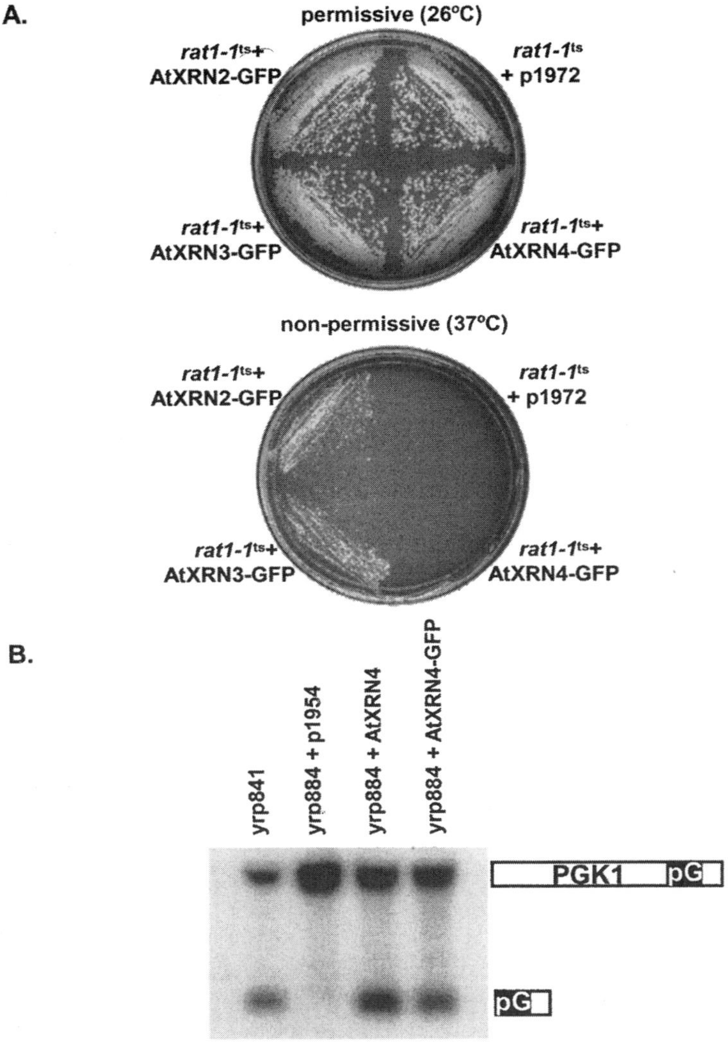

FIG. 4. Fusion of GFP to the C terminus of the AtXRNs does not affect their activities in yeast. (A) Overnight cultures of the *rat1-1*^{ts} strain transformed with p1972, and p1972 containing each of the AtXRNs, were streaked on SD plates as described in Complementation of *rat1-1*^{ts} and yielded results comparable to those of the AtXRNs without GFP fusions. (B) Northern blot analysis of the *PGK1* poly(G) reporter RNA in *xrn1*Δ cells transformed with p1954, p1954 containing AtXRN4, and p1972 containing AtXRN4–GFP. The blot was probed with the oligonucleotide complementary to the *PGK1* poly(G) repoter RNA. The structure of the poly(G) reporter, and the structure of the poly(G)-stabilized intermediate generated by Xrn1p, AtXRN4, and AtXRN4–GFP, are shown on the right-hand side. Again, the results are comparable with and without fusion to GFP.

Localization of XRN–GFP Proteins in Yeast Cells

The *rat1-1*^{ts} strain is transformed with a plasmid for expression of the XRN–GFP fusion, using standard methods. GFP without a fusion, expressed from p1972 (Fig. 3), can be used as a control as this protein is distributed uniformly across cells and is not localized preferentially to the nucleus or cytoplasm.[26] The growth of the transformants should be carried out in a manner similar to that used for the *rat1-1*^{ts} complementation studies described above. After dilution with SD medium to an OD_{600} of approximately 1, the cells are grown at 26° for an additional 4 hr. To stain nuclear DNA, 2,6-diamidino-2-phenylindole (DAPI; Sigma, St. Louis, MO) is added to a final concentration of 0.5 μg/ml, for the final 30 min of the 4-hr growth. It is sometimes difficult to obtain adequate staining of yeast nuclear DNA with DAPI. Staining can be enhanced by the addition of 20% (v/v) ethanol and incubation for approximately 5 min. This treatment does not appear to affect the intracellular localization or distribution of the AtXRN–GFP fusions. However, it leads to decreased intensity of GFP fluorescence, and a rapid quenching of GFP fluorescence for some AtXRN–GFP fusion proteins. For proteins whose expression in yeast is low, the decrease in GFP fluorescence due to ethanol treatment may make it difficult to obtain representative images of protein localization before GFP fluorescence decreases below detection. Optimization of ethanol concentration or incubation times may be required to obtain the best results for individual XRN–GFP fusions.

AtXRN–GFP Localization Studies in Plant Cells: Reiteration of Results of *rat1-1*^{ts} Complementation

The *rat1-1*^{ts} complementation studies indicated that the AtXRNs are likely targeted differentially in yeast and that *rat1-1*^{ts} complementation was a reliable predictor of nuclear targeting in yeast. To gain insight into the potential function of the AtXRNs in plant cells, and to determine whether *rat1-1*^{ts} complementation and localization studies were relevant to the intracellular localization of the AtXRN proteins in plants, localization of the AtXRN–GFP fusion proteins in plant cells was determined. These experiments employed transient expression in onion epidermal cell layers by particle bombardment, a system with several advantages for protein localization studies.[27] AtXRN2–GFP and AtXRN3–GFP, which could complement *rat1-1*^{ts}, were targeted to the nucleus in plant cells, while AtXRN4–GFP, which did not complement *rat1-1*^{ts}, was cytoplasmic.[9] These results indicate that complementation of *rat1-1*^{ts} gives a likely indication of the localization of XRN proteins in their native contexts.

[26] A. G. von Arnim, X. W. Deng, and M. G. Stacey, *Gene* **221,** 35 (1998).
[27] M. J. Vargona, R. J. Schmidt, and N. Raikhel, *Plant Cell* **4,** 1213 (1992).

Conclusions

There are potentially important differences between the activity and intracellular locations of XRNs of yeast and multicellular eukaryotes. We have described experiments that make use of complementation of yeast *xrn1* and *xrn2/rat1* mutants in combination with poly(G) tract studies, which allow for these differences to be addressed. The results of these studies can yield important insight into the possible function of XRN enzymes and can be used in conjunction with more detailed analysis of the XRN enzymes in their native contexts.

[23] 5′ → 3′-Exoribonuclease from Rabbit Reticulocytes

By LAWRENCE I. SLOBIN

Introduction

It has been demonstrated that 5′ → 3′-ribonucleases play an important role in the decay of mRNAs in yeast cells.[1] A 5′ → 3′-ribonuclease from yeast cells[2] named Xrn1p has been purified and cloned (see [20] in this volume[2a]). It has been implicated in a number of biological processes including mRNA degradation. A homolog of *XRN1* has been cloned from *Drosophila* (see [24] in this volume[2b]) and yeast (see [21] in this volume[2c]). The Xrn1p nuclease consists of a single polypeptide that degrades both RNA and DNA. It does not cleave RNAs that possess a 5′-terminal cap structure. Other 5′ → 3′-exoribonucleases have been partially purified from nuclei of mammalian cells.[3,4] Coutts and Brawerman found that extracts of mouse sarcoma 180 cells possess a 5′ → 3′-exoribonuclease that can cleave both capped and uncapped mRNA.[5,6] They observed that the initial cleavage of the substrate could occur at either the first, second, and probably the third phosphodiester linkage in some RNAs. The purification and characterization of a 5′ → 3′-exoribonuclease from rabbit reticulocytes that degrades capped and

[1] D. Muhlrad, C. J. Decker, and R. Parker, *Genes Dev.* **8,** 855 (1994).

[2] A. Stevens, *J. Biol. Chem.* **255,** 3080 (1980).

[2a] A. Stevens, *Methods Enzymol.* **342,** [20] 2001 (this volume).

[2b] I. V. Chernukhin, J. E. Seago, and S. F. Newbury, *Methods Enzymol.* **342,** [24] 2001 (this volume).

[2c] A. W. Johnson, *Methods Enzymol.* **342,** [21] 2001 (this volume).

[3] L. S. Lasater and D. C. Eichler, *Biochemistry* **23,** 4367 (1984).

[4] K. G. K. Murthy, P. Park, and J. L. Manley, *Nucleic Acids Res.* **19,** 2685 (1991).

[5] M. Coutts and G. Brawerman, *Biochim. Biophys. Acta* **1173,** 49 (1993).

[6] M. Coutts and G. Brawerman, *Biochim. Biophys. Acta* **1173,** 57 (1993).

0076-6879/01 $35.00

uncapped RNAs has been reported.[7] This report summarizes information about the reticulocyte enzyme.

Assay for Reticulocyte Exoribonuclease

Virtually any RNA greater than 10–15 bases long may be used for routine assay of the exoribonuclease. The author has not found any RNAs that are resistant to the action of the enzyme. The RNA substrate may be labeled internally with either ^{32}P- or ^{3}H-labeled nucleotides in an *in vitro* transcription reaction, or at the 3′ end with poly(A) polymerase and radiolabeled ATP. The assay described below employs a 382-nucleotide transcript complementary to the 3′ exon of murine c-*myc* RNA (Ambion, Austin, TX), which is labeled with [α-^{32}P]GTP (3000 Ci/mmol) in an *in vitro* transcription reaction. All solutions are made with nuclease-free deionized or distilled water. Standard 10-μl reactions contain 20 m*M* Tris-HCl, pH 7.4, at 37°, 60–70 m*M* KCl, 1 m*M* $MgCl_2$, acetylated bovine serum albumin (BSA, 100 μg/ml) (nuclease free), 3 m*M* dithiothreitol (DTT), human placental ribonuclease inhibitor (10^3 units/ml), 2–4 % (v/v) glycerol, and 5–25 n*M* [^{32}P]RNA plus added enzyme. Reactions are incubated for 20 min at 37° and terminated by the addition of 10 μl of a 0.3% (w/v) solution of poly(A) in H_2O. Other nucleic acids at comparable concentrations can be added at the end of the incubation. The time of incubation can be adjusted depending on the amount of enzyme added. Incubations at 30° also give satisfactory results. After addition of 50 μl of 3.5% (by volume) perchloric acid, tubes are placed on ice for 10 min. Precipitated RNA is removed by centrifugation in a microfuge for 10 min at 4°. An aliquot of the supernatant (50 μl) is taken for measurement of acid-soluble radioactivity in a scintillation counter. RNA degradation is a linear function of enzyme concentration up until apprximately 50% hydrolysis of the starting substrate. A more diagnostic assay for the enzyme, although far more time consuming, involves analysis of the RNA degradation products on 20% (w/v) acrylamide–8 *M* urea gels.[7] Because the exoribonuclease is highly processive (see below) only mononucleotide products are produced from internally labeled RNA substrates up to at least 2500 bases long. The author has never observed any intermediate degradation product from internally labeled substrates even at short reaction times.

Purification of Exoribonuclease

Reagents and Materials

Solid ammonium sulfate is added to protein solutions without any pH adjustments.

[7] S. Somoskeöy, M. N. Rao, and L. I. Slobin, *Eur. J. Biochem.* **237,** 171 (1996).

All buffers listed below contain 1 mM dithiothreitol, 0.1 mM EDTA, and either 1 mM $MgCl_2$ (buffers 1–7) or 1 mM magnesium acetate (buffers 8–11).

Buffer 1 : 20 mM K^+ HEPES (pH 7.4), 50 mM KCl
Buffer 2 : 20 mM K^+ HEPES (pH 7.4), 350 mM KCl
Buffer 3 : 20 mM K^+ HEPES (pH 7.4), 100 mM KCl
Buffer 4 : 20 mM K^+ HEPES (pH 7.4), 250 mM KCl
Buffer 5 : 20 mM K^+ HEPES (pH 7.4), 500 mM KCl
Buffer 6 : 50 mM potassium phosphate, pH 7.4
Buffer 7 : 350 mM potassium phosphate, pH 7.4
Buffer 8 : 20 mM Tris-HCl, pH 7.7 at 25°, 100 mM KCl, 1 mM
Buffer 9 : 20 mM Tris-HCl, pH 7.7 at 25°, 50 mM KCl
Buffer 10 : 20 mM Tris-HCl, pH 7.7 at 25°, 350 mM KCl
Buffer 11 : 20 mM Tris-HCl, pH 7.7 at 25°, 100 mM KCl, 20% (v/v) glycerol

Purification Procedures

All steps are performed at 4–6° unless indicated otherwise. The columns were equilibrated with starting buffer before use.

Step 1: Q-Sepharose Chromatography of Rabbit Reticulocyte Postpolysomal Supernatant Proteins. The postpolysomal supernatant proteins from 675 ml from rabbit reticulocyte lysates (Green Hectre Farms) were applied to a column (2.5 × 31 cm) of Q-Sepharose (Pharmaci, Piscataway, NJ) equilibrated with buffer 1. The column is washed with the same buffer until essentially all the hemoglobin is eluted. Bound proteins are then eluted with buffer 2. Eluted proteins (1.6 g) are fractionated with ammonium sulfate. The proteins precipitating between 40 and 70% saturated ammonium sulfate are recovered by centrifugation and dialyzed against buffer 6. Approximately 75% of the total nuclease activity in the postpolysomal supernatant is recovered.

Step 2: Hydroxyapatite Chromatography. The proteins fractionated by ammonium sulfate are applied to a column (2.5 × 20 cm) of hydroxyapatite (Ultrogel). After washing with buffer 6 until all the unbound proteins are eluted, the column is developed with a linear 1700-ml gradient between buffer 6 and buffer 7 at a flow rate of 2.5 ml/min. Fractions (20 ml) are collected and assayed. Approximately 60% of the ribonuclease activity is found in the flowthrough fraction (510 mg of protein) and 40% in the fractions eluting between 60 and 100 mM potassium phosphate (140 mg of protein). The activity bound to the hydroxyapatite column is concentrated by precipitation with 70% saturated ammonium sulfate and dialyzed against buffer 1.

Step 3: Heparin–Agarose Chromatography. The proteins from the previous step are applied to a column (2.5 × 8 cm) of heparin–agarose (Bio-Rad, Hercules, CA). After washing with buffer 1 to elute unbound proteins, the column is eluted in stepwise fashion with buffers 3–5. The ribonuclease activity distributes as follows:

10% of the activity in the flowthrough (30 mg of protein), 35% in buffer 3 (53 mg of protein), and 55% in buffer 4 (31 mg of protein). The proteins eluted with buffers 3 and 4 are pooled separately, concentrated by precipitation with 70% saturated ammonium sulfate, and dialyzed against buffer 8. The activities eluted with buffers 3 and 4 are purified further by the procedures described below. Both activities are processive exoribonucleases and display the same specificity toward various substrates. However, the enzyme eluting at the lower KCl concentration is significantly less pure. The results presented below refer to the exoribonuclease eluted with buffer 4.

Step 4: Superose-6 Chromatography. The nuclease activity eluted with buffer 4 in step 3 is applied to a column (1.6 × 62 cm) of Superose-6 Prep grade (Pharmacia). The column is calibrated with molecular weight standards before application of the nuclease. The column is developed with buffer 8. The majority of the nuclease activity (≥70%) is recovered in a peak that centers at about 150 kDa. The fractions in the activity peak are pooled (9.5 mg of protein).

Step 5: QMA MemSep Chromatography. The pooled fractions from the previous step are dialyzed against buffer 9 and chromatographed on a QMA MemSep 1000 cartridge (Millipore, Bedford, MA) equilibrated with buffer 9. The cartridge is eluted with a linear gradient (total volume of 100 ml), using buffers 9 and 10 at a flow rate of 3 ml/min. The majority of the protein (≅90%) as well as 70% of the activity is not bound to the column. However, there is also an activity peak that elutes between 50 and 100 m*M* KCl. The nuclease bound to the membrane (0.9 mg of protein) possesses a specific activity about four times greater than that of the flowthrough proteins. The use of several other quaternary amine ion exchangers in place of the MemSep cartridge does not materially improve the purification. The QMA-bound fractions containing the RNase activity are pooled and concentrated with an Amicon (Danvers, MA) Centriprep-10 concentrator before further chromatography.

Step 6: Superdex Chromatography. A Superdex 200 HR 10/30 column (Pharmacia) is equilibrated with buffer 11. The column is calibrated with molecular mass standards for size-exclusion chromatography (Bio-Rad). The pooled fractions from the previous step are chromatographed at room temperature, using a flow rate of 20 ml/hr; 0.33-ml fractions are collected. A symmetrical peak of activity centering on fraction 41 is observed (see Fig. 1) Fractions 40–43, comprising the most active fractions, are pooled and stored at −70° in small aliquots. A total of 103 μg of protein is recovered at a concentration of 65 μg/ml.

Using the procedure described above, the author has achieved an approximately 68,000-fold purification based on the nonhemoglobin proteins in the starting lysate. Data suggest that the yield of purified protein is low (<10%) for the following reasons: (1) The enzymatic activity of the ribonuclease present in the buffer 3 heparin–agarose eluate (see above) is purified by the procedures described. Although significantly less pure, its enzymatic properties are indistinguishable

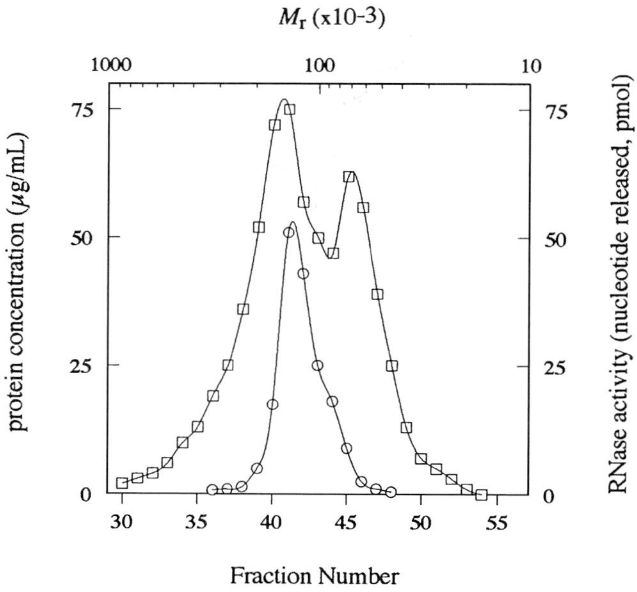

FIG. 1. Size-exclusion chromatography of exonuclease on Superdex 200 HR. Exonuclease (0.85 mg) after QMA chromatography (step 5) was chromatographed on a column of Superdex 200 HR. Each fraction was assayed for ribonuclease activity using 25 n*M* ^{32}P-labeled c-*myc* RNA. The column was calibrated with molecular size standards supplied by Bio-Rad. Fractions 40–43 were pooled and used for studies described here. (□) Protein concentration; (○) RNase activity (as nucleotides released). [Reproduced from S. Somoskeöy, M. N. Rao, and L. I. Slobin, *Eur. J. Biochem.* **237**, 171 (1996).]

from the more highly purified exoribonuclease; (2) the ribonuclease(s) that do not bind to the QMA MemSep cartridge are either similar or identical in properties to the purified enzyme. These observations raise the possibility that the enzyme described in this chapter exists in a variety of different forms. Alternatively, reticulocytes may possess a multiplicity of different processive $5' \rightarrow 3'$-exoribonucleases.

Physical Properties of Exoribonuclease

The exoribonuclease appeared to be reasonably stable when stored at $-70°$. No significant loss in activity was observed over a 3-month period. The enzyme was heat labile. Incubation for 10 min at 45° led to loss of about 90% of activity. The purified enzyme possessed a normal ultraviolet spectrum and appeared to be free of nucleic acids or nucleotides.

The purified enzyme contained three distinct polypeptide bands with apparent M_r values of 62,000, 58,000, and 54,000 (Fig. 2). The 62- and 58-kDa polypeptides showed the same staining density, suggesting a 1 : 1 stoichiometry. The staining

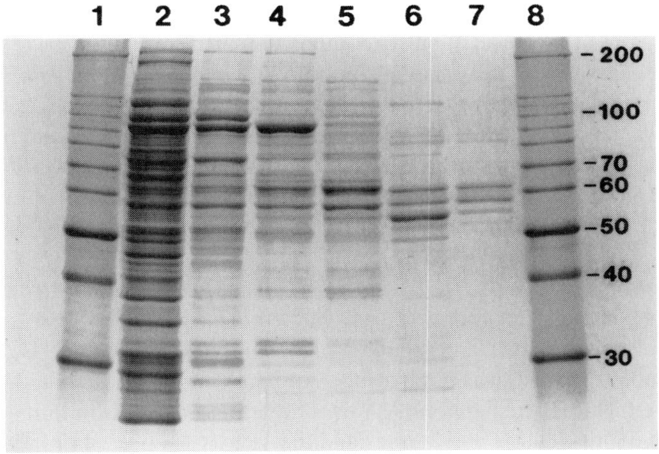

FIG. 2. SDS–PAGE of purified exonuclease. Approximately equal amounts of ribonuclease activity from pooled proteins at each step of purification were analyzed in a 10% (w/v) SDS–polyacrylamide gel as described.[7] The GIBCO-BRL 10-kDa protein ladder is shown in lanes 1 and 8. Proteins were visualized with a Coomassie Blue G stain. Lane 2, 350 m*M* KCl eluate of Q-Sepharose after precipitation of proteins with 40–70% saturated ammonium sulfate (step 1); lane 3, pool of proteins eluted from hydroxyapatite between 60 and 100 m*M* potassium phosphate (step 2); lane 4, proteins eluted from heparin–agarose at 250 m*M* KCl (step 3); lane 5, pool of activity after chromatography on Superose-6 (step 4); lane 6, proteins eluted from QMA MemSep 1000 between 50 and 100 m*M* KCl (step 5); lane 7, pool of fractions 40–43 after HPLC gel-exclusion chromatography on Superdex 200 HR (step 6, see Fig. 1). [Reproduced from S. Somoskeöy, M. N. Rao, and L. I. Slobin, *Eur. J. Biochem.* **237**, 171 (1996).]

of the 54-kDa band was distinctly less intense. The size of the native enzyme, as judged by gel filtration (Fig. 1), is intermediate between that of a dimer of 58- and 62-kDa polypeptides and a trimer of all three polypeptides. Ultraviolet cross-linking of a 5′-end-labeled RNA substrate with the purified enzyme produced a labeled product(s) whose size was consistent with one or more polypeptides of approximately 60 kDa.[7]

Enzymatic Properties of Exoribonuclease

Requirements for Activity

pH Optimum, Ionic Strength, and Metal Ions. The enzyme activity was little changed over the pH range 6.5–8.0. Activity was much the same in a variety of amine (Tris, HEPES), phosphate, and sulfonic acid (2-[(2-amino-2-oxyethyl) amino]ethanesulfonic acid, ACES) buffers at the same pH and ionic strength. Optimum activity was found between 60 and 90 m*M* KCl. However, the enzyme lost only about 50% of its activity when the KCl was lowered to 10 m*M* or raised to

200 mM. These findings were recorded for the substrate used for enzyme assay (see above). The possible effects of buffer composition on different substrate structures or on the processivity of the enzyme (see below) were not investigated. The Mg^{2+} dependence of the enzyme was investigated using capped c-*myc* RNA. The Mg^{2+} optimum was approximately 0.1 mM. An increase in Mg^{2+} concentration led to a loss of activity; the enzyme was about one-half as active at 0.8 mM as it was at 0.1 mM magnesium. At 4 mM Mg^{2+} the enzyme functioned at about 25% of its maximum rate. Mn^{2+} could not satisfy the metal ion requirement of the nuclease.

Substrate Specificity and Directionality of Exoribonuclease. In addition to the assay substrates, the following polynucleotides were hydrolyzed: poly(A), poly(U), poly(C), poly(G), and double-stranded poly(A):poly(U). DNA, both single and double stranded, was not hydrolyzed. Single-stranded DNA did, however, inhibit the enzyme. By using poly(A) it was determined that the minimal substrate size for the exonuclease was 10 bases and that activity appeared to increase with increasing polymer size up to a octadecamer. To determine the polarity of the exoribonuclease, an [^{3}H]poly(A) tail was added to capped ^{32}P-labeled c-*myc* and the rate of formation of acid-soluble ^{3}H and ^{32}P radioactivity was determined. The results of this experiment clearly demonstrated that degradation of the 5′ end of the substrate preceded hydrolysis at the 3′ end. These results are consonant with the view that the enzyme acts in a $5' \rightarrow 3'$ direction.

Products of Enzymatic Action. Only mononucleotide products are observed when a homogeneously labeled RNA substrate is hydrolyzed (Fig. 3B). Further analysis indicated that 5′-mononucleotide products are produced exclusively, for example, when c-*myc* RNA was synthesized with [α-^{32}P]GTP, 5′-GMP was the only hydrolysis product.[7] These observations strengthen the characterization of the enzyme as a highly processive exoribonuclease. Somewhat surprisingly, chimeric substrates containing deoxynucleotides at the 5′ end were effectively hydrolyzed. For example, the exonuclease cleaved the oligonucleotide $dA(A)_2(dA)_{12}$ to give the following digestion products in order of abundance: dA, dAAA, and dAA (Fig. 3A). The preferred cleavage site was clearly between the two ribo-containing nucleotides.[7] Chimeras containing up to 15 deoxynucleotides at the 5′ end were substrates provided that they were followed in the sequence by ribonucleotides. In all cases the oligodeoxynucleotide was released intact from the 5′ end of the chimeric substrate. A variety of cleavage events at the 5′ end were also observed when 5′-end-labeled poly(A) was used as a substrate: the predominant cleavage product was the dinucleotide pA_pA (Fig. 3C). Mono-, tri-, and tetranucleotide products can also be found. When a cap-labeled RNA substrate was hydrolyzed, the predominant products produced appeared to be (5′)m^{7}Gppp(5′)GpG and Gppp(5′)GpG[7]. Similar results were found by Coutts and Brawerman for a less purified enzyme from mouse sarcoma cells (see Introduction).

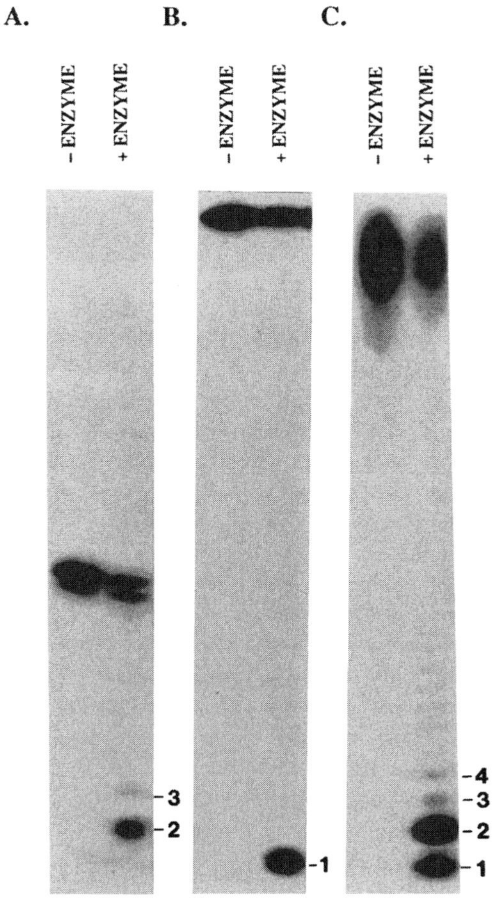

FIG. 3. Analysis of exonuclease cleavage products on polyacrylamide–urea gels. Standard reactions, containing 1 μl of purified enzyme and 100 fmol of the indicated substrate, were incubated for 20 min at 37°. An aliquot (2 μl) was taken for analysis by PAGE as described.[7] The following substrates were used: (A) [5′-^{32}P]dA(A)$_2$(dA)$_{12}$; (B) capped c-*myc* RNA uniformly labeled with [α-^{32}P]GTP; (C) [5′-^{32}P]poly(A). The numbers indicate the number of nucleotides in the hydrolysis products. [Reproduced from S. Somoskeöy, M. N. Rao, and L. I. Slobin, *Eur. J. Biochem.* **237,** 171 (1996).]

It has long been noted that the cap structure protects mRNA from degradation.[8,9] We have determined the kinetics of hydrolysis of capped and uncapped RNAs, using c-*myc* RNA containing either a 5′ cap or 5′-triphosphate terminated RNA. The cap structure was found to significantly alter both the K_m and V_{max} of the enzyme. Compared with the 5-triphosphate terminated RNA, the cap structure increased the K_m 4-fold and decreased the V_{max} 4-fold. No difference in kinetics was observed when substrates with triphosphate and monophosphate termini were compared. Apparently, RNA cap structure presents a significant impediment to the action of the exonuclease.

Inhibitors of Enzyme Action

A number of different agents inhibit enzyme action. EDTA was found to completely inhibit exonuclease activity when present at concentrations greater than twice the Mg^{2+} ion concentration in the assay mixture. *N*-Ethylmaleimide (NEM) at a concentration of 2.5 m*M* inhibited the release of mononucleotide product from uniformly labeled c-*myc* substrate by about 90%. It also inhibited the hydrolysis of 5′-end products from poly(A). However, examination of the effect of NEM on a small substrate, oligo(A), 7–17 bases long, gave a different result. We found that cleavage at the 5′ end of this substrate actually increased about 2-fold in the presence of NEM. Furthermore, the size distribution of 5′-end products was changed. AMP was not produced; instead, the products formed were di- and oligonucleotides. The enhanced activity of the exonuclease at the 5′ end of a substrate was also found with the chimeric substrate $dA(A)_2(dA)_{12}$. In addition, 2 m*M* NEM abolished the ultraviolet cross-linking of the enzyme to the 5′ end of a poly(A) substrate (see above). These observations suggest that NEM alters the way in which the enzyme binds to the 5′ terminus of a substrate. Treatment with the sulfhydryl reagent converts the exonuclease from a highly processive enzyme to one that acts only to cleave small substrates. These findings suggest that the enzyme possesses an upstream anchor site, altered by NEM treatment, which is essential for proper substrate binding and processive cleavage of large substrates.Upstream anchor sites have been proposed for other processive nucleases such as the 3′-exonucleases polynucleotide phosphorylase (PNPase),[10] and DNA exonuclease 1.[11] Evidence has been provided that such a site exists in ribonuclease II from *Escherichia coli*.[12]

Exonuclease action is also inhibited by nucleotide triphosphates (Table I). Hydrolysis of the terminal phosphate of the nucleotide does not appear to be involved in the inhibition because the nonhydrolyzable nucleotide ADPNP was about as effective as ATP. The presence of ADP in enzyme reaction mixtures had little, if any, effect (Table I). ATP and UTP appeared to be somewhat more effective as inhibitors than GTP and CTP. Examination of the concentration dependence of the effect of ATP showed that 50% inhibition was achieved at approximately

[8] Y. Furuichi, A. LaFiandra, and A. J. Shatkin, *Nature (London)* **266,** 235 (1977).

[9] K. Shimotohno, Y. Kodana, J. Hashimoto, and K. Miura, *Proc. Natl. Acad. Sci. U.S.A.* **74,** 2734 (1977).

[10] T. Godefroy, *Eur. J. Biochem.* **14,** 222 (1970).

[11] R. S. Brody, K. G. Doherty, and P. D. Zimmerman, *J. Biol. Chem.* **261,** 7136 (1986).

[12] V. Cannistraro and D. Kennell, *J. Mol. Biol.* **261,** 7136 (1994).

TABLE I
EFFECT OF NUCLEOTIDES ON EXONUCLEASE ACTIVITY

Nucleotide[a]	RNase activity[b]
—	100
ATP	37.8 ± 2.2
CTP	60.8 ± 2.7
GTP	62.9 ± 2.1
UTP	38.1 ± 2.4
ADPNP	42.5 ± 2.5
ADP	84.2 ± 2.8

[a] Each of the indicated nucleotides was added as an Mg^{2+} complex at a concentration of 0.5 m*M*.

[b] Assay mixtures containing 1 μl of enzyme and 200 fmol of capped c-*myc* RNA uniformly labeled with [α-^{32}P]GTP were incubated for 20 min at 37°. Reactions were analyzed by PAGE and the amount of mononucleotide product formed was determined with the AMBIS radioanalytic imaging system.[9] Results are presented as the mean of two independent determinations. RNase activity in the control reaction, without added nucleotides, was taken as 100.

0.2 m*M* ATP. The presence of ATP decreased the V_{max} of the enzyme. Unlike the case with NEM inhibition, the presence of nucleotides did not have a significant effect on the nature of the products produced from the 5′ end of a substrate. There was some indication, based on experiments with substrates differentially labeled at the 5′ and 3′ ends, that NTPs also interfere with enzyme processivity. However, it should be noted that regardless of the inhibitors employed (NTPs or NEM), we have never observed reaction intermediates (oligo- or polynucleotides) when uniformly labeled substrates were assayed.

Inhibition of Exonuclease Activity by eIF-4F

One of the ways in which the degradation of an mRNA may be modulated is by its association with one or more protein synthesis factors. One of the factors that regulates the initiation process is eIF-4F, a complex of three polypeptides that specifically binds to the 5′-cap structure of mRNAs.[13,14] To determine whether this factor modified exonuclease activity we examined the degradation of an internally labeled capped RNA in the presence of varying amounts of purified eIF-4F. The results, shown in Fig. 4, demonstrate that eIF-4F is a potent exonuclease

[13] N. R. Webb, V. J. Chari, G. DePhillis, J. W. Kozarich, and R. E. Rhoads, *Biochemistry* **23,** 177 (1984).

[14] W. Rychlik, P. R. Gardner, T. C. Vanaman, and R. E. Rhoads, *J. Biol. Chem.* **261,** 71 (1986).

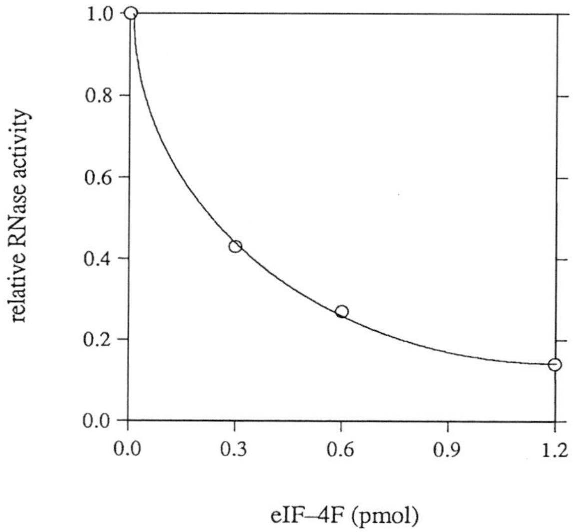

FIG. 4. eIF-4F inhibits exonuclease activity. Standard reaction mixtures contained 150 fmol of capped c-*myc* RNA that was uniformly labeled with [α-^{32}P]GTP and eIF-4F as indicated. They were incubated for 10 min at room temperature. Nuclease (2 μl) was then added and the reaction mixtures were incubated for 20 min at 37°. Acid-soluble radioactivity was determined as described in text. In the uninhibited reaction (no eIF-4F) approximately 90 fmol of the substrate was hydrolyzed to mononucleotides.

inhibitor. A 2-fold molar excess of eIF-4F over capped RNA substrate inhibited mononucleotide formation by about 50% and a 10-fold excess of factor inhibited by about 90%. The substrate contained a small amount (<10%) of uncapped 5′ ends that may have been hydrolyzed in the presence of eIF-4F. At the concentration of eIF-4F used in these experiments the factor had no effect on the hydrolysis of uncapped RNA. The inhibitory activity of eIF-4F supports the hypothesis that the enzyme functions exclusively as a 5′ → 3′-exoribonuclease.

eIF-4E, the cap-binding subunit of eIF-4F, appeared to be a less effective inhibitor. For example, we observed about 20% inhibition of exonuclease activity with 10 pmol of eIF-4E, using the same assay conditions described for eIF-4F. Another protein synthesis factor reported to bind at or near the 5′-cap structure as well as within the 5′-untranslated region of mRNA is eIF-2.[15] However, it required a 50-fold molar excess of eIF-2 over capped RNA to achieve a 50% inhibition of exonuclease activity; 90% inhibition was observed with a 300-fold excess of the factor. These observations indicate that if the exonuclease is involved in mRNA degradation *in vivo,* eIF-4F could function to retard a 5′- to -3′ decay process.

[15] H. A. A. van Heugten, A. A. M. Thomas, and H. O. Voorma, *Biochimie* **74,** 463 (1992).

[24] *Drosophila* 5′ → 3′-Exoribonuclease Pacman

By IGOR V. CHERNUKHIN, JULIAN E. SEAGO, and SARAH F. NEWBURY

Introduction

Early development in multicellular organisms is programmed by maternal gene products that are transported into the egg before fertilization. The initial coordinates of the body plan in the fertilized embryo are entirely dependent on maternal products, as no zygotic transcription occurs. Therefore up to a particular stage of development (cellularization in *Drosophila,* 2.5 hr after fertilization), the developmental program is entirely dependent on posttranscriptional control, in particular RNA localization, translation, and RNA stability. Of these three controlling events, the least is known about the control of RNA stability.

Analysis of early development in *Drosophila* has shown that RNA localization, control of translation, and mRNA stability are intimately linked.[1-3] Generally, translational repression leads to degradation of an RNA, and failure of an RNA to localize correctly also leads to its degradation. For example, the binding of Smaug protein to unlocalized *nanos* RNA causes it to be translationally repressed and then degraded.[4] After the initial coordinates of the developing embryo have been determined, the RNAs responsible for early development are specifically degraded and embryonic transcription commences.

Work on the yeast *Saccharomyces cerevisiae* has identified many ribonucleases and associated factors that control mRNA decay, RNA splicing, and rRNA processing. In yeast, it has been shown that 3′ → 5′ degradation/processing of RNA requires the exosome, which is a complex of at least 10 proteins.[5,6] Degradation of RNA in a 5′ → 3′ direction occurs by initial decapping of the mRNA, followed by 5′ → 3′ degradation of the RNA by Xrn1p. The two decapping proteins (Dcp1p and Dcp2p) and Xrn1p have been shown to be complexed to seven Lsm proteins, which are likely to form a ring encircling the RNA.[7-10] Because many of these

[1] D. Curtis, R. Lehmann, and P. D. Zamore, *Cell* **81,** 171 (1995).
[2] A. Jacobsen and S. Pelz, *Annu. Rev. Biochem.* **65,** 693 (1996).
[3] P. M. Macdonald and C. A. Smibert, *Curr. Opin. Genet. Dev.* **6,** 403 (1996).
[4] C. A. Smibert, J. E. Wilson, K. Kerr, and P. M. Macdonald, *Genes Dev.* **10,** 2600 (1996).
[5] C. Allmang, E. Petfalski, A. Podtelejnikov, M. Mann, D. Tollervey, and P. Mitchell, *Genes Dev.* **13,** 2148 (1999).
[6] P. Mitchell, E. Petfalski, A. Shevchenko, M. Mann, and D. Tollervey, *Cell* **91,** 457 (1997).
[7] T. Dunckley and R. Parker, *EMBO J.* **18,** 5411 (1999).
[8] C. A. Beelman, A. Stevens, G. Camponigro, T. E. LaGrandeur, L. Hatfield, D. M. Fortner, and R. Parker, *Nature (London)* **382,** 642 (1996).
[9] S. Tharun, W. He, A. E. Mayes, P. Lennertz, J. D. Beggs, and R. Parker, *Nature (London)* **404,** 515 (2000).

METHODS IN ENZYMOLOGY, VOL. 342
0076-6879/01 $35.00

proteins are conserved in higher eukaryotes, it is likely that the mechanism of RNA processing/degradation is similar, although components of the complexes may vary between species.

To understand the role of RNA stability in development, a number of approaches can be used. Once the genes encoding particular ribonucleases or associated factors have been identified, then expression of the RNA during development can be determined. In addition, a portion of the protein can be expressed and antibodies raised so that the spatial and temporal distribution of the encoded protein can be determined. We have used these techniques to show that the $5' \rightarrow 3'$-exoribonuclease Pacman is differentially expressed during development.[11] In a genetically tractable organism such as *Drosophila,* it is then often possible to identify a mutation in the gene of interest (FlyBase, *http://flybase.bio.indiana.edu.* or *http://fly.ebi.ac.uk:7081/*). Alternatively, the gene of interest can be ectopically expressed at particular developmental stages or in different tissues and the consequences of misexpression analyzed.[12]

Biochemical methods provide an alternative and complementary approach to the understanding of the mechanisms of action of these enzymes. The mechanisms whereby ribonucleases target and degrade RNA are not well understood and the way in which these enzymes may interact with each other or with other components of the cell is not at all clear. Biochemical characterization of these enzymes is essential in order to determine their active sites, the regions of the proteins that recognize RNA, and the domains of the proteins that interact with each other. The understanding of the ways in which these ribonucleases interact with each other and their target RNAs will not only shed light on the mechanism of gene regulation but is also likely to be crucial in understanding the link between translation, localization, and RNA stability.

This chapter concentrates on the methods we have used to express a *Drosophila* recombinat $5' \rightarrow 3'$-exoribonuclease, purify the protein, and analyze its activity *in vitro.* We offer these protocols as a starting point for investigators wishing to perform similar experiments on similar proteins from higher organisms.

Identification of Ribonucleases and Associated Factors

A number of approaches can be used to identify ribonucleases and associated factors in multicellular organisms. For example, if a particular region of RNA is known to be necessary and sufficient to promote its own degradation,[13] then this

---

[10] E. Bouveret, G. Rigaut, A. Shevchenko, M. Wilm, and B. Seraphin, *EMBO J.* **19,** 1661 (2000).

[11] D. D. Till, B. Linz, J. E. Seago, S. J. Elgar, P. E. Marujo, M. Elias, J. A. McClellan, C. M. Arraiano, J. E. G. McCarthy, and S. F. Newbury, *Mech. Dev.* **79,** 51 (1998).

[12] P. P. D'Avino and C. S. Thummel, *Methods Enzymol.* **194,** 565 (1999).

[13] A. Bashirullah, S. R. Halsell, R. L. Cooperstock, M. Kloc, A. Karaiskakis, W. W. Fisher, W. Fu, J. K. Hamilton, L. D. Etkin, and H. D. Lipshitz, *EMBO J.* **18,** 2610 (1999).

RNA sequence can be used as a "bait" in cross-linking analysis to identify interaction proteins. However, we have used the "candidate gene" approach, which has proved fruitful with the increasing information available through genomic databases and also because of the evolutionary conservation between ribonucleases. This approach is particularly useful where genetic screening would be difficult because the genes involved have pleiotropic effects and/or are functionally redundant.

Many ribonucleases or ribonuclease motifs show evolutionary conservation from the bacteria *Escherichia coli* through to humans. In a remarkable database analysis, where the amino acid sequences of prokaryotic ribonucleases such as RNase II were used to identify similar genes in other organisms, many 3′ → 5′-ribonucleases were identified or predicated in a range of eukaryotes.[14] However, for some ribonucleases, such as RRP4, homologs are not found in prokaryotes but are nevertheless well conserved from yeast to humans.[6] This conservation means that identifying ribonucleases in the organism of choice (where sequence data are available) is comparatively straightforward. In our experience it is advisable to search using conserved motifs and avoid regions with repeated motifs [such as tetratricopeptide repeat (TPR) motifs]. Searching the *Drosophila* database not only reveals the genomic location of the gene of interest, but also the expressed sequence tags (ESTs) at that location. ESTs are sequences at the 5′ ends of cDNA clones: now that the genomic sequence of *Drosophila* has been completed they have been mapped to their genomic locations. The sequences of these cDNA clones have been arranged in groups (clots) and care must be taken to try to identify the longest EST as many are truncated at either the 5′ or 3′ end. Usually, the homology between ribonucleases is such that the full-length clone can be identified from sequence comparison with, for example, the human homolog. This cDNA can then be obtained from Research Genetics (Huntsville, AL; http://www.resgen.com). If the full-length EST is not available then it may be necessary to experimentally identify the 5′ end of the gene by primer extension or similar techniques.

Expression and Purification of Pacman

Subcloning and Expression

In vitro characterization of the *S. cerevisiae* 5′ → 3′-exoribonuclease Xrn1p and the mouse homolog mXrn1p has usually been performed by overexpression of the protein in yeast and then purification by biochemical methods.[15–17] However, we wished to express the entire *Drosophila* 5′ → 3′-exoribonuclease Pacman in the bacteria *E. coli* and purify it by affinity chromatography. This method has the

[14] I. S. Milan, *Nucleic Acids Res.* **25,** 3187 (1997).

[15] A. Johnson and R. D. Kolodner, *J. Biol. Chem.* **266,** 14046 (1991).

[16] A. Stevens, *J. Biol. Chem.* **255,** 3080 (1980).

[17] V. I. Bashkirov, J. A. Solinger, and W. D. Heyer, *Chromosoma* **104,** 215 (1995).

advantage that it allows rapid purification of protein and that the same method can be applied to the preparation of truncated or otherwise mutated proteins for further analysis or for raising antibodies.

A number of plasmid vectors have been constructed to express eukaryotic proteins of interest in bacteria. The vector used should be a multicopy plasmid and include a selectable marker, a "tag" for purification [typically a hexahistidine (His_6) tag], and a promoter that can be induced when required. The latter feature is particularly useful when the overexpression of the protein of interest is toxic to the bacterial cell. We have used the His_6 tag vector pET28a (Novagen, Madison, WI), which has a T7*lac* promoter that can be induced by isopropyl-β–D-thiogalactopyranoside (IPTG). The vector also carries the natural promoter and coding sequence of the *lac* repressor (*lacI*), orientated so that the T7*lac* and *lacI* promoters diverge. When this type of promoter is used in DE3 lysogens to express target genes, the *lac* repressor acts both at the *lacUV5* promoter in the host chromosome to repress transcription of the T7 RNA polymerase gene by the host polymerase and at the T7*lac* promoter in the vector to block transcription of the target gene by any T7 RNA polymerase that may be made. Therefore, in the absence of IPTG, expression from the plasmid is repressed.

In the pET series of vectors, it is typically arranged so that the cDNA is subcloned into the vector, in-frame, as an *Nde*I–*Not*I fragment. It is usually necessary to introduce a suitable *Nde*I site at the start codon by polymerase chain reaction (PCR). Because of the large size of the *pacman* cDNA, we cloned *pacman* into pET28b by a three-way ligation (i.e., a 1300-nucleotide *Nde*I–*Not*I fragment generated by PCR, a 3400-bp *Hin*dIII–*Not*I fragment, and the vector cut with *Nde*I–*Not*I) to reduce the possibility of introducing mutations by PCR. In our experience, inserts of this size in an *E. coli* plasmid can be unstable and it is prudent to check the plasmid by sequencing and restriction enzyme analysis before proceeding further. It is also easier to transform the initial ligation mix into a host strain such as JM109, DH5α, or HB101 before proceeding to transform the correct construct into the expression host BL21(DE3).

Expression of Protein in Escherichia coli BL21(DE3)

The best way to determine the right conditions for protein expression is to perform all the procedures in mini- or midiscale. We found that full-length Pacman protein expression could be sensitive to conditions such as the time for culturing cells before and after induction, and the concentration of the inducer. The more truncated (smaller) forms of protein occurred along with protein degradation when the cell culture was grown for grown too long or with an increased concentration of IPTG.

1. Inoculate a 5-ml culture of LB selective medium from a BL21(DE3) colony carrying the correct construct and grow at 37° overnight.
2. Inoculate 500 ml of LB containing the appropriate antibiotic with 5 ml of the overnight culture. Incubate for 3 hr at 37° on an orbital shaker.

3. Collect a 10-ml aliquot of cells (noninduced sample).

4. Add IPTG at concentration of 0.5 m*M* (chosen experimentally), add 0.1 m*M* phenylmethylsulfonyl fluoride (PMSF), and incubate the culture for a further 2 hr.

5. Centrifuge the cells at 5000 rpm for 5 min at 4° and wash the pellet with phosphate-buffered saline (PBS) containing 1 m*M* PMSF.

6. Freeze the cell pellet at −20°.

7. Lyse the cells by adding buffer A [20 m*M* HEPES-KOH (pH 7.0), 3 *M* urea, 0.5 *M* NaCI, 10 m*M* 2-mercaptoethanol, 10–20 m*M* imidazole (concentration determined experimentally), cooled to 4°] directly to the frozen pellet and vortex it until the cells are completely resuspended. Freeze the lysate and store overnight at −20°.

8. Check an aliquot of the cell lysate samples collected from induced and non-induced cultures for protein expression by electrophoreses on a sodium dodecyl sulfate (SDS)–polyacrylamide gel and staining with Coomassie blue or by Western blotting (see below) with an antibody against Pacman protein or a monoclonal antibody against the His_6 tag portion of the protein (Sigma, St. Louis, MO) (see Fig. 1A).

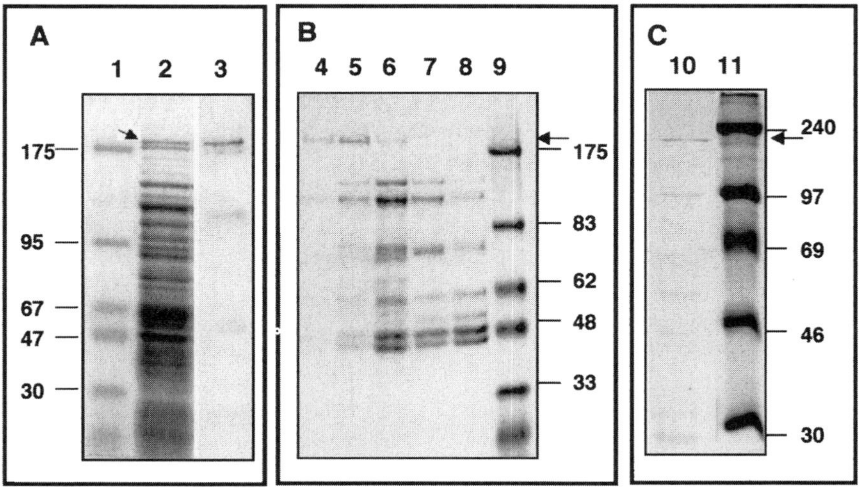

FIG. 1. (A) Detection of Pacman protein in bacterial and *Drosophila* extracts. Lane 1, molecular mass marker (in kilodaltons); lane 2, total bacterial lysate from *E. coli* expressing full-length Pacman, stained with Coomassie blue. Full-length protein (184 kDa) is marked with an arrow; lane 3, Western blot of total bacterial lysate, probed with the Pacman antibody. Control experiments show that other bands are cleavage products (Pacman protein is susceptible to cleavage by endogenous proteases). (B) Lanes 4–8, fractions collected from an S200 gel-filtration column. The eluted proteins were stained with Coomassie blue. Pacman protein is marked with an arrow; lane 9, molecular mass marker (in kilodaltons); (C) gel showing the purified Pacman that was used in the band-shift assays; lane 10, Purified protein stained with Coomassie blue; lane 11, molecular mass marker (in kilodaltons).

Protein Purification on Hexahistidine Tag Column

1. Charge the His_6 tag medium (Novagen) with Ni^{2+} according to the manufacturer instructions and prepare a 2 × 3 cm column.
2. Equilibrate the column with buffer A [20 m*M* HEPES-KOH (pH 7.0), 3 *M* urea, 0.5 *M* NaCl, 10 m*M* 2-mercaptoethanol, 10–20 m*M* imidazole (optimum concentration determined experimentally)]. Apply the cell lysate to the column at a flow rate of 1 ml/min. Wash the column with buffer A untill no protein can be detected in the flowthrough.
3. Elute the affinity-bound proteins with a gradient of 0.02–0.5 *M* imidazole. Collect the fractions and take aliquots. Check for the presence of Pacman protein by electrophoresis on a SDS–polyacrylamide gel with subsequent staining with Coomassie blue or by Western blotting.

S-200 gel Filtration

Further purification of the protein is performed by Sephacryl S-200 HR gel filtration on a 75 × 2.5 cm column equilibrated with buffer B [20 m*M* HEPES-KOH (pH 7.0), 0.5 *M* KCl, 10% (v/v) glycerol, 20 m*M* 2 mercaptoethanol, 2 m*M* EDTA, 1 m*M* PMSF].

1. After Ni^{2+} chromatography, load the fraction containing Pacman protein onto the column.
2. Perform the column chromatography at a flow rate of 0.5 ml/min. Collect 5-ml fractions and analyze aliquots of these for the presence of protein, as described for His_6 tag chromatography (see Fig. 1B and C).
3. Dialyze the fractions containing Pacman protein against storage buffer [20 m*M* HEPES-KOH (pH 7.0), 0.3 *M* KCl, 40% (v/v) glycerol, 2 m*M* EDTA, 20 m*M* 2-mercaptoethanol]. If necessary, concentrate the protein by standard procedures.
4. Make small aliquots of the dialyzed protein solution. The protein can be stored at −20° for short-term storage or at −70° for long-term storage.

In our hands, full-length Pacman protein is stable and functional after purification on the His_6 tag column. However, after subesequent purification it tends to become cleaved to give a 115-kDa protein that is, however, functionally active. It is possible that Pacman protein is complexed during its initial purification to cellular factors that stabilize Pacman against cleavage activity and that these are removed in subsequent purification steps.

Detection of Nucleic Acid Binding and Exonuclease Activity

The advantage of this method is that it gives relatively quick and easy purification of the protein using an organism (*E. coli*) that is familiar to most molecular biologists. The potential disadvantage is that the purified protein may not have exonuclease activity because it is not correctly modified posttranslationally. The activity of the homologous protein *S. cerevisiae* Xrn1p is often measured by filter binding assays.[15,18] However, we chose to measure activity by band-shift analysis, as binding and exonuclease digestion of nucleic acids can then be readily visualised and quantified. Because the active 5′ → 3′-exoribonuclease will rapidly degrade labeled nucleic acids, it is necessary to stall the processive nuclease so that the product can be easily detected. In a number of experiments, it has been shown that yeast Xrn1p is stalled by poly(G) tracts *in vivo* and *in vitro*.[19,20] In addition, mouse Xrn1p and *S. cerevisiae* Xrn1p have been shown to apparently bind to G_4 tetraplex, probably because the nuclease is stalled by the stable G-quartet structure.[21,22] Because mouse Xrn1p is highly homologous to Pacman, we reasoned that G_4 tetraplex would provide a suitable target nucleic acid for the activity assays. It is likely that this target would be a suitable substrate for analysis of the activity of many eukaryotic processive ribonucleases.

Preparation of G_4 Tetraplex

1. Use DNA or RNA oligonucleotides that have previously been shown to form a G_4-tetraplex structure *in vitro*.[21–23] We have used the oligonucleotide 5′-TATGGGGAGCTGGGGAAGGTGGGATTT-3′ and the oligonucleotide 5′-TGGACCAGACCTAGCA-3′ as competitors.
2. Dissolve the G_4 oligonucleotides in 10 μl of TE to 0.5–100 μM. Heat to 90° for 60 sec, and then chill on ice and spin briefly. Add 10 μl of TE plus 2 *M* KCl. In our experience the G_4 tetraplex forms spontaneously on the addition of this buffer.
3. Electrophorese the labeled G4 tetraplex on a 1× TBE (50 m*M* Tris, 50 m*M* boric acid, 1 m*M* EDTA)–6% (w/v) polyacrylamide gel. Detect the labeled tetraplex by staining with ethidium bromide and remove the slice containing the tetraplex with a scalpel blade. Crush the gel slice in 3 volumes of elution buffer (0.5 *M* ammonium acetate, 1 m*M* EDTA, pH 8.0) and incubate overnight at 37°. Centrifuge at 10,000*g* for 10 min at 4°, remove the supernatant, chloroform extract the DNA, and then concentrate it by ethanol precipitation.
4. Label the oligonucleotide at the 5′ end. Add the following to a sterile microcentrifuge tube:

Nuclease-free water	~32 μl
Kinase buffer (10×; New England BioLabs, Beverly, MA)	5 μl

[18] J. A. Solinger, D. Pascolini, and W.-D. Heyer, *Mol. Cell. Biol.* **19,** 5930 (1999).
[19] D. Muhlrad, C. J. Decker, and R. Parker, *Genes Dev.* **8,** 855 (1994).
[20] T. L. Poole and A. Stevens, *Biochem. Biophys. Res. Commun.* **235,** 799 (1997).
[21] V. I. Bashkirov, H. Scherthan, J. A. Solinger, J.-M. Buerstedde, and W. D. Heyer, *J. Cell Biol.* **136,** 761 (1997).
[22] Z. Liu and W. Gilbert, *Cell* **77,** 1083 (1994).

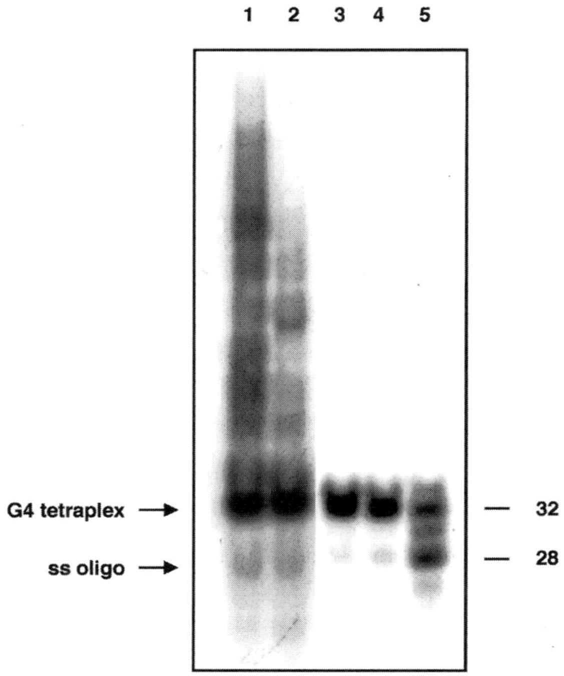

FIG. 2. Purification of the G_4 tetraplex. Lanes 1 and 2, annealed 5′-end-labeled G_4 oligonucleotides prior to gel purification; lanes 4 and 5, 5′-end-labeled G_4 tetraplex after gel purification; lanes 5, molecular weight markers (in nucleotides).

G_4 oligonucleotides	10 ng–1 μg)
[γ-^{32}P) ATP, 6000 Ci/mmol, 10 Ci/μl (Amersham Pharmacia, Piscataway, NJ)	5 μl
T4 polynucleotide kinase (New England BioLabs)	3 μl
Final volume	50 μl

5. Vortex the solution gently before incubation at 37° for 30 min. Remove the unincorporative radionucleotides from the labeled nucleic acids with a Sephadex G-50 spin column (Amersham Pharmacia) or similar system according to the manufacturer instructions. Check that the G_4 oligonucleotides have formed a higher order structure by gel electrophoresis (Fig. 2).

Alternatively, the G_4 oligonucleotides may be labeled first and then annealed to give the G-quartet structure.[23]

[23] D. Sen and W. Gilbert, *Methods Enzymol.* **211,** 191 (1992).

G_4 *Oligonucleotide-Binding Reaction*

1. The binding reaction is performed in a total volume of 20 μl. Add purified Pacman protein in a volume of 1–5 μl (~5–25 ng) to a mixture containing binding buffer [20 m*M* HEPES (pH 7.5) 100 m*M* KCl, 10% (v/v) [glycerol] 1 p*M* competitor oligonucleotide (5′-TGGACCAGACCTAGCA-3′), 1 μg of poly(dI-dC), and 200 cpm of 5′-end-labeled G_4 tetraplex oligonucleotides.
2. Mix the reagents gently and then centrifuge briefly at 10,000*g* for 5 sec at 4°.
3. Incubate at 4° for 30 min.
4. Electrophorese the reaction products on a 6% (w/v) polyacrylamide gel buffered with 1× TBE.
5. After electrophoresis is complete, dry the gel on a gel dryer and visualize the band shifts by autoradiography.

Nuclease Assay

Incubate the G_4 tetraplex with the enzyme as described as above, except that the binding buffer is supplemented with 3 m*M* $MgCl_2$ and the reaction is carried out at 25° for 30 min.

We have used the above-described method to show the full-length 184-kDa Pacman protein, purified from *E. coli* as described above, has both binding and exonuclease activity *in vitro* (Fig. 3).

We have also used Ni^{2+} chelate and ion-exchange chromatography to purify the N-terminal and C-terminal portions of the Pacman protein to 90% homogeneity for use in our functional assays.

Conclusions and Prospects

This chapter has focused on a method for rapid purification and analysis of the *Drosophila* ribonuclease pacman. The method given is likely to be generally applicable to the purification of other ribonucleases. By analysis of the activity of "wild-type" and "mutant" proteins *in vitro,* it should be possible to determine their mechanisms of action, their active domains, and perhaps to reconstitute an active "RNA degradation" complex *in vitro*. It will also be possible to determine the specificity of these ribonucleases for particular RNA sequences or secondary structures. In addition, it is now thought that RNA interference (RNAi), where double-stranded RNAs introduced into the cell promote repression of the homologous RNA, is achieved through specific degradation events,[24] perhaps involving one or more of the 5′ → 3′- or 3′→ 5′-ribonuclease complexes. An understanding of the mechanisms whereby interfering RNA promotes specific degradation of its target RNA may be possible by biochemical analysis *in vitro*. Biochemical work

[24] J. M. Bosher and M. Labouesse, *Nat. Cell Biol.* **2,** E31 (2000).

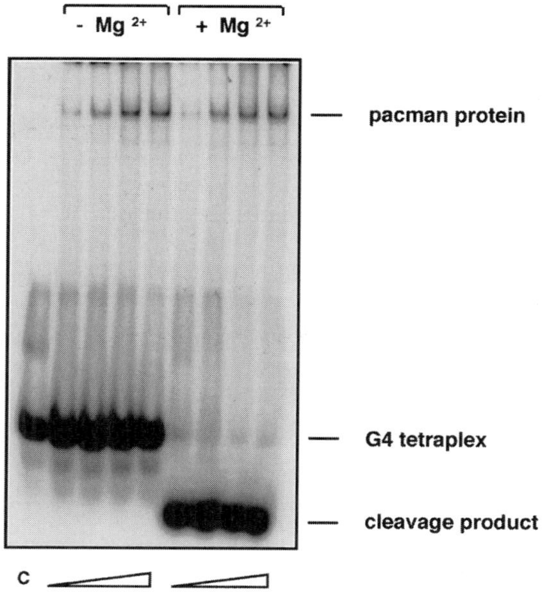

FIG. 3. Pacman protein binds to, and cleaves, G_4 tetraplex in the presence of magnesium ions. Full-length Pacman protein (184 kDa) was expressed in *E. coli* and purified with the His_6 tag system. Increasing amounts of protein were then incubated with G_4 tetraplex that had been labeled at the 5′ end. C, Control with no protein.

will also facilitate future structural analysis of the protein. In a few cases, progress has already been made in defining the catalytic domains of these RNA-binding proteins and their interactions with particular RNAs. For example, the yeast protein Xrn1p is known to have acidic N-terminal domains that are likely to contain the exonuclease, and Xrn1p has also been shown to be stalled by stable stem–loop structures.[17,18,20,25,26]

While the biochemical analysis of ribonucleases is likely to provide new information about their mechanisms of action, it is a particularly powerful approach when carried out in combination with genetic experiments. For example, analysis of mutant proteins generated *in vivo* can provide valuable information about the functions of these proteins.[26] Therefore these experiments, particularly in combination with genetic experiments, are likely to shed light on one of the least understood mechanisms of gene expression in eukaryotes.

[25] A. Holler, I. Bashkirov, J. A. Solinger, U. Reinhart, and W.-D. Heyer, *Eur. J. Biochem.* **231,** 329 (1995).

[26] A. M. Page, K. Davis, C. Molineux, R. D. Kolodner, and A. W. Johnson, *Nucleic Acids Res.* **26,** 3707 (1998).

[25] Purification of Poly(A)-Specific Ribonuclease

By ANDERS VIRTANEN, JAVIER MARTÎNEZ, and YAN-GUO REN

Introduction

The functional significance of the mRNA poly(A) tail has been debated over the years and several different functions have been assigned to it, for example, as a structural element influencing the translational efficiency, the mRNA half-life, and/or the intracellular transport of mRNAs. Poly(A) removal is an important step during mRNA decay and it has been established that mRNA degradation is initiated by degrading the mRNA poly(A) tail (reviewed in Ross[1]). Poly(A)-degrading nuclease activities have been studied in several eukaryotic systems (for reviews see Refs. 1–4). We defined a poly(A)-specific 3′-exonuclease in HeLa cell-free extracts and proposed a reaction pathway for mRNA poly(A) tail removal.[5,6] The HeLa cell activity was specific for degrading 3′-located poly(A) tails, required a 3′-located hydroxyl group, and released 5′-AMP as the mononucleotide product. The responsible nuclease activity has been purified, a molecular clone obtained, and the enzyme named poly(A)-specific ribonuclease (PARN).[7–9] The nuclease was initially called deadenylating nuclease (DAN).[7] PARN is present in at least two isoforms in *Xenopus* oocytes, one being 74 kDa in molecular size and the other 62 kDa.[8] The two forms differ in subcellular distribution in oocytes, the 62-kDa form being cytoplasmic while the 74-kDa form is nuclear. Two isoforms have also been identified in calf thymus cell-free extracts: a 74-kDa form 8 and a 54-kDa form. The 54-kDa form corresponds to a fragment of the full-length 74-kDa polypeptide.[9] It remains to be established whether the 54-kDa form is a proteolytic product of the 74-kDa polypeptide generated during purification or whether it represents a naturally occurring isoform. The 54-kDa form of PARN is an oligomeric (i.e., homotrimer) and highly processive poly(A)-specific exonuclease. Both isoforms of PARN interact with the mRNA 5′-end cap structure.[9–12] This interaction stimulates both the nuclease activity[9–11] and the processivity of

[1] J. Ross, *Microbiol. Rev.* **59,** 423 (1995).
[2] G. Caponigro and R. Parker, *Microbiol. Rev.* **60,** 233 (1996).
[3] P. Mitchell and D. Tollervey, *Curr. Opin. Genet. Dev.* **10,** 193 (2000).
[4] A. Virtanen and J. Åström, *Prog. Mol. Subcell. Biol.* **16,** 199 (1997).
[5] J. Åström, A. Åström, and A. Virtanen, *EMBO J.* **10,** 3067 (1991).
[6] J. Åström, A. Åström, and A. Virtanen, *J. Biol. Chem.* **267,** 18154 (1992).
[7] C. G. Körner and E. Wahle, *J. Biol. Chem.* **272,** 10448 (1997).
[8] C. G. Körner, M. Wormington, M. Muckenthaler, S. Schneider, E. Dehlin, and E. Wahle, *EMBO J.* **17,** 5427 (1998).
[9] J. Martinez, Y.-G. Ren, A.-C. Thuresson, U. Hellman, J. Åström, and A. Virtanen, *J. Biol. Chem.* **275,** 24222 (2000).

0076-6879/01 $35.00

degradation.[12] In this chapter we describe the purification of the 54-kDa PARN nuclease isoform from calf thymus cell-free extracts.[9] The same protocol can be used to purify PARN from HeLa cell-free extracts. A simple procedure to isolate the HeLa cell activity has been described.[5,6]

Assays and Substrates

To purify PARN we have established a dual assay strategy[5,6,9] that monitors both disappearance of the RNA substrate and appearance of the two reaction products, deadenylated RNA and released AMP. To detect disappearance of the substrate and the accumulation of the deadenylated RNA product polyadenylated RNA substrate [e.g., L3(A_{30})] radioactively labeled in the body of the RNA is used as the substrate during an *in vitro* deadenylation reaction. The reacted RNA is subsequently analyzed by analytical polyacrylamide gel electrophoresis. Release of AMP is investigated by *in vitro* deadenylation of L3(A_{30}) RNA substrate radioactively labeled in its poly(A) tail followed by detection of released mononucleotides by thin-layer chromatography (TLC). The strength of this dual assay strategy is that nucleases that degrade both the poly(A) tail and the RNA body of the substrate can be detected and excluded. An alternative assay based on detecting the release of the AMP mononucleotide has also been used.[7,13] In this assay radioactively labeled poly(A) is used as the substrate and the release of AMP is monitored as a trichloroacetic acid-soluble product.

Polyadenylated RNA substrate L3(A_{30}) can be prepared by using standard procedures for *in vitro* transcription (see, e.g., Åström *et al.*[5]). PARN is approximately five to six times more active if the RNA substrate is $m^7G(5')ppp(5')G$ capped during *in vitro* transcription.[9] It is important to calculate the specific activity of the synthesized RNA substrate in curies per millimole to be able to accurately quantitate the PARN activity.

Conditions for *in vitro* deadenylation have been described.[5,6,9] PARN requires monovalent cations and the optimal concentration is about 100 m*M*. The nuclease activity is higher in the presence of K^+ compared with Na^+. Divalent metal cations are essential, and Mg^{2+} is the preferred species, with an optimal concentration about 1 m*M*. The optimal pH is about 7. Standard conditions for *in vitro* deadenylation are 1–1.5 m*M* $MgCl_2$, 2.5% (w/v) poly vinyl alcohol (Mw 10,000; Sigma, St. Louis, MO), 100 m*M* KCl, 0.15 units of RNAguard (Amersham Pharmacia Biotech Piscataway, NJ), 5–20 fmol of RNA substrate, 20 m*M* HEPES–KOH (pH 7), 0.1 m*M* EDTA, 0.25 m*M* dithiothreitol (DTT), 10% (v/v)

[10] E. Dehlin, M. Wormington, C. G. Korner, and E. Wahle, *EMBO J.* **19,** 1079 (2000).

[11] M. Gao, D. T. Fritz, L. P. Ford, and J. Wilusz, *Mol. Cell* **5,** 179 (2000).

[12] J. Martinez, Y.-G. Ren, M. Ehrenberg, and A. Virtanen, *J. Biol. Chem.,* in press (2001).

[13] J. E. Lowell, D. Z. Rudner, and A. B. Sachs, *Genes Dev.* **6,** 2088 (1992).

glycerol.[6] A standard reaction volume is 15 or 25 μl and incubations are performed at 30° for 1 to 180 min. Approximately 0.1–0.4 unit of PARN activity should be used. During a standard purification procedure we use 1–6 μl of the obtained fraction to detect the nuclease activity and a standard incubation time is 30–60 min. We define 1 unit of deadenylation activity as the release of 1 μmol of AMP minute. Reactions are terminated and the reaction products are investigated either by purifying the RNA and subsequent electrophoresis in 10% (w/v) polyacrylamide (19:1, acrylamide–bisacrylamide)–7 *M* urea gels as previously described[5,6] or by one–dimensional (1-D) TLC analysis. Reactions to be investigated by polyacrylamide gel electrophoresis should be terminated by the addition of 150 μl of 0.5 *M* NaCl, 5 m*M* EDTA, 10 m*M* Tris-HCl (pH 7), 0.2% (w/v) sodium dodecyl sulfate (SDS), and glycogen (25 μg/ml). After terminating the reaction the reacted RNA substrate is purified by standard phenol extraction followed by ethanol precipitation. Reactions to be investigated by 1-D TLC should be terminated by the addition of EDTA to a final concentration of 50 m*M* and then directly subjected to 1-D TLC on polyethyleneimine (PEI)–cellulose F plates (Merck, Rahway, NJ), using 0.75 *M* KH_2PO_4, pH 3.5 (H_3PO_4) as the solvent. The liberated mononucleotide will migrate with the solvent while unreacted RNA substrate and deadenylated RNA products will stay at the loading spot.

Purification Procedure

We usually purify PARN in two steps. First we generate a partially purified fraction that is obtained after four chromatographic steps, carried out without freezing the selected fractions. The resulting Blue Sepharose fraction is reasonably stable and can be stored at −70° for at least 12 months. It is possible to freeze the sample at −70° after each fractionation step. To obtain homogeneously purified PARN the Blue Sepharose fraction is subjected to two consequtive steps of affinity purification. The resulting PARN activity is considerably less stable than the activity in the Blue Sepharose fraction.

All steps should be done at 4°. A typical protocol starting with a 400-ml crude ammonium sulfate fraction (14 g of protein) is given below. The given numbers for protein amount and fraction volumes are based on our experience and are useful for following each step of chromatography. A detailed example can be found in Martinez *et al.*[9] This procedure can be scaled up or down as long as the flow rates in centimeters per hour and the ratios between the diameter and the height of the matrices are kept the same. The PARN activity develops well during the Mono Q step if calf thymus is the starting material. If HeLa cell-free nuclear extracts are used PARN activity is easily detected already after the DEAE-Sepharose step. The partial purification protocol was initially established by using HeLa cell-free nuclear extract as the starting material.

Preparation of Cell-Free Extracts

All steps should be done at 4°. The preparation of the calf thymus cell-free extract is based on a procedure described by Wahle.[14] Calf thymus (3 kg) is cut into pieces and homogenized in an approximately equal volume of buffer 1 [50 m*M* Tris-HCl, 10 m*M* K_3PO_4, 1 m*M* EDTA, 10% (v/v) glycerol, 50 m*M* KCl, 0.1 m*M* DTT at pH 7.9], using a Waring blender (50 sec at low speed and 50 sec at high speed). Solid material is precipitated by centrifugation in a Sorvall (Newtown, CT) GSA rotor at 16,000*g* for 60 min at 4°. The obtained crude extract is filtered through a testsieve mesh 7-normal (Pascal, London, UK) and subsequently $NH_4(SO_4)_2$ (25% saturation) is added to the supernatant at 0.134 g/ml. The extract is stirred on ice for 2 hr and the precipitate is removed by centrifugation in a Sorvall GSA rotor at 16,000 *g* at 4° for 60 min. The obtained supernatant is supplemented with more $NH_4(SO_4)_2$ (45% saturation), 0.115 g/ml, and then treated as for the previous precipitation step. The pellet collected after the second centrifugation is dissolved in 2–4 volumes of buffer D [20 m*M* HEPES–KOH, 100 m*M* KCl, 1.5 m*M* $MgCl_2$, 0.2 m*M* EDTA, 0.5 m*M* DTT, 20% (v/v) glycerol at pH 8.2] and dailyzed for 10 hr at 4°, using dialysis tubing with a molecular weight cutoff at 6,000–8000 (Spectra/Por 1; Spectrum, Rancho Dominguez, CA). After dialysis the 25–45% ammonium sulfate fraction is frozen in liquid nitrogen and stored at −70°. Starting with 3.0 kg of calf thymus the obtained extract (~1000 ml) contains approximately 100 g of protein. The subsequent ammonium sulfate fraction contains approximately 14 g of protein in 400 ml. The protein concentration is determined by using the Bio-Rad (Hercules, CA) protein assay kit with bovine gamma globulin as reference.

Partial Purification of Poly(A)-Specific Ribonuclease

The ammonium sulfate fraction is added to DEAE-Sepharose CL-6B (Amersham Pharmacia Biotech) ion-exchange medium (350-ml packed matrix) equilibrated with buffer D (see above). Protein is allowed to bind to the matrix under slow stirring for 30 min at 4°. Unbound material is removed by washing the matrix with buffer D three times. The suspended matrix is recovered by centrifugation (Sorvall H4000 rotor at 800 rpm for 3 min) after each washing step. The washed matrix is then packed in a column (diameter, 70 mm) and washed (2 column volumes) with buffer D at a flow rate of 12 cm/hr. The column is eluted by two salt steps (buffer D supplemented with 0.17 and 1.0 *M* KCl) and eluted protein from each step is collected. The nuclease activity elutes in the 0.17 *M* KCl step fraction (1.7 g of protein, 450 ml) and is subsequently dialyzed against buffer D for 10 hr.

The dialyzed fraction is further purified by heparin–Sepharose CL-6B (Amersham Pharmacia Biotech) chromatography, using a column with a bed volume

[14] E. Wahle, *J. Biol. Chem.* **266,** 3131 (1991).

*RNase I.** RNase I* is purified from *E. coli*.[15] One unit is the activity needed to eliminate all full-length molecules in a 20-μl reaction containing 1 μg of poly(A) (~300 nucleotides) in 60 sec at 37°.

Mung Bean Nuclease. Mung bean nuclease is from Boehringer Mannheim (Indianapolis, IN), with units defined by the vendor.

Generation of Specific 3′-End Groups. Poly(C) is digested with various agents to generate specific 3′-ended molecules with a large fraction of products in the 40- to 60-nucleotide size range.[14] A portion is partially digested with RNase I* to give 2′,3′-cyclic phosphate ends. Oligonucleotides are separated from the enzyme by batch fractionation on S-Sepharose, with the oligonucleotides eluting in 10 m*M* morpholin ethanesulfonic acid (MES) buffer, pH 5.8, and RNase I* strongly bound. Another portion of poly(C) is digested with 1 *N* KOH for 5 min at 37° and then treated in 0.1 *N* HCl to convert the cyclic phosphate P ends to 2′-phosphate or 3′-phosphate ends[16]; we have found that this treatment produces an approximately equal amount of 2′- and 3′-phosphate ends.[15] Another portion is partially digested with pancreatic RNase A. RNase A also generates 2′,3′-cyclic phosphate ends as initial products, but unlike RNase I*, they are rapidly converted to 3′-phosphate ends.[17]

Oligonucleotide Substrates and Inhibitors. Preparative amounts of homopolymers are digested for 2 hr with a low level of *E. coli* RNase I[18] to give intermediate sizes.[14] The resulting oligonucleotides are immediately 5′-^{32}P labeled for 10 min in the T4 polynucleotide kinase reaction.[19] The products are fractionated by polyacrylamide gel electrophoresis (PAGE) and specific sizes, for example, 45 nucleotides, are cut out and eluted to use as starting substrates. Synthetic oligonucleotides are also purified by PAGE.[14]

Methods

Assays

Reaction rates measure the amount of mononucleotide released in a defined time.[14] The assay volume is usually 100 μl and includes 50 m*M* NH_4HCO_3–1 m*M* $MgCl_2$, pH 7.2, homopolymer, and enzyme. In view of its heat lability, assays are usually carried out at 23° and kept as short as possible. However, there is significant stabilization of enzyme by substrate, even at 37 or 45°, which minimizes loss of activity during assays. The reaction is terminated by heating at 100° for 1 min before diluting to 0.7 ml in 10 m*M* Tris and loading onto a Mono Q column

[15] V. J. Cannistraro and D. Kennell, *J. Bacteriol.* **173,** 4653 (1991).

[16] G. G. Brownlee, "Determination of Sequences in RNA" (T. S. Work and E. Work, eds.). American Elsevier, New York, 1972.

[17] H. Witzel and E. A. Barnard, *Biochem. Biophys. Res. Commun.* **7,** 289 (1962).

[18] J. Meador III and D. Kennell, *Gene* **95,** 1 (1990).

[19] V. J. Cannistraro, B. M. Wice, and D. E. Kennell, *J. Biochem. Biophys. Methods* **11,** 163 (1985).

(Pharmacia) driven either by a fast protein liquid chromatography (FPLC; Pharmacia) or model 2248 Pharmacia/LKB (Bromma, Sweden) high-performance liquid chromatography (HPLC) instrument. The gradient is from 0 to 2 *M* NaCl in 10 m*M* Tris. The mononucleotide elutes as a sharp peak at 0.22 *M* NaCl and is cleanly separated from the starting polymer, which elutes at~0.55 *M* NaCl. The extent of reaction is monitored by the peak height or area.

Identifying Enzyme Mechanism

Two different experimental approaches are presented to elucidate the mechanism of processive movement.[11] Both approaches examine intermediate stages of the reaction before dissociation of the enzyme–substrate complex. The RNase II reaction stops before the last rC residue of a 3′-$(rC)_n(dC)_m$ single-stranded oligonucleotide.[14] The enzyme stalls to give a transient enzyme–substrate complex. This characteristic of the enzymology provides an ideal means to analyze intermediate stages in the reaction. The first approach measures the extent to which a stalled complex acts as a competitive inhibitor of RNase II. If the association of the stalled intermediate complex were weaker, it would dissociate faster and be a less effective inhibitor of RNase II. Comparing the inhibition by stalled intermediates after various numbers of nucleotides are released indicates whether there is a progressive weakening of the enzyme–substrate complex with each hydrolytic event rather than an abrupt weakening of the complex just before dissociation.

The second approach examines the nucleotides in a stalled complex that are interacting with the enzyme. Various chemical and enzymatic probes have been tried, but the most useful is PDE I, a 3′-exonuclease.[11] Activities are used so that most PDE I molecules can bypass the altered, or interacting, nucleotide. In the case of many interacting nucleotides in a substrate, multiple bands are generated by the fraction of PDE I molecules that stop at each nucleotide affected by the RNase II interaction. By using a wide range of PDE I activities, bands can be seen from the 3′ end (at the lower activities) to the 5′ end (at higher activities) of the interacting substrate. Further studies will be necessary to identify the molecular basis for the PDE I blocks, but in these studies the observations are remarkably consistent with the kinetic results and are corroborated by the same kind of band generation in the reverse direction using PDE II, a 5′-exonuclease. Finally, besides consistency with the kinetics of linear oligomers, PDE I-generated banding patterns give a reasonable explanation for the observed interaction of RNase II with structured nucleic acids and their resultant vulnerability to endonucleases.

Results

Recognition Specificity of RNase II

3′-End Group. With the possible exception of 2′-phosphate, each of the following ends are present in the cell[14,15]: 3′-OH, 2′,3′-cyclic phosphate, and 3′-phosphate.

Ends with 3′-OH arise by the action of processing endo-RNases[20] or the exo-RNases RNase II and polynucleotide phosphorylase (PNPase). The 2′,3′-cyclic phosphate ends are the initial and major products of RNase I or RNase I* cleavages, because their conversion to 3′-phosphate ends by those ribonucleases is slow.[21] RNase M generates 3′-phosphate ends and has specificity similar to pancreatic RNase A,[22,23] and RNase A generates 3′-phosphate ends rapidly from the initial 2′,3′-cyclic phosphate intermediate.[17]

The rates of mononucleotide formation in the RNase II assay were used to measure the initiation rates with the different 3′-end groups at one concentration of substrate (6.6 μg/100 μl).[14]

$$\text{OH}/2',3'\text{-cyclic P}/3'\text{-phosphate}/2'\text{-phosphate} = 3.4/3.0/1.9/1.0$$

RNase II initiated almost twice as fast to 3′-OH as to 3′-phosphate ends. This difference probably results from the charge on the phosphate rather than from the phosphate itself, because the 2′,3′-cyclic phosphate reacted as fast as molecules with a 3′-OH end. At the same time the 2′-phosphate ends reacted half as fast as the 3′-phosphate and four times slower than the 3′-OH, indicating some positional effect of the phosphate. Nonetheless, the differences in rates for molecules with the three end groups known to be present in the cell are not large and all would be suitable substrates for RNase II.

Bond Cleavage Specificity Not for Ribose Moiety. Ribonuclease II is specific for RNA,[8,10] but can degrade 3′-terminal T–T bonds.[14] We compared 5′-$(rC)_n(dT)_{1-5}$ oligonucleotides and found that the rate of cleavage decreases progressively with the addition of each dT. Therefore, RNase II can cleave a –dT–dT-3′ bond if the nearby upstream nucleotides are ribonucleotides. Their distance from the terminal cleavage site may not be as sharply defined, but includes nucleotides 3 to 5 from the 3′ end. This region, which defines the specificity for RNA, is a discrete internal segment because residues further upstream can be deoxynucleotides without affecting cleavage of the 3′-end bonds.[14] The results were the same when deoxy(C) replaced deoxy(T).

Position of Anchor-Binding Site. It was concluded that both DNA exonuclease I[24,25] and polynucleotide phosphorylase[26] have two substrate-binding sites: the 3′ catalytic site and an upstream site. DNA oligonucleotide analogs are competitive inhibitors of RNase II,[14] suggesting binding at the active site. An upstream "anchor"-binding site[11,14] was identified with dC oligonucleotides of defined length as competitive inhibitors. $(dC)_{19}$ and $(dC)_{21}$ were weak inhibitors (Fig. 1a),

[20] V. J. Cannistraro and D. Kennell, *Eur. J. Biochem.* **213,** 285 (1993).
[21] P. F. Spahr and B. R. Hollingworth, *J. Biol. Chem.* **236,** 823 (1962).
[22] V. J. Cannistraro, M. N. Subbarao, and D. Kennell, *J. Mol. Biol.* **192,** 257 (1986).
[23] V. J. Cannistraro and D. Kennell, *Eur. J. Biochem.* **181,** 363 (1989).
[24] R. S. Brody, K. G. Doherty, and P. D. Zimmerman, *J. Biol. Chem.* **261,** 7143 (1986).
[25] R. S. Brody, *Biochemistry* **30,** 7072 (1991).
[26] T. Godefroy, *Eur. J. Biochem.* **14,** 222 (1970).

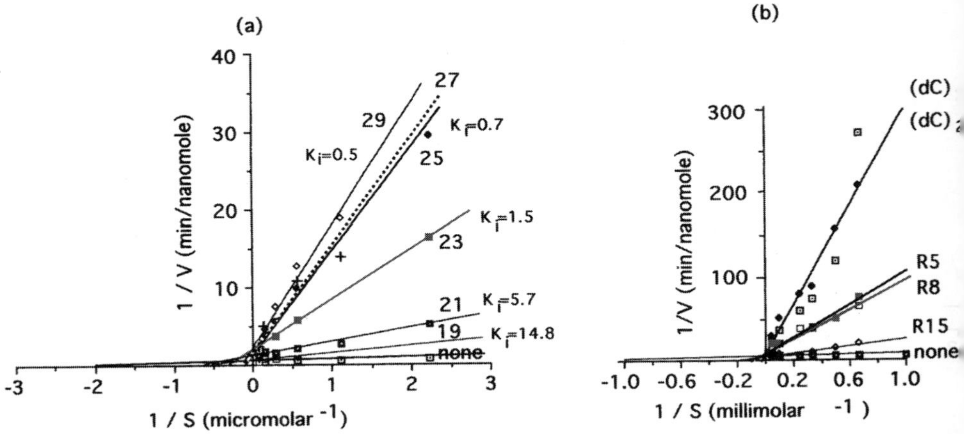

FIG. 1. Double-reciprocal plots of substrate concentration (S) versus reaction velocity (V) for degradation of poly(C) by RNase II in the presence of the designated oligonucleotide at 100 nM in a 100-μl reaction at 10°.[11] The "none" curves refer to control reactions (no DNA inhibitor). Each reaction was linear with time. (a) Oligonucleotides of $(dC)_n$ with the number (n) of residues indicated. (b) R5, R8, and R15 refer to the number of rC residues in the mixed RNA–DNA oligonucleotides.[11] Relative degrees of inhibition are derived from the relative slopes of the lines. The K_i values (nM) were derived from the equation $K_i = i/[(m_i/m_j) - 1]$, where i is the concentration of inhibitor (100 nM) and m_i and m_j are the slopes in the Lineweaver–Burk plot of inhibitor and no inhibitor lines, respectively. [Reprinted from V. J. Cannistraro and D. Kennell, *Biochim. Biophys. Acta* **1433,** 170 (1999), with permission from Elsevier Science.]

but addition of two more residues to give $(dC)_{23}$ increased the inhibition significantly and another two dC residues caused a further marked increase. Additional residues did not give a measurable increase. These results indicate that dC residues about 20 to 25 nucleotides from the 3′ end are important for RNase II binding.

Inhibition by dA oligonucleotides is at least 500 times less than that of dC oligonucleotides of corresponding size, with inhibition by $(dA)_{27}$ barely detectable.[14]

Specificity to Include Polarity of Binding. The association of substrate is further restricted by an obligatory direction of binding in the catalytic site.[14] The enormous difference in inhibition by $(dA)_n$ versus $(dC)_n$ made an experiment possible. Inhibition by (1) 5′-$(dA)_8(dC)_{25}$-3′ [$(dC)_{25}$ at 3′ end] was compared with that of (2) 5′-$(dC)_{25}(dA)_8$-3′[$(dA)_8$ at 3′ end]. If the 5′ end could bind as well as the 3′ end, then the two oligonucleotides should inhibit equally well. If the 3′ end binds preferentially (polarity effect), then oligonucleotide 1 should inhibit much better than oligonucleotide 2. The results showed a significant polarity (Fig. 2). The oligonucleotide with $(dC)_{25}$ at the 3′ end was more inhibitory than was the oligonucleotide with $(dA)_8$ at the 3′ end. The ratio of calculated K_i values was 4. Both inhibitor reactions gave the same V_{max} as with no inhibitor, showing they

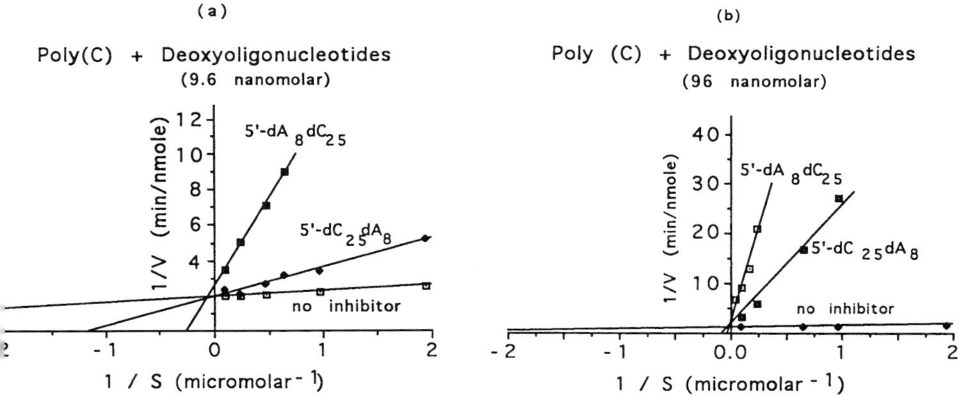

FIG. 2. Inhibition of RNase II degradation of poly(C) by the presence of $5'(dA)_8(dC)_{25}$ or $5'(dC)_{25}(dA)_8$ in a 100-μl reaction at 23°.[14] The reciprocal of concentration of poly(C) is plotted versus the reciprocal of reaction velocity. The concentration of inhibitor was 9.6 n*M* (a) and 96 n*M* in (b). [From V. J. Cannistraro and D. Kennell, *J. Mol. Biol.* **243,** 930 (1994), with permission of Academic Press.]

are competitive inhibitors. The results show that there is a directional polarity for positioning the substrate and suggest precise short-range recognition constraints.

Reaction Kinetics for RNA Homopolymers

Reaction velocities[14] at 23° for poly(A) and poly(U) were significantly faster than that of poly(C), with V_{max} values ~3.5 times faster (Fig. 3). The velocities are also shown for when the reactions occur in the presence of 20% (v/v) ethanol. At lower concentrations of poly(A), ethanol slows the reaction significantly, but the rate increases with concentration to give the same V_{max}. The intersection on the abscissa gave a ~16-fold higher K_m (67×10^{-7} *M* compared with 4×10^{-7} *M*), indicating that ethanol reduces significantly the affinity of RNase II for poly(A).

The results with poly(C) were quite different. At all concentrations of poly(C), ethanol increases the reaction rate similarly, by ~2.5-fold, to give a parallel line. In this case, even though the intercepts on the abscissa differ, ethanol probably has little effect on the binding to poly(C) because the kinetics of inhibition by $oligo(dC)_{33}$ with or without 20% (v/v) ethanol are the same.[14]

The strongest attractive forces between nucleic acids and proteins are provided by ionic bonds.[27] These bonds are weakened by the presence of salt. Salt had no significant effect on the reaction velocities of poly(A) but was a competitive inhibitor of poly(C). The V_{max} was the same but the K_m was much higher in salt

[27] W. Saenger, "Principles of Nucleic Acid Structure." Springer-Verlag, New York, 1984.

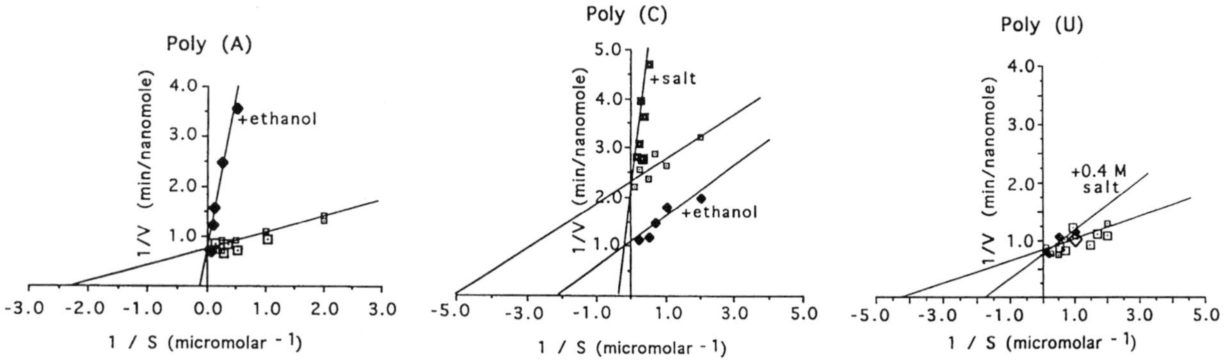

FIG. 3. Double reciprocal plots of substrate concentration (*S*) versus reaction velocity (*V*) for degradation by RNase II of poly(A), poly(C), and poly(U) at 23°.[14] *S* represents micromolar substrate, assuming 300 nucleotides per polymer, and *V* represents nanomoles of mononucleotide released per minute in a 100-μl reaction. KCl was present at 200 m*M* ["+ salt" lines in the poly(C) panel; larger squares in poly(A) and poly(U) panels] and 400 m*M* in the poly(U) panel, as indicated. Ethanol was at 20% (v/v) in the designated reactions, and the unmarked lines included reactions with no addition (small squares). The reactions were linear with time beyond the time chosen (2 or 4 min) for each concentration point shown. [From V. J. Cannistraro and D. Kennell, *J. Mol. Biol.* **243,** 930 (1994), with permission of Academic Press.]

(33×10^{-7} *M*). KCl at 0.2 *M* reduced the binding affinity of RNase II for poly(C) by about 16-fold. The results with poly(U) differed from those with either poly(A) or poly(C). The V_{max} and K_m values were unaffected by the presence of either 20% (v/v) ethanol or 0.2 *M* KCl.

The V_{max} values for poly(C) in ethanol, and for poly(A) or poly(U) with or without ethanol, were all approximately the same. However, in the absence of ethanol, poly(C) was degraded ~3.5 times slower at saturating levels of substrate. The following experiments were performed to understand the basis for these observations.

Poly(C) Degradation Occurring by Processive Leaps

When RNase II degrades poly(C) at low temperatures, the degradation is discontinuous.[14] The distances between stops are fairly constant at 10 to 12 nucleotides except during the terminal stage of the reaction, when the oligomer is <10 to 15 nucleotides (Fig. 4). Poly(A) and poly(U) degradation does not show discontinuities at discrete intervals until the substrate reaches the 10- to 15-mer size. Chase experiments with oligo(dC) inhibitors showed that most, or all, of the intense bands result from dissociation.[14] When added just before the enzyme, oligo(dC) inhibitors completely block initiation. When oligo$(dC)_{17}$ was added during degradation of excess poly(C), the band pattern did not change with time. This result shows that RNase II is dissociating, rather than pausing, to give the strong bands every 12 nucleotides.

Evidence that Nucleotides at Anchor Site Remain Bound during Reaction of Poly(C)

RNase II stalls at the last rC of single-stranded 3′-$(rC)_n(dC)_m$ oligonucleotides.[11,14] While the rates of association of oligonucleotides to RNase II are relatively fast, even for smaller ones, the dissociation rates are orders of magnitude slower (Fig. 5). The more residues released, the faster the stalled complex dissociates, resulting in a progressively weaker complex to inhibit RNase II activity (Fig. 1). Mixed RNA–DNA oligonucleotides with <12 rC residues remain associated with RNase II sufficiently long so that the intermediate complexes can be studied. A range of PDE I activities was used to footprint nucleotides interacting with RNase II. One set of oligonucleotides included

3′-$(dC)_{25}(dA)_8$ (no cleavable bonds)
3′-dT$(rC)_2(dC)_{15}(rC)_{22}$-5′ (two cleavable bonds)
3′-dT$(rC)_8(dC)_9(rC)_{22}$-5′ (eight cleavable bonds)

(At the time of these experiments the *in vitro* synthesis of the RNA-containing oligonucleotides required a starting 3′-DNA nucleotide, but as noted, RNase II degrades a 3′-dT–rC– – bond). In the absence of RNase II association, PDE I

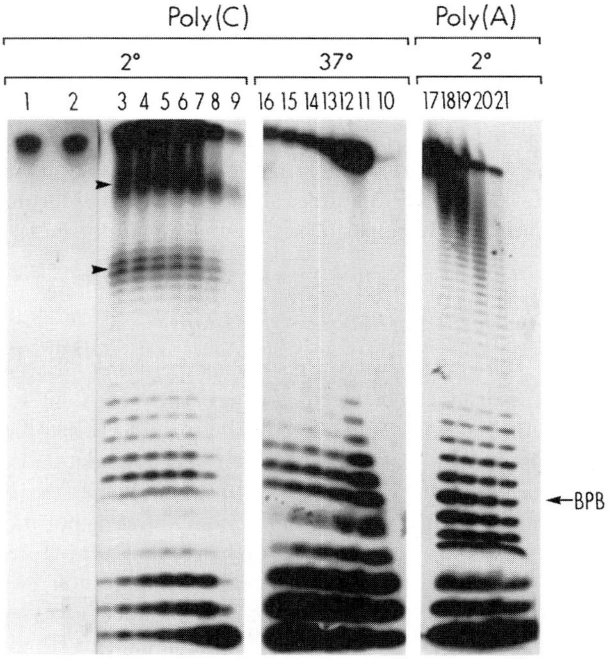

FIG. 4. Discrete periodic stops are seen only when RNase II degrades single-stranded poly(C) at low temperatures. Lanes 3–9 show the degradation of [5′-^{32}P]poly(C) (45-nucleotide length) by limiting RNase II in a 20-μl reaction at 2° for 10, 20, 30, 40, 60, 120, and 300 sec, respectively. Lane 1, no RNase II; lane 2, 0.1% (w/v) SDS added just before RNase II in a 5-min reaction: Lanes 10–16, degradation of [5′-^{32}P]poly(C) (45-nucleotide length) at 37° with samples taken 7 sec apart from 0 (lane 10) to 42 sec (lane 16); lanes 17–21, degradation of poly(A) (45-nucletide length) at 2° for 0, 20, 40, 60, and 120 sec, respectively. The film for lanes 10–21 was overexposed to detect any stops. Reactions were stopped by adding SDS to 0.1% (w/v) and products were separated by electrophoresis through a 20% (w/v) polyacrylamide gel. The bromphenol blue marker is shown (BPB). The lowest band is dimer C. The arrowheads point to the more intense bands at ~33- and ~21-nucleotide lengths. The former length is more obvious in original films but is seen here in lane 8. [From V. J. Cannistraro and D. Kennell, *J. Mol. Biol.* **243,** 930 (1994), with permission of Academic Press.]

degraded single-stranted oligonucleotides of RNA, DNA, or RNA–DNA processively to mononucleotides, that is, no intermediates were observed (Fig. 6c). With RNase II bound to 3′-$(dC)_{25}(dA)_8$, the lowest PDE I activities gave bands corresponding to nucleotides near the 3′ end. Higher activities showed a progressive shift of bands to smaller sizes as the PDE I degraded further through the DNA–RNase II complex. There was a continuous ladder of bands. This result suggests that every nucleotide of a dC oligonucleotide (and presumably rC oligonucleotide) forms an initial association with RNase II (Fig. 6a). In contrast, both RNA–DNA oligonucleotides (after cleavage of either two or eight bonds)

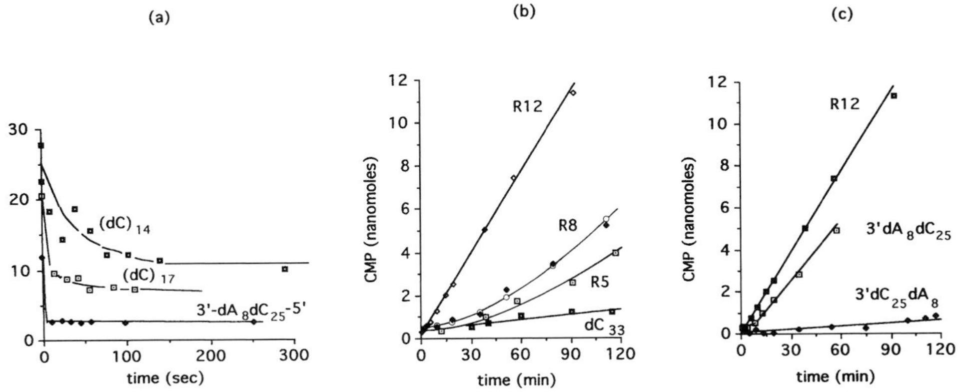

FIG. 5. The rates of association and dissociation of various oligonucleotides from RNase II at 10°. (a) Association rates: At time 0 a DNA oligonucleotide was added to 200 units of RNase II in 200 μl of Tris [0.1 μg of $(dC)_{17}$ or 3′-$(dA)_8(dC)_{25}$ and 1.0 μg of $(dC)_{14}$ (to have measurable inhibition)]. At the indicated times 20 μl was removed and added to 100 μl of RNase II reaction mix containing 150 μg of poly(C) and incubated for an additional 40 sec. Reactions were stopped by boiling and the amount of CMP was measured. (b and c) Dissociation rates: 0.1 μg of oligonucleotide (~100 n*M*) was added to 250 units of RNase II in 240 μl of Tris, and the reactions were initiated at 5 min by combining with RNase II reaction mix plus 150 μg of poly(C) (15m*M*) in 1.2 ml. The half-life equals the time when the rate of poly(C) degradation equals half the rate with no inhibitor (when half the complex has dissociated). The expected curve for dissociation of the RNase II–oligonucleotide complex (X, nanomoles of complex) with a $t_{1/2}$ of 60 min is shown (○) along with the observed values (◆). The rate of poly(C) degradation (r) is proportional to the amount of free RNase II; $r = K(X_0 - X)$ nmol of CMP/min, where $X = X_0$ at $t = 0$. Because the dissociation reaction is first order, $dX/dt = -kX$ and $r = KX_0(1 - e^{-kt})$. Integrating with time gives the nanomoles of CMP released (c) at $t.c = KX_0[t + (e^{-kt}/k) - 1/k]$ to give the expected points shown. The observed values and fitted curves for R12 (◇), R5 (□), and $(dC)_{33}$ (■) are also shown. [Reprinted from V. J. Cannistraro and D. Kennell, *Biochim. Biophys. Acta* **1433,** 170 (1999), with permission from Elsevier Science.]

gave a band-free gap between the 3′-proximal nucleotides and a set of 3′-distal residues (Fig. 6b and c). Catalysis by RNase II leads to dissociation of nucleotides between the catalytic and anchor sites. The gap size between these enzyme sites decreases with the number of residues removed. The remaining residues at the 3′ end (the first four to six nucleotides) must be interacting with the catalytic site and nucleotides 15 to ~25 from the 3′ end to the 5′ anchor site of the stalled RNase II.

The nucleotides bound to the anchor site of RNase II were the same whether 2 or 8 residues were released and ended at ~25 nucleotides from the original 3′ end. This number agrees closely with the most distal nucleotide associated with RNase II (nucleotide 25) predicted from the inhibition as a function of oligonucleotide length (Fig. 1). Note that the original nucleotides bound to the anchor site before degradation were the same nucleotides after degradation whether two or eight

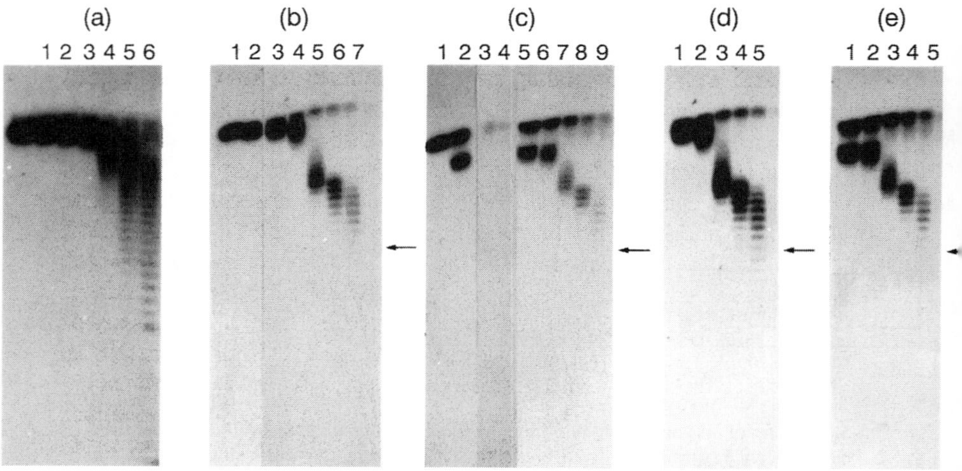

FIG. 6. Nucleotides bound to the 5′ anchor site of RNase II remain fixed during progressive removal of residues from the 3′ end. Nucleotides associated with RNase II were identified by blocked PDE I degradation of enzyme–5′-^{32}P-labeled oligonucleotide complexes. The ^{32}P-labeled oligonucleotides (~0.02 μg) were degraded for 60 sec by 40 units of RNase II in 20 μl at 2°. The resulting complexes were then degraded for an additional 30 sec by the indicated units of PDE I at 2° and the resulting 5′-^{32}P-labeled oligonucleotides separated by PAGE. (a) RNase II-bound DNA oligonucleotide 5′-^{32}P-$(dA)_8(dC)_{25}$ before (lane 1) and after degradation with 0.0001, 0.0002, 0.0004, 0.001, and 0.002 unit of PDE I, respectively (lanes 2–6). (b) Mixed RNA–DNA oligonucleotide 5′-^{32}P-$(rC)_{22}(dC)_{15}(rC)_2$dT (R2): lane 1, no treatment; lane 2, RNase II only; lanes 3–7, RNase II-bound oligonucleotide with 0.00004, 0.0001, 0.0002, 0.0004, and 0.001 unit of PDE I, respectively. (c) Mixed RNA–DNA oligonucleotide 5′-^{32}P-$(rC)_{22}(dC)_9(rC)_8$dT (R8): lane 1, no treatment; lane 2, RNase II only; lanes 3 and 4, 0.00002 and 0.0004 unit of PDE I only, respectively; lanes 5–9, RNase II-bound oligonucleotide with 0.00002, 0.00004, 0.0001, 0.0002, and 0.0004 unit of PDE I, respectively. (d) and (e) show the same gels as those in (b) and (c), respectively, with the films exposed longer to show the lower bands more clearly. The times between loading of consecutive samples account for the curvature of the full-length bands. The arrows identify nucleotide 25 from the 3′ end. [Reprinted from V. J. Cannistraro and D. Kennell, *Biochim. Biophys. Acta* **1433,** 170 (1999), with permission from Elsevier Science.]

bonds were hydrolyzed. If the substrate had "slid" through the enzyme, those bands would represent much smaller oligonucleotides.

3′ Binding Site Necessary to Maintain Stable Complex

The 5′ anchor site interaction by itself is not sufficient to maintain a stable substrate association. 3′-dT$(rC)_{15}(dC)_{16}$-5′-^{32}P was bound to RNase II, in the absence of Mg^{2+} and NH_4^+ to prevent reaction, before treatment with RNase A, which degraded the $(rC)_{15}$ segment. The complex was then separated from free substrate in a native polyacrylamide gel. The loss of the associated 3′ nucleotides by RNase

A degradation resulted in rapid dissociation of its distal [5′-^{32}P]DNA fragment.[11] Therefore, binding of substrate at the 5′ anchor site alone is not sufficient to maintain a stable complex.

RNase II Binding to Nucleic Acids with Secondary Structures

RNase II is single-strand specific.[11] However, it can attach to closed duplex structures. Equimolar concentrations of different structured DNA oligonucleotides were compared as inhibitors of RNase II. The strongest binding is to stem–loop structures that have free 3′ and 5′ arms. A strong stem–loop structure with blunt ends did not inhibit (not shown). However, with both single-stranded 3′ and 5′ arms the strong duplex is an effective inhibitor (Fig. 7a). With only a 3′ arm it was less inhibitory, with a K_i value ~2.5 times higher, while with only a 5′ arm it gave almost no inhibition. A weaker stem–loop structure with 3′ and 5′ single-strand arms (Fig. 7b) was a better inhibitor than its counterpart with the strong duplex; its K_i value was close to that of single strands of dC. With this weaker stem, eliminating the 5′ single strand had no observable effect, while eliminating the 3′ arm again resulted in almost no inhibition.

RNase II binding requires a 3′-ended single strand, and when the duplex is relatively unstable, the oligonucleotide can bind as well as does a linear molecule of dC, and a 5′ single-stranded arm does not add binding strength. This result suggests that anchor site binding to the 5′ side of the duplex can occur, perhaps due to duplex "breathing." However, when the duplex was thermodynamically stable, a 5′ single-stranded sequence contiguous with the duplex improved the binding by providing a needed anchor site.

RNase II Binding Sites on Oligonucleotides with Secondary Structures

The segment of nucleotides interacting with the anchor site of RNase II is a function of the oligonucleotide structure.[11] RNase II was bound to a weak stem–loop [5′-^{32}P]DNA oligonucleotide with 3′ and 5′ single-stranded arms and the complex probed by PDE I (Fig. 8a). Bands at the top of the gel correspond to the first four to six nucleotides at the 3′ end catalytic site of RNase II. However, unlike linear DNA, which gave a continuous series of bands from 1 to 25 nucleotides (Fig. 6a), there was a large gap of ~20 nucleotides before four or five bands that corresponded to nucleotides primarily on the 5′ proximal duplex strand. The three strongest bands represented T residues at positions 27, 29, and 32 from the 3′ end. These bands could reflect the much better binding of RNase II by oligo(dT) or oligo(dC) compared with oligo(dA)[14] or to the preferred spatial position of the hairpin to the 3′ catalytic site.

This latter explanation was shown to be the case when only purines were on the same 5′-sided strand of the duplex and the bands were in the same approximate

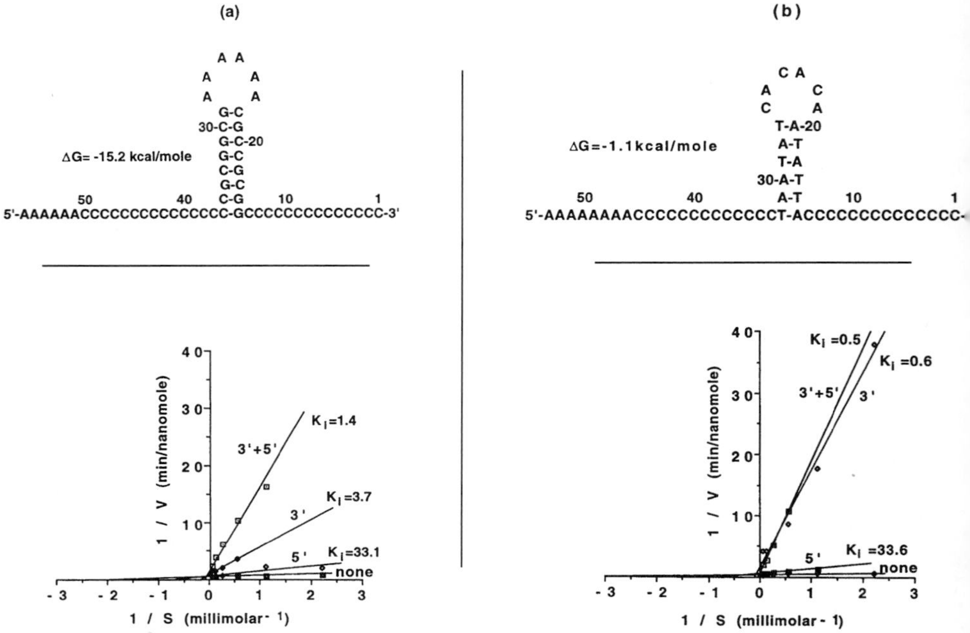

FIG. 7. The inhibition of RNase II by DNA oligonucleotides with strong (a) or weak (b) duplex structures and with 5′ and/or 3′ single strands. Shown are the structures with both 5′ and 3′ single strands. The free energy values (ΔG) are estimated by the gcg mfold program (version 8; GCG, Madison, WI). *Bottom:* Lineweaver–Burk plots for each oligonucleotide with its derived inhibitor constant (K_i) (see formula in Fig. 1) and controls (no inhibitor). The strong duplex without 5′ or 3′ single strands did not inhibit, that is, its points would fall on the "none" curves. Each reaction was linear for the 10 min at 23° with 94 n*M* oligonucleotide and 40 units of RNase II in 100 μl of RNase II reaction mix with the poly(C) concentration shown. Curves labeled 3′ in the lower frames are the same structure but missing the 5′ single strand arm and those labeled 5′ lack the 3′ arm. [Reprinted from V. J. Cannistraro and D. Kennell, *Biochim. Biophys. Acta* **1433,** 170 (1999), with permission from Elsevier Science).]

position (Fig. 8b). Therefore, the nucleotides bound to the anchor site are not dictated by the positions of pyrimidines but rather by their spatial distance from the catalytic site. An oligonucleotide with a longer 3′ single-strand arm (19 residues rather than 14), binds ~15 to 27 nucleotides from the 3′ end with residues on that 3′ single-strand arm and the same strand a short distance into the duplex (Fig. 8c).

Although the weak stem–loop without any 3′ single strand was a weak inhibitor of RNase II (Fig. 7), a DNA–protein complex could accumulate presumably from stem "breathing." If RNase II was bound only to a completely denatured molecule, every nucleotide would then be associated with the enzyme as occurs with $(dC)_{33}$

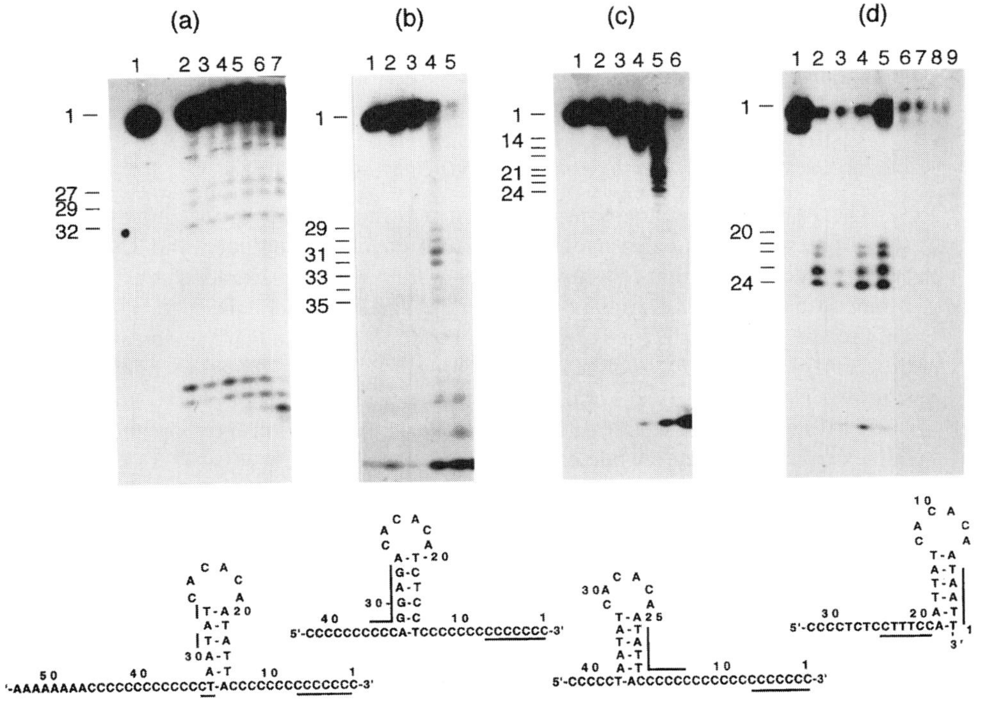

FIG. 8. PDE I degradation of RNase II-bound DNA oligonucleotides with secondary structures. Forty units of RNase II was bound to <0.02 μg of ^{32}P-labeled oligonucleotide at 23°. After 60 sec, the indicated amount of PDE I was added for an additional 30 sec before adding SDS/glycerol/dyes and running, on polyacrylamide gels. (a) A weak stem–loop structure; Lane 1, oligonucleotide alone reacted with 0.001 unit of PDE I; lanes 2–7, RNase II-bound oligonucleotide reacted with 0.00002, 0.00005, 0.0001, 0.0002, 0.0005, and 0.001 unit of PDE I, respectively. (b) A strong stem–loop structure with the pyrimidine-rich strand of the duplex on the strand opposite that in (a): Lane 1, oligonucleotide alone; lane 2, RNase II-bound oligonucleotide; lane 3, oligonucleotide reacted alone with 0.02 unit of PDE I; lanes 4 and 5, RNase II-bound oligonucleotide treated with 0.005 and 0.01 unit of PDE I, respectively. (c) A weak stem–loop DNA structure with a longer 3′ single strand: Lane 1, RNase II-bound oligonucleotide; lane 2, oligonucleotide reacted with 0.005 unit of PDE I; lanes 3–6, RNase II-bound oligonucleotide reacted with 0.0005, 0.001, 0.002, and 0.005 unit of PDE I, respectively. (d) A weak stem–loop with a single strand only on the 5′ end: Lane 1, RNase II-bound oligonucleotide; lanes 2–5, RNase II-bound oligonucleotide reacted with 0.0001, 0.0002, 0.0005, and 0.001 unit of PDE I, respectively; lanes 6–9, oligonucleotide alone reacted with the same units of PDE I as in lanes 2–5. The numbers on the side of each panel indicate the number of nucleotides from the 3′ end. The solid lines in the structures indicate the 3′ and 5′ nucleotides that correspond to the gel bands. Bands near the bottom of the gels are oligonucleotides resulting from near-limit digests. [Reprinted from V. J. Cannistraro and D. Kennell, *Biochim. Biophys. Acta* **1433,** 170 (1999), with permission from Elsevier Science.]

(Fig. 6a). However, the anchor site bands corresponded to nucleotides on the single strand proximal to the duplex on its 5′ side (Fig. 8d), separated by a large gap from the catalytic site bands, suggesting that RNase II is binding to a partially open stem structure, as is likely with the other structures in Fig. 8.

Disruption of Duplex Structures by RNase II Binding

The PDE I footprinting was consistent with nucleotides in the duplex of a stem–loop structure interacting with the anchor site of RNase II to inhibit RNase II activity. Such interactions could weaken or disrupt a double-stranded structure.[11] Mung bean nuclease is a stringent single-strand-specific endonuclease recognizing only relatively long stretches of single strands.[27a] It did not cleave bonds in three differently structured DNA oligonucleotides, even in their small single-strand loops (Fig. 9). The single-strand arms were not cut because it cannot cleave stretches of dC. However, with RNase II bound, cleavages occurred in the loop and both duplex strands even at low nuclease activities.

RNase I* of *E. coli* is encoded by the *rna* gene for the periplasmic RNase I,[15,18] and it appears to be a cytoplasmic precursor to RNase I.[28] Unlike RNase I, RNase I*, like RNase YI* of yeast,[29] does not make nicks in duplex RNA. We tested the binding of RNase II (in the absence of NH_4^+ and Mg^{2+}) for enhanced activity of RNase I* to degrade 5S rRNA. Without RNase II, RNase I* cleaved only the most vulnerable bonds identified earlier with RNase YI*[27a] When added together to 5S rRNA, many new bands were seen. The initial cleavages by RNase I* probably provided binding sites for RNase II that led to alteration of adjacent secondary structures for further RNase I* cleavages. Similar cooperativity between the two enzymes was seen with the degradation of a 62-nucleotide fragment of 5S RNA.[11]

Cycling of Reactive RNase II On/Off-Structured RNA

Under reactive conditions, RNase II might, or might not, dissociate when further degradation is blocked at a duplex or other structure.[11] A large, 110-nucleotide RNA containing all but the first 10 nucleotides from the 5′ end of the 5S rRNA gave a molecule with extensive secondary structures but with a 3′ end of 10 single-stranded nucleotides. RNase II was bound to this RNA in the absence of Mg^{2+} and NH_4^+ (Fig. 10, lane 2). Added oligonucleotide $(dC)_{25}$ prevents RNase II-substrate association,[14] but the competitor did not cause release of the substrate (Fig. 10, lane 4). However, when added during catalysis, approximately 90% of the substrate became free (Fig. 10, lane 3). Without competitor, almost all of the reacted substrate remained associated with RNase II (Fig. 10, lane 5) suggesting that the

[27a] V. J. Cannistraro and D. Kennell, *Methods Enzymol.* **341,** 118 (2001).
[28] S. K. Srivastava, V. J. Cannistraro, and D. Kennell, *J. Bacteriol.* **174,** 56 (1992).
[29] V. J. Cannistraro and D. Kennell, *Nucleic Acids Res.* **25,** 1405 (1997).

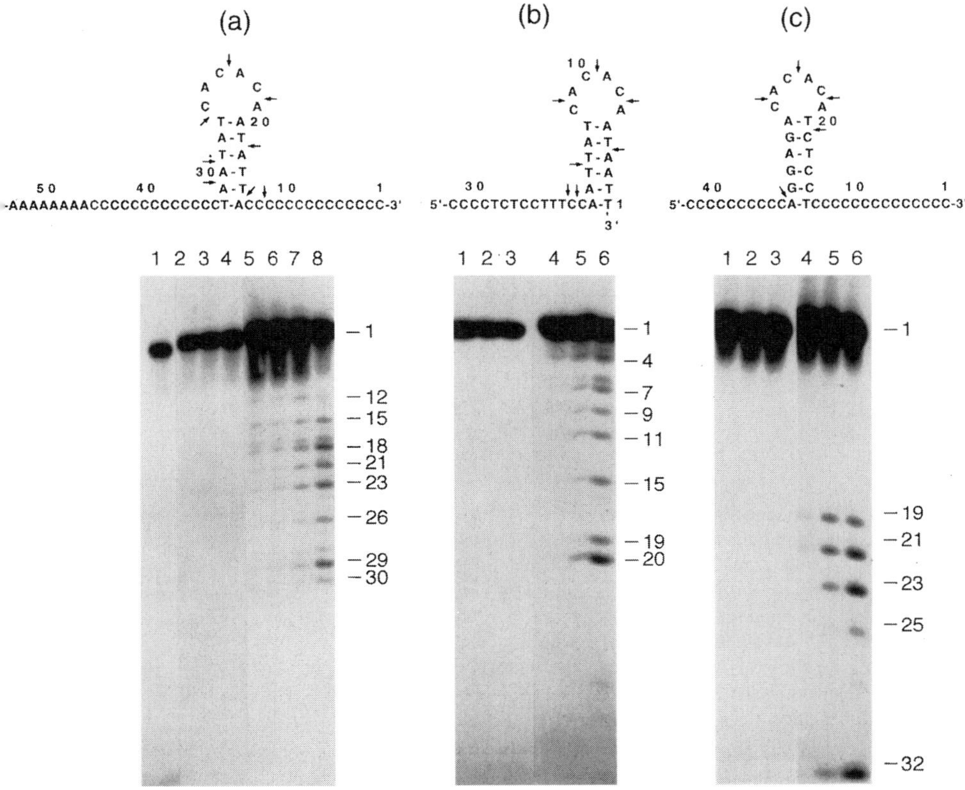

FIG. 9. RNase II binding to an oligonucleotide with secondary structures makes phosphodiester bonds accessible to a single-strand-specific endonuclease. RNase II (40 units) was added to the [5′-^{32}P]DNA oligonucleotide (~0.02 μg) in 50 mM Tris-HCl, pH 7.0, at 23° in 20 μl. After 60 sec, mung bean nuclease (nuclease) was added with $ZnCl_2$ (to give 1 mM) and incubated for an additional 60 sec before chilling to 0° and adding urea to 6 M plus SDS/glycerol/dyes before PAGE. Controls were only nuclease treated. (a) A weak AT stem: Lanes 1–4, free oligonucleotide and lanes 5–8, RNase II-bound oligonucleotide. Each set was reacted with 5, 10, 20, and 50 units of nuclease, respectively. (b) An AT stem missing the single-stranded 3′ arm: Lanes 1–3, free oligonucleotide and lanes 4–6, RNase II-bound oligonucleotide. Each set was reacted with 5, 10, and 20 units of nuclease, respectively. (c) A strong duplex structure: Lanes 1–3, free oligonucleotide and lanes 4–6, RNase II-bound oligonucleotide. Each set was reacted with 10, 20, and 50 units of nuclease, respectively. The numbers on the side indicate the number of nucleotides from the 3′ end. Arrows point to the observed cleavage sites. [Reprinted from V. J. Cannistraro and D. Kennell, *Biochim. Biophys. Acta* **1433,** 170 (1999), with permission from Elsevier Science.]

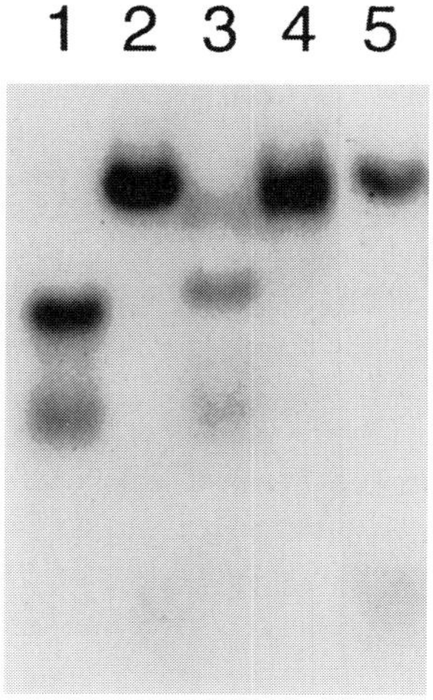

FIG. 10. Association/dissociation cycling by RNase II to a large structured oligonucleotide during its degradation by the enzyme. Forty units of RNase II was added to ~0.02 μg of the 5′-^{32}P-labeled 110-nucleotide fragment of the 5S rRNA missing the first 10 nucleotides of the 5′ end at 23° in a volume to give 20 μl/lane. The samples were run on a 5% (w/v) nondenaturing polyacrylamide gel. Oligonucleotide alone (lane 1); RNase II plus oligonucleotide in Tris–EDTA for 60 sec (lane 2) followed by Mg^{2+} plus NH_4^+ and 1μg of $(dC)_{25}$ for 60 sec (lane 3) or followed by 1 μg of $(dC)_{25}$ only (lane 4); RNase II plus oligonucleotide with Mg^{2+} plus NH_4^+ for 60 sec (lane 5). [Reprinted from V. J. Cannistraro and D. Kennell, *Biochim. Biophys. Acta* **1433,** 170 (1999), with permission from Elsevier Science.]

RNase II–oligonucleotide complex not only dissociates when it stops at a duplex, but is in equilibrium between association and dissociation.

Discussion

Dissociation and Slower Reaction Rate Dependent on Secondary Structure of Substrate

Single-strand homopolymers exist as stacked, or randomly coiled, structures. Low temperatures favor the helical structure; poly(C) has six nucleotides per turn,[30–32] but there is not a sharp transition from one form to the other.[33] The stacking is largely additive, rather than cooperative,[27] and under the conditions reported here, a given polynucleotide has regions of both structures. Hydrophobic

[30] G. D. Fasman, C. Lindblow, and L. Grossman, *Biochemistry* **3,** 1015 (1964).
[31] S. Arnott, R. Chandrasekaran, and G. W. Leslie, *J. Mol. Biol.* **106,** 735 (1976).
[32] M. S. Broido and D. R. Kearns, *J. Am. Chem. Soc.* **104,** 5207 (1982).
[33] W. C. Johnson, Jr., *Methods Biochem. Anal.* **31,** 61 (1985).

interactions are believed to account for most of the forces stabilizing the stacked coil[27] and become weaker in organic solvents such as ethanol. Poly(C) shows a marked hyperchromic effect from higher temperature or from addition of ethanol, and dissociations at discrete stop sites become less prominent.[14] At the same time, the reaction rate increases to equal that of the other homopolymers. Thus, both agents (temperature and ethanol) favor the random coil structure and each eliminates dissociations at discrete sizes and increases the reaction velocity of poly(C).[14]

Poly(C) Degradation Providing Unique System to Study Processivity Mechanism of RNase II

Initial studies used homopolymers to study RNase II binding to single-stranded RNA to avoid any effects of duplex structures, because cellular or random RNA sequences of >25 residues have a high probability for duplex formations.[34] Of the homopolymers, poly(G) forms multistranded complexes[35] and is not a substrate. Although poly(U) shows strong binding, like poly(C),[14] it has a random coil structure even at low temperatures[27] and did not show discrete gel bands when degraded by RNase II.[14] With an open site on the enzyme (Fig. 8), anchor site binding of poly(U) might include a wide range of nucleotide sets from the 3′ end, so that dissociations would give an undetectable "blur" of gel bands (Fig. 11A, inset). While the pyrimidine polymers are associated to RNase II by strong ionic bonds, poly(A) oligomers are bound much more strongly to a hydrophobic column[14] and the weak hydrophobic association to RNase II is weakened by ethanol.[14] In the absence of the strong anchor site binding, poly(A) must "slide" through the RNase II channel even though it possesses helix–coil structures at low temperatures.

The combination of strong binding and an ordered structure could account for the dissociations at discrete intervals of ~12 nucleotides, in turn accounting for the slower reaction of poly(C) compared with poly(A) or poly(U). These characteristics of helical poly(C) provide a system with which to study the RNase II reaction mechanism that, for the reasons given, is not apparent with other substrates.

Model for RNase II Reaction Mechanism

A model for the processive reaction of RNase II, when conditions favor the stacked single-strand structure, is shown in Fig. 11A.[11] Poly(C) is bound by ionic bonds to a catalytic site associated with the first few 3′ nucleotides and an anchor site for binding nucleotides ~15 to 25 from the 3′ end. The latter remain bound while the 3′ end is threaded through the catalytic site as the end nucleotides are cleaved off to generate a progressively decreasing strength of association. Binding of both the 3′-end and anchor site nucleotides is necessary to maintain the

[34] D. Kennell, *in* "Maximizing Gene Expression" (W. W. Reznikoff and L. Gold, eds.), p. 101. Butterworths, Stoneham, Massachusetts, 1986.

[35] W. Saenger, J. Riecke, and D. Suck, *J. Mol. Biol.* **93,** 529 (1975).

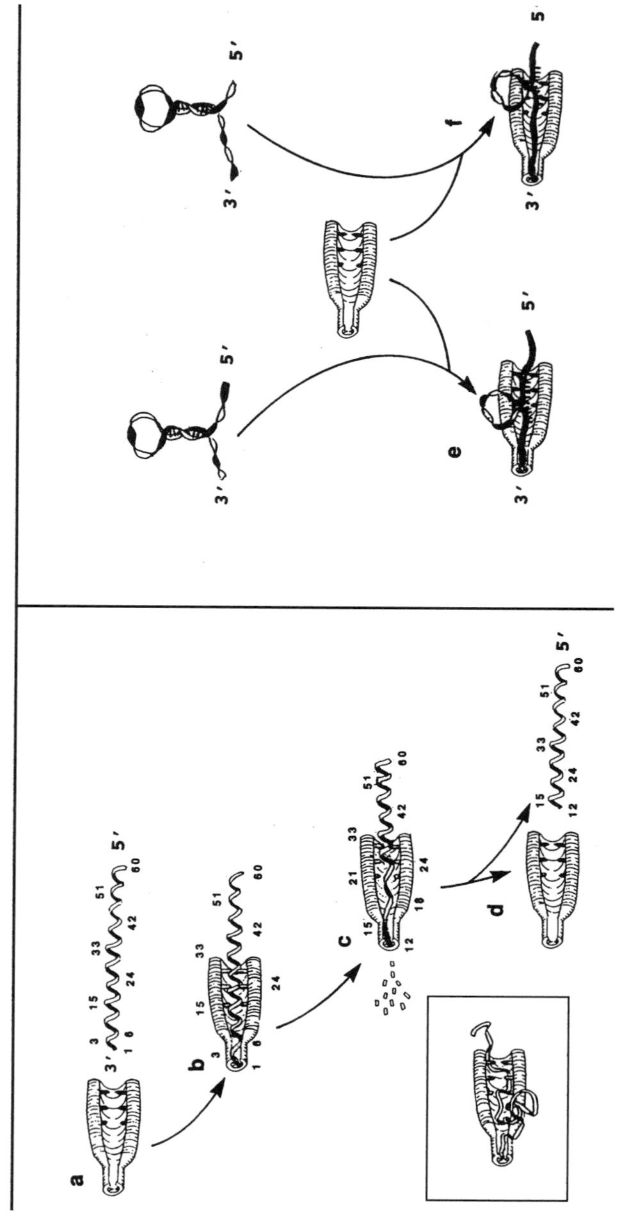
A
B
a
b
c
d
e
f
3′
5′

association. After ~12 nucleotides, the enzyme–substrate dissociates. Figure 11 (inset) shows one of many possible configurations of a polynucleotide when it is a random coil, e.g., poly(U), or poly(rC)$_{60}$ at higher temperatures or in ethanol. Poly(A) has a stronger helix–coil than poly(C),[27] but binding to the anchor site is by weak hydrophobic interactions. Without strong anchor site binding, poly(A) threads through the enzyme without periodic dissociations. However, all substrates start dissociating with each cleavage when they become so small that the anchor site cannot be filled by nucleotides. In this model the energy for progression is provided by the pulling force on the substrate at the catalytic site.

RNase II can bind to nucleic acids with secondary structures, but optimal binding requires both single-stranded 3′ and 5′ arms. 3′-Exonuclease footprinting showed that anchor site binding was not limited to the first 27 nucleotides from the 3′ end as with single-stranded oligonucleotides of C (Fig. 6), or to single-stranded segments in loops or arms, but rather was determined by their spatial distance to the 3′ catalytic site of RNase II (Fig. 8). Interaction of duplex nucleotides can weaken or disrupt the double-stranded stem and make it more vulnerable to single-strand nucleases. Both DNA and RNA complexes are made susceptible to single-strand nucleases, corroborating this interpretation.

FIG. 11. (A) A model for the processive reaction of RNase II. The RNA is poly(rC)$_{60}$. The structure of RNase II is not known but is depicted to be 60 to 70 Å in the longer dimension, and the reaction is occurring below the T_m for the helix ↔ random coil transition. (a) shows the free enzyme with a narrow groove that encompasses three or four nucleotides at the 3′ end of the substrate. The hydrolytic activity is shown by the two teeth near the 3′ end. The strong anchor site spans 12 or more nucleotides and is depicted by the transverse bar teeth. The numbers designate the nucleotide, starting with nucleotide 1 at the 3′ end. (b) shows the substrate just before cleavage of nucleotide 1. The 3′ end is attracted into the catalytic site, which weakens the stacking forces to make the first four or five nucleotides extended. The specificity of RNase II for ribonucleotide bonds may be defined at the binding sites for nucleotides 3 to 5.[14] The RNA is held fixed in place by the strong anchor site. (c) shows the reaction just when the first eleven 5′-CMP residues have been released. With the release of each nucleotide, the new 3′ end is attracted into the catalytic site, causing further extension of the helix that is still rigidly held in place by the anchor site. (d) The cycle is repeated but with the first 11 nucleotides missing from the substrate. *Inset:* One of many possible configurations of a polynucleotide when it is a random coil, e.g., poly(U), or poly(rC)$_{60}$ at higher temperatures. Variable numbers of nucleotides between catalytic and anchor site binding would lead to a random size distribution of released oligonucleotides and not give discrete gel bands. (B) A model of the binding of nucleic acids with secondary structures. The binding site of RNase II is open to allow positioning of a complex structure. The hydrogen bonds (–) between strands of the duplex are shown. They are also drawn on the separated duplex strands to identify the position of the disrupted duplex as a result of the anchor site binding. (e) depicts an oligonucleotide similar to the oligonucleotide in Fig. 8b. The anchor site binds to the single-stranded arm proximal to the duplex on its 5′ side and duplex nucleotides on that same strand. (f) depicts the oligonucleotide in Fig. 8c with a longer 3′ single-stranded arm. The anchor site binds to the single-stranded arm proximal to the duplex but on its 3′ side and duplex nucleotides on that same strand. [Reprinted from V. J. Cannistraro and D. Kennell, *Biochim. Biophys. Acta* **1433,** 170 (1999), with permission from Elsevier Science.]

However, RNase II cannot bind to a duplex structure itself, as shown by the absence of binding to a strong G–C duplex without a 3′ single-stranded arm. Rather, the anchor site is bound to the duplex in a secondary stage of the association and the binding is to only one of the two strands. The duplex strand selected could be specified by the location of the single strand able to bind part of the anchor site. The remaining contiguous nucleotides on that same strand of the duplex are probably recruited during the open phases of duplex "breathing." The more prolonged open phase of a weaker DNA duplex could account for its stronger binding and greater inhibition of RNase II activity.

The association of RNase II with a structured nucleic acid is not a static relationship but involves dissociation/reassociation cycling. Successful anchor site binding to duplex nucleotides of a molecule must be kinetically dependent on the simultaneous occurrence of its duplex breathing and association with enzyme.

Section III

Ribonuclease Complexes

[27] *Escherichia coli* RNA Degradosome

By AGAMEMNON J. CARPOUSIS, ANNE LEROY, NATHALIE VANZO, and VANESSA KHEMICI

Introduction

The *Escherichia coli* RNA degradosome is a nucleolytic complex containing four integral components: ribonuclease E (RNase E), polynucleotide phosphorylase (PNPase), RNA helicase B (RhlB), and enolase.[1-4] This complex was discovered during efforts to purify and characterize RNase E, which is a single-strand-specific endoribonuclease involved in the processing of ribosomal RNA and the degradation of messenger RNA.[5-8] RNase E is a large, multidomain protein containing an N-terminal catalytic site, a central arginine-rich RNA-binding domain, and a C-terminal region that serves as the scaffold on which the other components of the RNA degradosome assemble.[9,10] A "minimal" degradosome containing RNase E, PNPase, and RhlB can be reconstituted by a simple mixing protocol using purified components.[11] That procedure is found in [28] in this volume.[11a]

In this chapter we describe a large-scale purification yielding milligram amounts of the RNA degradosome, an assay for measuring RNase E activity, and a small-scale method for preparing extracts enriched in the RNA degradosome. We conclude with some general comments and a brief discussion about how the purification could be improved.

General Methods

The classic *Guide to Protein Purification* has been useful in the development of the procedures described below.[12] It should be consulted for the use of lysozyme

[1] C. P. Ehretsmann, A. J. Carpousis, and H. M. Krisch, *Genes Dev.* **6,** 149 (1992).
[2] A. J. Carpousis, G. Van Houwe, C. Ehretsmann, and H. M. Krisch, *Cell* **76,** 889 (1994).
[3] B. Py, C. F. Higgins, H. M. Krisch, and A. J. Carpousis, *Nature (London)* **381,** 169 (1996).
[4] A. Miczak, V. R. Kaberdin, C. L. Wei, and S. Lin-Chao, *Proc. Natl. Acad. Sci. U.S.A.* **93,** 3865 (1996).
[5] S. N. Cohen and K. J. McDowall, *Mol. Microbiol.* **23,** 1099 (1997).
[6] A. J. Carpousis, N. F. Vanzo, and L. C. Raynal, *Trends Genet.* **15,** 24 (1999).
[7] G. A. Coburn and G. A. Mackie, *Prog. Nucleic Acid Res. Mol. Biol.* **62,** 55 (1999).
[8] P. Régnier and C. M. Arraiano, *Bioessays* **22,** 235 (2000).
[9] N. F. Vanzo, Y. S. Li, B. Py, E. Blum, C. F. Higgins, L. C. Raynal, H. M. Krisch, and A. J. Carpousis, *Genes Dev.* **12,** 2770 (1998).
[10] V. R. Kaberdin, A. Miczak, J. S. Jakobsen, S. Lin-Chao, K. J. McDowall, and A. von Gabain, *Proc. Natl. Acad. Sci. U.S.A.* **95,** 11637 (1998).

0076-6879/01 $35.00

to break bacterial cells, the use of detergents in protein purification, the fractionation of protein mixtures with ammonium sulfate, the quantification of protein by the method of Lowry, and techniques for chromatography, sedimentation, and sodium dodecyl sulfate–polyacrylamide gel electrophoresis (SDS–PAGE). We suggest *Molecular Cloning: A Laboratory Approach* for procedures such as the amplification of DNA by polymerase chain reaction (PCR), *in vitro* transcription using bacteriophage T7 RNA polymerase, the purification of RNA by phenol extraction and ethanol precipitation, and the analysis of RNA by denaturing gel electrophoresis.[13]

Materials

For the large-scale purification of the RNA degradosome, we routinely use AC21, which is a derivative of *E. coli* MC1061. The construction of AC21 and its growth have been described.[2] Briefly, cells from a 16-liter culture are collected by centrifugation, yielding a paste (160–200 g) that is formed into two tablets, weighed, and stored at −70°. We use AC21 because it is the wild-type control for a series of isogenic strains harboring mutations in the gene encoding RNase E. We have also used MC1061 grown to high density in an automated fermentor. We believe that most commonly used laboratory *E. coli* strains should be compatible with the procedures described below.

Double-distilled or Milli-Q water, treated with diethyl pyrocarbonate (0.1%, w/v) and autoclaved, is used throughout. Diethyl pyrocarbonate, aprotinin, leupeptin, pepstatin A, phenylmethylsulfonyl fluoride (PMSF), DNase I, proteinase K, dithiothreitol (DTT), glycerol, Triton X-100, and Genapol X-080 are purchased from Fluka (Ronkonkoma, NY); hen, egg white lysozyme is from Sigma (St. Louis, MO). We have purchased these reagents from the same source over a 5-year period and have obtained consistent results. Nevertheless, this does not imply that reagents of comparable quality from other suppliers cannot be used. All other reagents are of the highest purity available, taking care to avoid traces of heavy metals, proteases, and nucleases.

Aprotinin (2 mg/ml), DTT (0.5 *M*), leupeptin (0.8 mg/ml), and proteinase K (20 mg/ml) are prepared in water; pepstatin A (0.8 mg/ml) in methanol; DNase I (2 mg/ml) in 50% (v/v) glycerol–0.001 *N* HCl. These reagents are stored in aliquots at −20°.

[11] G. A. Coburn, X. Miao, D. J. Briant, and G. A. Mackie, *Genes Dev.* **13,** 2594 (1999).

[11a] G. A. Mackie, G. A. Coburn, X. Miao, D. J. Briant, and A. Prud'homme-Genereux, *Methods Enzymol.* **342,** [28] 2001 (this volume).

[12] M. P. Deutscher (ed.), *Methods Enzymol.* **182** (1990).

[13] J. Sambrook, E. F. Fritsch, and T. Maniatis, "Molecular Cloning: A Laboratory Approach," 2nd Ed. Cold Spring Harbor Laboratory Press, Cold Spring Harbor, New York, 1989.

Triton X-100 (20%, v/v), Genapol X-080 (20%, v/v), and PMSF (0.1 *M* in 2-propanol) are prepared and stored at room temperature. These reagents should not be autoclaved. PMSF, which is toxic, should be handled with care.

Lysozyme, stored as a powder at 4°, is added directly to the lysis buffer in the large-scale purification. For the small-scale preparation of extracts, a fresh stock in water (50 mg/ml) is prepared just before use. In our hands, frozen stocks are unreliable in the procedures described below.

Concentrated stocks of nuclease-free bovine serum albumin (BSA, fraction V; Roche, Nutley, NJ) and yeast RNA (Roche) are prepared as follows. BSA (2 g) is acetylated in a 100-ml reaction as described[14] and then dialyzed exhaustively against 10 m*M* Tris-HCl (pH 8.3), 1 m*M* EDTA, 10 m*M* NaCl, heated to 80° for 30 min, and stored in aliquots (4°). Yeast RNA (10 g) is suspended with stirring at room temperature in 100 ml of 25 m*M* Tris-HCl (pH 7.5), 120 m*M* sodium acetate, 5 m*M* EDTA, 1% (w/v) sodium dodecyl sulfate (SDS), extracted with phenol and then phenol–chloroform, and precipitated with 2 volumes of ethanol (−20°). The RNA is precipitated with ethanol three more times by suspending in 10 m*M* Tris-HCl (pH 7.5), 1 m*M* EDTA, 100 m*M* NaCl (TEN) and adding a 1/10 volume of 3 *M* sodium acetate and 2 volumes of ethanol, and then dialyzed exhaustively against TEN and stored in aliquots (−20°). The concentration is determined by UV absorption, using an E_{280} of 0.63 (BSA) or an E_{260} of 20 (yeast RNA) as the extinction coefficient for a 1-mg/ml solution.

Purification of RNA Degradosome

Solutions

Lysis buffer (LB): 50 m*M* Tris-HCl (pH 7.5), 200 m*M* NaCl, 5% (v/v) glycerol, 3 m*M* EDTA, 1 m*M* DTT, 1 m*M* PMSF, aprotinin (2 μg/ml), pepstatin A (0.8 μg/ml), leupeptin (0.8 μg/ml), lysozyme (1.5 mg/ml)

DNase buffer (DB): 50 m*M* Tris-HCl (pH 7.5), 200 m*M* NaCl, 5% (v/v) glycerol, 40 m*M* magnesium acetate, 1 m*M* DTT, 1 m*M* PMSF, aprotinin (2 μg/ml), pepstatin A (0.8 μg/ml), leupeptin (0.8 μg/ml), 0.8% (v/v) Triton X-100, DNase I (40 μg/ml)

Suspension buffer (SB): 50 m*M* Tris-HCl (pH 7.5), 200 m*M* NaCl, 10% (v/v) glycerol, 0.5% (v/v) Genapol X-080, 1 m*M* EDTA, 1 m*M* DTT, 1 m*M* PMSF, aprotinin (2 μg/ml), pepstatin A (0.8 μg/ml), leupeptin (0.8 μg/ml)

Buffer G (BG): 10 m*M* Tris-HCl (pH 7.5), 10% (v/v) glycerol, 0.5% (v/v) Genapol X-080, 1 m*M* EDTA, 1 m*M* DTT, 1 m*M* PMSF, aprotinin (2 μg/ml), pepstatin A (0.8 μg/ml), leupeptin (0.8 μg/ml), BG is used with NaCl and for certain steps it is supplemented with BSA (specified in the protocol)

[14] N. Gonzalez, J. Wiggs, and M. J. Chamberlin, *Arch. Biochem. Biophys.* **182,** 404 (1977).

These buffers are prepared just before use with stirring at room temperature and then chilled on ice. They should not be stored.

Procedure

All steps are performed on ice or in a cold room, as rapidly as possible. With an automated fast protein liquid chromatograph (FPLC) it should be possible to perform the SP-Sepharose chromatography the evening of the first day of the purification and set up the glycerol gradients to run overnight the second day. After low-speed centrifugation (S-30 fraction), the preparation can be held on ice for a few hours. For longer periods it should be stored frozen (−70°). On thawing, any precipitate that forms should be removed by low-speed centrifugation.

The procedure is compatible with 50 to 150 g of cells. The volumes indicated below are for 100 g of cells. In the initial steps, the proportion of buffer to cell mass is critical and the volumes should be adjusted carefully. For example, a cell mass of 92 g should be suspended in 276 ml of LB, followed by 138 ml of DB and 108 ml of 5 *M* ammonium chloride.

Lysis. The night before starting the purification, a tablet of cells (−70°) is moved to −20°. The following steps, which need not be done in a cold room, are performed by keeping the receptacle for a Waring blender chilled in an ice bath except when actually homogenizing the suspension. The frozen cells are transferred to a thick plastic bag, broken into pea-size pieces with a plastic mallet, and transferred to a 1-liter glass receptacle. Add 300 ml of LB; grind at low speed (18,000 rpm) for 1 min and then at high speed (22,000 rpm) for 1 min. Incubate for 30 min. Homogenize for 1 min at high speed. Add 150 ml of DB and homogenize for 1 min at low speed. Incubate for 30 min. Add 112.5 ml of 5 *M* ammonium chloride. Homogenize for 1 min at low speed. Incubate for 30 min.

30,000g Supernatant (S-30). Clarify the lysate by low-speed centrifugation in a Sorvall (Newtown, CT) GSA rotor (13,000 rpm, 4°, 1 hr).

200,000g supernatant (S-200). Prepare a high-speed supernatant by centrifugation in a Beckman (Fullerton, CA) Ti60 rotor (45,000 rpm, 4°, 2 hr). Divide the S-30 fraction into thick-walled polycarbonate centrifuge tubes (uncapped, 22 ml/tube). Because the volume of the S-30 fraction is usually 400 to 500 ml, access to three Ti60 rotors is required to treat the entire S-30 fraction at once. We prefer using uncapped tubes because in our hands it is simpler, faster, and cleaner than systems relying on capping or sealing. When decanting the high-speed supernatant, avoid the flocculent material that forms a soft layer above the hard, transparent pellet containing the ribosomes.

Ammonium Sulfate Precipitation (ASP-40). The S-200 fraction is fractionated by slowly adding ammonium sulfate (40% saturation, 0.226 g/ml) with constant stirring in an ice bath. After the last addition, continue stirring for another 20 min and then collect the precipitate by centrifugation in a Sorvall GSA rotor

(13,000 rpm, 4°, 20 min). Decant the supernatant and drain the tubes by inverting them on a paper towel. Suspend the pellets in SB (1/4 volume of the S-200 fraction). Because the ASP-40 fraction is applied directly to an ion-exchange column, it is important to properly drain the tubes and use the indicated volume of SB.

SP-Sepharose Chromatography (SP-Sph). Equilibrate a 24-ml column (16 × 120 mm) of SP-Sepharose (Pharmacia, Piscataway, NJ) with 50 ml of BG (250 m*M* NaCl) at a flow rate of 1 ml/min. Load the ASP-40 fraction at 0.2 ml/min and then wash with 20 ml of BG [400 m*M* NaCl, BSA (0.5 mg/ml)]. Elute with 30 ml of BG [600 m*M* NaCl, BSA (0.5 mg/ml)] at the same flow rate, collecting 0.8-ml fractions. A small peak of UV-absorbing material should elute with the step to 600 m*M* NaCl. The peak fractions can be stored individually or pooled for glycerol gradient sedimentation if further purification is necessary.

Glycerol Gradient Sedimentation (G-Grd). In a tube for a Beckman SW40 rotor, form a 10-ml linear 15–40% (v/v) glycerol gradient in BG (500 m*M* NaCl). Layer 3 ml of the SP-Sph fraction on top and separate by centrifugation (40,000 rpm, 4°, 16 hr). Collect from the bottom, taking 40 drops per fraction (approximately 26 fractions). Most of the degradosome should sediment in a broad band to near the middle of the tube. The lower molecular weight material, including the carrier BSA used in the SP-Sepharose chromatography, remains at the top. The peak fractions are pooled, aliquoted, and stored at −20° (short term) or −70° (long term). The aliquots can be repeatedly frozen and thawed with no apparent loss of RNase E activity.

Assay of RNase E Activity

The substrate used to assay RNase E activity has been described.[2] 9Sa RNA is a truncated form of 9S RNA, which is a precursor of 5S rRNA (Fig. 1). We use

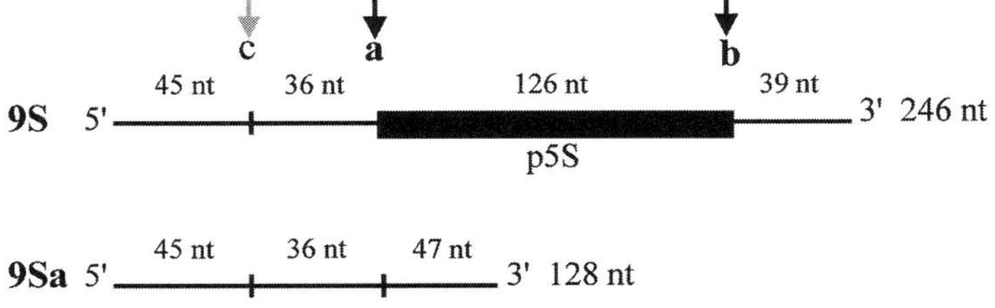

FIG. 1. The 9S precursor of 5S rRNA and the 9Sa derivative have both been used to measure RNase E activity. A 126-nucleotide precursor of 5S rRNA (p5S), which is 6 nucleotides longer then the mature 5S rRNA, is derived from the 9S precursor by RNase E cleavages at the a and b sites (solid arrows). Cleavage at the c site is detected at high concentrations of enzyme (shaded arrow). Pilot experiments showed that the rate of cleavage at the a site was comparable in comparisons between the 9S and 9Sa substrates. We prefer using the 9Sa substrate because the analysis of its digestion by denaturing gel electrophoresis is simpler than that of the 9S substrate.

the digestion the 9Sa RNA as a measure of RNase E activity, and define 1 unit of activity as the amount of enzyme needed for complete cleavage at the a site under the conditions described in this section. In this assay, we use a large excess of competitor yeast RNA (0.2 mg/ml, final). Under these conditions, the rate of cleavage depends on the concentration of the competitor RNA and is independent of the concentration of radioactive substrate. Although a unit as defined here is arbitrary (i.e., not based on a chemical rate), the assay detects activity in crude fractions and allows the comparison of activity between different steps in the purification. We use 30° as the standard temperature because it permits activity measurements of temperature-sensitive mutants of RNase E. The wild-type enzyme can be assayed at 37°.

Solutions

Enzyme buffer (EB): 10 m*M* Tris-HCl (pH 7.5), 500 m*M* NaCl, 5% (v/v) glycerol, 0.5% (v/v) Triton X-100, 1 m*M* EDTA, 1 m*M* DTT, BSA (0.5 mg/ml)

Substrate buffer (SB): 10 m*M* Tris-HCl (pH 7.5), 5 m*M* $MgCl_2$ yeast RNA (0.25 mg/ml)

Proteinase K (PK): 10 m*M* Tris-HCl (pH 7.5), 10 m*M* EDTA, 0.2% (w/v) SDS, proteinase K (0.2 mg/ml)

Formamide–urea–dye mix (FUD): To 20 ml of formamide (deionized) add with stirring at room temperature 5 g of urea (ultrapure), 0.25 ml of 0.5 *M* EDTA, 0.25 ml of 10× TBE, 0.125 ml of 1% (w/v) xylene cyanol, and 0.125 ml of 1% (w/v) bromphenol blue.

EB, SB, and PK are prepared just before use. The FUD mix is stored in aliquots at −20°.

Procedure

[^{32}P]UTP-labeled 9Sa RNA is synthesized by transcription with T7 RNA polymerase of a DNA template prepared by PCR amplification.[2] A detailed procedure is beyond the scope of this chapter. Sambrook *et al.*[13] should be consulted for a general description of these procedures.

The following steps are performed on ice unless otherwise noted. Enough substrate for all the reactions is prepared by diluting [^{32}P]UTP-labeled 9Sa RNA into SB (0.5–1.0 μCi/ml, final). Fractions from the large-scale purification are diluted 5-fold (lysate, S-30, S-200), 10-fold (ASP-40), or 20-fold (S-Sph, G-Grd) into EB. The enzyme is then further diluted in a series of 2-fold steps by mixing equal volumes of enzyme and EB, and changing the pipette tip after each dilution. Assemble the reactions on ice by mixing 2 μl of enzyme with 8 μl of substrate and then incubating at 30° for 30 min. Quench the reactions on ice, add 10 μl of PK, and incubate at 50° for 10 min. Cool the reactions to room temperature, add

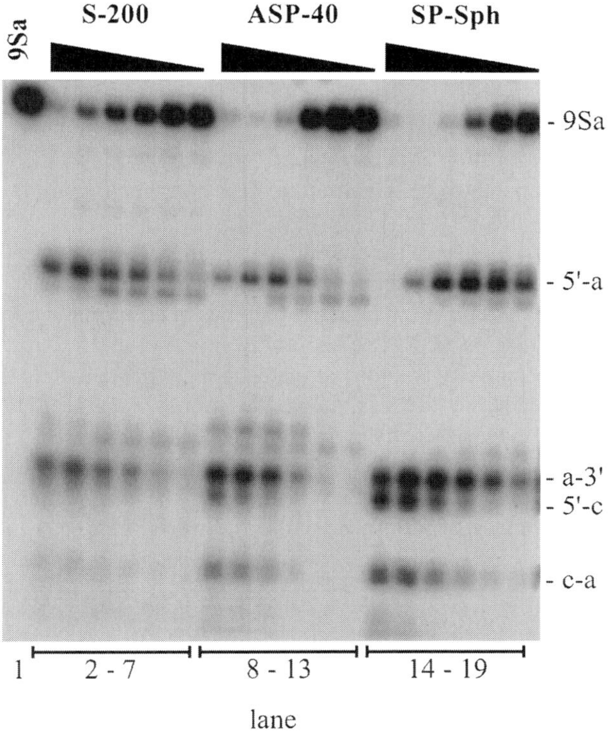

FIG. 2. Autoradiogram of a denaturing polyacrylamide gel [9% (w/v), 37.5 : 1, 7 *M* urea, 1× TBE showing the assay of RNase E activity in the S-200, ASP-40, and SP-Sph fractions from a large-scale purification of the degradosome (see Table I). The fractions were 2-fold serially diluted as described in text. The amount of total protein in the assay ranged from 6.4 to 0.20 μg (lanes 2 to 7), from 1.6 to 0.05 μg (lanes 8 to 13), and from 64 to 2 ng (lanes 14 to 19). Lane 1 is a control in which the 9Sa substrate was incubated under the same conditions and processed with the other samples. To the right is shown the position of the 9Sa substrate and the 5′-a, a-3′, 5′-c, and c-a products.

20 μl of FUD, heat at 85° for 5 min, and then quench on ice. A sample (20 μl) of each reaction is separated by electrophoresis on denaturing polyacrylamide gels. The gels are dried and exposed on X-ray film at −70° with an intensifying screen.

The autoradiograph in Fig. 2 shows the analysis of the S-200, ASP-40, and SP-Sph fractions from a large-scale purification. By careful visual inspection, it is possible to estimate 1 unit of activity. For example, in lanes 14–19 (SP-Sph fraction), partial digestion was observed in lanes 17–19 whereas in lane 16 the 9Sa substrate is completely cleaved at the a site and some of the 5′-a product is further digested by cleavage at the c site. Taking the amount of protein assayed in lane 16 (16 ng) as 1 unit of activity gives a specific activity of 62.5 units/μg of protein (62,500 units/mg). It is admittedly more difficult to make this determination with a crude fraction (S-200) because there is some undigested 9Sa substrate even at the highest concentration of protein. This leads to a systematic underestimation of RNase E activity that is apparently due to inhibition in the crude fractions. Whether the inhibitor(s) interferes directly with RNase E or acts indirectly, perhaps by sequestering the 9Sa substrate, is not known. An additional complication in assaying the crude fractions is that the yield of products does not appear to be quantitative. This could be due to degradation by other nucleases.

Table I shows an analysis of the amount of protein, activity, and specific activity in each step of a large-scale purification. Note that the activity increases between the lysate and ASP-40 fraction. This effect, which has been reproduced many times, is likely due to inhibition of RNase E activity in the crude fractions (discussed above). Because the activity is underestimated, we cannot calculate the yield and extent of purification in the initial fractions. We estimate that the yield of RNase E in the ASP-40 fraction is 40% on the basis of Western blot analysis. This gives an 8-fold enrichment in the ASP-40 fraction. The overall purification summarized in Table I shows a 3% yield with a 310-fold enrichment. These values seem reasonable given that the abundance of RNase E has been estimated as 1000 molecules per cell.[15]

TABLE I
RNase E ACTIVITY DURING PURIFICATION OF *Escherichia coli* RNA DEGRADOSOME[a]

Fraction	Protein (mg)	Activity (units)	Specific activity	Yield[b] (%)	Enrichment (fold)
Lysate	18,000	620,000	34	—	—
S-30	9,800	470,000	48	—	—
S-200	6,200	1,600,000	260	—	—
ASP-40	900	2,300,000	2,600	40	8
SP-Sph	9.2	670,000	73,000	12	230
G-Grd	1.8	180,000	100,000	3	310

[a] Preparation starts with 100 g of cells. The amount of protein was determined by the method of Lowry, using a BSA standard. The activity was determined as described in text. The specific activity is expressed as units per milligram of protein. The fractions are lysate, total lysate before centrifugation; S-30, 30,000*g* supernatant; S-200, 200,000*g* supernatant; ASP-40, ammonium sulfate pellet (40% saturation); SP-Sph; chromatography on SP-Sepharose; and G-Grd, sedimentation on a glycerol gradient. Data presented are a composite of two preparations because we rarely assayed the lysate and S-30 fractions during the later stages of developing this procedure.

[b] The yield in the S-30 and S-200 fractions cannot be determined because the activity is inhibited in the crude fractions (see text). The 40% yield in the ASP-40 fraction is an estimation based on Western blotting.

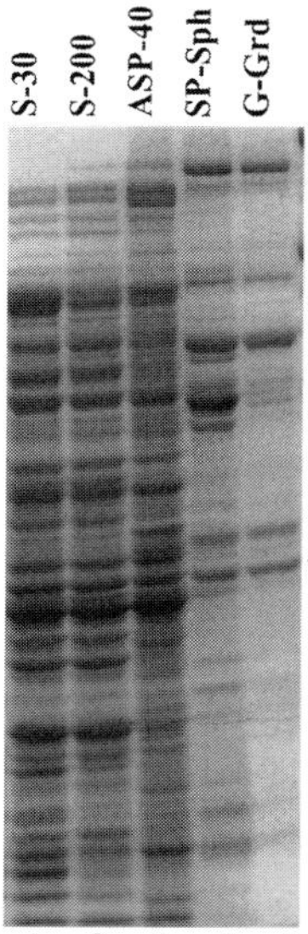

FIG. 3. An SDS–polyacrylamide gel (9%, w/v) stained with Coomassie blue showing protein from various fractions of the large-scale purification. The amount of protein loaded in lanes 1 to 5 was, respectively, 74, 56, 32, 9.2, and 4.8 μg. To the right is indicated the position of RNase E (Rne), a proteolysis product of RNase E (Rne*), PNPase (Pnp), RhlB, and enolase (Eno). Also indicated is a region of the gel where trace amounts of DnaK and polyphosphate kinase (Ppk) have been identified. RNase E is a 118-kDa protein that migrates at 180 kDa in SDS–PAGE because of the unusual amino acid composition of the C-terminal half of the protein. The molecular masses of the other major components by SDS–PAGE are PNPase (85 kDa), RhlB (50 kDa), and enolase (48 kDa).

Figure 3 shows an SDS–PAGE analysis of the fractions from a large-scale purification. To the right is shown the position of RNase E (Rne) and the three other major components of the RNA degradosome: Pnp (PNPase), RhlB, and Eno (enolase). Rne* is a proteolysis product in which part of the C-terminal region of the protein has been removed. RNase E is known to be sensitive to proteolysis.[2,16] Despite extensive precautions to minimize proteolysis, we have not been able to eliminate the Rne* product from the highly purified preparations of the RNA degradosome. Nevertheless, we believe that this proteolysis occurs during the purification because Rne* is not detected by Western blotting when total lysates are prepared by directly breaking the cells with heat in SDS–PAGE loading buffer.

[15] M. Kido, K. Yamanaka, T. Mitani, H. Niki, T. Ogura, and S. Hiraga, *J. Bacteriol.* **178,** 3917 (1996).
[16] E. A. Mudd and C. F. Higgins, *Mol. Microbiol.* **9,** 557 (1993).

Just below Pnp there are several proteins in the 60- to 70-kDa range that are present in trace amounts in the G-Grd fraction. Among these are DnaK and polyphosphate kinase (Ppk), which have been identified by protein sequencing and for which there is evidence suggesting a physical association with the RNA degradosome.[9,17]

In Fig. 3 RNase E, which migrates as one of the largest proteins in *E. coli,* is clearly visible in the ASP-40 fraction. The only other protein identified in this region of the gel is MukB, which is 6-fold less abundant than RNase E.[15] We have occasionally monitored the purification of the RNA degradosome, particularly during the later steps in the procedure, following RNase E by SDS–PAGE.

Preparation of Extracts on Small Scale

A partial purification of [^{35}S]methionine-labeled RNA degradosome was originally performed for immunoprecipitation experiments during the initial characterization of the RNA degradosome.[2] Variations of this protocol have been described in subsequent work.[9,15] We describe here our current procedure for the small-scale preparation of an ASP-60 fraction containing the RNA degradosome. The protocol is convenient for handling several cultures simultaneously. We use 60% ammonium sulfate to ensure that mutant forms of the RNA degradosome are efficiently precipitated. However, 40% ammonium sulfate is sufficient when working with wild-type cells.

Solutions

Lysis buffer (LB*): 50 m*M* Tris-HCl (pH 7.5), 100 m*M* NaCl, 5% (v/v) glycerol, 3 m*M* EDTA, 1 m*M* DTT, 1 m*M* PMSF, aprotinin (2 μg/ml), pepstatin A (0.8 μg/ml), leupeptin (0.8 μg/ml), lysozyme (1.5 mg/ml)

DNase buffer (DB*): 50 m*M* Tris-HCl (pH 7.5), 100 m*M* NaCl, 5% (v/v) glycerol, 30 m*M* magnesium acetate, 1 m*M* DTT, 1 m*M* PMSF, aprotinin (2 μg/ml), pepstatin A (0.8 μg/ml), leupeptin (0.8 μg/ml), 0.4% (v/v) Genapol X-080, DNase I (20 μg/ml)

Buffer G (BG*): 10 m*M* Tris-HCl (pH 7.5), 5% (v/v) glycerol, 0.5% (v/v) Genapol X-100, 1 m*M* EDTA, 1 m*M* DTT, 1 m*M* PMSF, aprotinin (2 μg/ml), pepstatin A (0.8 μg/ml), leupeptin (0.8 μg/ml)

Procedure

All steps should be performed on ice or in a cold room unless otherwise indicated. A 100-ml culture of *E. coli* is grown to the end of log phase and quenched on ice. The cells are collected by centrifugation and washed in 10 ml of 50 m*M* Tris-HCl, pH 7.5. Carefully drain the tube and weight it to determine the cell mass

[17] E. Blum, B. Py, A. J. Carpousis, and C. F. Higgins, *Mol. Microbiol.* **26,** 387 (1997).

(by subtracting the weight of the empty tube). As in the large-scale procedure, the proportion of buffer volume to cell mass is important. The volumes indicated here are for 100 mg of cells. The actual volumes should be adjusted in proportion to the cell mass as described above. Suspend the cells in 1 ml of LB*. Freeze at −70°. The preparation can be stored indefinitely at this point. Thaw on ice. The preparation should become viscous. Add 0.5 ml of DB* and incubate at 4° with gentle rocking for 30 min. Check that the viscosity has been reduced. If not, continue the incubation for another 15 min. Add 0.375 ml of 5 *M* ammonium chloride and continue with gentle rocking for another 30 min. The following centrifugation steps are performed with a Beckman TLA100.3 rotor. Prepare a 30,000*g* supernatant (24,000 rpm, 4°, 1 hr), and then a 200,000*g* supernatant (61,000 rpm, 4°, 1 hr). Add ammonium sulfate (60% saturation, 0.361 g/ml) and mix by inversion (manually) until the salt dissolves, and then incubate with gentle rocking for another 15 min. Collect the precipitate by centrifugation (24,000 rpm, 4°, 30 min). Carefully decant the supernatant and drain the tube thoroughly by inverting it on a paper towel. Suspend the pellet with 0.30 ml of GB* (300 m*M* NaCl). Store at −20° (short term) or −70° (long term).

General Comments

Proteolysis

The proteolysis of RNase E during the purification is a significant problem. The N-terminal and C-terminal halves of the protein are separated by a long proline-rich region and there are other proline-rich stretches in the C-terminal half of the protein. We believe that these regions, which could serve as "hinges" between different domains, render RNase E particularly sensitive to proteolysis during purification. It should be noted that proteolytic fragments containing the N-terminal catalytic domain are active.[2,16] Several *E. coli* ribonuclease activities that have been described previously, such as RNase F, RNase K, and RNase N, could be degradation products of RNase E.[18] This is now believed to be the case for RNase K.[19] However, this conclusion needs to be reevaluated in light of the characterization of RNase G (CafA), an *E. coli* endoribonuclease that is a homolog of RNase E.[20,21]

Detergent

We initially developed the lysis step with Triton X-100, and then switched to Genapol X-080 in the latter steps of the purification because Triton, which

[18] C. P. Ehretsmann, A. J. Carpousis, and H. M. Krisch, *FASEB J.* **6,** 3186 (1992).

[19] U. Lundberg, O. Melefors, B. Sohlberg, D. Georgellis, and A. von Gabain, *Mol. Microbiol.* **17,** 595 (1995).

[20] Z. Li, S. Pandit, and M. P. Deutscher, *EMBO J.* **18,** 2878 (1999).

[21] M. R. Tock, A. P. Walsh, G. Carroll, and K. J. McDowall, *J. Biol. Chem.* **275,** 8726 (2000).

absorbs strongly in the UV, is incompatible with monitoring column chromatography steps. Genapol has properties similar to Triton but does not absorb in the UV. In the preparation of the lysate, the proportion of buffer volume to cell mass is critical because of the use of nonionic detergent. On occasion, during the ammonium sulfate fractionation, the precipitate floats instead of sediments. When this happens, we stop the purification and begin again. This problem is probably due to a high detergent-to-protein ratio, possibly because of inefficient lysis. In our hands, Genapol seems less prone to interference with the ammonium sulfate precipitation. The protocol for small-scale extracts has been optimized with this detergent.

Ionic Strength

For both the purification of the RNA degradosome and the preparation of extracts, we add ammonium chloride (1 M, final) to the lysate before centrifugation. This step is important for the efficient extraction of RNase E. At lower ionic strength, RNase E pellets with the fast sedimenting material even in the presence of detergent. The loss in low salt suggests an ionic interaction with components of the cell wall or the ribosomes. This is perhaps not surprising considering that the C-terminal half of RNase E contains several highly charged regions.

Carrier Bovine Serum Albumin

In the SP-Sepharose step we wash the column and elute the protein in a buffer containing BSA (0.5 mg/ml) to reduce the loss of activity that is often seen when the protein concentration drops below 3 mg/ml during chromatography. BSA, prepared as described above, is a relatively inexpensive carrier that is compatible with a variety of chromatography media. We recommend its use whenever possible.

Stability

In numerous trials, we attempted to optimize stability. RNase E purified as described here can be heated in the enzyme buffer (EB) to 45° for 10 min with little or no loss of activity. It can also withstand large dilutions. We believe that the high salt (500 m*M* NaCl), the detergent [0.5% (v/v) Triton X-100], and the carrier BSA (0.5 mg/ml) are important components of the EB. A variety of chromatographic media such as DEAE-cellulose, hydroxyapatite, and heparin–agarose were abandoned because their use resulted in unacceptable losses of activity.

Proteinase K

In the assay of RNase E activity, the samples are prepared for denaturing gel electrophoresis by degrading the protein with proteinase K. In our hands, this step is essential because protein interferes with the electrophoresis by retarding the

RNA. This effect is observed even with highly purified degradosome, suggesting that the interference is due to one of the components of the complex. RNase E, PNPase, and RhlB are all known to bind to RNA. An alternative to proteinase K is to extract the samples with phenol and precipitate with ethanol, but this is more tedious and it increases error due to losses during the treatment.

Rethinking the Procedure

Polyethyleneimine

Many researchers will find the preparation of supernatants by high-speed centrifugation tedious, particularly with large volumes. The major component removed by high-speed centrifugation is the ribosome. It is possible to use polyethyleneimine (PEI) as an alternate treatment to remove bulk nucleic acid, including the ribosomes. In fact, we have developed a small-scale procedure with a PEI precipitation step that was used for immunoprecipitation experiments.[9] Nevertheless, we are concerned that trace amounts of PEI could interfere with the activity of the degradosome. Thus, we recommend caution when using PEI, especially in preparations for the study of enzymatic activity.

Gel-Permeation Chromatography

The other cumbersome procedure is the use of glycerol gradient centrifugation in the last step of the purification. There has been a report in which this step was replaced by gel-permeation chromatography on Bio-Gel A-1.5m (Bio-Rad, Hercules, CA).[22] We have made several trials using Sephacryl S-300 HR. In principle, gel-permeation chromatography in a high ionic strength buffer containing detergent is an interesting alternative to the glycerol gradients. However, precautions should be taken to prevent dilution, which can lead to the inactivation of RNase E. In our hands, the use of carrier BSA during gel-permeation chromatography helps to maintain activity.

Acknowledgments

Early work on the purification of RNase E was supported by the Swiss National Science Foundation (31-30936.91). A. J. C. thanks R. E. Epstein for encouragement during work at the University of Geneva, and G. Van Houwe for technical assistance. Our research in Toulouse is supported by the Centre National de la Recherche Scientifique (CNRS), with additional funding from the European Union, the Cancer Research Association (ARC), the Midi Pyrénées Region, and the Fundamental Microbiology Program of the Ministry of Education (MENRT).

[22] G. A. Coburn and G. A. Mackie, *J. Mol. Biol.* **279,** 1061 (1998).

[28] Preparation of *Escherichia coli* Rne Protein and Reconstitution of RNA Degradosome

By GEORGE A. MACKIE, GLEN A. COBURN, XIN MIAO, DOUGLAS J. BRIANT, and ANNIE PRUD'HOMME-GENEREUX

Introduction

The endoribonuclease RNase E was discovered and characterized by D. Apirion and colleagues.[1,2] It was initially described as an activity required for the penultimate step in the maturation of 5S rRNA, the processing of a 9S precursor to a 126-residue pre-5S RNA. It has since emerged as a major player in both rRNA processing and mRNA decay.[3–5] RNase E activity is readily demonstrated in crude extracts, but purification and characterization of the enzyme proved unexpectedly difficult.[5] Two approaches to this challenge ultimately succeeded. In the first, described by Carpousis *et al.* (see [27] in this volume[5a]), RNase E activity was purified as part of a larger complex, the RNA degradosome.[6–8] In the second, the complete *rne* gene was cloned and sequenced, permitting overexpression of its product, the Rne protein.[9] The Rne protein is surprisingly large (1061 amino acid residues) and acidic (p*I* 5.4). In purified form, it manifests endonucleolytic activity identical to that of crude RNase E in the absence of other components of the RNA degradosome.[9]

The discovery that a number of key components of the RNA-processing and decay apparatus are organized into a multicomponent complex, the RNA degradosome,[6–8] begs the question of the function of the individual components. One avenue of investigation relies on reconstitution of the RNA degradosome from purified components.[10] Ultimately, this should permit the dissection of the roles of the individual components and the assembly *in vitro* of complexes prepared from mutant proteins. This chapter describes methods for the purification of Rne, the key scaffold in the assembly process, and for the reconstitution of active degradosomes from individually purified enzymes.

[1] B. K. Ghora and D. Apirion, *Cell* **15,** 1055 (1978).

[2] T. K. Misra and D. Apirion, *J. Biol. Chem.* **254,** 11154 (1979).

[3] J. G. Belasco, *In* "Control of Messenger RNA Stability" (J. G. Belasco and G. Brawerman, eds.), pp. 3–12. Academic Press, San Diego, California, 1993.

[4] O. Melefors, U. Lundberg, and A. von Gabain, *In* "Control of Messenger RNA Stability" (J. G. Belasco and G. Brawerman, eds.), pp. 53–70. Academic Press, San Diego, California, 1993.

[5] G. A. Coburn and G. A. Mackie, *Prog. Nucleic Acids Res. Mol. Biol.* **62,** 55 (1999).

[5a] A. J. Carpousis, A. Leroy, N. Vanzo, and V. Khemini, *Methods Enzymol.* **342,** [27] 2001 (this volume).

[6] A. J. Carpousis, G. Van Houwe, C. Ehretsmann, and H. M. Krisch, *Cell* **76,** 889 (1994).

0076-6879/01 $35.00

Preparation of Crude RNase E

Buffers and Solutions

M9ZB (per liter): Combine 10 g of BDH (Poole, UK) peptone from casein [Difco (Detroit, MI) Bacto-tryptone or Humco Sheffield (Texarkana, TX) NZ amine can be substituted], 6 g of Na_2PO_4, 3 g of KH_2PO_4, and 1 g of NH_4Cl; sterilize by autoclaving and supplement with sterile $MgSO_4$ to 1 m*M* and $CaCl_2$ to 0.1 m*M*

NZYCM (per liter): Combine 10 g of BDH peptone (or equivalent), 5 g of yeast extract, and 2.5 g of Difco Casamino Acids; sterilize by autoclaving and add sterile $MgSO_4$ to 1 m*M*

Isopropyl-β-D-thiogalactopyranoside (IPTG): 0.1 *M* solution prepared by filter sterilization

Buffer A: Combine 50 m*M* Tris-HCl, 10 m*M* $MgCl_2$, 60 m*M* NH_4Cl, and 0.5 m*M* EDTA; adjust to pH 7.8 with HCl sterilize by autoclaving, and supplement with either 6 m*M* 2-mercaptoethanol or 0.5 m*M* dithiothreitol (DTT) and with protease inhibitors (see text)

Buffer D: Combine 20 m*M* Tris-HCl, 1 m*M* $MgCl_2$, 20 m*M* NH_4Cl, and 0.1 m*M* EDTA; adjust to pH 8.0, sterilize by autoclaving, and supplement with DTT, glycerol, and protease inhibitors as above

Buffer G: Combine 50 m*M* Tris-HCl (pH 8.0), 6 *M* guanidine hydrochloride, 150 m*M* NaCl, 20% (v/v) Glycerol, and 0.1 m*M* EDTA; supplement with DTT to 1 m*M*

Buffer R: Combine 50 m*M* Tris-HCl, pH 8.0 (or 50 m*M* HEPES–KOH, pH 7.6), 150 m*M* NaCl, 20% (v/v) Glycerol, and 0.1 m*M* EDTA; supplement with DTT to 1 m*M*

Reconstitution buffer: 20 m*M* Tris-HCl (pH 7.5), 1 m*M* $MgCl_2$, 20 m*M* KCl, 1.5 m*M* DTT, 10 m*M* sodium phosphate, pH 7.5 (optional), 3 m*M* ATP (neutralized) (optional)

Formamide–dye buffer: Deionized formamide, 0.5× TBE buffer (see below), 0.01% (w/v) xylene cyanol FF, 0.01% (w/v) bromphenol blue

Sodium dodecyl sulfate (SDS) running buffer: 50 m*M* Tris base, 190 m*M* Glycine, 0.5 m*M* neutralized EDTA, 0.1% (w/v) SDS

TBE running buffer: 90 m*M* Tris base, 90 m*M* boric acid, 2 m*M* EDTA (disodium salt)

[7] B. Py, C. F. Higgins, H. M. Krisch, and A. J. Carpousis, *Nature* (*London*) **381,** 169 (1996).
[8] A. Miczak, V. R. Kaberdin, C.-L. Wei, and S. Lin-Chao, *Proc. Natl. Acad. Sci. U.S.A.* **93,** 3865 (1996).
[9] R. S. Cormack, J. L. Genereaux, and G. A. Mackie, *Proc. Natl. Acad. Sci. U.S.A.* **90,** 9006 (1993).
[10] G. A. Coburn, X. Miao, D. J. Briant, and G. A. Mackie, *Genes Dev.* **13,** 2594 (1999).

Growth of Cultures

The following applies to recombinant strains. To ensure the retention of plasmids, starter cultures must not be overgrown. A loopful or single colony of the appropriate strain (e.g., strain GM402 [BL21(DE3) containing pGM102][9]) is inoculated into 10 ml of supplemented M9ZB medium containing ampicillin (50 mg/liter) or carbenicillin (25 mg/liter). Two serial 100-fold dilutions are made from this culture into identical 10-ml aliquots of the same medium. Cultures are grown overnight with shaking at 30°.

A 1 : 100 dilution of a turbid, but not saturated, overnight culture is used to initiate growth of larger cultures in supplemented M9ZB medium containing carbenicillin. It is critical to maintain vigorous aeration during growth. For this reason we use the largest flasks available and add no more than one-tenth its nominal volume (e.g., 200 ml of culture in a 2-liter flask). Cultures are grown at 29–30° with vigorous shaking. Growth is followed by turbidity (e.g., absorbance at 600 nm). When cultures reach an A_{600} of 0.4, IPTG is added to 1 m*M*. After 15 min, the cultures are diluted with an equal volume of warm NZYCM containing carbenicillin (50 mg/liter) and growth is continued at 29–30° with vigorous shaking for up to 5 hr.

At harvest, each flask of culture is swirled for 1 min in a slurry of ice and water to ensure rapid chilling. All subsequent manipulations are performed in a cold room or at 4°. Cultures are transferred to cold, tared centrifuge bottles and the cells are harvested by centrifugation for 15 min at 5000 rpm in a Beckman (Fullerton, CA) JA-10 rotor or its equivalent. The supernatants are discarded and the cell pellets are completely suspended with one-fourth the original culture volume of cold buffer A containing 6 m*M* 2-mercaptoethanol. The suspension is pooled in a single bottle and the cells are harvested by centrifugation as described above. The supernatant is discarded and any remaining liquid is carefully wiped from the neck of the bottle. The pellet is weighed (yields are about 4–5 g/liter) and is suspended with 3.5 ml/g wet weight of buffer A containing 0.1 m*M* DTT and 7.5% (v/v) glycerol. Care should be taken to obtain a smooth suspension without excessive shearing. The suspended cells can be processed further or frozen quickly and stored at −70°. Frozen cells should be processed within a few days.

Preparation of S-30 Extracts

If frozen, cell suspensions are thawed in cold water in a cold room and then transferred to a chilled Aminco (SLM Aminco, Rochester, NY) French pressure cell. The suspension is passed twice through the cell at a pressure of 8000 lb/in^2. Higher pressures appear to cause the loss of RNase E activity. The lysate is supplemented with fresh DTT to 0.1 m*M*, with DNase I (1–2 U/ml; Sigma, St. Louis, MO), and with a 1-μg/ml concentration of each of the following protease inhibitors: leupeptin, aprotinin, and pepstatin (all obtained from Sigma). Phenylmethylsulfonyl fluoride (PMSF; Sigma-Aldrich, Milwaukee, WI) can also be added but is relatively ineffective in the buffers used. The lysate is left on ice for 10 min, and then clarified by centrifugation for 45 min at 15,000 rpm in a Beckman JA-20 rotor or equivalent. The supernatant (S-30) is saved and the volume is determined. The pellet is discarded. Inclusion bodies, a problem with some truncated forms of the Rne protein, will be in this pellet if formed.

Preparation of AS-26 Fraction

The clarified crude extract is diluted 4-fold with buffer A containing 0.1 m*M* DTT and 7.5% (v/v) glycerol and the cocktail of protease inhibitors enumerated above. While the lysate is stirred on ice, enzyme-grade $(NH_4)_2SO_4$ is added gradually over a period of 15 min to reach a concentration of 26% (w/v) or about 45% of saturation. It is not necessary to adjust the pH. After stirring for 15 min to ensure equilibration, the suspension is centrifuged for 20 min at 12,000 rpm in a Beckman JA-20 rotor. The supernatant is decanted and the pellet is suspended with 1 ml of buffer A supplemented with DTT and protease inhibitors as described above for each gram (wet weight) of the original cell pellet.

The suspended ammonium sulfate pellet is dialyzed at 4° against two changes of 100 volumes of buffer D containing 60 m*M* NH_4Cl, 0.1 m*M* DTT, and 10% (v/v) glycerol for 4–5 hr in total. The dialyzed material, the AS-26 fraction, is divided into suitable portions (e.g., 1 ml) and quick frozen before storage at −70°. A small portion should be retained for the determination of protein concentration (typically 4–7 mg/ml), for assaying RNase E activity, and for examining the quality of the preparation and the level of expression of Rne by SDS polyacrylamide gel electrophoresis (PAGE). Characterization of this preparation has been described.[11]

Purification of Rne

Preparative Gel Electrophoresis

The following is an expansion of a published method.[9] We typically employ a preparative slab gel (GIBCO-BRL, Gaithersburg, MD) with interior dimensions of 170 × 150 × 1.5 mm. The lower separating gel contains 5.5% (w/v) acrylamide (49 : 1 acrylamide : bisacrylamide), 0.375 *M* Tris-HCl (pH 8.8), 5% (v/v) glycerol, 0.1% (w/v) SDS. The upper stacking gel contains 4.5% (w/v) acrylamide (49 : 1 acrylamide : bisacrylamide), 0.06 *M* Tris-HCl (pH 6.8), 0.5 m*M* neutralized EDTA and 0.1% (w/v) SDS. The running buffer is described above. A sample containing 3 mg of the AS-26 fraction is diluted with an equal volume of 120 m*M* Tris-HCl (pH 6.8), 3% (w/v) SDS, 100 m*M* DTT, 0.5 m*M* EDTA, 10% (v/v) glycerol, and traces of bromphenol blue. The sample is boiled for 5 min and applied to the gel. The initial voltage is set at 85 V until the tracking dye has entered the separating gel; the voltage is then raised to 150–180 V and the run is continued until the dye has exited the gel for 10 min. After the apparatus is disassembled, a vertical

[11] G. A. Mackie, *J. Bacteriol.* **173,** 2488 (1991).

slice is removed from one or both edges of the gel and stained with 0.1% (w/v) Coomassie blue in H_2O. Staining in an acidic fixing solution is also possible, but there is significant shrinking and swelling of the gel, making subsequent alignment difficult. The remainder of the gel is left adhering to one of the plates, wrapped with cellophane (e.g., Saran Wrap), and stored at 4°. After destaining, the guide strip is used to identify the position of the Rne band. We find it helpful to lay the plate with the gel on a sheet of finely ruled paper (e.g., graph paper) in order to align the guide strip and the main gel. The region of the main gel containing the Rne band is excised with a razor blade and cut into small cubes no more than 2 mm in any dimension.

Electroelution of the Rne protein can be accomplished in several ways. We have used a Bio-Rad (Hercules, CA) electroelution apparatus (model 422) with success. Before use the glass tubes and fritted disks are autoclaved in H_2O previously treated with 0.1% (w/v) diethyl pyrocarbonate (Sigma-Aldrich) while the rubber bungs are soaked in 0.1% (w/v) SDS at 60° for 60 min and rinsed with sterile H_2O. The dialysis caps are pretreated as recommended by the manufacturer. In this method, the cubed gel pieces are placed in a tube to a depth of about 1 cm above a fritted disk. The tube is filled with running buffer (as described above) and placed in the Bio-Rad apparatus. Elution is performed at 9–10 mA per tube for up to 7 hr at ambient temperature. The lower reservoir is stirred and buffer is continuously recirculated with a peristaltic pump, taking precautions not to short-circuit the current path. Bubbles tend to accumulate under the bottom of the dialysis caps. They can be removed with a stream of buffer delivered carefully from a curved disposable pipette.

After electroelution, the glass tube and frit are slipped carefully from the rubber bung and dialysis cap. This will cause foaming of the eluate, which can be suppressed with 2–5 μl of 2-butanol. The eluates are removed with a pipettor and transferred to a Corex centrifuge tube. The dialysis caps are rinsed sequentially with 0.3 ml of running buffer, which is pooled with the eluates. DTT is added to 1 m*M* and the eluted protein is precipitated by the addition of 4.5 volumes of acetone. SDS will precipitate immediately, but the suspension is left at −15°. The precipitated material is recovered by centrifugation at 12,000 rpm for 60 min in a Beckman JA-20 rotor. The supernatant is removed and the pellet is washed three times with a total of 20 ml of 80% (v/v) acetone containing 1 m*M* DTT. The final pellet is "dried" by allowing any acetone to evaporate at ambient temperature for 10 min.

Renaturation of Rne

The pellet containing Rne (as an SDS salt) is dissolved at room temperature in a minimal volume of buffer G supplemented with 1 m*M* DTT. Ideally, to avoid excessive dilution, no more than 0.5 ml of buffer should be used. This material is warmed to 37° for up to 60 min. In the original method, the dissolved material is then diluted 20- to 40-fold in buffer R supplemented with 1 m*M* DTT, 0.1 m*M* PMSF, and a 1-μg/ml concentration each of leupeptin, aprotinin and pepstatin.

The intent is to reduce the concentration of guanidine hydrochloride rapidly to less than 0.25 *M* in order to promote refolding of the denatured protein.[12] As an alternative, we have dissolved the Rne–SDS pellet with 1.0 ml of buffer G, warmed it as described above, and diluted this 3-fold with buffer R so that the guanidine hydrochloride concentration is reduced to 1.5 *M*. Refolding is achieved during dialysis (below). In either case, the diluted Rne is dialyzed against two changes of 100 volumes each of buffer D containing 60 m*M* NH_4Cl, 0.1 m*M* DTT, and 10% (v/v) glycerol at 4° for at least 20 hr in total. The dialyzed material is concentrated 10- to 20-fold in a centrifugal concentrator [e.g., a Millipore (Acton, MA) Centricon-30]. The quality of the eluted protein and its approximate concentration can be determined by analysis on analytical SDS–PAGE, using proteins of known concentration [e.g., bovine serum albumin (BSA) or β-galactosidase] as standards. Examples for two different preparations have been published.[9,10] This method has also been applied successfully to soluble deletions of Rne.[10] Methods for assaying RNase E activity[9,11] are described elsewhere.

Nondenaturing Purification of Rne

Virtually all of the more recent strategies employ overexpressing strains to facilitate purification, in some cases with purification "tags."[8] Taraseviciene and co-workers[13] prepared extracts from an overexpressing strain by sonication and fractionated the crude extract by centrifugation at 200,000*g* in the presence of 0.4 *M* $(NH_4)_2SO_4$ (S-200). This step presumably releases Rne from ribosomes or membrane fragments. Enrichment of the Rne protein in the S-200 was achieved by precipitation with $(NH_4)_2SO_4$ to 40% of saturation. Protein in the pellet was fractionated on a Toyopearl gel-filtration column. Subsequent immunoaffinity purification on tandem Sepharose columns charged with nonimmune and anti-Rne antisera, respectively, yielded a preparation of Rne that visually appears to be >60% homogeneous. Some of the smaller protein species present in the purified material may represent proteolytic breakdown products. This procedure did result in a significant loss of activity, however, mostly in the immunoaffinity step.

We have developed an independent alternative to the foregoing procedure. The principal difficulty is that only a portion, about 40%, of the overexpressed Rne protein behaves as a "soluble" protein, while the majority is either aggregated or in a complex. Cultures of GM402 are grown and extracts prepared as described above. The crude extract is made 0.5 *M* in NH_4Cl and centrifuged for 60 min at 200,000*g* to release Rne,[6] which is concentrated by precipitation with $(NH_4)_2SO_4$ to 40% of saturation. This material is "soluble," but elutes from a Bio-Gel A 1.5M column in the flowthrough, ahead of β-galactosidase and RNA polymerase. Better results, however, are obtained by passing material from the ammonium sulfate pellet over

[12] D. A. Hager and R. R. Burgess, *Anal. Biochem.* **109,** 76 (1980).

[13] L. Taraseviciene, S. Naureckiene, and B. E. Uhlin, *J. Biol. Chem.* **269,** 12167 (1994).

a Hi-Trap Affi-Gel Blue column (Pharmacia Amersham Biotech, Piscataway, NJ). Rne is applied to this column in 10 m*M* HEPES, 1 m*M* EDTA, 2.5% (v/v) glycerol, 0.25% (v/v) Genapol X-080, 0.1 m*M* DTT, 0.1 m*M* PMSF, pH 7.6. Rne is retained on the column and is eluted with a linear gradient of KCl (0–3 *M* in the loading buffer) at about 2 *M* KCl. Although not homogeneous, this fraction is almost 8-fold more active than the ammonium sulfate pellet from which it is derived.

A significant portion of the Rne in the crude extract, up to 60%, is not soluble after extraction with NH_4Cl and remains in the 200,000*g* pellet unless a detergent such as 0.25% (v/v) Genapol X-080 (Fluka, Ronkonkoma) is included in the extraction buffer. The 200,000*g* supernatant in such an example is also fractionated with ammonium sulfate and passed over an Affi-Gel Blue column as described above. Roughly half the applied Rne flows through this column and half is retained, eluting at about 2 *M* KCl as described above. The yield of Rne can be increased by applying the material in the flowthrough to a second Affi-Gel Blue column. The major contaminants in the fractions retained on Affi-Gel Blue are of low molecular weight and presumably could be removed by gel filtration.

Reconstitution of RNA Degradosome

Purification of Components

The degradosome contains roughly stoichiometric quantities of Rne (likely a dimer), polynucleotide phosphorylase (PNPase; minimally a trimer, but possibly in two copies), enolase (likely a dimer), and RhlB (unknown subunit structure).[7,8,15a] It contains lesser amounts of other proteins, including polyphosphate kinase.[14] The purification of Rne is described above. Purifications of PNPase[15] and RhlB[10] from recombinant strains overexpressing these enzymes have also been described. Purification tags are not employed to facilitate purification and the effect of their presence on reconstitution has not been determined. A purification of enolase has been developed in the laboratory of B. Luisi (Department of Biochemistry, Cambridge University, Cambridge, UK).

Reconstitution

A method for reconstituting an active "minimal" degradosome has been described.[10] In outline, incubations contain a suitable buffer, an RNA substrate, ATP and/or sodium phosphate, and recombinant proteins. This mixture is incubated and samples are withdrawn to assay the disappearance of substrate. Alternatively,

[14] E. Blum, B. Py, A. J. Carpousis, and C. F. Higgins, *Mol. Microbiol.* **26,** 387 (1997).

[15] G. A. Coburn and G. A. Mackie, *J. Mol. Biol.* **279,** 1061 (1998).

[15a] G.-G. Liou, W.-N. Jane, S. N. Cohen, N.-S. Lin, and S. Lin-Chao, *Proc. Natl. Sci. U.S.A.* **98,** 63 (2001).

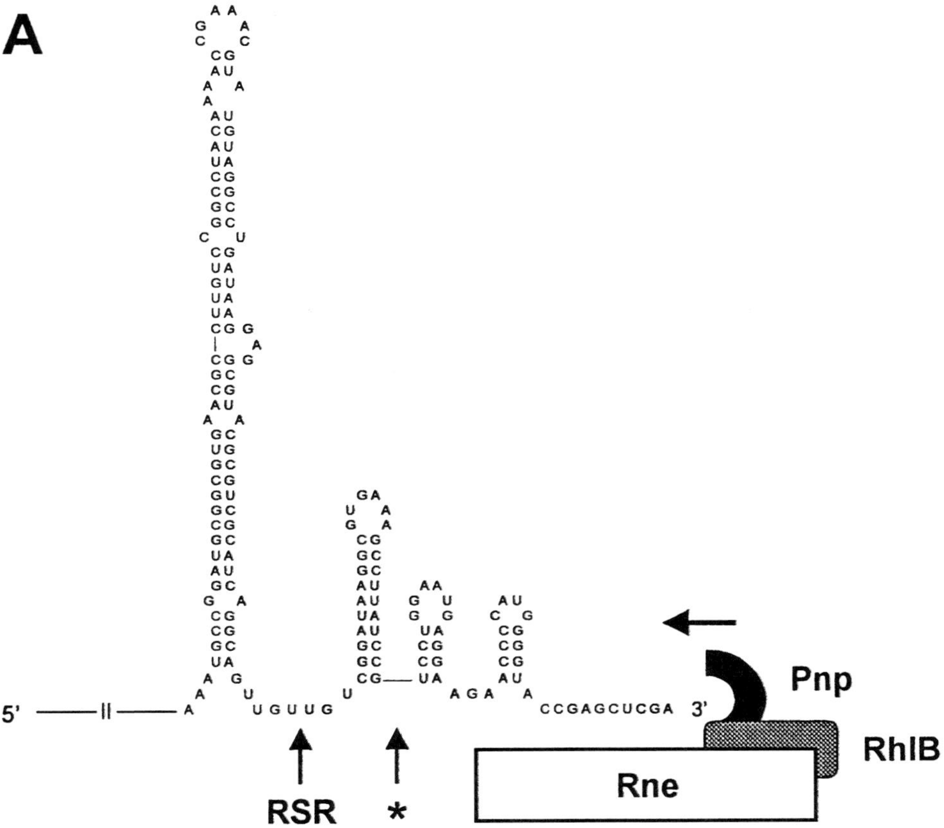

FIG. 1. The secondary structures of substrates for assaying degradosome activity. (A) The 375-nucleotide *malEF* RNA from the maltose operon.[7,10,17] The 3′ attack of a minimal degradosome is shown schematically (Rne, open rectangle; PNPase, filled semicircle; RhlB, shaded rectangle). RSR, REP (repetitive extragenic palindrome)-stabilized RNA (due to stalling of PNPase at the base of the stable hairpin);*, an intermediate stalled product.[10] (B) The 210-nucleotide *rpsT*/180–poly(A)$_{30}$ RNA.[10,18] The attack of a minimal degradosome on the 3′-poly(A) extension is shown schematically as in (A).

the formation of protein–protein complexes can be assessed by coimmunoprecipitation.

The standard assay for activity following reconstitution is based on that described by Py *et al.*[7] The buffer for reconstitution (see above) contains much lower ionic strengths than the standard buffer for assaying RNase E activity.[11] Sodium phosphate and ATP are not necessary for the physical reconstitution of complete or partial degradosomes. The assay of coupled RNA helicase–polynucleotide

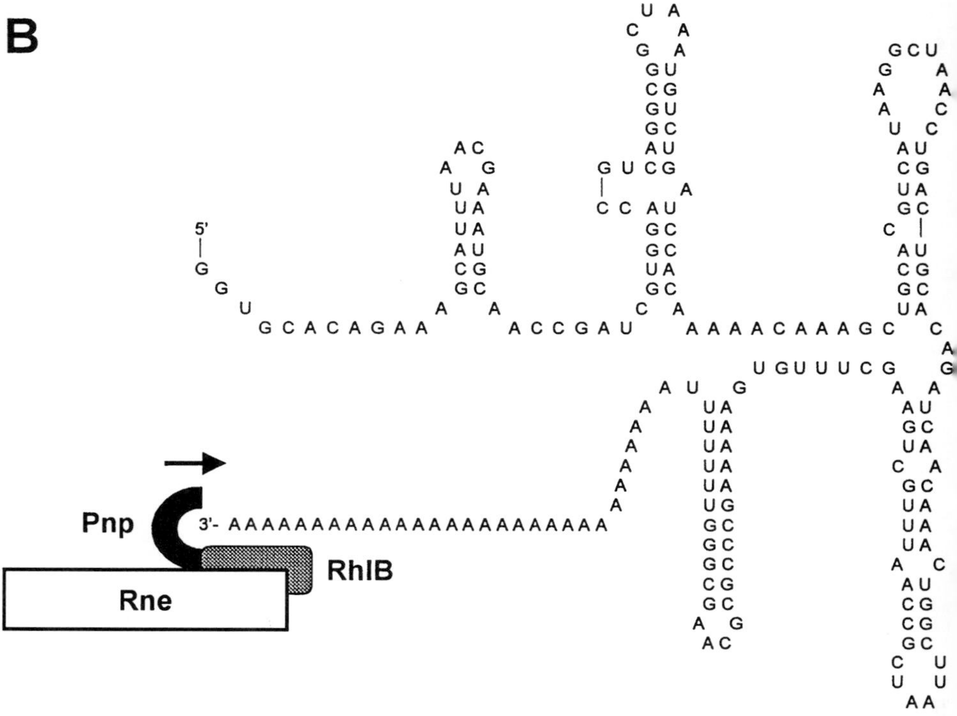

FIG. 1. (*continued*)

phosphorylase activity requires that substrates possess a single-stranded 3′ extension.[10,16] Figure 1[7,10,17,18] illustrates two suitable substrates: (Fig. 1A) a 375-nucleotide RNA containing the *malEF* intercistronic region[7,10] or (Fig. 1B) a 210-nucleotide RNA containing 180 nucleotides from the 3′ end of the *rpsT* mRNA extended by a 30-nucleotide poly (A) tail [*rpsT*/180–poly$(A)_{30}$].[10] Methods for preparing these RNAs have been described.[11] A typical reconstitution (40 μl) is assembled at 4° in a microcentrifuge tube by adding water, buffer, and cofactors first, followed by the RNA substrate to a final concentration of 20 n*M*. Proteins are added to the incubation in the following amounts per 40 μl: Rne, 30 ng (∼250 fmol of monomer); PNPase, 60 ng (∼250 fmol of trimer); RhlB, 30 ng (∼300 fmol of putative dimer). Incubations are performed at 30°. For the assessment

[16] E. Blum, A. J. Carpousis, and C. F. Higgins, *J. Biol. Chem.* **274,** 4009 (1999).

[17] R. S. McLaren, S. F. Newbury, G. S. C. Dance, H. C. Causton, and C. F. Higgins, *J. Mol. Biol.* **221,** 81 (1991).

[18] G. A. Mackie, J. L. Genereaux, and S. K. Masterman, *J. Biol. Chem.* **272,** 609 (1997).

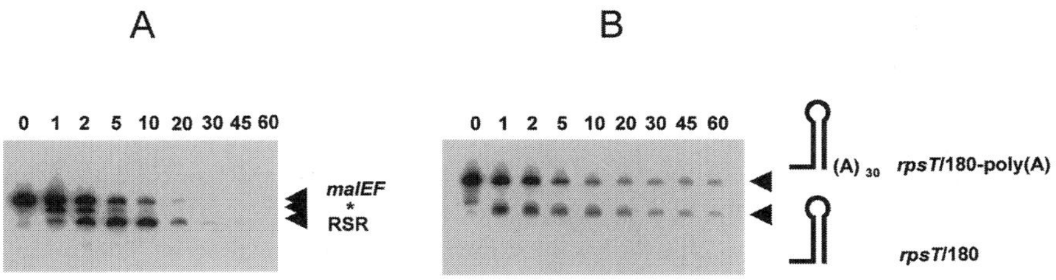

FIG. 2. Activity of a minimal reconstituted degradosome against (A) the 375-nucleotide *malEF* RNA or (B) the 210-nucleotide *rpsT*/180–poly$(A)_{30}$ RNA. Reconstitutions were performed as described in text and samples of 4 μl were withdrawn at the time in minutes (indicated above each panel). Samples were denatured in buffered formamide, separated on 6% (w/v) polyacrylamide gels containing 8 *M* urea, and visualized by phosphorimaging. The triangles in the right margins of each panel denote the initial substrate and various intermediates; the asterisk refers to the "star" intermediate in Fig. 1A. [These data are taken from G. A. Coburn, X. Miao, D. J. Briant, and G. A. Mackie, *Genes Dev.* **13,** 2594 (1999), with permission.]

of physical reassociation by coimmunoprecipitation, the quantities of protein and the volume of incubation should be increased by a factor of 25. Details of the procedures for immunoprecipitation are outlined elsewhere.[10,19]

Reconstituted activity is assessed by monitoring the ATP and phosphate-dependent disappearance of the full-length substrate on 6% (w/v) polyacrylamide gels containing 8 *M* urea[7,10] in TBE buffer. Samples, typically 4 μl, are withdrawn from the assay at the desired time, mixed with 12 μl of formamide–dye buffer, and boiled for 60 sec before separation. The gels are run until the xylene cyanol dye has migrated about 75% of the length of the gel for *rpsT* RNA substrates or until the dye has run off for *malEF* RNA substrates. Gels are fixed briefly in 5% (v/v) acetic acid containing 5% (v/v) ethanol, rinsed in H_2O, and dried before exposure. Typical examples of such assays are shown in Fig. 2. In the case of the *malEF* RNA, PNPase alone will shorten the substrate by about 35 residues to yield a 340-nucleotide species termed the "RSR."[7,17] Further degradation of this intermediate to mononucleotides and limit oligonucleotides is dependent on ATP and the presence of the RhlB helicase (Fig. 2A). Likewise, with the *rpsT* RNA (Fig. 2B), PNPase alone will shorten the poly(A) tail on the substrate to yield an ~180-nucleotide intermediate, but degradation past the terminal stem–loop is dependent on ATP and RhlB.[10] As an alternative to electrophoretic analysis, activity could be monitored by release of ribonucleoside 5′-diphosphates.

The method for reconstituting a minimal degradosome described here yields macromolecular complexes whose formation is dependent on Rne and whose activity depends on PNPase, RhlB, phosphate, and ATP.[10] To this extent, the

[19] N. F. Vanzo, Y. S. Li, B. Py, E. Blum, C. F. Higgins, L. C. Raynal, H. M. Krisch, and A. J. Carpousis, *Genes Dev.* **12,** 2770 (1998).

reconstituted complex recapitulates the properties of unfractionated degradosomes.[7,10] However, the stoichiometry of the reconstituted complexes has not yet been determined accurately. Nonetheless, the development of a method for the reconstitution of the degradosome should open the path to a systematic investigation of the role of each of the components in the assembly process and to elucidating the role of Rne in activating the helicase activity of RhlB and coupling PNPase activity to that of RhlB. The methods described here have already shown that the N-terminal portion of Rne harboring the endonuclease activity and the S1 RNA-binding motif is not essential for reconstitution or RhlB-dependent PNPase activity.[10]

Acknowledgments

Previously unpublished work described in this chapter was supported by a grant from the former Medical Research Council of Canada and its successor, the Canadian Institutes of Health Research.

[29] Purification of Yeast Exosome

By PHILIP MITCHELL

Introduction

Most of the ribonucleases described in this volume are structurally distinct and function independently. However, many of the $3' \rightarrow 5'$-exoribonucleases found in eukaryotic cells are assembled together in a single multienzyme complex known as the exosome. The exosome from the budding yeast *Saccharomyces cerevisiae* is the best characterized, although homologous complexes are found in other eukaryotes including humans,[1,2] and it is the yeast complex with which this chapter is concerned. Remarkably, 10 of the 11 components of the yeast exosome are either demonstrated to have $3' \rightarrow 5'$-exoribonuclease activity, or are putative $3' \rightarrow 5'$-exoribonucleases on the basis of sequence homology.[1–4] Notably, different components of the exosome act with either processive or distributive kinetics and function by either hydrolytic or phosphorolytic mechanisms (for a review, see van Hoof and Parker[5]).

Genetic analyses in yeast have shown that the exosome functions in the 3′-end maturation of diverse stable RNAs, including 5.8S rRNA,[1,2] small nuclear RNAs

[1] P. Mitchell, E. Petfalski, A. Schevchenko, M. Mann, and D. Tollervey, *Cell* **91,** 457 (1997).
[2] C. Allmang, E. Petfalski, A. Podtelejnikov, M. Mann, D. Tollervey, and P. Mitchell, *Genes Dev.* **13,** 2148 (1999).
[3] S. Mian, *Nucleic Acids Res.* **25,** 3187 (1997).
[4] K. T. D. Burkard and J. S. Butler, *Mol. Cell. Biol.* **20,** 604 (2000).

0076-6879/01 $35.00

(snRNAs) and small nucleolar RNAs (snoRNAs).[6,7] The exosome has additional functions in pre-rRNA processing, where it degrades the 5′ external transcribed spacer (5′ ETS) fragment, as well as aberrant RNAs that arise from errors in the order of pre-rRNA-processing events.[8–10] The exosome also functions in mRNA metabolism, degrading nonspliced pre-mRNAs in the nucleus,[11] deadenylating mRNAs,[12] and degrading mRNA after loss of the poly(A) tail.[13]

The exosome has been routinely purified from extracts of *S. cerevisiae* by immunoaffinity chromatography using an epitope-tagged version of the Rrp4p component, consisting of two copies of the *z* IgG-binding domain of protein A from *Staphylococcus aureus* fused to the N terminus of Rrp4p (denoted *zz*-Rrp4p).[14] Purification was initially performed with the aim of identifying proteins associated with Rrp4p and analyzing their function in particular RNA-processing reactions. For this reason, the protocols are based on Western analyses, rather than an enzyme purification strategy. It should be noted that analyses are complicated by the observations that the exosome is heterogeneous, both with respect to structure and enzymatic activity.[2,15] Approximately 1 μg of exosome complex is obtained per liter of culture (equivalent to a crude extract containing about 100 mg of total protein). Enzyme assays can, however, be performed on the IgG–agarose-bound fraction of crude lysates[14] and such assays exhibit negligible contaminant nuclease or phosphatase activities.

Preparation of Cell Extracts

Materials

Yeast strain P49,[1] or a related strain
Beckman (Fullerton, CA) J-25 centrifuge with JLA10.500 and JA25.50 rotors, or equivalent
Beckman GS-6R benchtop centrifuge, or equivalent
Glass beads (425- to 600-μm diameter; Sigma, St. Louis, Mo)

---

[5] A. van Hoof and R. Parker, *Cell* **99,** 347 (1999).

[6] C. Allmang, J. Kufel, G. Chanfreau, P. Mitchell, E. Petfalski, and D. Tollervey, *EMBO J.* **18,** 5399 (1999).

[7] J. Kufel, C. Allmang, G. Chanfreau, P. Petfalski, D. L. J. Lafontaine, and D. Tollervey, *Mol. Cell. Biol.* **20,** 5415 (2000).

[8] J. de la Cruz, D. Kressler, D. Tollervey, and P. Linder, *EMBO J.* **17,** 1128 (1998).

[9] C. Allmang, P. Mitchell, E. Petfalski, and D. Tollervey, *Nucleic Acids Res.* **28,** 1684 (2000).

[10] N. I. T. Zanchin and D. S. Goldfarb, *Nucleic Acids Res.* **27,** 1283 (1999).

[11] C. Bousquet-Antonelli, C. Presutti, and D. Tollervey, *Cell* **102,** 765 (2000).

[12] P. Mitchell and D. Tollervey, unpublished data (1998).

[13] J. S. Jacobs Anderson and R. Parker, *EMBO J.* **17,** 1497 (1998).

[14] P. Mitchell, E. Petfalski, and D. Tollervey, *Genes Dev.* **10,** 502 (1996).

[15] P. Mitchell, unpublished data (1999).

Solutions

YPD medium: 2% (w/v) Bacto-peptone, 1% (w/v) yeast extract, 2% (w/v) glucose

TMN-150 buffer (1 liter, chilled): 10 m*M* Tris-HCl (pH 7.6), 5 m*M* $MgCl_2$, 150 m*M*

NaCl, 0.1% (v/v) nonidet P-40 (NP-40)

Phenylmethylsulfonyl fluoride (PMSF; Sigma): 0.1 *M* stock in 2-propanol

Protocol

The strain P49, which expresses *zz*-Rrp4p, is grown overnight in YPD at 30°. One milliliter of a freshly saturated preculture inoculated into a 1-liter culture volume will reach a suitable harvesting $OD_{600\ nm}$ of 1–2 in about 14 hr. Six cultures can be conveniently performed in parallel. The cells are harvested by centrifugation at 4000 rpm for 10 min at 4°, using a Beckman JLA10.500 rotor. All subsequent steps are performed on ice with prechilled solutions, unless stated otherwise.

The pellets from four 500-ml bottles are resuspended in 20 ml of TMN-150 buffer, transferred to a 50-ml screw-cap polypropylene tube, and recentrifuged at 3000 rpm for 10 min at 4° in a benchtop centrifuge. Typically, about 3 g of wet cell pellet is obtained per liter of culture. The pellets are resuspended in 1 volume of TMN-150 buffer and an equal volume of sterile, diethyl pyrocarbonate (DEPC)-treated glass beads is added. PMSF is added to a final concentration of 0.5 m*M* and the cell–glass bead mixture is vortexed 10 times for 30 sec, with 1-min intervals on ice. The cell debris is pelleted as described above, and the supernatants from the three concomitant lysates are pooled. At this point it is advantageous to subject the pellets to a further round of lysis. The combined crude extracts are cleared by centrifugation in a Beckman JA25.50 rotor at 16,000 rpm (30,000*g*) for 20 min at 4°. This low-speed spin (LSS) fraction is then transferred to a fresh 50-ml screw-cap tube (Greiner, Frickenhausen, Germany), glycerol is added to a final concentration of 10% (v/v), and the material is stored at −80°.

For analytical scale preparations, 200-ml cultures are harvested directly in four 50-ml tubes. After resuspending the combined pellets in 20 ml of TMN-150, the cells are centrifuged as described above and taken up in 1 ml of TMN-150. Lysis in the presence of 1 ml of glass beads is performed, as described above. After pelleting the cell debris and glass beads, the supernatant is transferred to a 2.2-ml Eppendorf tube and centrifuged at 15,000 rpm in a benchtop microcentrifuge for 5 min. The cleared supernatant is transferred to 1.5-ml Eppendorf tubes and glycerol is added to 10% (v/v). Store at −80°.

Glycerol Gradient Analysis of Exosome Complexes

Glycerol gradient analysis enables the resolution of the exosome complex from free *zz*-Rrp4p. Although not routinely used for preparative purposes, it is the most suitable method available for assaying the physical status of the complex. The yeast exosome has an estimated sedimentation coefficient of 14S.[1]

Materials

LSS extract from strain P49
Gradient-forming device
Beckman XL-100 ultracentrifuge and SW40 rotor, or equivalent
Peroxidase–anti-peroxidase (PAP) antibody (Sigma)
Sodium dodecyl sulfate–polyacrylamide gel electrophoresis (SDS–PAGE) unit
Western blot transfer unit

Solutions

TMN-150 buffer (50 ml) containing glycerol at 10 and 30% (v/v)
Buffers for SDS–PAGE, Western transfer, and immunoblot analysis

Protocol

Prepare TMN-150 buffer containing glycerol at 10 and 30% (v/v) final concentration. A 12-ml gradient of the two solutions is poured, using a suitable gradient-forming device, taking care to avoid frothing at low volumes. Samples of 0.1–0.2 ml are loaded onto each gradient and centrifugation is performed at 36,000 rpm in an SW40 rotor for 24 hr. Samples containing 10% (v/v) glycerol can be conveniently diluted with 1 volume of TMN-150 before loading. After centrifugation, the gradient is divided into 0.5-ml fractions. Aliquots (10 μl) of each fraction are resolved by SDS–PAGE through a 10% (w/v) acrylamide gel. After Western transfer, the *zz*-Rrp4p protein is decorated with PAP antibody (Sigma) and detected by enhanced chemiluminescence (ECL), using commercial detection reagents (Amersham, Arlington Heights, IL) according to the manufacturers instructions. The exosome complex peaks at about fraction 16; free *zz*-Rrp4p peaks at about fraction 4 (counting from the top of the gradient).

Exosome Purification

A construct incorporating a cleavage site for the TEV protease between the *zz* tag and the Rrp4p protein has been made,[16] allowing Rrp4p-containing complexes to be recovered from the immunoaffinity-purified material in a native form.

[16] P. Mitchell and D. Tollervey, unpublished data (2000).

Materials

LSS extract from strain P246
IgG–Sepharose 6FF (Pharmacia, Piscataway, NJ)
Econo purification column (2 ml; Bio-Rad, Hercules, CA)
TEV protease (GIBCO-BRL, Gaithersburg, MD)

Solution

TMN-150 buffer (1 liter)

Protocol

A 100-μl packed column of IgG–Sepharose (Pharmacia) is prepared by pouring sufficient gel suspension into a disposable 2-ml Econo purification column (Bio-Rad) and washed with 10 ml of TMN-150 buffer. The LSS fraction from strain P246 is passed twice through the column under gravitational pressure. After washing the column with 100 ml of TMN-150, the IgG–Sepharose column is drained, the outflow is capped, and the resin is resuspended in 200 μl of TMN-150 containing 20 units of TEV protease. After incubation overnight at 4°, the reaction mixture is drained into an Eppendorf tube and the IgG–Sepharose matrix is washed with another 200 μl of TMN-150. The combined eluates are made up to 10% (v/v) glycerol and stored at −80°.

Resolution of Distinct Exosome Complexes

Two major forms of the exosome complex, which differ by the presence or absence of the Rrp6p component,[2] are readily resolved by ion-exchange chromatography. Enzymatic analyses of subfractions of the exosome population have also revealed distinct enzymatic properties.[15]

Materials

LSS extract from strain P49
DEAE, Mono Q, and Mono SP Sepharose resin (Pharmacia)
IgG–Sepharose 6FF (Pharmacia)
Beckman GS-6R benchtop centrifuge, or equivalent
Econo low-pressure chromatography column (20 ml; Bio-Rad)
Econo disposable column (2 ml; Bio-Rad)

Solutions

TMN-0 buffer: 10 m*M* Tris-HCl (pH 7.6), 5 m*M* $MgCl_2$, 0.1% (v/v) NP-40
TMN-150, TMN-200, TMN-320, and TMN-500 buffer; TMN-0 buffer containing NaCl at the concentration (millimolar) indicated

Protocol

The LSS fraction is diluted with 0.5 volume of TMN-0 buffer, giving a final concentration of 100 m*M* with respect to Na^+. This material (~90 ml) is batch-bound in two 50-ml screw-cap tubes to 40 ml of prewashed DEAE-Sepharose at 4° for 1 hr. Pellet the resin by centrifugation at 2000 rpm in the Beckman benchtop centrifuge. Retain the supernatant on ice. After washing the DEAE-Sepharose resin three times with 30 ml of TMN-100 for 5 min, bound material is eluted three times with 30 ml of TMN-300. The pooled eluates, which include free *zz*-Rrp4p and the exosome fraction lacking Rrp6p, are diluted with TMN-0 buffer to a final concentration of 100 m*M* Na^+ and passed through a 10-ml Mono Q Sepharose FF (Pharmacia) column packed in an Econo low-pressure chromatography column (Bio-Rad). After washing with 5 bed volumes of TMN-100, a fraction containing free *zz*-Rrp4p is eluted in 50 ml of TMN-200. The exosome fraction lacking Rrp6p is subsequently eluted in 50 ml of TMN-320. The DEAE-Sepharose supernatant, which includes the Rrp6p-containing exosome fraction, is passed through a 10-ml Mono SP Sepharose column prewashed in TMN-100. After washing, the bound material is recovered by elution in 50 ml of TMN-500. The two resolved forms of the complex are then purified by immunoaffinty chromatography over IgG–Sepharose, as described above.

To enrich for the exosome activity, the LSS fraction is passed through a 2-ml Mono SP Sepharose FF column packed in a disposable column that has been prewashed with TMN-150. The bound material is washed twice with 5 ml of TMN-150 and eluted with 8 ml of TMN-500. Glycerol is added to the eluate to 10% (v/v) final concentration and the eluate is stored at −80°. This fraction, which is stable for at least several months at −80° and can undergo multiple rounds of freeze–thawing without obvious loss of activity, yields highly active enzyme on immunoaffinity purification.

Chromatography over Mono Q Sepharose FF resolves a processive form of the complex from the distributive activity. To recover this fraction from the enriched activity, 250 μl of Mono SP Sepharose FF eluate is diluted with 1 ml of TMN-150 (final Na^+ concentration of 220 m*M*) and batch-bound to 50 μl of prewashed Mono Q Sepharose FF beads in an Eppendorf tube for 15 min at 4°. Alternatively, this fraction can be purified directly from LSS extracts by binding at 150 m*M* NaCl. After centrifugation at low speed in a benchtop microcentrifuge, the supernatant containing the processive activity is transferred to a clean tube and kept on ice. The resin is washed three times with 1 ml of TMN-150 and the bound material is recovered by elution in 250 μl of TMN-500.

Preparation of RNA Substrates for Assaying Exosome Activity *in Vitro*

An *in vitro* assay has been developed for analyzing exosome activity, which utilizes a short, 5′-end-labeled RNA substrate. Degradation products are subsequently analyzed by PAGE. A stock of unlabeled RNA is prepared and an aliquot is freshly labeled and purified for each set of experiments.

Materials

pBS cloning vector (Stratagene, La Jolla, CA)
*Xba*I restriction enzyme
T3 RNA polymerase, polynucleotide kinase
Alkaline phosphatase (calf intestine phosphatase; New England BioLabs, Beverly, MA)
Fluorescence thin-layer chromatography (TLC) plates and UV light source for UV shadowing
[γ-^{32}P]ATP (>5000 Ci mmol^{-1}; Amersham)
PAGE unit

Solutions

RNA extraction buffer: 10 m*M* Tris-HCl (pH 7.6), 1 m*M* EDTA, 150 m*M* NaCl, 0.1% (w/v) SDS
TBE buffer for PAGE

Protocol

A 33-nucleotide-long RNA substrate is transcribed from *Xba*I-linearized pBS, using T3 RNA polymerase according to standard protocols. The incubation mix is diluted to 200 μl and treated with 1 U of alkaline phosphatase at 37° for 60 min. NaCl is added to 150 m*M* and the mixture is extracted three times with an equal volume of phenol–chloroform. The RNA is recovered by ethanol precipitation in the presence of 10 μg of glycogen carrier and purified by electrophoresis through a 12% (w/v) polyacrylamide gel; the full-length RNA runs close to the xylene cyanol marker. After detection by UV shadowing, the full-length RNA is excised with a scalpel in a minimal piece of acrylamide gel. The gel slice is crushed with a clean plastic tip in an Eppendorf tube and the RNA is eluted into 200 μl of extraction buffer. After 4 hr of shaking at room temperature on an Eppendorf (Hamburg, Germany) Thermomixer the supernatant is drawn off and the extraction step repeated. Glycogen carrier (10 μg) and 2 volumes of ethanol are added to the combined eluates and the RNA is precipitated at −20° for 1 hr. After centrifugation at 15,000 rpm in a benchtop microcentrifuge for 20 min, the pellet is washed with 150 μl of 70% (v/v) ethanol and air dried for 10 min. The purified RNA is dissolved in 20 μl of H_2O and stored at −20°.

For labeling, 1 μl of dephosphorylated RNA is incubated with [γ-^{32}P]ATP and polynucleotide kinase according to standard protocols. After heat inactivation at 70° for 10 min, an equal volume of gel loading dye [95% (v/v) formamide, 20 m*M* EDTA, 0.5% (w/v) xylene cyanol, 0.5% (w/v) bromphenol blue] is added and the labeled RNA is gel purified as described above. The purified full-length RNA is taken up in 100 μl of H_2O and stored at $-20°$.

In Vitro Assay for Exosome Activity

Assays of exosome activity have routinely been performed while the complex is still attached to the IgG–agarose resin. Immunoaffinity chromatography can be performed on crude lysates without the need for prior purification, or on isolated subfractions of the complex. As controls, mock affinity chromatography purifications are performed on lysates from yeast expressing nontagged Rrp4p.

Materials

LSS extract from P49 and control strain
IgG–agarose beads (Sigma)
RNasin (Promega, Madison, WI)
5′-Labeled RNA substrate
Thermomixer (Eppendorf)
PAGE unit

Solutions

TMN-150 buffer
Reaction buffer: 20 m*M* Tris-HCl (pH 7.6), 50 m*M* KCl, 5 m*M* $MgCl_2$, bovine serum albumin (BSA, 100 μg/ml), 10 m*M* dithiothreitol (DTT)
TBE buffer for PAGE

Protocol

IgG–agarose beads (20 μl) prewashed with TMN-150 buffer are mixed with 100 μl of LSS extract in an Eppendorf tube and made up to 1 ml with TMN-150 buffer. Mix for 1 hr at 4°. Spin the mixture in a benchtop microcentrifuge at a low speed (2000 rpm) to prevent compression of the resin. Discard the supernatant and wash the pellet three times with 1 ml of TMN-150 buffer for 10 min at 4°. Resuspend once with 50 μl of reaction buffer, centrifuge, place on ice, and resuspend in 100 μl of reaction buffer. Add 1 μl of RNasin, followed by 1 μl of RNA substrate. Aliquot 20 μl of the mixture into each of five 1.5-ml centrifuge tubes, taking care to keep the beads in suspension by gently flicking the side of the tube. Transfer four tubes to an Eppendorf Thermomixer and incubate at 30° with shaking. The 0-min time point is stopped while still on ice by the addition of 20 μl of gel loading dye. The reactions in the four aliquots are stopped after 5, 10, 20, and 40 min.

When the time course is completed, incubate the samples at 65° for 10 min and spin briefly in a microcentrifuge. Load 10 μl from each time point onto a

0.4-mm-thick 12% (w/v) polyacrylamide gel and run the gel until the bromphenol blue dye is about 10 cm from the wells. Such thin gels can be vacuum dried without adding glycerol or fixing; these gels tend not to stick directly to paper, so transfer the gel first to Saran Wrap and then lay paper on the reverse side before vacuum drying.

Acknowledgments

I acknowledge the support and encouragement given by David Tollervey, in whose laboratory this work was carried out. This work has been funded by the Wellcome Trust.

Section IV

Organellar Ribonucleases

[30] Genetic and Biochemical Approaches for Analysis of Mitochondrial Degradosome from *Saccharomyces cerevisiae*

By ANDRZEJ DZIEMBOWSKI and PIOTR P. STEPIEN

Introduction

Regulation of RNA turnover is one of the major mechanisms of controlling gene expression in yeast mitochondria. RNA degradation is carried out by the yeast mitochondrial degradosome (also known as mtEXO); a multiprotein enzyme complex showing NTP-dependent 3′→5′-exoribonuclease activity *in vitro*. It was first identified and purified by the group of H. P. Zassenhaus.[1] The enzyme is composed of three subunits with molecular masses of 75, 84, and 111 kDa, all of which are encoded in the nuclear genome and after translation in the cytoplasm are imported into the mitochondria. The *SUV3* and *DSS1* genes have been identified as encoding the 84- and 111-kDa subunits, respectively. The analysis of mutants indicates that the degradosome participates in turnover of different species of mitochondrial RNAs and may be involved in processing of both 3′ and 5′ ends of mitochondrial RNAs.

Biochemical Analysis

Studies of mitochondrial nucleases from wild-type yeast strains are hampered by the contaminating activity of the *NUC1* gene product. It has the activity of a single-stranded RNase, single- and double-stranded DNA endonuclease, and 5′-exonuclease of double-stranded DNA,[2] and it may participate in recombination of mitochondrial DNA.[3] Yeast strains bearing the knockout allele of the *NUC1* gene are viable and respiration competent,[4] and therefore are recommended for experiments with other nucleases.

Screening for RNase activity in nuc1p-deficient mitochondrial lysates resulted in identification and purification of the degradosome: an exoribonuclease with a native size of approximately 160 kDa.[1] Our data indicate that the enzyme is not membrane bound, because it can be purified from total cell extracts without the aid of detergents. Studies by Min *et al.*[1] showed that the purified enzyme is specific only for single-stranded RNA and the K_m for this substrate was found to be

[1] J. Min, R. M. Heuertz, and H. P. Zassenhaus, *J. Biol. Chem.* **26,** 7350 (1993).

[2] R. D. Vincent, T. J. Hofmann, and H. P. Zassenhaus, *Nucleic Acid Res.* **16,** 3297 (1998).

[3] H. P. Zassenhaus and G. Denninger, *Curr. Genet.* **25,** 142 (1994).

[4] H. P. Zassenhaus, T. J. Hofmann, R. Uthayashanker, R. D. Vincent, and M. Zona, *Nucleic Acids Res.* **16,** 3283 (1998).

0076-6879/01 $35.00

0.53 μg/ml. The activity is dependent on NTPs: both deoxy- and ribonucleoside triphosphates can stimulate RNA hydrolysis with K_m values ranging from 20 to 90 μM, while the nonhydrolyzable analogs were inactive. During the exoribonucleolytic reaction ATP is converted to ADP and PO_4, and approximately one or two molecules of ATP are required to release each nucleotide from the single-stranded RNA (ssRNA) substrate. RNA degradation performed by the purified enzyme is strictly polar and proceeds from the 3′ end to the 5′ end. The products of the reaction are 5′-nucleoside monophosphates. Sodium dodecyl sulfate–polyacrylamide gel electrophoresis (SDS–PAGE) analysis revealed that the enzyme is composed of three polypeptides of approximately 75, 90, and 110 kDa.

Genes Encoding Subunits of Yeast Mitochondrial Degradosome

So far only two yeast nuclear genes have been identified as encoding degradosome subunits. The *SUV3* gene was isolated[5] as a suppressor of the *SUV3-1* mutant,[6] using the technique of yeast colony Northern hybridization.[7] The *SUV3* gene was shown to be located on chromosome XVI; it encodes a protein of 737 amino acids with a predicted molecular mass of 84 kDa, which shows motifs characteristic of ATP-dependent RNA helicases from the DEAD/DExH family.[5] Orthologs of the *SUV3* gene were found in a variety of organisms, including the purple bacterium *Rhodobacter sphaeroides*, *Arabidopsis*, *Caenorhabditis*, *Drosophila*, and *Homo sapiens*[8,9]; analysis of the motifs indicates that the SUV3-like proteins form a distinct and highly conserved family of RNA helicases, containing the DEIQ box, and common to all eukaryotic organisms.[9] Direct evidence of the identity of the *SUV3* gene product with the degradosome subunit was obtained using Western blots.[10]

The second subunit of the degradosome complex is encoded by the yeast nuclear gene *DSS1*, located on chromosome XIII. The *DSS1* gene was isolated in our laboratory as a multicopy suppressor of an *SUV3* deletion and it encodes a 970-amino acid protein, with a predicted molecular mass of 111 kDa.[11,12] The

[5] P. P. Stepien, S. P. Margossian, D. Lansman, and R. A. Butow, *Proc. Natl. Acad. Sci. U.S.A.* **89,** 6813 (1992).

[6] H. Conrad-Webb, P. P. Perlman, H. Zhu, and R. A. Butow, *Nucleic Acids Res.* **18,** 1369 (1990).

[7] P. P. Stepien and R. A. Butow, *Nucleic Acids Res.* **18,** 380 (1989).

[8] A. Dmochowska, P. Stankiewicz, P. Golik, P. P. Stepien, E. Bocian, I. Hansmann, and E. Bartnik, *Cytogenet. Cell Genet.* **83,** 84 (1998).

[9] A. Dmochowska, K. Kalita, M. Krawczyk, P. Golik, K. Mroczek, J. Lazowska, P. P. Stepien, and E. Bartnik, *Acta Biochim. Pol.* **46,** 155 (1999).

[10] S. Margossian, H. Li, H. P. Zassenhaus, and R. A. Butow, *Cell* **84,** 199 (1996).

[11] A. Dmochowska, P. Golik, and P. P. Stepien, *Curr. Genet.* **28,** 108 (1995).

[12] A. Dziembowski, M. Malewicz, M. Minczuk, P. Golik, A. Dmochowska, and P. P. Stepien, *Mol. Gen. Genet.* **260,** 108 (1998).

DSS1-encoded protein shows motifs characteristic of bacterial RNase II, and it is homologous to the fungal Cyt4 protein, which is involved in RNA processing, splicing, and turnover in *Neurospora crassa*; and the *vacB* gene product from *Shigella flexneri*, which encodes RNase R and appears to be necessary for efficient translation of plasmid-encoded mRNA. Evidence of the identity of the *DSS1* gene product with the degradosome subunit was obtained by tagging the protein with a hexahistidine (His_6) tag, and by subsequent detection of the activity of mtEXO eluted from Ni columns (our unpublished data, 2000).

The yeast nuclear gene encoding the 75-kDa subunit has not been yet identified.

Phenotypes of *SUV3* and *DSS1* Mutants

Three different mutations in genes encoding degradosome subunits have been tested: knockouts of the *SUV3* or *DSS1* gene, and a missense mutant of the *SUV3* gene. A plethora of effects could be observed, which in most cases were identical for *SUV3* and *DSS1* mutants. The observed mutant phenotypes indicate deregulation of turnover and processing of mitochondrial RNAs and can be divided into several groups.

The major effect of either *SUV3* or *DSS1* gene knockout is the lack of mtEXO activity.[10,12] The remaining phenotypes seem to be a consequence of this fact. In experiments in which yeast strains harbored mitochondrial genomes with introns, the most characteristic mutant phenotype was strong accumulation of excised group I introns[6] and instability of intron-containing transcripts.[10,13,14] Changes in RNA stability were also detected when intronless mitochondrial genomes were present in our mutant strains—the levels of *COB1* mRNA and 16S rRNA were low.[12]

The second group of mutant phenotypes involves changes in processing of precursors of mitochondrial RNAs (mtRNAs): the generation of a proper 3′-end terminus by cleavage of *fit1* mRNA close to the conserved dodecamer sequence was affected in *SUV3* mutants.[10] We have found that in yeast strains harboring intronless mitochondrial genomes both *SUV3* and *DSS1* knockouts gave rise to large extensions from the 3′ end for the 21S rRNA and from the 5′ end for 16S rRNA, *COB1*, *VARI*, and *ATP6/8* mRNAs (our unpublished results, 2000).

As a consequence of these disturbances in RNA turnover and processing, yeast cells of strains disrupted for *SUV3* or *DSS1* genes are respiration incompetent, mitochondrial translation is stopped, and mitochondrial genomes are quickly lost.[12]

[13] P. P. Stepien, L. Kokot, T. Leski, and E. Bartnik, *Curr. Genet.* **27,** 234 (1995).
[14] P. Golik, T. Szczepanek, E. Bartnik, P. P. Stepien, and J. Lazowska, *Curr. Genet.* **28,** 217 (1995).

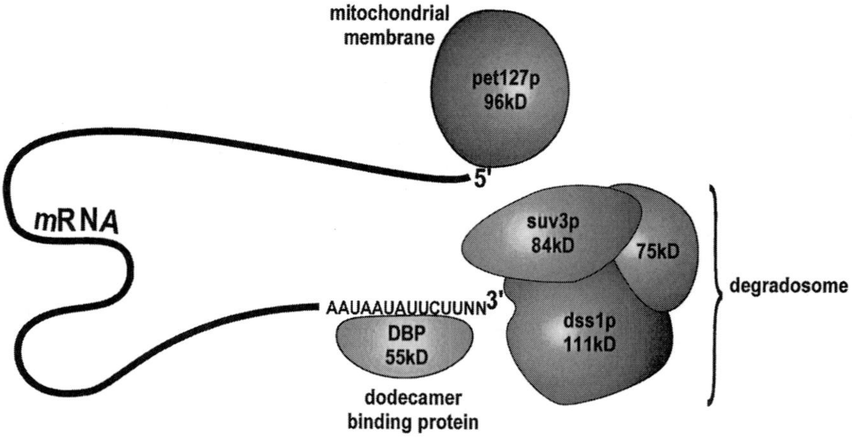

FIG. 1. Model of interactions of yeast mitochondrial degradosome with mRNA.

Model of Degradosome Function in Yeast Mitochondria

The yeast mitochondrial degradosome (mtEXO) participates in turnover and processing of RNA and its activity is indispensable for proper functioning of mitochondria. In contrast to the nucleus, in mitochondria the regulation of gene expression occurs primarily at the posttranscriptional level. The main mechanisms involve processing of RNA precursors and regulation of RNA stability and translatability. Yeast mitochondrial RNAs have different half-lives: intergenic regions generated during processing of the primary transcripts and spliced group I introns have rapid turnover, while mRNA and ribosomal RNA are much more stable. Therefore some molecular mechanisms must recognize specific structures in RNA or RNA–protein complexes and regulate RNA turnover.

In yeast mitochondria mRNAs are not polyadenylated. Mature 3′ ends are generated by endonucleolytic cleavage two nucleotides downstream from the conserved dodecamer element 5′-AAUAAUAUUCUU-3′.[15] A 55-kDa protein that binds with high affinity to the dodecamer sequence has been purified,[16,17] but the gene encoding this polypeptide has not been identified yet. The dodecamer-binding protein (DBP) may be involved in regulation of mRNA stability: *in vitro* experiments have shown that it protects RNA from degradation by the NTP-dependent exoribonuclease.[16]

It seems that the turnover of yeast mitochondrial RNAs is regulated by the interplay of the dodecamer-binding protein and the degradosome (Fig. 1). In wild-type

[15] T. J. Hofmann, J. Min, and H. P. Zassenhaus, *Yeast* **9,** 1319 (1993).

[16] J. Min and H. P. Zassenhaus, *Mol. Cell. Biol.* **13,** 4167 (1993).

[17] H. Li and H. P. Zassenhaus, *Biochem. Biophys. Res. Commun.* **261,** 740 (1999).

strains the 3′ ends of mature mRNAs are protected from the exoribonucleoytic action of the degradosome. Unprotected excised intergenic regions and introns are subject to quick degradation. For ribosomal RNAs, the binding of ribosomal proteins is most probably responsible for the protection.

The biochemical activity of the mitochondrial degradosome is probably similar to degradosome complexes described for bacteria. The presence of RNA helicases in complexes degrading RNA is necessary for unwinding the stem–loop structures that impede the progress of exoribonucleases. RNA helicases may also play a role in the recognition of RNA features that modulate the decay rates (for reviews see Refs. 18 and 19).

Mutations in the *SUV3* or *DSS1* gene deregulate the activity of the degradosome: RNA molecules that are of low abundance in wild-type strains, such as the omega intron, accumulate to high levels. On the other hand, other RNAs (but not all) are degraded. This indicates that besides the NTP-dependent exoribonuclease activity another RNA-degrading enzyme exists; it cannot be excluded that the third, so far unknown, degradosome subunit also has ribonucleolytic activity.

The activity of the mitochondrial degradosome is not restricted to intron turnover as was suggested previously.[20] Most experiments in our laboratory were performed with yeast strains containing intronless mitochondrial genomes and the results show that the degradosome complex is necessary for proper metabolism of mtRNA in general.

Mutations in the *SUV3* or *DSS1* gene affect processing events on both 3′ and 5′ ends of mtRNA. This suggests that turnover and processing of mitochondrial transcripts could be linked. Moreover, it seems possible that there are functional interactions between 3′ and 5′ ends of mitochondrial RNA: we have found that both *SUV3* and *DSS1* gene disruptions can be suppressed by higher expression of the yeast nuclear gene *PET127*,[21] a membrane-bound protein involved in 5′-end processing of mtRNAs.[22] Furthermore, a mutation within the 5′ untranslated region of the mitochondrial *COB1* gene resulting in lack of expression of cytochrome *b* protein can be supressed by a missense mutation within the *DSS1* gene.[23] Although these observations support our hypothesis, further studies are needed; in particular, the yeast genes encoding the 75-kDa subunit of the degradosome and the dodecamer-binding protein must be identified.

[18] J. S. J. Anderson and R. Parker, *EMBO J.* **17,** 1497 (1998).
[19] A. J. Carpousis, N. F. Vanzo, and L. C. Raynal, *Trends Genet.* **15,** 24 (1999).
[20] S. P. Margossian and R. A. Butow, *Trends Biochem. Sci.* **21,** 392 (1996).
[21] T. Wegierski, A. Dmochowska, A. Jablonowska, A. Dziembowski, E. Bartnik, and P. P. Stepien, *Acta Biochim. Polon.* **45,** 935 (1998).
[22] G. Wiesenberger and T. D. Fox, *Mol. Cell. Biol.* **17,** 2816 (1997).
[23] W. Chen, M. A. Islas-Osuna, and C. L. Dieckmann, *Genetics* **151,** 1315 (1999).

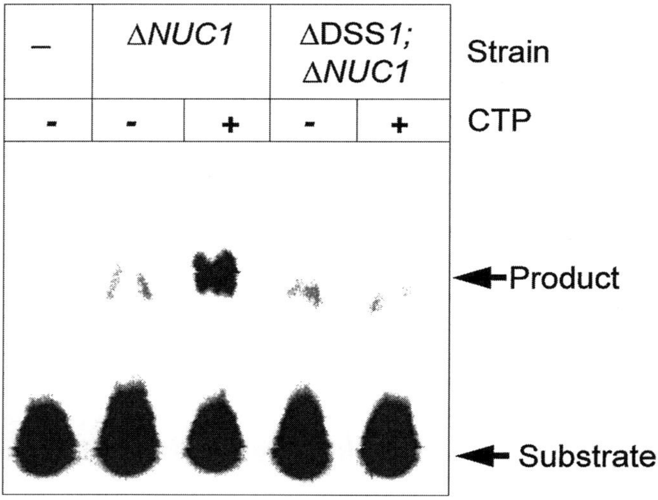

FIG. 2. *In vitro* activity of NTP-dependent exoribonuclease in mitochondrial extracts from control strain (Δ*NUC*) and Δ*DSS1* disruptant in an isogenic yeast strain that also carries the *NUC1* disruption. The assay was carried out as described by Min *et al.*[1] The first lane represents the control strain with no extract added. [Reprinted with the kind permission of Springer-Verlag: Fig. 1 in A. Dziembowski, M. Malewicz, M. Minczuk, P. Golik, A. Dmochowska, and P. P. Stepien, *Mol. Gen. Genet.* **260,** 108 (1998).]

Exoribonuclease Activity Assays

The assay for NTP-dependent $3' \rightarrow 5'$-exoribonuclease activity is based on the fact that the reaction product (monoribonucleosides) migrates at the front of the chromatogram and the substrate remains at the origin. The substrate for assays is uniformly [^{32}P]UTP-labelled RNA obtained by *in vitro* transcription, and is incubated with the enzyme in an appropriate buffer with or without NTPs. Exoribonuclease reaction products are developed on a thin-layer chromatography (TLC) polyethyleneimine (PEI) plate and analyzed on a PhosphorImager radioactivity scanner (Molecular Dynamics, Sunnyvale, CA) (Fig. 2). NTPase activity of the enzyme can be measured under the same reaction conditions as for exoribonuclease but using [γ-^{32}P] nucleotide triphosphate and cold RNA.

Preparation of Extracts

The NTP-dependent exoribonuclease activity of the mitochondrial degradosome can be detected in crude mitochondrial lysates. Yeast strains with an inactivated *NUC1* gene must be used in such assays, otherwise the strong ribonucleolytic activity of nuc1p will make the assay impossible. In our hands it has been necessary to purify mitochondria by Percoll gradient centrifugation. We routinely follow the procedure described by Riezman *et al.*[24] with minor modifications. Yeast cells are grown in YPD medium [1% (w/v) yeast extract, 2% (w/v) Bacto-peptone, 2% (w/v) glucose] to the early logarithmic phase (OD 1–1.5) and are converted to spheroplasts and homogenized in a Potter glass homogenizer. Mitochondria are isolated by differential centrifugation and further purified on a Percoll step gradient.

Isolated mitochondria are resuspended in lysis buffer [20 m*M* Tris-HCl (pH 8.0), 0.5 *M* KCl, 5 m*M* $MgCl_2$, 1 m*M* EDTA, 2 m*M* dithiothreitol (DTT), 1 m*M* phenylmethylsulfonyl fluoride (PMSF), 15% (v/v) glycerol, 2% (v/v) Triton X-100] to a protein concentration of approximately 10 mg/ml, incubated for 30 min in ice, and subsequently centrifuged at 20,000*g* for 15 min at 4°. The supernatant is collected, frozen in liquid nitrogen, and stored at −80°.

Preparation of Substrate

For *in vitro* transcription we use a commercial kit based on T7 RNA polymerase (e.g., Promega, Madison, WI). Several different linearized plasmids containing the T7 RNA polymerase promoter are used as templates (e.g., Bluescript KS plasmid linearized by *Afl*III restriction enzyme). The length of RNA is between 100 and 300 bases. After the reaction the RNA is phenol extracted and purified by centrifugation in a mini Quick-Spin DNA column (Boehringer Mannheim/Roche, Indianapolis, IN).

Exoribonuclease Reaction Conditions

Reactions are usually carried out essentially as described by Min *et al.*[1] with minor modifications.

1. Combine the following:

   Tris-HCl (pH 8.0), 10 m*M*
   $MgCl_2$, 10 m*M*
   KCl, 25 m*M*
   Dithiothreitol (DTT), 1 m*M*
   ^{32}P-Labeled RNA, 10 ng
   CTP, 1 m*M*

The same mixture reaction without CTP is used as a control.

2. After addition of 10 μg of mitochondrial protein extract or purified mitochondrial degradosome the total volume is 10 μl.

[24] H. Riezman, R. Hay, S. Gasser, G. Daum, G. Schneider, C. Witte, and G. Schatz, *EMBO J.* **2,** 1005 (1983).

3. The reaction mixture is incubated for 30 min at 30°.

4. The reactions are stopped by the addition of EDTA to a final concentration of 20 m*M*, or are analyzed directly.

5. A portion (usually 1.5 μl) of the reaction mixture is spotted on a TLC polyethyleneimine (PEI) plate (Merck, Rahway, NJ) and developed in 1 *M* formic acid, 0.5 *M* LiCl. The plates are dried and analyzed in a PhosphorImager radioactivity scanner.

Purification of Mitochondrial Degradosome Complex

Purification of the native degradosome complex was first reported by Min *et al.*[1] The procedure starts with 3 kg of yeast and requires isolation of mitochondria and subsequent column chromatography: on heparin–Sepharose, QAE-cellulose, Affi-Gel Blue, octyl-agarose, and ATP–agarose. Partial purification of the complex, employing a His_6 tag fusion, has been reported by Margossian *et al.*[10] In our hands the best method turned out to be the tandem affinity chromatography (TAP method) described by Rigaut *et al.*[25] This approach is based on affinity chromatography of protein A and calmodulin-binding peptide (TAP tag) fused to the C terminus of selected protein. These two tags are separated by a TEV protease cleavage site. The protein complex containing the fusion protein is initially bound to an IgG matrix, and after washing is released from the matrix by TEV protease cleavage. The complex is then purified by chromatography on calmodulin-coated beads. We isolated the mitochondrial degradosome, using a yeast strain with the TAP tag fused to the C terminus of the Suv3 protein. The major advantage of this method is that it needs much less yeast, does not require isolation of mitochondria, and yields a pure, active, three-protein complex in only two steps (Fig. 3). The procedure takes only 2 days.

All procedures for the TAP method are described in detail at *www.embl-heidelberg.de/ExternalInfo/seraphin/Tap.html*. Here we describe the procedure optimized for degradosome isolation.

Construction of Yeast Strain Expressing TAP-Tagged SUV3 Gene

A gene fusion was constructed by attaching the TAP tag to the C terminus of the *SUV3* gene by *in vivo* homologous recombination, using the *URA3* gene as a selection marker. We use polymerase chain reaction (PCR)-based genomic tagging, using plasmid pBS1539 as a template[25,26] and the following primers:

[25] G. Rigaut, A. Shevchenko, B. Rutz, M. Wilm, M. Mann, and B. Seraphin, *Nat. Biotechnol.* **17,** 1030 (1999).

[26] O. Puig, B. Rutz, B. G. Luukkonen, S. Kandels-Lewis, E. Bragado-Nilsson, and B. Seraphin, *Yeast* **14,** 1139 (1998).

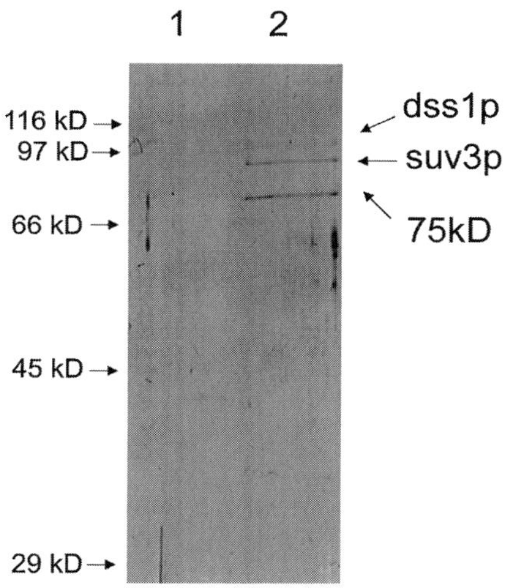

FIG. 3. TAP purification of the mitochondrial degradosome. A silver-stained SDS–polyacrylamide gel. Lane 1, control purification from wild-type strain without TAP tag, lane 2, mitochondrial degradosome subunits purified from yeast strain containing TAP-tagged *SUV3* gene.

SUVTAPp: ATGTTAAGGCTTTCAAGCAGTGTAGCTGTTTCTATTT ACAGGATATATTACGACTCACTATAGGG
SUVTAP1: GGTTCTACAAGAGGTCACCTCTCATCTTCGAGAAGA AGATTGCGTACATCCATGGAAAAGAGAAG

Addition of a 21-kDa TAP tag to Suv3 protein gives no phenotype. The yeast strain containing this construct remains respiration competent and grows on nonfermentable carbon sources at the same rate as the wild-type strain. After purification of the mitochondrial degradosome the Suv3 protein has a 5-kDa TAP tag composed of calmodulin-binding peptide, spacer regions, and a TEV protease cleavage site.

Preparation of Total Yeast Extract

Extracts are prepared in the same buffers as described by Rigaut *et al.*,[25] but cells are disrupted by agitation with glass beads instead of passage through a French press.

Buffers

Buffer A: 10 m*M* K–HEPES (pH 7.9), 10 m*M* KCl, 1.5 m*M* $MgCl_2$, 0.5 m*M* DTT, 0.5 m*M* PMSF, 2 m*M* benzamidine, 1 m*M* leupeptin, 2 m*M* pepstatin A, 4 m*M* chymostatin, 2.6 m*M* aprotinin 500

Buffer D: 20 m*M* K–HEPES (pH 7.9), 50 m*M* KCl, 0.2 m*M* EDTA (pH 8.0), 0.5 m*M* DTT, 20% (v/v) glycerol, 0.5 m*M* PMSF, 2 m*M* benzamidine

Procedure

1. The yeast strain containing an *SUV3* gene tagged with the TAP tag is grown overnight in 4 liters of YPD medium to a cell density of 1.4–1.8. Cells are harvested and washed once in deionized water.
2. Cells are resuspended in an equal volume of buffer A and transferred to a 150-ml Corex tube. An equal volume of acid-washed glass beads (0.5 mm) is then added, and the cells are homogenized by shaking through a distance of about 50 cm for 5 min (30-sec intervals separated by 1 min of cooling in ice). The homogenization step is critical because not only the cell wall but also the mitochondria must be disrupted. The efficiency of extraction can be easily checked by Western blotting, using peroxidase–anti-peroxidase (PAP) (Sigma, St. Louis, MO) as an antibody.[25] We usually obtain 60% extraction efficiency.
3. After homogenization the extract is transferred to a new tube and 2 *M* KCl is added to final concentration of 0.2 *M*.
4. The extract is centrifuged in an SW 41 Ti Beckman (Fullerton, CA) rotor at 20,500 rpm for 30 min at 4°. Supernatant is transferred to a new tube and centrifuged in the same rotor at 35,000 rpm for 1 hr and 24 min. The supernatant is collected.
5. Extract is dialyzed for 3 hr in 2 liters of buffer D.
6. Dialyzed extract is frozen in liquid nitrogen and stored at −80°.

Purification

Purification is performed exactly as described by Rigaut *et al.*[25] (*www.embl-heidelberg.de/Externallnfo/seraphin/TAP.html*). The only difference is that for the first step a 2-fold larger volume of IgG–agarose bead suspension is used. Western blot analyses have shown that under conditions modified in this way a higher portion of mtEXO binds to the IgG–agarose beads.

Buffers

IPP150: 10 m*M* Tris-HCl (pH 8.0), 150 m*M* NaCl, 0.1% (v/v) NP-40

TEV cleavage buffer: 10 m*M* Tris-HCl (pH 8.0), 150 m*M* NaCl, 0.1% (v/v) NP-40, 0.5 m*M* EDTA, 1 m*M* DTT

IPP150 calmodulin-binding buffer: 10 m*M* 2-mercaptoethanol, 10 m*M* Tris-HCl (pH 8.0), 150 m*M* NaCl, 1 m*M* magnesium acetate, 1 m*M* imidazole, 2 m*M* $CaCl_2$, 0.1% (v/v) NP-40

IPP150 calmodulin elution buffer: 10 m*M* 2-mercaptoethanol, 10 m*M* Tris-HCl (pH 8.0), 150 m*M* NaCl, 1 m*M* magnesium acetate, 1 m*M* imidazole, 2 m*M* EGTA, 0.1% (v/v) NP-40

Preparation of Beads

For each 10 ml of extract 400 μl of IgG–agarose bead suspension is used in the first step (Sigma) and 200 μl of calmodulin beads suspension is used for the second step (Stratagene, La Jolla, CA). Beads are washed with 5 ml of IPP150 (IgG beads) or with 5 ml of IPP150 calmodulin-binding buffer (Calmodulin beads) in small columns.

Binding of Mitochondrial Degradosome to IgG Beads. The extract is supplemented with Tris-HCl (pH 8.0) to a final concentration of 10 m*M*, NaCl to a final concentration of 150 m*M*, and NP-40 to a final concentration of 0.1% (v/v). The extract is transferred to a column with washed IgG beads (400 μl per 10 ml of extract) and rotated for 2 hr at 4°. Plugs are removed and the column is drained by gravity flow.

The column is washed with 30 ml of IPP150 for every 10 ml of extract and with 10 ml of TEV cleavage buffer.

Release of Yeast Mitochondrial Degradosome by Tobacco Etch Virus Protease Cleavage

The bottom of the column is closed and 1 mL of tobacco etch virus (TEV) protease cleavage buffer and approximately 100 units of TEV protease (GIBCO-BRL, Gaithersburg, MD) are added per 10 ml of extract. The column is closed and incubated with rotation for 2 hr at 16° and then the flowthrough solution is collected. TEV cleavage buffer (200 μl) is added to collect the solution from the column dead volume.

Purification by Chromatography on Calmodulin-Coated Beads

1. Three volumes of calmodulin-binding buffer is added to the eluate. The EDTA coming from the TEV cleavage buffer is titrated by addition of 3 μl of 1 *M* $CaCl_2$ per 1 ml of IgG eluate.
2. The solution is transferred to the column containing calmodulin beads (200 μl per 10 ml of extract).
3. The column is closed and rotated at 4° for 1 hr. The column is then drained by gravity flow and washed with 30 ml of IPP150 calmodulin binding buffer per 10 ml of extract. Purified enzyme is eluted in five fractions of 200 μl with IPP150 calmodulin elution buffer.
4. The isolated mitochondrial degradosome can be directly used for activity assays because the elution buffer does not contain components inhibiting its activity. It is important to dilute it in exonuclease reaction buffer at least 5-fold because of high salt concentration in elution buffer. From 40 g of yeast wet weight we obtain about 50 ng of active three-protein mitochondrial degradosome complex. The results shown in Fig. 3 indicate that in the purified enzyme preparation, no contaminating protein bands can be detected.

Conclusions

Despite intensive research, many questions about the yeast mitochondrial degradosome remain unanswered. Efforts to identify the third gene encoding the degradosome subunit are currently in progress in our laboratory. Successful purification of relatively large amounts of the active enzyme should enable us to study its interactions with RNA and other proteins. A combination of genetical and biochemical approaches will increase our understanding of how the system of mtRNA turnover is regulated.

Acknowledgments

This work was supported by the Polish State Committee for Scientific Research, Grant No. 6P04A 01818, and by the Polish–French Center for Biotechnology of Plants. We thank Ewa Bartnik for helpful comments, Michał Mińczuk, and Michał Koper for help in manuscript preparation, and Bertrand Seraphin for plasmids containing TAP tag.

[31] Direct Sizing of RNA Fragments Using RNase-Generated Standards

By BARBARA SOLLNER-WEBB, JORGE CRUZ-REYES, and LAURA N. RUSCHÉ

Electrophoretic migration of RNA is affected by its terminal OH or P character to a sizable and variable extent, considerably more than is DNA. If not accounted for when sizing RNA, errors of up to two nucleotides can arise, even when using sequencing ladders prepared from RNA with the same sequence. Many cellular cleavage events of interest generate 3′-OH termini, yet these termini are not formed when using the standard methods of preparing RNA sequencing ladders, which all generate 2′,3′-cyclic P and 5′-OH termini. In contrast to RNases generally used for sequence analysis, nuclease P1 generates 3′-OH and 5′-P termini, but it has been considered sequence nonspecific. Fortunately, conditions have been found in which nuclease P1 cleaves in a highly G-specific manner and different conditions in which it cleaves in an A-preferential manner, allowing convenient preparation of the desired sequencing ladders.

The Common Approaches to RNA Sequence Analysis

Studies of RNA nucleolytic events as described in this volume generally involves precisely defining where the cleavages occur by determining the exact size or end location of the product RNA. Of interest are numerous different

0076-6879/01 $35.00

endonucleolytic, 3′→5′-exonucleolytic, and 5′→3′-exonucleolytic events that serve in the maturation of rRNAs, tRNAs, small nucleolar RNAs (snoRNAs), etc., as well of mRNAs by RNA editing (e.g., *see* Ref. 1) and in probing of RNA structure. Several methods are used for RNA sizing, electrophoretically resolving either the RNA itself, a reverse transcriptase-generated DNA copy of the RNA (for 5′-end analysis), or a DNA molecule that had been primed with the RNA (for 3′-end analysis). However, with *in vitro* processing reactions, where the input RNA is end labeled and of known sequence and the cleavage fragments are generally $\leq$200 nucleotides in length, the most straightforward approach is to directly compare the electrophoretic migration of the resultant RNA fragment with sequencing ladders prepared from the same input RNA, resolved in adjacent gel lanes. These ladders serve to align to the known sequence of the input RNA; they need not determine the complete sequence *de novo*. Depending on whether the input RNA is 5′- or 3′-end-labeled, the upstream or downstream cleavage product is analyzed.

The key to accurate RNA size determination is to select appropriate marker ladders that not only are from the same sequence RNA but also have comparable termini. Cleavage agents commonly used to generate such ladders are sequence-specific RNases, including RNase T1 (which cleaves after G residues) and RNase A (which cleaves after C and U residues), and alkali, which cleaves after any residue. These agents all generate 5′-OH and 2′,3′-cyclic P termini, the latter only slowly converting to 3′-P termini.[2–8] Alternatively, chemical protocols can be used to cleave RNA in a sequence-preferential manner,[9] generating 5′-P termini and 3′ termini bearing an abasic cleaved ribose. If the experimental RNA has the same terminal P or OH character as the sizing markers, migration can be directly compared. However, if the terminal moieties are not the same, their charge differences affect relative electrophoretic migration.

The Sizing Ambiguity

Experiments show that the presence of a terminal P versus a terminal OH can significantly affect electrophoretic migration of otherwise identical RNA

[1] B. Sollner-Webb, L. N. Rusché, and J. Cruz-Reyes, *Methods Enzymol.* **341,** 101 (2001).

[2] J. Cruz-Reyes, K. Piller, L. Rusché, M. Mukherjee, and B. Sollner-Webb, *Biochemistry* **37,** 6059 (1998).

[3] F. Ausubel, R. Brent, R. Kingston, D. Moore, J. Seidman, J. Smith, and K. Struhl, "Current Protocols in Molecular Biology", Vol. 1, pp. 3.1–3.2. John Wiley & Sons, New York, 1989.

[4] H. Donis-Keller, A. Maxam, and W. Gilbert, *Nucleic Acids Res.* **4,** 2527 (1977).

[5] G. Knapp, *Methods Enzymol.* **180,** 192 (1989).

[6] Y. Kuchino and S. Nishimura, *Methods Enzymol.* **180,** 154 (1989).

[7] I. Schildkraut, *in* "Nucleases" (S. Linn, R. Lloyd, and R. Roberts, eds.), 2nd Ed., pp. 469–483. Cold Spring Harbor Laboratory Press, Cold Spring Harbor, New York, 1993.

[8] T. Uchida and F. Egami, *in* "The Enzymes" (P. Boyer, ed.), Vol. 4, 3rd Ed., pp. 205–223. Academic Press, New York, 1971.

TABLE I
ELECTROPHORETIC MIGRATION OFFSET CAUSED BY DIFFERENT TERMINI ON RNA OR DNA

Termini being compared	Charge difference	Migration offset (nt)	
		If in RNA[a]	If in DNA[b]
5′-P vs 5′-OH	−2	~0–1[c]	~1/2[d]
3′-P vs 3′-OH	−2	~1/4–2[c]	~1/2[d]
2′,3′-cyclic P vs 3′-OH	−1	~1/4–2[c]	NA
3′-P vs. 2′,3′-cyclic P	−1	~0[c]	NA

Abbreviations: nt, Nucleotide; NA, not applicable.

[a] Faster migration of an RNA bearing the indicated phosphorylated terminus, relative to an otherwise identical RNA bearing the second noted terminus.

[b] Faster migration of a DNA bearing the indicated phosphorylated terminus, relative to an otherwise identical DNA bearing the indicated OH terminus.

[c] Determinations are from Cruz-Reyes *et al.*[2]

[d] Determinations are from Sollner-Webb and Reeder.[10]

(summarized in Table I[2,10]). Unlike DNA, where terminal P differences alter migration by $\leq 1/2$ nucleotide[10], with RNA the migration offsets can be considerably larger and more variable.[2] In standard sequencing gels analyzing 5′-end-labeled RNA, molecules bearing a 3′-OH can migrate anywhere from ~0 to >2 nucleotides slower than otherwise identical molecules bearing a 3′-P. Figure 1A presents typical electrophoretic offsets resulting from different 3′ moieties. It shows an RNase T1 sequence ladder of 5′-labeled RNA in which the 3′ termini either bear their original 2′,3′-cyclic P moieties or were subsequently converted to 3′-P or 3′-OH moieties. Surprisingly, the ladder with 3′-OH termini migrates substantially slower than those with 2′,3′-cyclic P or 3′-P termini, which migrate similarly. The offset caused by 3′-OH versus 2′,3′-cyclic P termini varies from barely discernible to somewhat over 2 nucleotides in different electrophoretic runs and even throughout the length of a single gel, according to features not yet understood. Figure 1B and C illustrates different electrophoretic offsets in different electrophoretic runs; this variability prevents rigorous size determination of RNA when using common sizing standards with different 3′ moieties. Because of this large and variable electrophoretic offset, to avoid errors of up to 2 nucleotides in size assignment of an experimental RNA, it is important that a sequencing ladder have the same 3′-terminal character.

Electrophoretic offsets are also an issue when molecules differ in 5′-terminal character (i.e., using 3′-end-labeled RNAs to analyze the downstream fragment generated by an experimental cleavage). Molecules bearing a 5′-OH migrate slower

[9] D. Peattie, *Proc. Natl. Acad. Sci. U.S.A.* **76,** 1760 (1979).

[10] B. Sollner-Webb and R. Reeder, *Cell* **18,** 485 (1979).

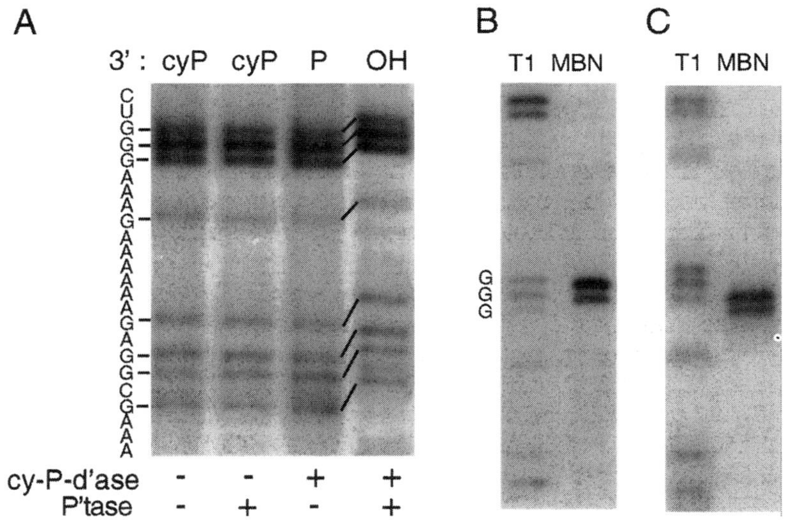

FIG. 1. Relative mobility of RNAs with 2′,3′-cyclic P, 3′-P or 3′-OH termini. (A) After 5′ labeling with phosphatase-resistant [^{35}S]thiophosphate, RNA was treated with RNase T1 (lanes 1) and in addition with phosphatase (P'tase; lane 2), 2′,3′-cyclic phosphodiesterase (cy-P-d'ase; lane 3), or 2′,3′-cyclic phosphodiesterase followed by phosphatase (lanes 4). Corresponding bands are indicated by lines, shown from ~70 to ~90 nucleotides in length. These data demonstrate that sequencing ladders generated by RNase T1 retain their 2′,3′-cyclic P termini (cyP) and migrate considerably faster than the same RNA with 3′-OH termini but similarly to otherwise identical RNA with 3′-P termini. In many electrophoretic runs, faster migrating fragments show greater offsets than the slower migrating fragments, but in other runs the reverse is observed. (B and C) Two separate sequencing gels, both 6% (w/v) polyacrylamide, analyzing 5′-^{32}P-labeled RNA treated with RNase T1 (T1, yielding a ladder with 2′,3′-cyclic P termini) or mung bean nuclease (MBN, which cleaves this RNA at two adjoining positions, yielding 3′-OH termini). The G triplet is ~90 nucleotides in length. [Data are from Cruz-Reyes *et al.*[2]]

than otherwise identical molecules bearing a 5′-P terminus, with the offset ranging from barely discernible to somewhat over 1 nucleotide in different electrophoretic runs. Because the magnitude of this offset also is not predictable, sequencing ladders should have the same terminal 5′-OH or 5′-P character as the experimental RNA to avoid a potential sizing error of 1 nucleotide.

Resolving the RNA Sizing Ambiguity

The previous sections have summarized that many RNA nucleolytic events of interest generate 3′-OH and 5′-P termini, yet to analyze the upstream fragment arising from such cleavages, there traditionally have not been protocols available to generate the needed sequencing ladders bearing 3′-OH termini. This section will summarize a convenient solution. One RNase long recognized to form

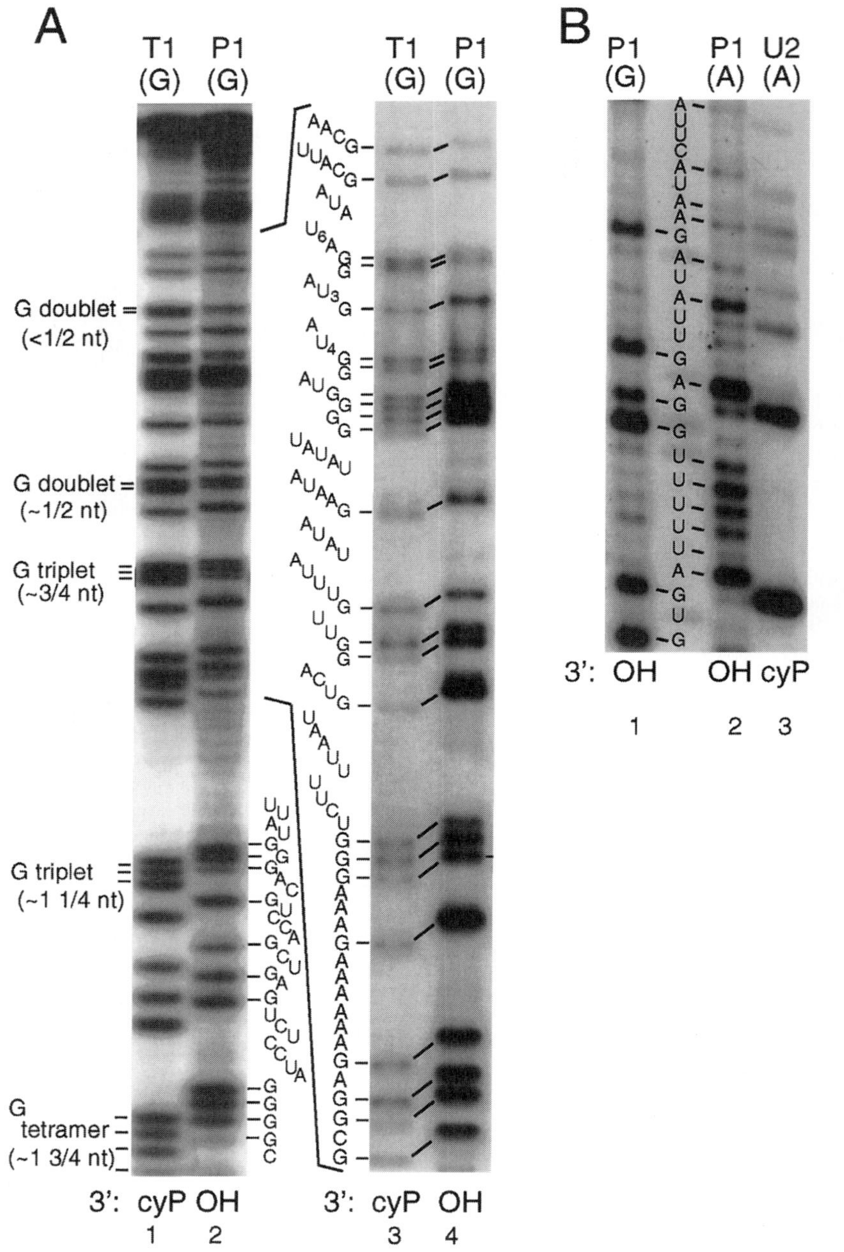

FIG. 2. Nuclease P1 can generate G-specific or A-preferential sequencing ladders and yields 3′-OH and 5′-P termini. 5′-^{32}P-labeled RNAs were treated with nuclease P1 under conditions where the cleavage is selective for occurring just 3′ off G residues [P1 (G)] or just 3′ off A residues [P1 (A); these conditions also cleave 3′ off U residues], with RNase T1 [T1 (G), which cleaves after G residues], or with RNase U2 [U2 (A), which cleaves just 3′ of A residues].

3′-OH and 5′-P termini is nuclease P1. However, nuclease P1 is generally accepted as relatively sequence nonspecific (e.g., Refs. 11 and 12). Nonetheless, under basic conditions (pH ~9), especially in the presence of urea, nuclease P1 was found to cleave specifically after G residues.[2] Under these conditions, nuclease P1 generates a sequence ladder that looks much like an offset version of an RNase T1 ladder (Fig. 2A). Furthermore, under acidic conditions (pH ~5), nuclease P1 was found to cleave most preferentially after A residues, but also after U residues.[2] In such reactions, the strong nuclease P1 bands form a ladder that looks much like an offset version of an RNase U2 (or RNase T2) ladder (Fig. 2B). Thus, mapping nucleolytic events that yield 3′-OH and 5′-P termini, relative to the known sequence of the substrate RNA, can be accomplished with nuclease P1 to generate a comparable marker ladder. If there are G residues near the experimental cleavage site, we recommend using the high-pH (G) conditions, but if not the low-pH (A + U) conditions form a favorable marker ladder. To identify the migration position of every residue, a ladder generated by alkali and an additional ladder generated by RNase T1 or T2 can be used to determine the migration offset. Alternatively, a ladder generated by nuclease P1 at neutral pH, where it cleaves more sequence nonspecifically, can be used. In this manner, nuclease P1 can serve to generate marker ladders which unambiguously determine the cleavage site of an experimental RNase that generates 3′-OH and 5′-P termini.

Enzymatic Treatments and Gel Electrophoresis

Enzymatic digestions can be carried out for 5–15 min at 50°, using $\sim 5 \times 10^4$ cpm of substrate RNA plus 1 μg of tRNA as carrier in 10 μl. G-specific conditions for nuclease P1 are 5–10 units of enzyme in 20 m*M* sodium P or sodium citrate (pH ~9), 6.5 *M* urea, 1 m*M* $ZnCl_2$. [Xylene cyanol (0.04%, w/v) can be included, allowing loading directly on the gel after denaturation at 95° for 2 min.] A-selective conditions for nuclease P1 are ~0.01 unit of enzyme in 20 m*M* sodium acetate (pH 5), 0.5 m*M* zinc acetate, and 12% (w/v) glycerol; this reaction should be treated with phenol and ethanol precipitated before gel loading. G-specific ladders by RNase T1 can be generated in many different buffers, although more homogeneous cleavage at G residues is favored on RNA denaturation, such as using 0.5–1 unit of enzyme in 33 m*M* sodium citrate (pH 5), 1.5 m*M* EDTA, 9.5 *M* urea, and can be loaded on the gel directly. Electrophoresis can be in Tris–borate–EDTA (pH 8.3) buffer, using 4–20% (w/v) polyacrylamide–8 *M* urea gels, depending on the size of the fragments.[13]

[11] M. Silberklang, A. Gillum, and U. L. RajBhandary, *Nucleic Acids Res.* **12,** 4091 (1977).

[12] U. RajBhandary, *FASEB J.* **39,** 2815 (1980).

[13] J. Sambrook, E. Fritsch, and T. Maniatis, "Molecular Cloning: A Laboratory Manual." Cold Spring Harbor Laboratory Press, Cold Spring Harbor, New York, 1989.

[32] *Chlamydomonas reinhardtii* as a Model System for Dissecting Chloroplast RNA Processing and Decay Mechanisms

By CLARE SIMPSON and DAVID STERN

I. Introduction

A. *Role of RNA Processing and Decay in Regulating Chloroplast Gene Expression*

RNA processing in chloroplasts includes mRNA 5′- and 3′-end processing, intron splicing, and intercistronic cleavages of polycistronic messages, as well as typical tRNA and rRNA processing. These posttranscriptional steps, along with changes in RNA stability, have received considerable attention for two reasons. First, changes in chloroplast gene expression during chloroplast biogenesis or in response to environmental signals have much more often been shown to be posttranscriptional rather than transcriptional. Second, genetic studies designed to identify nuclear mutants defective in chloroplast function have nearly exclusively identified posttranscriptional defects. These data, along with *in vitro* systems developed to dissect the relevant mechanisms, have been the subject of a number of reviews.[1–6]

B. *Comparison of Chlamydomonas with Other Model Systems*

The molecular genetics of chloroplast RNA metabolism has been primarily elaborated in three model systems: *Chlamydomonas,* maize, and *Arabidopsis.* These analyses have been complemented by *in vitro* assays in *Chlamydomonas* as well as in spinach and tobacco. Work in spinach, initially focused on transcription initiation,[7] has profited from the simplicity of isolating purified chloroplasts, whereas tobacco was the first vascular plant for which stable chloroplast transformation was established.[8] In terms of genetic screens, isolation of chloroplast

[1] M. Goldschmidt-Clermont, *Int. Rev. Cytol.* **177,** 115 (1998).
[2] A. Barkan and D. B. Stern, Chloroplast mRNA processing: Intron splicing and 3′-end metabolism. *In* "A Look beyond Transcription: Mechanisms Determining mRNA Stability and Translation in Plants" (J. Bailey-Serres and D. R. Gallie, eds.), p. 162. American Society of Plant Physiologists, Rockville, Maryland, 1998.
[3] J.-D. Rochaix, *Plant Mol. Biol.* **32,** 327 (1996).
[4] R. A. Monde, G. Schuster, and D. B. Stern, *Biochimie* **82,** 573 (2000).
[5] A. Barkan and M. Goldschmidt-Clermont, *Biochimie* **82,** 559 (2000).
[6] W. Zerges, *Biochimie* **82,** 583 (2000).
[7] W. Gruissem, B. M. Greenberg, G. Zurawski, D. M. Prescott, and R. B. Hallick, *Cell* **35,** 815 (1983).
[8] Z. Svab, P. Hajdukiewicz, and P. Maliga, *Proc. Natl. Acad. Sci. U.S.A.* **87,** 8526 (1990).

METHODS IN ENZYMOLOGY, VOL. 342
0076-6879/01 $35.00

biogenesis mutants is straightforward in haploid *Chlamydomonas,* as well as in diploid maize and *Arabidopsis.* Of these three, however, only *Chlamydomonas* has also been used extensively for *in vitro* assays. In light of these general comments, we review briefly the merits of each system.

1. Maize. The use of maize to study chloroplast biogenesis has been described in this series.[9] Its primary benefits are simple screens based on high chlorophyll fluorescence or leaf color, and the viability of homozygous mutant seedings for several weeks due to seed reserves, yielding useful amounts of material for mechanistic studies. Mutants have been characterized that affect a variety of posttranscriptional processes.[10–13] Maize also has certain disadvantages. For example, although transposon-tagged mutations are easy to obtain, cloning the inactivated nuclear gene can be somewhat tedious. In addition, mutants must be maintained as heterozygotes. Like other monocots, purified chloroplasts are difficult, but by no means impossible, to isolate in sufficient quantities to develop *in vitro* systems. Finally, the large maize nuclear genome is unlikely to be completely sequenced in the foreseeable future.

2. Arabidopsis. Like their maize counterparts, *Arabidopsis* chloroplast biogenesis mutants are seedling lethal and homozygotes must be maintained on sucrose, which itself may profoundly affect chloroplast metabolism.[14] *Arabidopsis* has the additional disadvantage of limiting material, especially in mutants, due to lack of seed reserves. These aspects are substantially counterbalanced by genomics resources, enabling rapid cloning of genes identified through mutational screens, although the number of cloned genes confirmed to participate in chloroplast RNA processing or stability is still limited.[15,16] In addition, the complete genome sequence offers an unparalleled opportunity to identify potential factors of interest. Indeed, up to 14% of all *Arabidopsis* proteins may be plastid targeted,[17] and gene families such as that encoding the potential regulatory PPR proteins have been identified[18] and will certainly be targets of reverse genetics. We view these functional genomic approaches as complementary to maize and *Chlamydomonas,* which have their own genetic and biochemical attributes.

3. Chlamydomonas. As a haploid, unicellular organism, *Chlamydomonas* has several advantages over its vascular plant counterparts (for review see Grossman[19]).

[9] A. Barkan, *Methods Enzymol.* **297,** 38 (1998).
[10] D. McCormac and A. Barkan, *Plant Cell* **11,** 1709 (1999).
[11] D. G. Fisk, M. B. Walker, and A. Barkan, *EMBO J.* **18,** 2621 (1999).
[12] B. D. Jenkins, D. J. Kulhanek, and A. Barkan, *Plant Cell* **9,** 283 (1997).
[13] A. Barkan, *Plant Cell* **5,** 389 (1993).
[14] J. Sheen, L. Zhou, and J. C. Jang, *Curr. Opin. Plant Biol.* **2,** 410 (1999).
[15] J. Meurer, C. Grevelding, P. Westhoff, and B. Reiss, *Mol. Gen. Genet.* **258,** 342 (1998).
[16] Y. Y. Yamamoto, P. Puente, and X. W. Deng, *Plant Cell Physiol.* **41,** 68 (2000).
[17] *Arabidopsis* Genome Initiative, *Nature (London)* **408,** 796 (2000).
[18] I. D. Small and N. Peeters, *Trends Biochem. Sci.* **25,** 46 (2000).
[19] A. R. Grossman, *Curr. Opin. Plant Biol.* **3,** 132 (2000).

Recessive mutations are readily revealed, and a lack of gene duplication relative to the ancient tetraploid genomes of maize[20] and *Arabidopsis*[21] reduces issues related to gene redundancy.[22] Its small size and rapid generation time facilitate large-scale screens, and nonphotosynthetic mutants are fully viable if provided a reduced carbon source. *Chlamydomonas* photosynthesis mutants also are viable in the dark, which allows light-sensitive strains to be maintained.

From a genomics point of view, *Chlamydomonas* is the only genetically tractable system in which both the nuclear and chloroplast genomes can be routinely transformed. Although this limitation is being overcome in vascular plants, the workload will still be far less in *Chlamydomonas.* A number of selectable marker and reporter gene options are available (see below), and progress in using antisense technology has been reported.[23] The *Chlamydomonas* Genetics Center, a repository for strains and clones, can be accessed at *http://www.biology.duke.edu/chlamy/,* and the relational database ChlamyDB is at *http://ars-genome.cornell.edu/cgi-bin/WebAce/webace?db=chlamydb.* The completion of the 204-kb chloroplast genome sequence has been achieved in the authors' laboratory (GenBank accession numbers AF396929), and sequencing efforts by the *Chlamydomonas* Genome Project and the Kazusa DNA Research Institute (Chiba, Japan) have deposited more than 55,000 expressed sequence tags (ESTs) from normalized libraries in GenBank. Physical mapping progress and other nuclear genome information has been reviewed.[24] A screenable bacterial artificial chromosome (BAC) library filter is available from Incyte Genomics (Palo Alto, CA; *http://www.incyte.com/*).

C. RNA Processing and Stability Factors in Chlamydomonas Chloroplasts

1. Factors Inferred from Mutant Studies. The functions of wild-type genes can be deduced from mutant phenotypes. A good example is a large group of nuclear genes whose products are required for *trans*-splicing of the *psaA* mRNA,[25] one of which has been cloned.[26] Other typical mutant phenotypes include rRNA maturation defects,[27] instability or blocked translation of individual mRNAs (see reviews cited above), or suppression of these phenotypes from reversion screens.

2. Mechanisms Inferred from Chimeric Gene Studies. For RNA stability factors in particular, it has been possible to locate the *cis* element associated with RNA instability in mutants, using reporter genes. In general, this has revealed that

[20] B. S. Gaut and J. F. Doebley, *Proc. Natl. Acad. Sci. U.S.A* **94,** 6809 (1997).
[21] T. J. Vision, D. G. Brown, and S. D. Tanksley, *Science* **290,** 2114 (2000).
[22] A. M. Settles, A. Baron, A. Barkan, and R. A. Martienssen, *Genetics* **157,** 349 (2001).
[23] M. Schroda, O. Vallon, F.-A. Wollman, and C. F. Beck, *Plant Cell* **11,** 1165 (1999).
[24] P. A. Lefebvre and C. D. Silflow, *Genetics* **151,** 9 (1999).
[25] M. Goldschmidt-Clermont, J. Girard-Bascou, Y. Choquet, and J.-D. Rochaix, *Mol. Gen. Genet.* **223,** 417 (1990).
[26] K. Perron, M. Goldschmidt-Clermont, and J.-D. Rochaix, *EMBO J.* **18,** 6481 (1999).
[27] S. P. Holloway and D. L. Herrin, *Plant Cell* **10,** 1193 (1998).

nuclear factors act on the 5′ untranslated region (UTR), and furthermore that they protect transcripts from a net 5′ → 3′-exonuclease activity (reviewed in Refs. 4 and 5). However, at least one factor acting at the 3′ UTR has been identified.[28] As expected, all translation factors tested have been shown to interact with the 5′ UTR.

3. Biochemical Data. Mechanistic details can be approached through *in vitro* studies, which in *Chlamydomonas* have been used to study mRNA 3′ processing[29] and the specificities and intraorganellar localization of a variety of RNA-binding activities.[30–32] In addition, cloning of stability factor genes has revealed through cell fractionation that these proteins exist in RNA-containing complexes.[33,34] The relevant techniques are outlined in Section II.G.3.

II. Methods

A. *Creation of Nuclear Mutants by Insertional Mutagenesis*

1. Overview. Insertional mutagenesis by glass bead-mediated transformation[35] has been widely used to create null mutants that are tagged by the transforming DNA, which integrates apparently randomly into nuclear DNA. Its efficiency makes it effective for any screen in which a null mutation would confer a useful phenotype, typically the loss (acetate dependence) or recovery (acetate independence) of photosynthesis. Integrated DNA is stably maintained in the nuclear genome, although expression of both homologous and heterologous sequences may be subject to silencing when selection is removed, particularly those derived from foreign genes.[36] In addition, transforming DNA insertion is often accompanied by deletions and/or rearrangements, which must be taken into account in strategies to recover sequences flanking the locus of interest.

2. Choice of Selectable Marker and Strain. The endogenous *NIT1* and *ARG7* genes have been widely used as nuclear mutagens, by complementing *Chlamydomonas* auxotrophs unable to reduce nitrate and synthesize arginine, respectively. Both markers are used as large genomic clones and the *ARG7* gene has extensive repetitive DNA, which are disadvantages for gene-tagging purposes. Plasmid rescue is the most efficient method of cloning DNA at the integration site. It requires a single, simple insertion of the transforming DNA, and the *Escherichia coli* replicon

[28] H. Levy, K. L. Kindle, and D. B. Stern, *J. Biol. Chem.* **274,** 35955 (1999).

[29] D. B. Stern and K. L. Kindle, *Mol. Cell. Biol.* **13,** 2277 (1993).

[30] A. Danon and S. P. Y. Mayfield, *EMBO J.* **10,** 3993 (1991).

[31] W. Zerges and J. D. Rochaix, *J. Cell Biol.* **140,** 101 (1998).

[32] F. Ossenbühl and J. Nickelsen, *Mol. Cell. Biol.* **20,** 8134 (2000).

[33] E. Boudreau, J. Nickelsen, S. D. Lemaire, F. Ossenbuhl, and J.-D. Rochaix, *EMBO J.* **19,** 3366 (2000).

[34] F. E. Vaistij, E. Boudreau, S. D. Lemaire, M. Goldschmidt-Clermont, and J.-D. Rochaix, *Proc. Natl. Acad. Sci. U.S.A.* **97,** 14813 (2000).

[35] K. L. Kindle, *Proc. Natl. Acad. Sci. U.S.A.* **87,** 1228 (1990).

[36] H. Cerutti, A. M. Johnson, N. W. Gillham, and J. E. Boynton, *Plant Cell* **9,** 925 (1997).

sequences must be functional. This method has been used successfully to clone *NIT1*-tagged alleles.[37,38] However, it is more often the case that bacterial sequences become uncoupled from the *Chlamydomonas* marker during integration, which requires the construction and screening of plasmid libraries for clones containing sequences present in the transformation vector, to derive associated flanking DNA. The *ARG7* gene frequently integrates as multiple, rearranged copies, creating large deletions at the integration site[39] (our unpublished results). Large deletions have also been characterized at *NIT1* insertion sites.[37] This complicates molecular analysis of transformants, and may necessitate chromosome walking from the locus of the cloned, flanking DNA.

Two newer constructs have been designed to address these issues. One is an *ARG7* shuttle marker intended to facilitate plasmid rescue in *E. coli;* the *Chlamydomonas RBCS2* promoter was fused to the *ARG7* cDNA rather than the longer and repetitive genomic clone, and a promoter was added to select for complementation of *E. coli argH* mutants.[40] Another is the dominant marker *ble,* which confers resistance to phleomycin or zeomycin, and can be used in any genetic background.[41] The first cassette contained the *Streptoalloteichus hindustanus ble* coding region fused to the *Chlamydomonas RBCS2* 5′ and 3′ UTRs. Versions of the *ble* cassette such as pSP108 also confer resistance in *E. coli,* driven by the *lac* promoter.[41] Striking increases in transformation efficiency have been achieved by insertion of two endogenous introns, and rare restriction sites have been introduced to facilitate analysis of transformants.[42] These are discussed in detail at *http://www.ucl.ac.uk/biology/ble1.htm,* where requests for clones can be made. Although *NIT1, ARG7,* and *ble* are most often used, other *Chlamydomonas* genes have also been used as selectable markers to rescue their cognate mutants, for example, *NIC7* (nicotinamide requirement), *THI10* (thiamine requirement),[43] and *CRY1* (cryptopleurine and emetine resistance).[44] In principle, any transformant population could be screened for photosynthetic mutants, as long as the transformation marker did not rescue an initial defect of this nature.

The following genomic clones are available from the *Chlamydomonas* Genetics Center: *NIT1* genomic (P-387), *NIT1* genomic, tagged (P-640); *ARG7* (P-389). A tag derived from ϕX174, added to an *ARG7* intron, allows introduced copies of

[37] L. W. Tam and P. A. Lefebvre, *Genetics* **135,** 375 (1993).

[38] K. Inoue, B. W. Dreyfuss, K. L. Kindle, D. B. Stern, S. Merchant, and O. A. Sodeinde, *J. Biol. Chem.* **272,** 31747 (1997).

[39] N. J. Gumpel, L. Ralley, J. Girard-Bascou, F.-A. Wollman, J. H. Nugent, and S. Purton, *Plant Mol. Biol.* **29,** 921 (1995).

[40] A. H. Auchincloss, A. I. Loroch, and J.-D. Rochaix, *Mol. Gen. Genet.* **261,** 21 (1999).

[41] D. R. Stevens, J.-D. Rochaix, and S. Purton, *Mol. Gen. Genet.* **251,** 23 (1996).

[42] V. Lumbreras, D. R. Stevens, and S. Purton, *Plant J.* **14,** 441 (1998).

[43] P. J. Ferris, *Genetics* **141,** 543 (1995).

[44] J. A. Nelson, P. B. Savereide, and P. A. Lefebvre, *Mol. Cell. Biol.* **14,** 4011 (1994).

ARG7 to be distinguished from the endogenous copy.[45] This clone is available from the authors. Many strains carrying *nit1, cw15,* and *arg7*[46] mutations are available from the *Chlamydomonas* Genetics Center, and suitable choices are discussed in Kindle.[47]

3. Glass Bead Nuclear Transformation Protocol. This widely used method and its alternatives have been reviewed in detail in an earlier volume.[47] Here we present methods geared to restoring photoautotrophy to a photosynthesis mutant, or to carrying out random mutagenesis with several of the markers described above. In any situation except with *NIT1,* the use of cell wall-deficient strains carrying the *cw15* mutation is recommended, although pretreatment of nondeficient strains with autolysin is another option. Introduction of the *cw15* marker is done by standard genetic techniques.[48] All manipulations should be carried out in a sterile hood.

1. Grow the recipient strain in nonselective medium [Tris–acetate–phosphate (TAP) for nonphotosynthetic strains; TAP supplemented with arginine at 100 μg/ml for *arg7* strains (arginine can be autoclaved during TAP preparation)]; SGII.NH_4 for *nit1* strains; TAP for *ble* transformation,[48] until cells reach early log phase (density of 1–2 $\times$ 10^6/ml). Pellet cells at moderate speed (e.g., 5000 rpm for 5 min in a GSA-type rotor).
2. Concentrate cells 100-fold by resuspending the pellet in selective medium (TAP-minimum, TAP without arginine, SGII.NO_3, or TAP medium for *ble* transformations), and shake at room temperature for 2–4 hr. Starving *nit1* cells for nitrogen allows efficient transformation of walled cells without addition of autolysin, probably because gametogenesis is induced and autolysin is released. This step also increases *NIT1* transformation efficiency.[35]
3. If transforming a walled strain, treat cells with gamete autolysin for approximately 45 min, instead of performing step 2. Resuspend the cell pellet from step 1 in 1/25 to 1/100 the original volume in autolysin. Spin and wash twice with selective medium before transformation, and then resuspend at 1/100 original culture volume and transform immediately. [Prepare autolysin as described in Kindle.[47] Test the efficacy of autolysin treatment by counting duplicate samples vortexed with and without detergent in a hemacytometer. *cw15* cells lyse in a 1% (v/v) solution of the nonionic detergent Nonidet P-40 (NP-40), whereas walled cells are unaffected.]

[45] N. J. Gumpel, J.-D. Rochaix, and S. Purton, *Curr. Genet.* **26,** 438 (1994).

[46] Somewhat confusingly, *arg2* and *arg7* are mutant alleles of the same gene, *ARG7*. Thus either can be complemented by transformation with the wild-type *ARG7* plasmid. The two mutant alleles will complement each other in diploids, and thus *arg2/arg7* has an Arg^+ phenotype.

[47] K. L. Kindle, *Methods Enzymol.* **297,** 27 (1998).

[48] E. H. Harris, "The *Chlamydomonas* Sourcebook: A Comprehensive Guide to Biology and Laboratory Use." Academic Press, San Diego, California, 1989.

4. Wash 0.5-mm glass beads with concentrated sulfuric acid, rinse with purified H_2O until the pH is neutral, allow to dry, and then weigh 300-mg aliquots into glass culture tubes and bake for approximately 2 hr at 204°).

5. Work in lots of a few tubes. To 300 μl of cells, aliquoted into 15-ml Corning (Acton, MA) tubes, add 100 μl of 20% (w/v) polyethylene glycol (PEG) 8000 (Sigma, St. Louis, MO) (autoclaved, must be freshly prepared periodically), and 1 μg of linearized DNA (increasing the amount of DNA increases the frequency of multiple insertions). Add 300 mg of sterile glass beads and vortex for 15–20 sec at top speed. Pipette cells immediately onto solid selection medium and spread gently with a sterile plastic BAC loop (Fisher Scientific, Pittsburgh, PA).

6. A 24-hr recovery period in TAP is required after *ble* transformations. Wash cells once with TAP to remove the PEG, and then resuspend in fresh TAP and allow 24 hr of growth. Plate on TAP containing phleomycin (2 μg/ml; Sigma) or zeomycin (Zeocin, 4–5 μg/ml; Invitrogen, Carlsbad, LA). Zeocin is less expensive and is therefore recommended.

4. Electroporation Protocol. A high-efficiency nuclear transformation method for cell wall-deficient strains has been developed.[49] Significant features of this method include addition of sucrose and carrier DNA, preincubation of cells at 10–20°, and mixing electroporated cells with a starch suspension to improve plating efficiency. Although this method requires determination of optimal electroporator settings and temperature for any particular strain, this is justified by the high transformation efficiencies, up to 2×10^5 transformants per microgram of DNA, which will facilitate large-scale insertional mutagenesis efforts. The *ARG7* and *NIT1* selectable markers give 10-fold higher efficiencies than the *ble* gene (K. Shimogawara, personal communication, 2001). Strains carrying the *cw15* and *cwd* mutations were tested,[49] and the method also works with walled strains treated with autolysin.[19] A modified protocol is provided here, adapted for use with the Bio-Rad (Hercules, CA) Gene Pulser (K. Shimogawara, personal communication, 2001).

1. Grow cultures to early log phase, and chill on ice before addition of a 1 : 2000 dilution of 10% (v/v) Tween 20. Harvest cells by centrifugation at 800*g* for 5 min at 4°, and resuspend in TAP containing 40 m*M* sucrose, at a final density of $1–4 \times 10^8$ cells/ml. Add 10 μg of linearized plasmid DNA and 200 μg of denatured salmon sperm DNA per milliliter of cell suspension. Place 40 μl of the suspension in a disposable 1-mm electroporation cuvette. Place the cuvettes in a water bath at the appropriate preincubation temperature (see below for comments on optimization). Cells must be electroporated within 1 hr of preincubation.

[49] K. Shimogawara, S. Fujiwara, A. Grossman, and H. Usuda, *Genetics* **148,** 1821 (1998).

2. Apply a 2.2-kV/cm electric pulse (±10–20% for optimization), using a Bio-Rad Gene Pulser, with capacitance set at 25 μF and resistance at infinity (∞).[50] These settings should give a 5-msec time constant.

3. After electroporation, incubate the cuvettes in a 25° water bath for at least 5 min, but no longer than 60 min, and plate 0.5–50 μl of cell suspension by adding 1 ml of starch suspension and spreading on a 9-cm petri plate. Use media appropriate to the selectable marker, as described in the glass bead protocol above. *ble* transformants require a 24-hr recovery in TAP liquid medium before plating with the starch suspension. Prepare the starch suspension by sequential washes of corn starch in distilled water and ethanol. Store washed starch in 70% (v/v) ethanol. Before transformations, subject the starch to repeated washes in TAP–sucrose, and then resuspend at 20% (w/v) in TAP–sucrose, and add PEG 8000 to 0.4% (w/v).

4. *Optimization:* Strains have different optimal preincubation temperatures, between 10 and 20°[49]; 15° is recommended as an empirical starting temperature (K. Shimogawara, personal communication, 2001). Strains also require different electric conditions for optimal transformation rates, which should fall into the range of 1.8–2.3 kV/cm.

B. Creating Mutants by Chloroplast Transformation

Chlamydomonas chloroplasts are transformed with a gene gun (Bio-Rad Biolistic PDS-1000/He system) to bombard cells spread on petri plates with DNA-coated tungsten particles. Transformation occurs via homologous recombination, which will introduce any foreign sequence into the genome if flanked by chloroplast sequences. Circular plasmids are generally used so that when a double crossover occurs, only the sequences between the flanking DNA, and not vector sequences, are introduced. Transformants are initially heteroplasmic (mixture of transformed and untransformed genomes), but after three or four passages under selection, the ~80 chloroplast genomes per cell become identical (homoplasmic). Chloroplast transformation allows gene inactivation, site-directed mutagenesis, and the introduction of chimeric constructs into intergenic regions. In each case, the recipient *Chlamydomonas* strain required and the design of the transformation cassette will be somewhat different, as described in Sections II.B.1, 2, and 4.

1. Transformation Method. Biolistic transformation is a standard procedure[51] and therefore not described in detail here. 5-Fluoro-2′-deoxyuridine (FdUrd) is a

[50] Double the volume of cell suspension to 80 μl and capacitance to 50 μF when using the Gene Pulser II.

[51] J. E. Boynton, N. W. Gillham, E. H. Harris, J. P. Hosler, A. M. Johnson, A. R. Jones, B. L. Randolph-Anderson, D. Robertson, T. M. Klein, K. B. Shark, and J. C. Sanford, *Science* **240,** 1534 (1988).

thymidine analog used to reduce chloroplast DNA copy number from 8–18% to 1–2% of total cellular DNA, presumably by inhibition of thymidylate synthase.[48,52] FdUrd (Sigma), can be added to 0.5 m*M* when inoculating the transformation culture. Make a 0.5 *M* stock in H_2O, filter sterilize, and store at $-20°$. This optional step is used to increase transformation efficiency. The use of FdUrd as a chloroplast DNA mutagen is described in Section II.C.3.

Transformation efficiency decreases after cells reach early log phase ($1–2 \times 10^6$ cells/ml), so counting cells is critical. Transformant colonies should be visible under a microscope in 4–5 days, or in 1 week by eye. Control plates of cells bombarded with particles lacking DNA should always be included to monitor spontaneous mutations, especially with antibiotic selection. Tungsten particles occasionally become oxidized, so preparation of a new batch of particles, and/or of spermidine, should be considered if transformation efficiency drops. Transformants should be streaked for single colonies and serially passaged under selection until polymerase chain reaction (PCR), Southern blots, or other screening protocols indicate that they are homoplasmic. The technique of cotransformation, where unlinked selectable and screenable markers are cobombarded, is applicable to both chloroplast and nuclear transformation. The efficiency of obtaining cotransformants carrying both the sequence under selection, and the second sequence of interest, is higher when the total size of the DNA to be introduced is limited. For chloroplasts, appropriate flanking sequences must be present in each transformation cassette.

a. Selection for Restoration of Photoautotrophy. Selection for restoration of photoautotrophy is powerful with low background, although specific recipient strains are required in each case. Recovery of photosynthesis through transformation of an *atpB* deletion strain was the first successful selection,[51] and the *atpB* downstream region remains a popular site for the insertion of reporter genes, which can be detected by screening after selection for photosynthesis. The *tscA* gene and the *petA–petD* intergenic region have been used in the same way.[53,54] When regulatory sequences of a chloroplast gene are being studied, for example, in the 5′ or 3′ UTRs, some introduced mutations will not disrupt gene function. If a strain is available carrying a deletion or other mutation in the gene of interest, it can be transformed with the modified copy of the gene, which serves as the selectable marker.[55,56]

[52] K. L. Kindle, K. L. Richards, and D. B. Stern, *Proc. Natl. Acad. Sci. U.S.A.* **88,** 1721 (1991).

[53] M. Goldschmidt-Clermont, Y. Choquet, J. Girard-Bascou, F. Michel, M. Schirmer-Rahire, and J.-D. Rochaix, *Cell* **65,** 135 (1991).

[54] W. Sakamoto, N. R. Sturm, K. L. Kindle, and D. B. Stern, *Mol. Cell. Biol.* **14,** 6180 (1994).

[55] Y. Choquet, D. B. Stern, K. Wostrikoff, R. Kuras, J. Girard-Bascou, and F.-A. Wollman, *Proc. Natl. Acad. Sci. U.S.A.* **95,** 4380 (1998).

[56] D. C. Higgs, R. S. Shapiro, K. L. Kindle, and D. B. Stern, *Mol. Cell. Biol.* **19,** 8479 (1999).

b. Antibiotic Resistance. Mutations in genes encoding 16S rRNA have provided selectable markers conferring resistance to spectinomycin[52] and streptomycin.[48,57] A reporter gene can be inserted adjacent to the rDNA, or introduced at an unlinked site by cotransformation.[58] The bacterial *aadA* gene confers resistance to spectinomycin and streptomycin,[59] and is available in several versions from the *Chlamydomonas* Genetics Center, differing in the flanking 5′ and 3′ UTRs. It has been widely used as a linked marker for site-directed mutagenesis of endogenous genes, as well as to discern sites of action of nuclear gene products. An additional advantage is that even when fused to *Chlamydomonas* sequences, *aadA* is active enough in *E. coli* that recombinant plasmids can be selected with ampicillin plus spectinomycin (20 μg/ml). In *Chlamydomonas,* apply selection by bombarding cells on plates containing spectinomycin (100 μg/ml). Select for homoplasmic transformants by streaking for single colonies several times on plates containing spectinomycin[56] at $\geq$400 μg/ml (the resistance level will depend on the promoter and 5′ and 3′ UTRs). Spontaneous mutations in the 16 rRNA gene will also confer spectinomycin resistance, and therefore the use of "no DNA" controls is essential.

aphA-6, which encodes aminoglycoside phosphotransferase, allows selection on kanamycin or amikacin.[60] The availability of a second dominant, portable marker will facilitate serial transformation of the chloroplast genome, although two marker recycling strategies have previously been proposed.[61] Direct drug selection is not possible for 24 hr after *aphA-6* transformation. Spread cells onto autoclaved nylon filters laid on TAP plates before particle bombardment. After 24 hr in dim light, transfer filters to TAP plus kanamycin (65 μg/ml). Streak colonies onto TAP plus kanamycin (100 μg/ml). *aadA* is recommended as the first choice of selectable marker, because selection for spectinomycin resistance can be applied immediately after particle bombardment. *aphA-6* is recommended for secondary transformation of spectinomycin-resistant strains.

2. Reverse Genetics. The inactivation of any chloroplast gene not required for cell viability can be achieved by chloroplast transformation, and many examples have been published for both *Chlamydomonas* and tobacco. Insert an *aadA* cassette into the gene to be disrupted, or use it to fully replace the coding region either in or out of phase. Include 1–2 kb of endogenous sequences on each side of the *aadA* cassette for efficient targeting. On transformation and selection for spectinomycin resistance, homologous recombination will disrupt the target gene. All wild-type genomes will be lost after an adequate period of selection, and the

[57] S. M. Newman, J. E. Boynton, N. W. Gillham, B. L. Randolph-Anderson, A. M. Johnson, and E. H. Harris, *Genetics* **126,** 875 (1990).

[58] X. Chen, K. Kindle, and D. Stern, *EMBO J.* **12,** 3627 (1993).

[59] M. Goldschmidt-Clermont, *Nucleic Acids Res.* **19,** 4083 (1991).

[60] J. M. Bateman and S. Purton, *Mol. Gen. Genet.* **263,** 404 (2000).

[61] N. Fischer, O. Stampacchia, K. Redding, and J.-D. Rochaix, *Mol. Gen. Genet.* **251,** 373 (1996).

resultant phenotype can be studied. In the case of essential[62] genes, however, it will not be possible to obtain homoplasmic transformants. One way around this problem is to weaken rather than disrupt the target gene. An example is *clpP*, whose function could be studied after mutagenesis of the translation initiation codon.[63]

3. Site-Directed Mutagenesis. cis elements in 5′and 3′ UTRs or introns can be studied by site-directed mutagenesis. The desired mutations are introduced by cotransformation, using *aadA* as the linked selectable marker. If the mutations in the target gene are subtle, it can be useful to introduce a silent restriction fragment length polymorphism (RFLP) to aid in PCR screening.[64] As mentioned above, the *aadA* gene can be introduced into virtually any site in the genome, although in some cases it may generate novel cotranscripts that may or may not be desirable. A number of examples using this strategy have been published.[56,63,65,66]

4. Chimeric Reporter Constructs. The two commonly used reporters in the *Chlamydomonas* chloroplast are the bacterial genes *uidA,* encoding β-glucuronidase (GUS), and *aadA,* which is also used extensively as a selectable marker. These reporters have the convenient property of visual assays, either colorimetric (GUS) or colony formation (*aadA*), and quantitative fluorometric (GUS) or enzyme (*aadA*) assays are also available. The functionality of 5′ and 3′ UTRs is determined by expressing the *uidA* and *aadA* coding regions under their control. The use of reporters allows the precise definition of *cis*-acting sequences mediating RNA metabolism and decay, particularly in the context of nuclear mutant backgrounds. Reporter genes are also used to create chimeric transcripts with synthetic phenotypes. For example, a chloroplast transformant was created to express chimeric 5′ UTR *rbcL*:*uidA*:3′ UTR *psaB* transcripts, which are 15 times less stable in light than in darkness.[67] Cotransformation is used to integrate reporter constructs into the chloroplast genome, as described above, for example, with selection being for photosynthesis and subsequent screening for *uidA* or *aadA*.

[62] "Essential" may either mean essential for cell viability, or essential for expression of the selectable marker cassette. One way this can be tested is by removing heteroplasmic transformants from selection and letting the genomes sort out stochastically. If only wild-type genomes are recovered, then the disrupted gene is essential for cell viability.

[63] W. Majeran, F.-A. Wollman, and O. Vallon, *Plant Cell* **12,** 137 (2000).

[64] X. Chen, K. L. Kindle, and D. B. Stern, *Plant Cell* **7,** 1295 (1995).

[65] J. Nickelsen, M. Fleischmann, E. Boudreau, M. Rahire, and J.-D. Rochaix, *Plant Cell* **11,** 957 (1999).

[66] R. G. Drager, D. C. Higgs, K. L. Kindle, and D. B. Stern, *Plant J.* **19,** 521 (1999).

[67] M. L. Salvador and U. Klein, *Plant Physiol.* **121,** 1367 (1999).

5. Poly(G) Tract. A sequence of 18 consecutive guanosine residues, the poly(G) tract, first used in yeast,[68] has been used as a tool to trap unstable transcripts and RNA-processing intermediates in the *Chlamydomonas* chloroplast. The initial use was to trap chloroplast transcripts that were unstable because of deletion of a 3′ UTR stem–loop structure, most likely by impeding a 3′→5′-exonuclease activity[69] that had been previously implicated in 3′ mRNA processing.[29] The poly(G) tract has also been used to prevent degradation of unstable transcripts in several nuclear mutant backgrounds. The poly(G) tract was inserted into the 5′ UTR of the genes of interest, at different positions relative to the mature mRNA 5′ end. The poly(G) was mapped to the 5′ ends of accumulating transcripts, and was interpreted as impeding 5′ → 3′-exonucleases.[65,66,70,71]

C. Obtaining Suppressors of Nonphotosynthetic Mutants

As mentioned above, the dispensable nature of photosynthesis in *Chlamydomonas* provides the opportunity to select for nucleus- or chloroplast-encoded second-site suppressors of nonphotosynthetic mutants. Because the initial mutants themselves may be due to either chloroplast (e.g., 5′ UTR) or nuclear mutations, there are four possible combinations (nuclear suppressor of chloroplast mutation, nuclear suppressor of nuclear mutation, etc.). Each of these combinations has been obtained in one or more studies. In general, chloroplast suppressors will identify *cis* elements or locally interactive sequences (e.g., base pairing within a UTR), whereas nuclear suppressors will identify additional *trans*-acting factors. All the classic advantages of suppressor screens apply to *Chlamydomonas,* for example, to identify interacting partners, or to obtain nonlethal alleles where null alleles would not survive or would have no phenotype. Once phenotypic reversion is obtained, genetic crosses must be used to find whether the mutation is a reversion, intragenic suppression, or extragenic, and whether it is in the nuclear or chloroplast genome. Pairwise crosses are used to determine whether different suppressors are allelic or unlinked.

1. Spontaneous Mutants. Isolation of spontaneous mutants allows recovery of either chloroplast and nuclear mutations, and is less likely to result in cells with multiple mutations, which complicate genetic analysis and may reduce fertility. Grow cultures to log phase (2.5×10^6 cells/ml), and then pellet cells and resuspend in 10 ml of minimal medium (lacking acetate) at 2.5×10^8 cells/ml. Spread 1 ml (2.5×10^8 cells) on each of ten 15-cm-diameter plates, after adding 3 ml of sterile

[68] C. J. Decker and R. Parker, *Genes Dev.* **7,** 1632 (1993).
[69] R. G. Drager, M. Zeidler, C. L. Simpson, and D. B. Stern, *RNA* **2,** 652 (1996).
[70] R. G. Drager, J. Girard-Bascou, Y. Choquet, K. L. Kindle, and D. B. Stern, *Plant J.* **13,** 85 (1998).
[71] F. E. Vaistij, M. Goldschmidt-Clermont, K. Wostrikoff, and J.-D. Rochaix, *Plant J.* **21,** 469 (2000).

0.7% (w/v) agarose (melted and kept at 45°) to each milliliter of cells. Allow the agar to harden, and then place the plates in high light to select for phenotypic revertants.

2. UV Mutagenesis. UV light is an effective mutagen, but may also generate sterile strains due to multiple mutations. Therefore, the dosage should be calibrated as described below. In our laboratory, a UV Stratalinker 1800 (Stratagene, La Jolla, CA) is used (a more powerful 2400 model is also available); the 1800 is programmed to deliver 6.6×10^4 μJ at 254 nm (a germicidal UV lamp can also be used).[48,72] Before mutagenesis, it is critical to establish the optimal dose, rather than relying on previously published values, or the LED reading of the instrument. Grow a liquid culture to 2.5×10^6 cells/ml in TAP, in dim light. Plate serial dilutions of cells on TAP, and then irradiate with a range of doses. Lids must be removed during irradiation. Plate serial dilutions of untreated cells from the same culture to determine the plating efficiency. The correct treatment will result in 5–10% cell survival, when plating efficiency is taken into account.

For large-scale mutagenesis, grow 1 liter of cells to 2.5×10^6 cells/ml in TAP in dim light (wrap the flask loosely in foil). Harvest the cells by centrifugation at room temperature. Resuspend the pellet in 10 ml of minimal medium at 2.5×10^8 cells/ml. Spread 1-ml aliquots onto each of ten 15-cm-diameter TAP-minimum plates, as described in Section II.C.1. Once the agar has hardened, remove the lids and irradiate with 6.6×10^4 μJ in the Stratalinker. Immediately place the plates in total darkness for 24 hr to prevent photoreactivation, and then move them to high light. Plating efficiency and viability of the cells used for mutagenesis should be monitored by plating untreated cells from the same culture on TAP.

3. Selection for Chloroplast Suppressors Using 5′-Fluoro-2′-deoxyuridine. In addition to the ability of FdUrd to reduce chloroplast DNA copy number, it is mutagenic to chloroplast DNA in stationary phase and during gametogenesis. In contrast, in other systems such as mammalian tumor cell lines, FdUrd is mutagenic only during the S phase of the cell cycle[73] and in vascular plants, it is ineffective probably because of uptake problems. FdUrd treatment has resulted in point mutations,[74] deletions, duplications,[48] and rearrangements[75] of the chloroplast genome, including a partial chloroplast suppressor of a nuclear mutation.[76]

[72] K. Shimogawara, D. D. Wykoff, H. Usuda, and A. R. Grossman, *Plant Physiol.* **120,** 685 (1999).

[73] T. S. Lawrence, M. A. Davis, H. Y. Tang, and J. Maybaum, *Int. J. Radiat. Biol.* **70,** 273 (1996).

[74] W. Zerges, J. Girard-Bascou, and J.-D. Rochaix, *Mol. Cell. Biol.* **17,** 3440 (1997).

[75] D. C. Higgs, R. Kuras, K. L. Kindle, F.-A. Wollman, and D. B. Stern, *Plant J.* **14,** 663 (1998).

[76] J.-D. Rochaix, M. Kuchka, S. Mayfield, M. Schirmer-Rahire, J. Girard-Bascou, and P. Bennoun, *EMBO J.* **8,** 1013 (1989).

Grow a culture of the strain to be mutagenized to midlog phase (5×10^6 cells/ml), and inoculate into TAP containing filter-sterilized FdUrd at 0.5–1 m*M*, to obtain a starting density of 1×10^4 cells/ml. Cultures can be divided into smaller flasks or tubes to recover independent mutations, and grown to stationary phase (10^7 cells/ml), followed by an additional 24-hr wait before plating[48] on minimal medium in bright light (E. Harris, personal communication).

D. Mutant Analysis

The protocols outlined below are germane to secondary screens of mutants or suppressors, or for the analysis of transformants. General principles are presented, although the steps needed will vary depending on the particulars of an experiment and not surprisingly, different laboratories use different protocols. Where multiple methods exist, we present ones that have worked well in our hands.

1. High Chlorophyll Fluorescence Screens. The original and simplest screen for photosynthesis defects is simply to determine the acetate requirement by replica plating a strain on TAP and minimal medium. In suppressor screens or for complementing PS^-mutants, plating on minimal medium is the selection. However, the number of genetic loci wherein mutations might cause acetate dependence is large, and many such mutants may not have a direct effect on chloroplast gene expression. For this reason, researchers using maize and *Arabidopsis* as well as *Chlamydomonas* have developed screens for high chlorophyll fluorescence (HCF). HCF results from the disruption of photosynthetic electron transport, because in its absence absorbed light energy is emitted as fluorescence rather than being used for photosynthesis. Excitement of chlorophyll with an appropriate wavelength of light will result in much higher fluorescence in photosystem I (PSI), photosystem II (PSII) or cytochrome b_6/f complex mutants relative to wild-type cells. For maize and *Arabidopsis* a hand-held longwave UV lamp is typically used.[77,78] For *Chlamydomonas,* however, it is possible to screen plates containing 100 or more colonies using excitation from below and monitoring fluorescence with a filtered CCD camera.[79] Although this device is not commercially available, the published design can be replicated by most machine shops. Screens are useful both for primary mutant isolation, for example, from insertional mutagenesis transformations, and also for tracking segregation of an HCF mutation after genetic crosses.

Because HCF screening is limited to electron transport mutants, mutations affecting the ATP synthase, ribulose-bisphosphate carboxylase (Rubisco) and some other genes will be missed. If the researcher wishes to target these, either acetate

[77] D. Miles, *Methods Enzymol.* **69,** 3 (1980).

[78] J. Meurer, K. Meierhoff, and P. Westhoff, *Planta* **198,** 385 (1996).

[79] P. Bennoun and D. Béal, *Photosynthesis Res.* **51,** 161 (1997).

screening or enrichment[80,81] will be required. On the other hand, mutations that indirectly affect electron transport (e.g., translation apparatus) may appear in HCF screens.

2. Total Protein Isolation. Protein subunits of the major photosynthetic complexes accumulate stoichiometrically. Thus, in most cases the entire complex will fail to accumulate if the synthesis or stability of one subunit is significantly affected. Therefore, immunoblot analysis of total proteins, using antibodies raised against selected complex subunits, allows lesions responsible for nonphotosynthetic phenotypes to be rapidly assigned to the photosynthetic complex (but not necessarily the precise gene) affected. Isolate total proteins from 3 ml of cells grown in TAP to midlog phase (5×10^6 cells/ml). Resuspend the cell pellet in 100 μl of sodium dodecyl sulfate (SDS) sample buffer [50 m*M* Tris (pH 6.8), 5% (v/v) 2-mercaptoethanol, 2% (w/v) SDS, 0.1% (w/v) bromphenol blue, and 10% (v/v) glycerol], and boil for 5 min. Remove insoluble material by centrifugation, and heat the soluble material to 65° before subjecting 10 μl to SDS–polyacrylamide gel electrophoresis (PAGE).[58] Variations may aid in the extraction and analysis of hydrophobic membrane proteins.[82]

3. Examination of RNA Phenotypes

a. Isolation of Total RNA. Use 1 ml of TRI reagent (Molecular Research Center, Cincinnati, OH) to isolate total RNA from a maximum of 10–15 ml of cells grown in TAP to midlog phase (5×10^6 cells/ml). Five to 10 μg of total RNA is sufficient for RNA filter hybridization experiments. Follow the manufacturer's instructions, with these modifications:

1. Add TRI reagent to the cell pellet, mix by pipetting, and incubate at 55° for 5–10 min.
2. Add 0.2 ml of chloroform, vortex at top speed for 15 sec.
3. After the final 70% (v/v) ethanol wash suggested by manufacturer, add a second 70% (v/v) ethanol wash.
4. Wash the pellet with 100% ethanol.
5. Drain the ethanol; do not air dry. Immediately resuspend gently in buffer or water. If necessary, incubate at 55° for 5–10 min.
6. Store samples or aliquots at −80°.

b. Run-on Transcription. A number of nuclear *Chlamydomonas* mutants that fail to accumulate individual chloroplast transcripts have been recovered. To differentiate between a transcription versus RNA stability defect, run-on transcription

[80] H. S. Shepherd, J. E. Boynton, and N. W. Gillham, *Proc. Natl. Acad. Sci. U.S.A.* **76,** 1353 (1979).
[81] G. W. Schmidt, S. Matlin, and N.-H. Chua, *Proc. Natl. Acad. Sci. U.S.A.* **74,** 610 (1977).

was performed. This technique entails the elongation in the presence of [^{32}P]UTP of previously initiated transcripts in permeabilized cells, and is also referred to as run-off transcription, or RNA pulse labeling. The resultant labeled RNA is used as a probe against blotted plasmid digests or dot/slot blots, and the signal will be proportional to the density of RNA polymerase molecules on the gene of interest and, theoretically, to the relative uridine content of the genes. If hybridized blots are not treated with ribonuclease, the relative sizes of target fragments, which must be in excess over the probe, will not be a factor. To determine that the targets are in excess, hybridize 3-fold dilutions of the probe to separate blots, and ensure that the relative transcription rates are identical. The length of labeling is also important. While we recommend 15 min, shorter pulses will be less affected by differential transcript stability, at the cost of diminished probe activity. The freeze–thaw permeabilization method[83] also labels nuclear and mitochondrial transcripts, so these may be assayed concurrently.[84] However, the signals from chloroplast transcripts are typically much stronger.

METHOD. Grow 5- to 50-ml cultures to stationary phase (1×10^7 cells/ml), harvest the cells, and wash twice in 10 m*M* HEPES (pH 7.5), 250 m*M* sucrose, 150 m*M* KCl, 0.4 m*M* phenylmethylsulfonyl fluoride (PMSF), and 1 m*M* EDTA. Pellet the washed cells and add buffer to obtain a final density of 5×10^7 cells/30 μl of buffer. Store in 30-μl aliquots at -80°. On the day of use, thaw 30-μl aliquots on ice, then freeze at -80° or on dry ice and rethaw three times.[69] Add 30 μl of 2× transcription buffer [1 *M* sucrose, 60 m*M* $MgCl_2$, 50 m*M* HEPES (pH 7.5), 15 m*M* dithiothreitol (DTT), 50 m*M* NaF, 0.25 m*M* GTP, 0.5 m*M* ATP, 0.25 m*M* CTP, 250 μCi of [^{32}P]UTP, 80 units of RNasin (Promega, Madison, WI)].[85] Leave on ice for 1 min, and then label at 26° for 15 min. Terminate the reaction by adding 10 μl of 20% (w/v) SDS and 130 μl of H_2O. Phenol–chloroform extract twice, and then precipitate the nucleic acids with ethanol. Wash the pellet with 70% (v/v) ethanol and resuspend in 100 μl of H_2O. Gently denature labeled RNA if desired, and then hybridize to immobilized DNA fragments representing the genes of interest.

c. In Vivo RNA Pulse–Chase. Rates of chloroplast transcript turnover, and thus actual half-lives, can be determined after the incorporation of ortho[^{32}P] phosphate during a 10-min pulse in phosphate-starved cells. This is followed by a chase of unlabeled orthophosphate. Deplete the cells of phosphate by growth in TAP containing 20 μM phosphate for at least four or five generations, until early log

[82] R. Kuras, C. de Vitry, Y. Choquet, J. Girard-Bascou, D. Culler, S. Buschlen, S. Merchant, and F.-A. Wollman, *J. Biol. Chem.* **272,** 32427 (1997).

[83] G. Gagné and M. Guertin, *Plant Mol. Biol.* **18,** 429 (1992).

[84] The older method of toluene permeabilization has been used successfully as an alternative to the freeze–thaw method [M. Guertin and G. Bellemare, *Eur. J. Biochem.* **96,** 125 (1979)]; however, only chloroplast transcripts are labeled.

[85] W. Sakamoto, K. L. Kindle, and D. B. Stern, *Proc. Natl. Acad. Sci. U.S.A.* **90,** 497 (1993).

phase (1–2 × 10^6 cells/ml) is reached. Pellet the cells and resuspend at a density of 2 × 10^7 cells/ml in TAP lacking phosphate, and then shake for 30 min. Add 40 μCi of ortho[^{32}P]phosphate per milliliter, allow to incorporate for 20 min with vigorous shaking at 25 μE m^{-2} sec^{-1} irradiance. Terminate the pulse with unlabeled orthophosphate added to 2 m*M* and then wash the cells once with TAP medium. Incubate with illumination during the chase.[86] Alternatively, terminate the pulse with medium containing 13.6 m*M* orthophosphate, and allow the chase to proceed in this medium.[87] Harvest 1 × 10^8 cells at each time point and isolate RNA immediately. Hybridize labeled RNA to immobilized DNA fragments of genes of interest.

4. Genetic Analysis. Standard methods for genetic manipulation are described in detail in *The Chlamydomonas Sourcebook.*[48] Briefly, perform genetic crosses to determine whether the mutation of interest is in the nuclear or chloroplast genome. Traits conferred by the chloroplast genome should be studied further by sequencing of the relevant genes, or may arise through genome rearrangements.[75] Determine through backcrossing to a wild-type strain whether the phenotype is conferred by a single nuclear mutation. More sophisticated techniques are available to assign mutations a map position on the 17 known chromosomes.[48]

For insertional mutants, to determine whether the phenotype of interest is tightly linked to the transformation cassette, cross the mutant to a suitable tester strain. For example, cross a nonphotosynthetic *ble* transformant to a wild-type (*ble*^{s}) strain; cross an *ARG7* transformant to an *arg2* or *arg7* strain. Examine the cosegregation of the mutant phenotype with DNA sequences derived from the cassette in transformant progeny genomes, using traditional DNA filter hybridization techniques, and/or with phenotypes conferred by the transformation vector such as phleomycin resistance or arginine prototrophy. Investigate cosegregation either in tetrad or random progeny. Because these mutants may have poor fertility, prior outcrossing to a wild-type strain may be necessary to obtain sufficient progeny. Overall, it can be expected that 30–50% of insertional mutants will have transforming DNA genetically linked to the phenotype, similar to results obtained with mutants arising in T-DNA-transformed *Arabidopsis* populations.

5. DNA Protocols

a. Rapid DNA Minipreparation for PCR of Chloroplast Genes. We use a rapid lysis procedure to obtain total DNA for PCR of the chloroplast genome, used for screening chloroplast transformants. Scrape cells from a plate, or use a pellet derived from a small liquid culture. Resuspend in 20–50 μl of lysis buffer [100 m*M*

[86] H. Lee, S. E. Bingham, and A. N. Webber, *Plant Mol. Biol.* **31,** 337 (1996).
[87] M. L. Salvador, U. Klein, and L. Bogorad, *Plant J.* **3,** 213 (1993).

Tris (pH 8.0), 100 m*M* EDTA, 250 m*M* NaCl, 0.1% (w/v) SDS], depending on the pellet size. Dilute lysed cells 10-fold with H_2O, vortex briefly, and use 1 μl for PCR.

b. Total DNA Isolation. The following protocol is modified from Rochaix.[88] It can be scaled up for larger culture volumes.

1. Incubate the pellet from 10 ml of cells grown to stationary phase for 1 hr at 55° in 1 ml of proteinase K buffer [10 m*M* Tris (pH 8.0), 10 m*M* EDTA, 10 m*M* NaCl]; add SDS to 0.5% (w/v) and proteinase K to 200 μg/ml.
2. Phenol–chloroform extract once, then add 50 μg of RNase A and incubate for 15 min at 37°. Repeat the organic extraction.
3. Add 0.1 volume of 3 *M* sodium acetate, pH 4.8, and 0.5 volume of 2-propanol to the aqueous phase, and collect DNA by centrifugation. Rinse the pellet with 70% (v/v) ethanol, and then drain and resuspend in 300 μl of TE.
4. Add 1 volume of 2 *M* NaCl, 20% (w/v) PEG 8000. Incubate on ice for >2 hr, or overnight at 4°.
5. Centrifuge for 30 min in a benchtop centrifuge at 4°, and wash the pellet two or three times with 70% (v/v) ethanol. Resuspend the pellet in 30 μl of TE.

E. Cloning Nuclear Genes Affected in Chloroplast RNA Metabolism Mutants

The genetic methods mentioned in the previous section are essential to define whether, after insertional mutagenesis, the nuclear locus of interest has been tagged by the transforming DNA. The requirements for a successful plasmid rescue approach were discussed above. If these conditions can be satisfied, that is, the insertion carries a selectable marker capable of expression in *E. coli,* proceed according to standard protocols.[37] Otherwise, construct plasmid libraries from DNA that has been size selected in an agarose gel, and identify clones containing sequences derived from the transformation vector.[89] Identify and sequence the *Chlamydomonas* flanking DNA in these clones, and use it to screen wild-type cosmid or BAC libraries. Attempt to rescue the mutant by complementation with genomic clones, using the glass bead protocol described above.

To complement recessive nuclear mutations created by other methods (e.g., UV mutagenesis), one approach is to rescue mutants by using an indexed wild-type cosmid library.[90] Genes encoding chloroplast RNA metabolism factors were

[88] J.-D. Rochaix, *Methods Enzymol.* **65,** 785 (1980).

[89] Southern blots carried out with different restriction enzymes and a vector probe from the transformation construct will reveal which digest will yield a useful size for cloning, that is, not too large to survive gel isolation but large enough to contain reasonable amounts of flanking *Chlamydomonas* DNA. It will be necessary to find single-copy sequences in this flanking DNA to proceed with gene isolation from the BAC library.

[90] H. Zhang, P. L. Herman, and D. P. Weeks, *Plant Mol. Biol.* **24,** 663 (1994).

cloned by using this[71] and other cosmid libraries.[26,33] Use of the high-efficiency electroporation method is recommended. Cloning genes where the phenotype is due to a dominant mutation or a second-site nuclear suppressor mutation[91] requires the construction of a genomic library from the mutant strain, which is transformed into the appropriate recipient strain. For example, if a photosynthetic revertant of a nonphotosynthetic strain carrying a mutation in a chloroplast gene is obtained, as in Higgs *et al.*,[56] a genomic library would be constructed from the suppressor strain, and used to rescue the original nonphotosynthetic strain.

F. Biochemical Approaches: In Vivo Methods

1. RNA Structure Determination. Chloroplast RNA secondary structure is of obvious importance in posttranscriptional gene regulation. Because computer predictive programs have limited reliability, direct analysis of RNA structure is a valuable complement. RNA secondary structure has been investigated in our laboratory by dimethyl sulfate (DMS) RNA modification.[56] DMS will readily infiltrate living cells and methylates cytosine and adenine residues (*in vivo* methylation), unless they are protected by base pairing or complexed to proteins.[92] These can be distinguished by treating purified total RNA (*in vitro* methylation), where only secondary structure would inhibit DMS modification. Primer extension is used to reveal sites of methylation, as methylated residues stop primer extension 1 nucleotide downstream of the methylated residue. As a control for stochastic pausing by reverse transcriptase, untreated RNA is used. The difficulty arises in deducing the actual secondary structure from the methylation patterns. RNA folding programs can be useful as well as evolutionary filtering, or even folding by manual inspection can be productive if only local pairing is assumed.

a. In Vivo. Resuspend the pellet derived from a 10-ml culture grown to 2×10^6 cells/ml in 1 ml of DMS buffer [10 m*M* Tris (pH 7.5), 10 m*M* $MgCl_2$, 3 m*M* $CaCl_2$], and add 5 μl of 7.9 *M* DMS (Aldrich, Milwaukee, WI). After a 5-min incubation at 25°, the reaction is terminated by adding 50 μl of 2-mercaptoethanol. Pellet the cells and extract RNA as described above. Twenty-five micrograms of RNA is used for primer extension. If a poor signal is obtained, it can often be improved by using a different primer.

[91] This applies only to dominant or semidominant suppressors. If the suppression is recessive, the only available approaches are map-based cloning or insertional mutagenesis of a heterozygous diploid [A. M. Preble, T. H. Giddings, Jr., and S. K. Dutcher, *Genetics* **157,** 163 (2001)], unless the suppressor is found to have a nonphotosynthetic phenotype in an otherwise wild-type background. In this case, it can be treated as any recessive mutant.

[92] DMS as well as other agents have been extensively used for ribosome and ribozyme structural analysis; our method was modified from one used in the Cech laboratory [A. J. Zaug and T. R. Cech, *RNA* **1,** 363 (1995)].

b. In Vitro. Add 75 μg of total RNA extracted from untreated cells to 200 μl of *in vitro* nuclease buffer [10 m*M* HEPES (pH 7.9), 230 m*M* KCl, 81 m*M* $MgCl_2$, 50 μM EDTA, 41 m*M* DTT, 8% (v/v) glycerol], and add 1 μl of 7.9 *M* DMS. After a 5-min incubation at 25°, the reaction is terminated by adding 50 μl of 2-mercaptoethanol. Precipitate the RNA and wash with ethanol to remove the DMS. Twenty-five micrograms of RNA is used for primer extension.

2. Use of Gene Expression Inhibitors. Chlamydomonas cells growing in liquid culture are an ideal model system in which to study the effects of certain inhibitors. Drugs are rapidly taken up by cells, and the immediate response can be determined. There are also no pleiotropic effects caused by wounding, for example. In general, short incubation times are recommended to reduce secondary effects. Strains that are resistant to particular inhibitors are available from the *Chlamydomonas* Genetics Center, and can be used to confirm the specific action of inhibitors tested.[93] The more costly inhibitors can be added directly to permeabilized chloroplasts or used in *in vitro* assays. Add inhibitors to cultures in early log phase. For convenience, divide cultures into aliquots, and harvest equal numbers of cells at desired time points after inhibitor addition.

a. Transcription Inhibitors. Actinomycin D is a nonspecific inhibitor of both nuclear and chloroplast transcription. It is added to logarithmically growing cultures to reveal transcript half-lives. Grow strain to 2–3 $\times 10^6$ cells/ml, and add actinomycin D to a final concentration of 50–100 μg/ml.[94,95] Tagetitoxin (Epicentre Technologies, Madison, WI) specifically inhibits the plastid-encoded *E. coli*-like RNA polymerase[96] and quantitatively inhibits chloroplast transcription in permeabilized *Chlamydomonas* cells (A. Zoeger and D. Stern, unpublished results, 2000); however, its cost precludes large-scale *in vivo* experiments. Rifampicin is another specific inhibitor of chloroplast transcription. It has been reported to inhibit the bulk of *Chlamydomonas* chloroplast transcription,[48,97] and has been used at 250 μg/ml[94]; however, *trnE* transcription was reported to be rifampicin insensitive.[98]

b. Translation Inhibitors. Cycloheximide inhibits translation on cytosolic ribosomes, and is typically added to cultures at 8–20 μg/ml[76,83,99,100] to study chloroplast RNA metabolism in the absence of *de novo* cytoplasmic protein

[93] These include cycloheximide, chloramphenicol, anisomycin, and other resistance markers (see ChlamyDB: *http://ars-genome.cornell.edu/cgi-bin/webAce/webace?db=chlamydb).*

[94] S. Hwang, R. Kawazoe, and D. L. Herrin, *Proc. Natl. Acad. Sci. U.S.A.* **93,** 996 (1996).

[95] N. N. Deshpande, Y. Bao, and D. L. Herrin, *RNA* **3,** 37 (1997).

[96] D. E. Mathews and R. D. Durbin, *J. Biol. Chem.* **265,** 493 (1990).

[97] S. J. Surzycki, *Proc. Natl. Acad. Sci. U.S.A.* **63,** 1327 (1969).

[98] D. Jahn, *Arch. Biochem. Biophys.* **298,** 505 (1992).

[99] R. Kawazoe, S. Hwang, and D. L. Herrin, *Plant Mol. Biol.* **44,** 699 (2000).

[100] J. F. Gera and E. J. Baker, *Mol. Cell. Biol.* **18,** 1498 (1998).

synthesis. Short (<2 hr) incubation times are recommended to limit the likely pleiotropic effects of a global block on the translation of chloroplast-destined proteins. Anisomycin can also be used.[48,99] Chloramphenicol and lincomycin are inhibitors of chloroplast translation. Inhibition of chloroplast translation may affect half-lives of chloroplast transcripts, if factors involved in decay are synthesized on chloroplast ribosomes, or if ribosome binding affects RNA stability directly. In one case, cells expressing a highly unstable *psaB* transcript were grown to midlog phase, and chloramphenicol at 100 μg/ml was added. Steady state levels of this transcript were significantly increased within 30 min.[86]

3. Effects of Light/Redox State on Chloroplast RNA Metabolism. Chlamydomonas is well suited to the study of the effects of light and redox state on chloroplast RNA metabolism. These are to some degree related because the redox poise of chloroplasts will very with light quantity and quality and thus the photosynthetic electron transport flux, or it can be manipulated experimentally.[101] The effect of light itself can be determined after placing cells under a light/dark cycle; however, this also synchronizes cell divisions. How to distinguish light from circadian effects is discussed below.

a. Studying Effects of Light and/or Redox State. Cells entrained by at least three cycles of a 12-hr light/12-hr dark regimen have been used to study group I intron splicing[95] and RNA degradation.[67] Cell division becomes synchronized under these conditions, provided that cultures are diluted daily to maintain density in early log phase.[87] Because the cells under investigation are synchronized, the effects of the cell cycle and circadian rhythm must be distinguished from the effects of light/redox changes, as discussed below. It is also necessary to exclude changes in transcription rates due to light, darkness, or inhibitors, by means of run-on transcription or transcription inhibitor experiments.[95]

To investigate whether changes in chloroplast RNA metabolism are caused by light stimulation of photosynthesis, the following inhibitors are used. *N'*-(3,4-Dichlorophenyl)-*N,N*-dimethylurea (DCMU) at 3–20 μM[67,95] inhibits electron transport from PSII to plastoquinone, and 2,5-dibromo-3-methyl-6-isopropyl-p-benzoquinone (DBMIB) at 1–10 μM[95] blocks transfer from plastoquinone to cytochrome b_6/f. DTT is a dithiol reductant, used at 5– 20 m*M*, whose specificity can be confirmed by using oxidized DTT as a control.[67] In contrast, the monothiol reductant 2-mercaptoethanol had much more limited effects in the latter study, which implicates an endogenous redox carrier with two sulfhydryl groups in redox regulation of RNA degradation. Interestingly, dithiol redox carriers have also been implicated in the light-regulated translation of chloroplast mRNAs.[102] The oxidant diamide also alters the redox state of cells, although incubation periods longer than 30 min were toxic to cells.[67]

[101] T. Pfannschmidt, A. Nilsson, and J. F. Allen, *Nature (London)* **397,** 625 (1999).
[102] T. Trebitsh, A. Levitan, A. Sofer, and A. Danon, *Mol. Cell. Biol.* **20,** 1116 (2000).

b. Cell Cycle. Cells entrained by at least three cycles of a light/dark regimen are required, as described above. Asynchronous cultures maintained at high density can provide a comparison with synchronous cultures diluted daily to keep density in early log phase.[87] Temperatures ranging from 23°[95] to 32°[67] have been used; 32° is a suitable growth temperature for rapid doubling of wild-type cultures during the light period (8-fold to 16-fold per day; U. Klein, personal communication, 2001), although nonphotosynthetic mutant strains are likely to become adversely affected at higher temperatures. To determine changes in RNA accumulation caused by the endogenous circadian control of chloroplast transcription, demonstrated for several genes,[94,99] perform parallel analyses of synchronized cultures transferred to continuous light or darkness.

G. Biochemical Approaches: In Vitro Methods

1. In Vitro Chloroplast RNA Processing and Degradation Assays

a. 3′ End Processing and Degradation. A spinach chloroplast stromal extract originally developed as an *in vitro* transcription system[103] was also found to be capable of efficient 3′ processing of synthetic tRNA and mRNA precursor transcripts, and rapidly degraded mRNAs lacking 3′ UTR inverted repeat sequences.[104] We later adapted this procedure to *Chlamydomonas* chloroplasts, and showed that correct 3′ processing of *atpB* precursor transcripts occurs *in vitro*.[29] Like the spinach extract, the *Chlamydomonas* proteins more rapidly degrade synthetic RNAs containing a poly(A) tail.[105] This extract may therefore be useful to study a variety of mRNA processing pathways. Because after chloroplast isolation the two protocols are identical, only the chloroplast isolation is described here. Note that yields will be substantially lower than for spinach, but several milligrams of protein may still be expected.

Isolation of intact chloroplasts is based on published procedures.[29,106] Subsequent steps are according to the published method.[103]

1. Grow a 4.8-liter culture of a strain carrying a cell wall-deficient mutation such as *cw15* to 2.5×10^6 cells/ml. Harvest cells and keep ice-cold.
2. During harvesting and washing of cells, prepare Percoll gradients (step 6).
3. Wash the cells in 20 m*M* HEPES, pH 7.5.
4. Resuspend at 10^8 cells/ml in breaking buffer (0.3 *M* sorbitol, 50 m*M* HEPES–KOH, 2 m*M* EDTA, 1 m*M* $MgCl_2$), plus 1% (w/v) bovine serum albumin (BSA).
5. Disrupt the cells in an ice-cold Kontes bomb (Kimble/Kuntes, Vineland, NJ; Yeda press). Pressurize 6 ml of cells to 35 lb/in^2 for 3 min.[107] Carefully bleed the disrupted cells into a cold 50-ml tube.

[103] W. Gruissem, B. M. Greenberg, G. Zurawski, and R. B. Hallick, *Methods Enzymol.* **118,** 253 (1986).
[104] D. B. Stern and W. Gruissem, *Cell* **51,** 1145 (1987).
[105] Y. Komine, L. Kwong, M. C. Anguera, G. Schuster, and D. B. Stern, *RNA* **6,** 1 (2000).
[106] W. R. Belknap, *Plant Physiol.* **72,** 1130 (1983).

6. Make 70 and 45% (v/v) Percoll solutions in breaking buffer plus 1% (w/v) BSA. Pour a step gradient with 4 ml of the 70% (v/v) solution carefully overlaid with 4 ml of the 45% (v/v) solution, in a 15-ml Corex tube.

7. Isolate intact chloroplasts by loading 3 ml of lysate from step 5 onto each step gradient.[108] Centrifuge at 11,500 rpm in an HB-4 or HB-6 swinging bucket rotor for 10 min.

8. Aspirate off the clear 3-ml sample layer and green film. Collect the band of intact chloroplasts at the 45%/70% interface with a transfer pipette. Unbroken cells will pellet, and broken chloroplasts will form a film at the top.

9. Dilute intact chloroplasts 6-fold in breaking buffer without BSA in 250-ml centrifuge bottles. Pellet by accelerating the GSA rotor to 5000 rpm, and then turning it off.

10. Resuspend the chloroplasts in 8 ml of buffer A [10 m*M* Tris (pH 7.9), 1 m*M* EDTA, 5 m*M* DTT].

b. Assays of 5′ UTR Stability. A chloroplast lysate from the mutant strain *nac2-26,* in which *psbD* transcripts are specifically destabilized *in vivo,*[109,110] preferentially degraded synthesized *psbD* 5′ UTR transcripts *in vitro,* compared with a wild-type lysate. Chloroplast isolation was substantially similar to the method described above, except a BioNebulizer (Glas-Col, **http://www.glascol.com**)[111] was used to break cells. Chloroplasts were lysed in a buffer containing 0.2% (v/v) Triton X-100, which appeared to release additional activities from membranes.

2. In Vitro RNA-Binding Assays. UV cross-linking of chloroplast lysates and chloroplast subfractions to *in vitro*-labeled transcripts has been used to identify numerous RNA-binding proteins.[112] The particular isolation/fractionation procedures will significantly affect the spectrum of proteins identified. Of particular note is a *Chlamydomonas* chloroplast fractionation procedure that allows the isolation of low-density membranes.[31] These are stably associated with RNA-binding proteins, and are physically associated with thylakoid membranes. Also, identification of RNA-binding activities can be complicated by comigration in gels, Mg^{2+},[31] redox status, phosphorylation, storage procedures,[32] and other factors. The reader

---

[107] The ideal breaking pressure will vary between strains. Once the first few gradients have been run, note the proportions of broken versus intact chloroplasts, and the amount of unbroken cells. In a good preparation most cells will be broken, and about 50% of the chloroplasts will be intact.

[108] Although we have tried other types of tubes, the 15-ml Corex tube provides the best separation and visibility.

[109] M. R. Kuchka, M. Goldschmidt-Clermont, J. van Dillewijn, and J.-D. Rochaix, *Cell* **58,** 869 (1989).

[110] J. Nickelsen, J. Van-Dillewijn, M. Rahire, and J.-D. Rochaix, *EMBO J.* **13,** 3182 (1994).

[111] C. M. Okpodu, D. Robertson, W. F. Boss, R. K. Togasaki, and S. J. Surzycki, *BioTechniques* **16,** 154 (1994).

is referred to the primary literature as well as other specialized references.[113] Gel mobility shift assays and RNA affinity chromatography have also been used to identify *Chlamydomonas* chloroplast proteins that bind to 5′ UTRs.[114,115] Many of the proteins identified by these methods have been implicated in translation, rather than RNA processing and decay.

3. Identification of High Molecular Weight Complexes Containing RNA Maturation Factors and RNA. Many RNA-processing factors exist in multiprotein complexes. Reports have also implicated such complexes in the metabolism of *Chlamydomonas* chloroplast transcripts.[33,71] While the reader is referred to the original literature for a detailed protocol,[33] the general strategy is described here. Soluble extracts are prepared by lysis of cells washed in a protein inhibitor mix. This is followed by sonication and ultracentrifugation to remove membranes. The supernatant is fractionated in a continuous 0.1–1.3 *M* sucrose gradient, or by size-exclusion chromatography. Aliquots of fractions are subjected to SDS–PAGE, immunoblotting, and the use of antibodies raised against the protein of interest. A significant finding was that RNA is present in fractions enriched for the protein factors under investigation.

III. Conclusions

Because posttranscriptional mechanisms are so important in chloroplasts, many methods have been developed or adapted to increase our understanding of them. For *Chlamydomonas,* genetic and transformation methods are well established, whereas biochemical approaches have been reported only more recently. We expect an explosion of new information in the coming years. In addition, as for other systems, genomics projects will increase the rate of gene cloning and also the pressure to develop, in parallel, proteomic and mechanistic assays.

[112] D. B. Stern, H. Jones, and W. Gruissem, *J. Biol. Chem.* **264,** 18742 (1989).

[113] G. Schuster and W. Gruissem, *In vitro* processing of chloroplast RNA and analysis of RNA–protein interactions by ultraviolet cross-linking. *In* "Methods in Plant Molecular Biology: A Laboratory Course Manual" (P. Maliga, D. F. Cashmore, A. R. Cashmore, W. Gruissem, and J. E. Varner, eds.), p. 209. Cold Spring Harbor Laboratory Press, Cold Spring Harbor, New York, 1995.

[114] A. Danon and S. P. Mayfield, *Science* **266,** 1717 (1994).

[115] A. Danon and S. P. Mayfield, *EMBO J.* **13,** 2227 (1994).

[33] Chloroplast mRNA 3′-End Nuclease Complex

By SACHA BAGINSKY and WILHELM GRUISSEM

Control of mRNA stability has been established as an important mechanism for the regulation of chloroplast gene expression. For example, during the development of etioplasts into photosynthetically active chloroplasts in higher plants, the relative transcription rates of many chloroplast DNA-encoded genes do not change significantly. Instead, the rapid accumulation of several chloroplast mRNAs is accompanied by increases in their relative half-lives.[1–5] These observations suggest that mRNAs in etioplasts are rapidly degraded but become stabilized during chloroplast development. The increase in RNA stabilization appears to be linked to specific processing events at the 5′ and 3′ ends of the respective mRNAs, but the exact mechanisms that link RNA processing and stability are still incompletely understood.

The RNA in the 3′ untranslated region (3′ UTR) of most plastid mRNAs is predicted to fold into a stem–loop structure, which stabilizes the mRNA by protecting it from nucleolytic degradation.[6] These stem–loop structures, however, differ widely in their predicted thermodynamic stabilities, which often do not correspond to the observed stability of the mRNAs. It has been argued, therefore, that the secondary structure alone cannot account for the differential mRNA stabilities found during chloroplast development.[6,7] Several proteins involved in chloroplast mRNA 3′-end processing have now been identified, among them a high molecular weight protein (HMW) complex with an apparent molecular mass of approximately 550 kDa. Although the complete subunit structure of the complex remains to be elucidated, one of its components is a 100-kDa RNA-binding protein with an enzymatic activity that can degrade RNA processively in a 3′→5′ direction. In the process, the enzyme produces nucleoside diphosphates, the typical product of phosphorolytic RNA degradation.[7] The protein resembles *Escherichia coli* polynucleotide phosphorylase (PNPase), which has an important role in prokaryotic mRNA decay. *In vitro,* the isolated chloroplast HMW protein complex alone completely degrades mRNA precursor molecules with extended 3′ ends. Thus, in order to process the mRNA 3′ end correctly, the HMW complex requires additional

[1] X. W. Deng and W. Gruissem, *Cell* **49,** 379 (1987).
[2] R. R. Klein and J. E. Mullet, *J. Biol. Chem.* **262,** 4341 (1987).
[3] P. Klaff and W. Gruissem, *Plant Cell* **3,** 517 (1991).
[4] W. Gruissem, *Cell* **27,** 161 (1989).
[5] W. Gruissem and J. T. Tonkyn, *Crit. Rev. Plant Sci.* **12,** 19 (1993).
[6] D. B. Stern and W. Gruissem, *Cell* **51,** 1145 (1987).
[7] R. Hayes, J. Kudla, G. Schuster, L. Gabay, P. Maliga, and W. Gruissem, *EMBO J.* **15,** 1132 (1996).

0076-6879/01 $35.00

proteins that must confer specificity to the processing reaction. Several additional chloroplast RNA-binding proteins (ctRNPs) have now been identified, including a group of proteins that range in molecular mass between 24 and 33 kDa. These nuclear-encoded ctRNPs interact with the HMW complex and appear to have a critical role in the regulation of mRNA stability in higher plant chloroplasts.[8–10]

Correct RNA processing *in vitro* requires the presence of the HMW complex, the ctRNPs, and at least one endonuclease. It is not known how these protein factors interact during chloroplast mRNA processing to establish a mature and stable 3′ end. In this chapter we discuss procedures for the purification of the proteins involved in chloroplast mRNA 3′-end processing, together with appropriate control assays to monitor the proteins throughout their purification. Figure 1 summarizes the protein fractionation protocol, and results are shown for representative fractionation experiments using EconoQ ion-exchange chromatography (Fig. 2), single-strand DNA affinity chromatography (Fig. 3), and gel-filtration chromatography (Fig.4). The protocol presented here should facilitate our understanding of chloroplast RNA processing and metabolism, but will also be of general utility in the purification of RNA-processing components.

Preparation of RNA Probe

The *petD* mRNA has been established as a useful model transcript with which to study RNA–protein interactions and processing at the 3′ end of RNA *in vitro*. The *petD* gene encodes subunit IV of the cytochrome b_6/f complex, which is directly involved in photosynthetic electron transport. During chloroplast development, *petD* mRNA is cleaved from the polycistronic *psbB-psbH-petA-petD* precursor to yield a monocistronic transcript that accumulates in the light,[11] but is rapidly degraded in the dark.[12] Processing at the 3′ end of the *petD* mRNA terminates the transcript in a stem–loop structure, which confers stability to the RNA. The correct processing at the 3′ end of the *petD* mRNA can be reproduced *in vitro* with a *petD* 3′-UTR RNA probe synthesized by transcription from a plasmid carrying the *petD* 3′ UTR.[6] The synthetic RNA contains 70 nucleotides of the coding region and extends 58 nucleotides 3′ proximal to the stem–loop, which itself is 46 nucleotides long. To produce the RNA substrate, the plasmid encoding the transcript is linearized by digestion with *Xba*I and purified by phenol–chloroform extraction. The transcription reaction is performed for 1 hr at 37°, using 0.5 μg of linearized plasmid as DNA template in a total volume of 20 μl containing 40 m*M*

[8] G. Schuster and W. Gruissem, *EMBO J.* **10,** 1493 (1993).
[9] M. Sugita and M. Sugiura, *Plant Mol. Biol.* **32,** 315 (1996).
[10] R. Hayes, J. Kudla, and W. Gruissem, *Trends Biochem. Sci.* **24,** 199 (1999).
[11] P. Westhoff and R. G. Herrmann, *Eur. J. Biochem.* **171,** 551 (1988).
[12] J. Kudla, R. Hayes, and W. Gruissem, *EMBO J.* **15,** 7137 (1996).

Tris-HCl (pH 8.0), 6 m*M* $MgCl_2$, 2 m*M* spermidine, 10 m*M* dithiothreitol (DTT), 0.5 m*M* each of GTP, CTP, and ATP, 0.1 m*M* UTP, 10 U of RNasin (Boehringer Mannheim/Roche, Indianapolis, IN), 10 U of T7 RNA polymerase (Boehringer Mannheim/Roche), and 100 μCi of [α-^{32}P]UTP (20 μCi/μl, 800 Ci/mmol). After transcription, the RNA products are isolated by ethanol precipitation (addition of 15 μl of 5 *M* ammonium acetate and 100 μl of ethanol, incubation for 30 min at $-20°$, followed by centrifugation at 10,000 rpm in an Eppendorf centrifuge). To improve the quality of the RNA template, it may be necessary to purify the full-length transcript by gel electrophoresis. For this purpose, the precipitated RNA is resuspended in formamide dye [deionized formamide, 0.1%(w/v) xylene cyanol, 0.1% (w/v) bromphenol blue], heated to 85° for 2 min, and loaded onto a denaturing acrylamide gel (minigel) [5%(w/v) acrylamide (19 : 1), 8 *M* urea in 0.5$\times$ TBE]. Gels are preelectrophoresed for 1 hr at 200 V in 0.5$\times$ TBE before sample loading. The sample is electrophoresed for 30 min at 200 V, after which the wet gel is covered with Saran Wrap and exposed to X-ray film. The full-length transcript (300 nucleotides) is excised and eluted from the gel matrix in an Eppendorf tube by adding 180 μl of diethyl pyrocarbonate (DEPC)-treated water, 20 μl of DEPC-treated 3 *M* sodium acetate (pH 5.5), 1 μl of DEPC-treated 0.25 *M* EDTA, and 1 μl of 20% (w/v) sodium dodecyl sulphate (SDS). The mixture is incubated overnight at 4°, after which the elution buffer is removed and transferred to a new Eppendorf tube, and the RNA is recovered by ethanol precipitation. The RNA template is then resuspended in 100 μl of DEPC-treated water and stored in 10-μl aliquots at $-80°$.

In Vitro Analysis of RNA-Binding Activities

Protein fractions can be analyzed for RNA-binding proteins that interact with an RNA template, using UV light cross-linking.[8] During irradiation of the nucleic acid–protein mixture with UV light, purine and pyrimidine radicals are generated that form covalent bonds with nearby amino acids. After UV light irradiation, RNA not protected by proteins is removed by digestion with RNases, and the protein mixture is then separated by SDS–polyacrylamide gel electrophoresis (PAGE).[13] Proteins covalently attached to radioactive nucleic acids are indirectly labeled and can be detected by autoradiography.

Using UV light cross-linking assays, partially fractionated chloroplast protein mixtures can be analyzed for proteins that bind the *petD* 3′-end mRNA. Approximately 5 μg of the chloroplast protein mixture [add buffer E (buffer composition is described below) to the protein fraction to give a total volume of 5 μl], 2 μl of RNA probe (50,000 cpm/μl), 3.75 m*M* $MgCl_2$, 2 m*M* DTT, and 10 m*M* KCl are combined in a total volume of 10 μl in an Eppendorf tube. The mixture is incubated on ice for 5 min and then placed in a UV-Stratalinker (Stratagene,

[13] U. K. Laemmli, *Nature* (*London*) **227,** 680 (1970).

La Jolla, CA) with the lids of the Eppendorf tubes opened. The RNA–protein mixture is then irradiated twice with 0.12 J. After addition of 2 μg of RNase A, the mixture is incubated at 37° for 20 min. The reaction is terminated by addition of 4 μl of SDS-sample buffer, the samples are heated for 5 min at 75°, and proteins are separated by 10%(w/v) SDS–PAGE.[13] After electrophoresis, the gel is dried and subjected to autoradiography. A typical RNA–protein cross-linking profile is shown in Fig. 2. In this assay a partially fractionated chloroplast protein fraction was subjected to UV cross-linking with the *petD* 3′-UTR RNA template and the interacting proteins were analyzed by SDS–PAGE. By comparison with molecular weight markers, RNA-binding proteins of 100, 55, and 41 kDa and the smaller ctRNPs of 33, 28, and 24 kDa can be detected (Fig. 2).

In Vitro Analysis of RNA-Processing Activities

Although the UV cross-linking assay is a useful tool with which to monitor RNA-binding proteins during a purification procedure, it does not provide information on the catalytic activities or function of the proteins in RNA processing. To analyze the RNA processing or degradation activity of isolated proteins or protein fractions, an *in vitro* RNA-processing assay is useful. In this assay, a radioactively labeled RNA probe (prepared as described above) is incubated with the protein fraction of interest. After incubation, the RNA is purified and analyzed by polyacrylamide gel electrophoresis (described previously) and autoradiography.

In the specific assay described here, chloroplast proteins are analyzed for processing activity of the *petD* 3′-end precursor RNA by combining the following: 5 μg of the isolated chloroplast protein fraction in buffer E (buffer composition is described below), buffer E to give a total volume of 5μl together with the chloroplast protein mixture, 2 μl of RNA substrate, 3.75 m*M* $MgCl_2$, 2 m*M* DTT, 10 m*M* KCl, and 1 m*M* sodium phosphate (pH 7) in a total volume of 10 μl in an Eppendorf tube. Samples are incubated for 30–60 min at room temperature. The reaction is terminated by addition of 90 μl of stop solution [6 *M* urea, 1% (w/v) SDS, 4.5 m*M* aurin tricarboxylic acid (Sigma, St. Louis, MO)] and the RNA is purified by phenol–chloroform extraction. The RNA products are precipitated by addition of 2 μl of 3 *M* sodium acetate (pH 5.5), 2 μg of *E. coli* tRNA, and 200 μl of ethanol, incubation at −20° for 30 min, and analyzed by RNA gel electrophoresis (as described previously) and autoradiography. RNA-processing activity can be monitored by the maturation of the 300-nucleotide precursor RNA to a smaller product of 209 nucleotides, which is stabilized for approximately 120 min (Fig. 5, lanes 1 and 2). The 3′ end of this *in vitro*-processed product has been previously shown to coincide with the 3′ end of the mature *petD* mRNA *in vivo*. In both cases, the 3′ end of the RNA terminates a few nucleotides downstream of the stem–loop structure (Fig. 5).

To monitor the quality of the chloroplast protein fractions, the processing reaction should be stopped at various time points (e.g., 2, 5, 10, 20, 40, and 60 min)

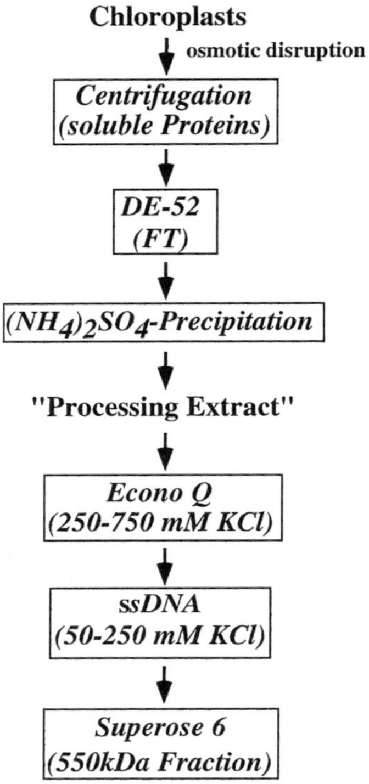

FIG. 1. Schematic presentation of the biochemical fractionation protocol developed for the analysis of proteins involved in chloroplast mRNA 3′-end processing and RNA degradation.

and RNA products should be analyzed. A high-quality chloroplast protein fraction containing RNA-processing activity should process the precursor in 1–3 min and stabilize the mature product for up to 120 min.

Fractionation of Proteins Involved in mRNA 3′-End Processing

The functional assay described above is useful for the identification of RNA-binding proteins and dissection of their interaction during chloroplast mRNA 3′-end processing. In this section, we discuss the purification of the protein factors and the functional characterization of their activities. A schematic diagram of the fractionation procedure is shown in Fig. 1.

Preparation of Chloroplast RNA-Processing Extract

Intact chloroplasts are isolated as described[14] from 500 g of washed leaf material, harvested from 3-week-old spinach plants (hydroponically grown under

[14] W. Gruissem, B. M. Greenberg, G. Zurawski, and R. B. Hallick, *Methods Enzymol.* **118,** 253 (1986).

short-day conditions). The leaf tissue is homogenized in a Waring blender in grinding solution (5× GM mix: 5 mM $Na_4P_2O_7$, 250 mM HEPES, 1.65 M sorbitol, 10 mM EDTA, 5 mM $MgCl_2$, 5 mM $MnCl_2$, adjusted to pH 6.8 with NaOH; diluted to 1× GM mix before use with water and DTT added to a final concentration of 10 mM) and filtered through two layers of cheesecloth and two layers of Miracloth. The filtrate is centrifuged for 15 min in a GS-3 rotor at 5000g, and the resulting pellet is resuspended in a small volume (plan for a final volume of 8 ml to load onto one Percoll gradient) of the grinding solution. The resuspended chloroplast solution is loaded onto a Percoll step gradient [Percoll step gradients: 20 ml of a 40% (v/v) Percoll solution is layered on a 7.5-ml 70% (v/v) Percoll solution].

PCBF: 100 ml of Percoll, 3 g of polyethylene glycol (molecular weight 8000), 1 g of bovine serum albumin (BSA), 1 g of Ficoll

Percoll solution (40%, v/v): 20 ml of 5× GM mix, 6 mg of reduced glutathione, 1 ml of 1 M, DTT, 40 ml of PCBF, H_2O to 100 ml

Percoll solution (70%, v/v): 20 ml of 5× GM mix, 6 mg of reduced glutathione, 1 ml of 1 M, DTT, 70 ml of PCBF, H_2O to 100 ml

The solution is centrifuged at 8000 g for 20 min at 4° in an HB-4 rotor. Intact chloroplasts that accumulate at the 40 to 70% interphase are fractionated and repurified by a second Percoll step gradient centrifugation as described above. The purified chloroplasts are washed three times with grinding solution (1× GM) and centrifuged to remove the Percoll. After the final centrifugation, the chloroplast pellet is resuspended in 6 ml of a buffer containing 10 mM Tris-HCl (pH7.9), 1 mM EDTA, and 5 mM DTT and lysed with 4 × 10 strokes in a glass homogenizer (chloroplasts are incubated for 5 min on ice between the strokes). After addition of 7 ml of a buffer containing 50 mM Tris-HCl (pH 7.9), 10 mM, $MgCl_2$, 2 mM DTT, 25% (w/v) sucrose, and 50% (v/v) glycerol, the lysed chloroplast mixture is fractionated with 2.3 ml of a saturated ammonium sulfate solution (approximately 4 M) and stirred slowly for 20–30 min at 4°. Thylakoid membranes and hydrophobic proteins are removed by centrifugation in a 70.iTi rotor at 50,000 rpm for 2 hr at 4°. The supernatant is passed over a column of preswollen DE-52 (5-ml column volume, 1-cm diameter, flow rate of 1.5 ml/min) equilibrated with at least 6 column volumes of chromatography buffer [25 mM Tris-HCl (pH 7.9), 0.42 mM EDTA, 2.8 mM DTT, 4.2 mM $MgCl_2$, 10% (w/v) sucrose, 20% (v/v) glycerol, and 0.5 M ammonium sulfate]. The flowthrough (FT) is recovered and its volume is determined. Finely ground ammonium sulfate powder is added to the FT solution (0.28 g/ml) and left to stir for 30 min at 4°. Precipitated proteins are collected by centrifugation at 25,000 rpm in a 70.iTi rotor for 25 min and resuspended in 1 ml of buffer E [20 mM HEPES–KOH (pH 7.9), 60 mM KCl, 12.5 mM $MgCl_2$, 0.1 mM EDTA, 2 mM DTT, and 5% (v/v) glycerol]. The protein solution is then dialyzed for 15 hr in 2 liters of buffer E on ice, and dialysis is continued for an additional

3 hr in fresh buffer E. This protein solution represents the chloroplast processing extract (Fig. 1), which can correctly process the *petD* mRNA 3′ end to a stable mature product (Fig. 5, lane 2).

EconoQ Chromatography

To identify the proteins that are involved in chloroplast 3′-end processing, it is necessary to fractionate the processing extract. As a first step, proteins can be fractionated by anion-exchange chromatography on EconoQ (Bio-Rad, Hercules, CA). This chromatography offers high protein binding capacity of the ion-exchange matrix and good reproducibility. In addition, the charge of a quaternary substituted amine is stable even at extreme pH ranges. It is critical for this type of chromatography, however, that protein solution to be fractionated on EconoQ be free of ammonium sulfate (see below), which interferes with the chromatographic behavior of proteins. The protein solution should also be cleared from possible precipitates by centrifugation at 14,000 rpm in an Eppendorf centrifuge for 15 min at 4° before column loading to prevent clogging of the column. The concentration of proteins in the solution should be approximately 10 mg/ml and should not significantly exceed this value. Before chromatography of the protein solution, a 5-ml column (prepacked; Bio-Rad) should be equilibrated with buffer E at a flow rate of 1 ml/min. This flow rate does not require high-pressure loading and can be achieved easily with a peristaltic pump. The protein solution (1ml) is loaded onto the column at the same speed, followed by extensive washing with 6–8 column volumes of buffer E (see above). Washing of the bound protein fraction is complete when the OD_{280} of the flowthrough is close to zero. The bound proteins are eluted with 20 ml of a linear salt gradient (60 m*M* to 1 *M* KCl) and collected in twenty 1-ml fractions, which are dialyzed against buffer E for 10 hr. Each fraction is then analyzed for protein content and RNA-binding activity by cross-linking to the *petD* RNA 3′-end probe.

A typical result of the EconoQ chromatography is shown in Fig. 2, which reveals several of the known RNA-binding proteins that are eluted from the EconoQ column. We previously identified the 100-kDa protein as a polynucleotide phosphorylase,[7] and the 24 and 28 ctRNPs as CS-type RNA-binding proteins.[8] The 41-kDa protein was identified as CSP41[15] by matrix-assisted laser desorption ionization time-of-flight mass spectrometry (MALDI/TOF) analysis (data not shown). Although the 55-kDa RNP (Fig. 2[16]) has not yet been isolated, it was reported previously to bind the *petD* 3′-end precursor with high affinity.[7] Only the 100-kDa species and ctRNPs clearly coelute in the same fractions, while most of the other RNA-binding proteins are separated efficiently in the salt gradient (Fig. 2). When

[15] J. J. Yang, G. Schuster, and D. B. Stern, *Plant Cell* **8,** 1409 (1996).

[16] M. Bradford, *Anal. Biochem.* **72,** 248 (1970).

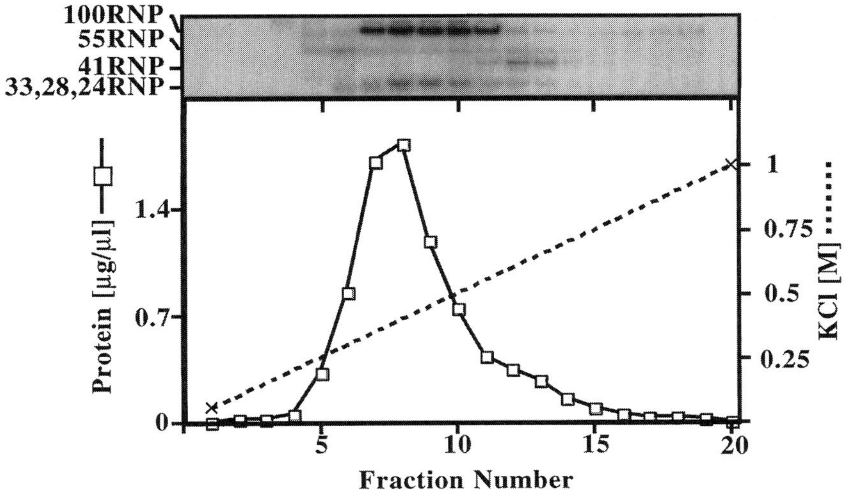

FIG. 2. EconoQ chromatography of the chloroplast RNA-processing extract. The graph shows the protein elution profile based on Bradford assays[16] and the salt gradient (*bottom*). RNA-binding activity of proteins in each fraction was determined by UV cross-linking assays using radioactively labeled RNA and SDS–PAGE analysis (*top*). The RNA-binding proteins previously identified by molecular cloning or biochemical fractionation are shown on the left.

analyzed for RNA-degradation/processing activity, fractions 7–10 correctly process the *petD* 3′-end precursor to a stable mature product (data not shown). For further purification, fractions 6–14 are pooled and dialyzed overnight in buffer E with 5 m*M* KCl. The pooled fractions correctly process the *petD* mRNA substrate and stabilize the product (Fig. 5, lane 3).

Single-Strand DNA Affinity Chromatography

Although anion-exchange chromatography is a convenient first step in the purification protocol for chloroplast RNA-processing proteins, other proteins not functionally related to plastid RNA processing (e.g., the abundant photosynthetic enzyme ribulose-1,5-bisphosphate carboxylase) are still present in the protein fractions. As a second purification step, single-strand DNA (ssDNA) is applied as an affinity matrix for the purification of nucleic acid-binding proteins. It should be noted, however, that assaying biological activity of proteins during protein fractionation by affinity chromatography is often less reproducible than during ion-exchange chromatography. To ensure reproducibility, it is important that all equipment be thoroughly cleaned and free from nuclease contamination, because this can affect the quality chromatography performance. Because DNA–protein interactions are dynamic processes, all work during ssDNA chromatography should be performed quickly at 4°, using prechilled buffers. The sample is applied to

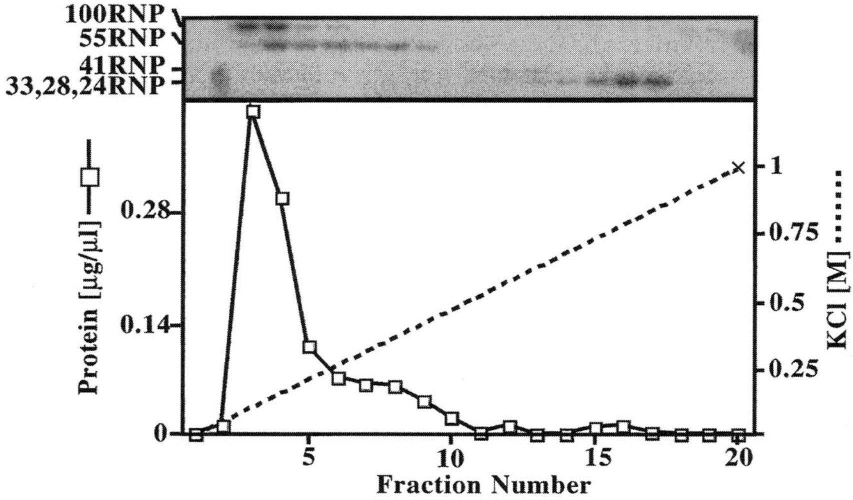

FIG. 3. Single-strand DNA affinity chromatography of pooled fractions obtained after EconoQ chromatography. The graph shows the protein elution profile based on Bradford assays (see Fig. 2) and the salt gradient (*bottom*). RNA-binding assays (*top*) were performed as described in Fig. 2.

a 2-ml column of ssDNA [1-cm column diameter, 3.5 mg of ssDNA per gram of cellulose (Sigma)] equilibrated with buffer E containing 5 m*M* KCl at a flow rate of 0.5 ml/min (peristaltic pump, 10 column volumes). The column is washed with ~10 column volumes of the equilibration buffer (buffer E). Bound proteins are eluted with a linear salt gradient from 5 m*M* to 1 *M* KCl, and collected in twenty 1-ml fractions that are subsequently dialyzed against buffer E (containing 60 m*M* KCl).

Figure 3 shows results from a ssDNA chromatography experiment. The 100-55; and 41-kDa RNA-binding proteins and the ctRNPs are eluted successively at increasing ionic strength of the elution buffer (Fig. 3). When fractions 2–17 containing all RNA-binding proteins are pooled, concentrated, and tested for the ability to process the *petD* 3′-UTR transcript, the protein mixture could correctly process the *petD* precursor RNA substrate to a stable product (Fig. 5, lane 4), although with a shorter half-life. Thus, the protein mixture obtained after ssDNA chromatography contains all components necessary for RNA processing. But because the RNA-binding proteins bind to ssDNA with different affinities, the loss of RNA stabilization capacity of the pooled fractions could be a consequence of changes in the stoichiometry of the proteins required for processing and stabilization.

Size-Exclusion Chromatography on Superose 6

As shown above, the chloroplast RNA-binding proteins can be efficiently separated by a combination of ion-exchange and affinity chromatography (Figs. 2 and 3). These results suggest that the proteins are not assembled into a stable

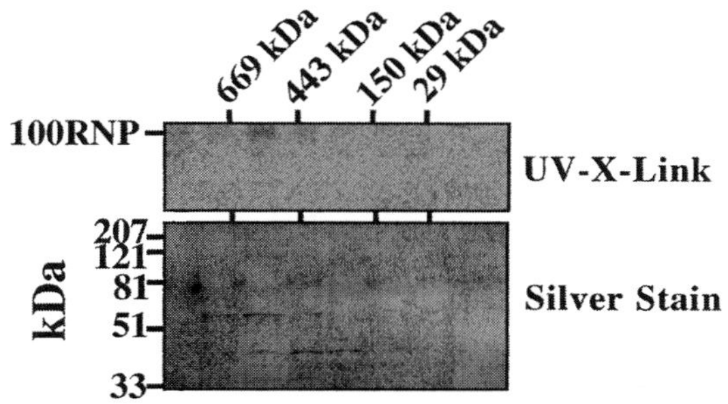

FIG. 4. Size-exclusion chromatography of protein fractions from ssDNA chromatography containing the 100-kDa PNPase. Proteins in individual fractions were analyzed by silver staining[17] (*bottom*) and assayed for RNA-binding activity (*top*). Size calibration of the Superose 6 column is shown (*top*) and was based on the following marker proteins: thyroglobulin, 669 kDa; ferritin, 443 kDa; aldolase, 150 kDa; chymotrypsinogen A, 29 kDa. Size markers for SDS–PAGE (*bottom*) are shown on the left.

complex. However, the 100-kDa PNPase has previously been found as part of a protein complex with an apparent molecular mass of approximately 550 kDa.[7] To isolate this chloroplast protein complex, the fractions containing PNPase activity after ssDNA affinity chromatography (fractions 2 and 3, Fig. 3) are pooled. The proteins in these pooled fractions are separated by size-exclusion chromatography on Superose 6. For this chromatography step, a size-calibrated prepacked Superose 6 column (Pharmacia, Piscataway, NJ) of 24-ml bed volume and 13 ($\pm$2)-μm bead size (Fig. 4) is operated at a pressure of 1–1.5 MPa. Buffer E, which is used in this procedure, should be filtered through a 0.2-μm pore size filter and the protein sample should be clarified by centrifugation at 10,000 rpm for 15 min at 4° before loading. To ensure constant pressure, columns are operated with a BioCad sprint chromatography system (PE Biosystems, Foster City, CA). The columns is first equilibrated with two column volumes of buffer E, followed by loading of the sample in a volume of 200–300 μl. If necessary, the sample volume can be reduced by using Centricon concentrator tubes (molecular weight cutoff, 3000; Amicor Danvers, MA). One-milliliter fractions spanning a molecular size range of approximately 669–29 kDa (as determined by marker proteins) are collected and subjected to UV cross-linking analysis. Separation quality and protein concentrations are determined by silver staining.[17]

Figure 4 shows the results from a Superose 6 gel-filtration chromatography experiment. A 100-kDa RNA-binding protein appears in a fraction with an apparent molecular mass of approximately 500–600 kDa. The protein in the corresponding silver-stained band at approximately 110 kDa has been identified as the 100-kDa PNPase by MALDI/TOF analysis (S. Baginsky, unpublished data, 2000). The silver-stained gel also reveals a small number of additional proteins that coelute

[17] C. R. Merril, D. Goldman, and M. R. van Keuren, *Methods Enzymol.* **96,** 230 (1983).

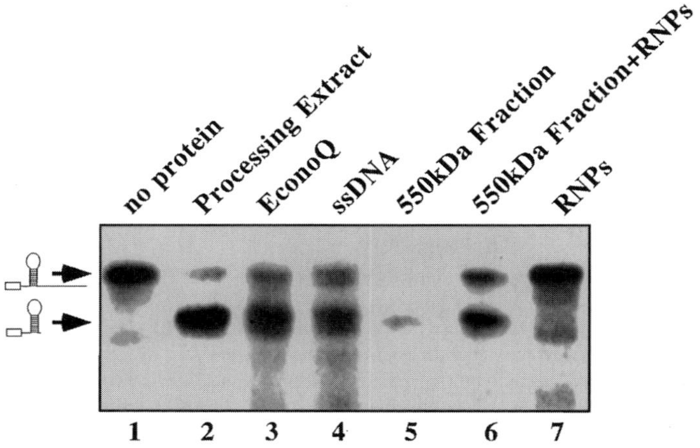

FIG. 5. RNA-processing activities of the protein fractions collected during different steps of the purification procedure. Representative results from an *in vitro*-processing experiment using the designated protein fractions and a radioactively labeled *petD* mRNA 3′-end precursor substrate RNA are shown. The substrate RNA was incubated for 45 min with no protein (lane 1), the complete chloroplast processing extract (lane 2), the pooled EconoQ fractions 6 to 14 (lane 3), the pooled ssDNA fractions 2 to 17 (lane 4), and the 550-kDa fraction after gel filtration (lane 5). The EconoQ and ssDNA fractions retain correct 3′-end processing activity, while the 550-kDa Superose 6 fraction containing PNPase effectively degrades the substrate. Correct 3′-end processing can be restored partially to the 550-kDa fraction by addition of a protein fraction enriched for ctRNPs (lane 6). This ctRNP protein fraction itself has no significant RNA 3′-end processing activity (lane 7), suggesting that the proteins may modify the activity of the 550-kDa PNPase complex.

with the PNPase, such as proteins of 55 and 48 kDa. Figure 5 shows that incubation of the *petD* mRNA precursor with the 550-kDa fraction results in nearly complete degradation of the substrate (Fig. 5, lane 5). This result suggests that gel-filtration chromatography separates from the protein complex-associated proteins that are required for correct processing and stability of the RNA substrate. Correct processing activity can be restored in part by reconstitution of the 550-kDa complex with a chloroplast protein fraction that has been enriched for the ctRNPs in the 24- to 41-kDa range (Fig. 5, lanes 6 and 7).

Concluding Remarks

The mechanisms that control mRNA processing and metabolism in higher plant plastids have important functions in the regulation of photosynthesis and developmental processes. The biochemical purification methods discussed in this chapter were developed to provide insights into the mechanisms and proteins involved in 3′-end processing, stabilization, and degradation of chloroplast RNAs.

The present model, which is based on the types of biochemical fractionation experiments described here, suggests that the processing of the *petD* mRNA 3′ end *in vivo* is initiated by an endonucleolytic cleavage approximately 10 to 15 nucleotides downstream of the stem–loop structure. This cleavage is followed by $3' \rightarrow 5'$ exonucleolytic degradation of the RNA to the base of the stem–loop structure that comprises the mature 3′ end. Additional RNA-binding proteins interact with the stem–loop structure to protect the 3′ end against further exonucleolytic attack.[10] Degradation of the *petD* RNA substrate is initiated by an endonucleolytic cleavage upstream of the stem–loop structure. A 67-kDa protein recognized by antibodies generated against the RNase E protein from *E. coli* has been shown to cleave the RNA upstream of the stem–loop in a site-specific manner.[7] The endonucleolytic cleavage products are subsequently polyadenylated and rapidly degraded by PNPase in the high molecular weight complex and or associated exonucleases.[10]

The association of a PNPase-like enzyme activity with a highmolecular weight protein complex in the chloroplast is reminiscent of the "degradosome" involved in prokaryotic mRNA metabolism (see [27] in this volume[18]). In *E. coli*, PNPase is assembled into a high molecular weight complex with at least four other proteins, including an ATP-dependent RNA helicase, RNase E, and enolase.[19] Although the association of the chloroplast PNPase with a 550-kDa complex is similar to *E. coli*, we have been unable to detect other proteins in the 550-kDa Superose 6 fraction by MALDI/TOF analysis that are similar to those found in the degradosome. It is possible, however, that a degradosome-like complex exists in the chloroplast as well, because an ATP-dependent RNA helicase can be detected in gel-filtration fractions corresponding to 800–900 kDa that also contain PNPase activity (not shown).

Together, the biochemical dissection of chloroplast mRNA processing and degradation described here and reviewed elsewhere[10] suggests that the mechanisms in plants involve enzymes similar to those found in prokaryotic cells. But fractionation of the chloroplast RNA-processing extract has also revealed several small, nuclear-encoded RNA-binding proteins (ctRNPs) not present in *E. coli* that appear to have important functions in chloroplast mRNA stability and regulation of nucleolytic activities. The development of novel fractionation protocols and reconstitution assays will now rapidly advance our understanding of chloroplast metabolism.

Acknowledgments

The authors thank Drs. Dina Phayre, Mohammed Bellaoui, Corina Marx, and Greg del Val for helpful discussions and for training in the use of the BioCad Sprint. S.B. was supported by a DFG-Postdoctoral fellowship (Ba1902/1-1).

[18] A. J. Carpousis, A. Leroy, N. Vanzo, and V. Khemici, *Methods Enzymol.* **342,** [27] 2001 (this volume).
[19] A. J. Carpousis, N. F. Vanzo, and L. C. Raynal, *Trends Genet.* **15,** 24 (1999).

[34] Chloroplast p54 Endoribonuclease

By KARSTEN LIERE, JÖRG NICKELSEN, and GERHARD LINK

Chloroplasts are the important sites of photosynthesis and major anabolic pathways within green plant cells. They contain their own genetic system and all enzymes required for gene expression both at transcriptional and posttranscriptional levels. Throughout biogenesis and function of the photosynthetic apparatus, there is a continuous demand for synthesis and replenishment of proteins that undergo turnover in response to changing environmental conditions.

Much of the regulation of chloroplast gene expression has been assigned to stability determinants at the level of RNA precursors, including interactions of sequence and/or structural elements with their cognate binding proteins. A number of these proteins are known to assemble at the 5′-untranslated region (UTR) of chloroplast mRNAs, protecting them from exonucleolytic degradation and forming ribonucleoprotein (RNP) complexes involved in translational control (see [32] in this volume,[1] and Refs. 2 and 3). Likewise, the function of proteins that bind to 3′-noncoding sequences of chloroplast transcripts has become increasingly well defined as a result of combined biochemical and molecular genetic approaches (see [33] in this volume [4]). Apart from criteria such as direct RNA binding versus participation of non-RNA-binding proteins in larger RNP complexes, both exonucleolytic and endonucleolytic activities have been characterized. Furthermore, posttranslational modifications have been identified to play an important role in the control of activity of certain 3′-RNA-binding proteins.[3]

One of the proteins known to be controlled by phosphorylation is a 3′-endoribonuclease, which was first found to specifically bind to a conserved U-rich sequence element (UUUAUCU) of chloroplast precursor transcripts.[5–7] This protein, which purifies as a monomeric polypeptide of an apparent molecular size of 54 kDa, was subsequently termed p54.[8] Furthermore, it was shown that p54 activity not only depends on the phosphorylation state of the enzyme, but also on its SH-group redox state mediated by glutathione *in vitro.*[8] These *in vitro* findings may have

[1] C. Simpson and D. B. Stern, *Methods Enzymol.* **342,** [32] 2001 (this volume).
[2] A. Danon and S. P. Mayfield, *EMBO J.* **13,** 2227 (1994).
[3] S. P. Mayfield and A. Cohen, *in* "A Look beyond Transcription: Mechanisms Determining mRNA Stability and Translation in Plants" (J. Bailey-Serres and D. R. Gallie, eds.), p. 174. American Society of Plant Physiologists, Rockville, Maryland, 1998.
[4] S. Baginsky and W. Gruissem, *Methods Enzymol.* **342,** [33] 2001 (this volume).
[5] J. Nickelsen and G. Link, *Nucleic Acids Res.* **17,** 9637 (1989).
[6] J. Nickelsen and G. Link, *Mol. Gen. Genet.* **228,** 89 (1991).
[7] J. Nickelsen and G. Link, *Plant J.* **3,** 537 (1993).
[8] K. Liere and G. Link, *Nucleic Acids Res.* **25,** 2403 (1997).

0076-6879/01 $35.00

implications for the *in vivo* regulation of chloroplast RNA precursors at their 3′ end. Here we present the purification and properties of the p54 endoribonuclease from mustard (*Sinapis alba*), which might turn out to be a key player in posttranscriptional regulation of chloroplast gene expression.

Purification and Characteristics of p54 Endonuclease

Purification

As a first step toward the purification of the p54 endonuclease, chloroplast protein extracts from mustard (*S. alba*) are prepared.[5,9] Although the procedure has been developed for mustard, it has been found suitable for other higher plant species as well without significant modifications. Briefly, chloroplasts from 1 kg of 4-day-old light-grown mustard seedlings are isolated by differential centrifugation and sucrose density gradient centrifugation. After lysis by osmotic shock, the insoluble material is sedimented by centrifugation and chloroplast proteins are precipitated from the supernatant after addition of ammonium sulfate to 55% saturation. Precipitates are resuspended in 20 ml of buffer E [30 m*M* Tris-HCl (pH 7.6), 0.1 *M* NaCl, 0.1 m*M* EDTA, 5 m*M* 2-mercaptoethanol, 20% (v/v) glycerol] and dialyzed against 2 liters of buffer E overnight.

In the purification procedure[7] this solution is passed over a 20-ml DEAE-cellulose (DE32; Whatman, Clifton, NJ) column, which has been equilibrated with 5 volumes of the same buffer at a flow rate of 0.5 ml/min. Under these conditions p54 is detected in the flowthrough fraction, which is adjusted to 0.25 *M* NaCl by adding an appropriate amount of salt under gentle stirring.

This fraction is then loaded onto an 8-ml column of heparin–Sepharose CL-6B (Amersham Pharmacia Biotech, Piscataway, NJ) equilibrated with buffer E containing 0.25 *M* NaCl at a flow rate of 0.5 ml/min. After washing with at least 3 volumes of this buffer proteins are eluted by applying a 60-ml linear 0.25–2 *M* NaCl gradient at a flow rate of 0.1 ml/min. The higher ionic strength of the loading buffer has been found to decrease the low-affinity binding of most chloroplast proteins to the polyanion heparin and, as a result, the purity of eluted proteins is considerably enhanced.

After dialysis of fractions against buffer E containing 0.1 *M* NaCl overnight, activity assays (see below) indicate that the p54 protein elutes between 0.3 and 0.7 *M* NaCl from the heparin–Sepharose column. Active fractions are pooled, adjusted again to 0.25 *M* NaCl, and passed over a 2-ml column of poly(U)–Sepharose 4B (Amersham Pharmacia Biotech) equilibrated with the same buffer at a flow rate of 0.5 ml/min. Most of the residual proteins fail to bind to the ribohomopolymer under these conditions and thus almost pure p54 is eluted from the column in a single

[9] J. Nickelsen, *in* "Plant Molecular Biology Manual" (S. B. Gelvin and R. A. Schilperoort, eds.), 2nd Ed., p. D6. Kluwer Academic Publishers, Dordrecht, The Netherlands, 1998.

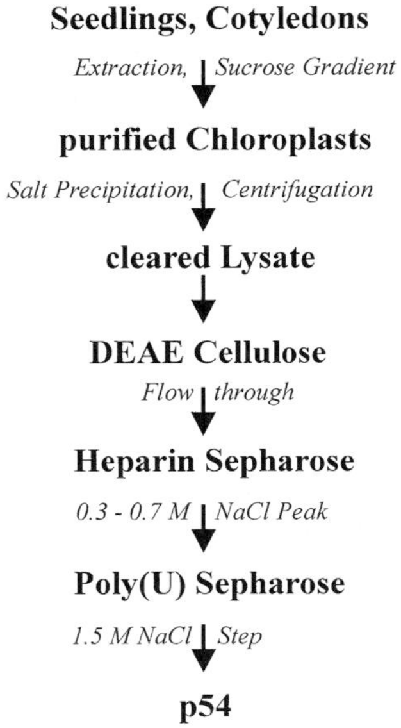

FIG. 1. Purification scheme of the p54 RNase from mustard chloroplasts.

step with 10 ml of buffer E containing 1.5 *M* NaCl. Fractions with RNA-binding activity are dialyzed overnight against 1 liter of buffer E containing 0.1 ml of NaCl and 50% (v/v) glycerol to increase protein concentration. After quick-freezing of samples in liquid nitrogen, they can be stored at −80° without significant loss of p54 activity. In Fig. 1, a summary scheme of the p54 purification[7] is given.

Activity assays that are routinely used during the purification include RNA gel retardation and UV cross-linking experiments to detect binding to 3′ RNA probes, as well as RNA processing (endonuclease) assays. The gel retardation and cross-linking assays have been described in detail[5,9] and are therefore not described here, while the RNA processing assay is outlined below.

Endonuclease Assay

The nucleolytic activity of p54 can be readily monitored after incubating radiolabeled *in vitro* transcripts from 3′ regions of chloroplast genes with chloroplast proteins under conditions used for assaying RNA-binding activity.[5,9] Ten-microliter aliquots of the reaction mixture are removed at various time points, at which times the reaction is immediately stopped by mixing with 100 μl of 1% (w/v) sodium dodecyl sulfate (SDS) solution containing 2 μg of *Escherichia coli* tRNA. After phenol–chloroform extraction and ethanol precipitation, processing products are electrophoretically separated on a 6% (w/v) denaturing urea–polyacrylamide gel and visualized by exposure to X-ray film.

To decide whether the detected processing products have been generated by exonucleolytic trimming or endonucleolytic cleavage, RNA probes are used that are end labeled at either the 5′ or the 3′ terminus.[10] Alternatively, processing products are analyzed by 5′- and 3′-end-labeled DNA probes of the same region by applying a nuclease S1-based approach, which has been used for the mapping of endonucleolytic p54 cleavage sites within mustard *trnK* and *rps16* 3′ RNAs.[5] Both RNA maturation products arising from an endonucleolytic cleavage event are mapped by either the 5′-or the 3′-end-labeled DNA probe. In contrast, an RNA 3′ end that has been generated through successive exonucleolytic degradation in a 3′→5′ direction is detected only with the 3′-end-labeled probe, but not with the probe labeled at the 5′ end.

Unlabeled *in vitro* transcripts (0.5 μg) are processed in a standard RNA-binding reaction for an appropriate time, which has been previously determined by using a radiolabeled probe as described above. Protein-free processing products are then hybridized with DNA probes covering the same chloroplast 3′ region and carrying a radiolabel at the 3′ or 5′ end. By using standard procedures,[11] the 3′ end label is introduced by filling in with Klenow enzyme and [α-^{32}P]dATP at an appropriate restriction site, producing 5′ protruding ends. The 5′ end is labeled after dephosphorylation with calf intestine phosphatase and subsequent phosphorylation by using polynucleotide kinase and [γ-^{32}P]ATP.[11] Processing products and DNA probes are hybridized overnight in 50 μl of piperazine-*N,N*′-bis(2-ethanesulfonic acid) (PIPES) buffer [40 m*M* PIPES (pH 6.4), 400 m*M* NaCl] containing 80% (v/v) formamide at 47°. S1 digestion buffer (500 μl) is added and RNA–DNA hybrids are treated for 30 min at 37° with 10 units of nuclease S1, which selectively digests single-stranded nucleic acids. Samples are mixed with 100 μl of 1 *M* Tris-HCl, pH 9.5, and 20 μl of 0.25 *M* EDTA and nucleic acids are phenol extracted and ethanol precipitated. Nuclease S1-resistant products are electrophoresed in a 6% (w/v) denaturing urea–polyacrylamide gel alongside DNA sequencing products as size markers. Signals are visualized by autoradiography.

Determination of p54-Binding Sites

Initially, p54 was identified as a chloroplast RNA-binding protein capable of specifically interacting with the 3′ regions of the *trnK, rps16,* and *trnH* primary transcripts.[5,6] The precise RNA sites recognized by p54 are mapped by an RNase T1 protection assay. Therefore, G residues of *in vitro* transcripts from the chloroplast 3′ regions are radiolabeled with [α-^{32}P]GTP by using a standard *in vitro* transcription procedure.[9] The RNA (2 ng) is incubated with 30 μg of chloroplast protein extract[9] in binding buffer, which is essentially buffer E without glycerol. After incubation for 15 min at 20°, 10 units of RNase T1 is added

[10] D. B. Stern, H. Jones, and W. Gruissem, *J. Biol. Chem.* **264,** 18742 (1989).

[11] J. Sambrook, E. F. Fritsch, and T. Maniatis, *in* "Molecular Cloning: A Laboratory Manual." Cold Spring Harbor Laboratory Press, Cold Spring Harbor, New York, 1989.

to the reaction mixture. Over a time period of 10 min this G-specific endonuclease cleaves the RNA probe after all G positions except those that are protected by the bound protein complex. Samples are then electrophoresed in a 5% (w/v) nondenaturing polyacrylamide gel (prepared in 0.5 *M* Tris-HCl, pH 8.8) at 25 mA. The electrophoresis buffer contains 25 m*M* Tris and 192 m*M* glycine. The RNase T1-resistant RNA–protein complex (T1R) is visualized after covering the wet gel with Saran Wrap and exposing it to X-ray film at 4° for an appropriate time. The complex is cut out of the gel, the gel piece is ground into small pieces, and the RNA component (T1R RNA) is eluted overnight at 4° in elution buffer [0.3 *M* sodium acetate (pH 7.0), 0.1% (w/v) SDS]. After ethanol precipitation, samples of the eluted RNA as well as of the original full-size *in vitro* transcript of the corresponding chloroplast 3′ region are each completely digested with 20 units of RNase T1 in 10 μl of water at 37° for 30 min.

Subsequent electrophoresis at 1500 V in 20% (w/v) denaturing urea polyacrylamide gels provides the sizes of the different RNA molecules. These include the T1R RNA, its G-specific digestion products, and the cleavage products of the full-size transcript; the latter serve as convenient marker fragments to assign T1R RNA cleavage products to distinct regions within the chloroplast 3′ region. The undigested T1R RNA provides a control to determine whether the sum of the cleavage product sizes is identical to its size. It should be noted that the applicability of the method is somewhat limited by the RNA region that is analyzed. If it contains a high proportion of G residues, it might be difficult to detect RNase T1 cleavage products of a characteristic length that allow an unambiguous assignment. However, as chloroplast intergenic regions usually are AU rich and contain only few guanosines, this should be a critical point only in rare cases.

By using this approach the p54-binding regions have been mapped within the *trnK* and *rps16* 3′-RNA regions. Both binding sites contain the 7-mer sequence UUUAUCU followed by a downstream stretch of multiple U residues. The functional significance of the detected elements has been confirmed by analyzing site-directed mutants of the 3′ regions in RNA-binding assays.[6]

Effects on p54 Activity of Changes in Phosphorylation and Redox State *in Vitro*

Phosphorylation is known to affect a number of chloroplast proteins involved in organellar gene expression, including transcription factors,[12] proteins of a 5′-RNA-binding complex implicated in translational control,[2,3] and the 3′-RNA-binding protein 28RNP.[13–15] Whereas these proteins were found to be inhibited by

[12] K. Tiller and G. Link, *EMBO J.* **12,** 1745 (1993).

[13] M. Kanekatsu, H. Munakata, K. Furuzono, and K. Ohtsuki, *FEBS Lett.* **335,** 176 (1993).

[14] I. Lisitsky and G. Schuster, *Nucleic Acids Res.* **23,** 2506 (1995).

[15] R. Hayes, J. Kudla, G. Schuster, L. Gabay, P. Maliga, and W. Gruissem, *EMBO J.* **15,** 1132 (1996).

TABLE I
POSTTRANSLATIONAL MODIFICATIONS AFFECTING ACTIVITY OF p54[a]

Modification	Pretreatment 1	Pretreatment 2	Effect
Phosphorylation	PKA		++
Dephosphorylation	CIAP		− − −
Reduction	GSH		− − −
Oxidation	GSSG		+
Phosphorylation/reduction	PKA	GSH	− − −
Phosphorylation/oxidation	PKA	GSSG	+++
Reduction/phosphorylation	GSH	PKA	− −
Oxidation/phosphorylation	GSSG	PKA	++
Dephosphorylation/reduction	CIAP	GSH	− − −
Dephosphorylation/oxidation	CIAP	GSSG	− −
Reduction/dephosphorylation	GSH	CIAP	− − −
Oxidation/dephosphorylation	GSSG	CIAP	− −

Abbreviations: PKA, protein kinase A; CIAP, calf intestinal alkaline phosphatase; GSH, reduced glutathione; GSSG, oxidized glutathione.

[a] The positive (+) or negative response (–) of p54 to the various treatments as detailed in Fig. 2 is indicated by the number of symbols.

phosphorylation and activated by dephosphorylation, the opposite is true for p54, for which phosphorylation resulted in elevated levels of both processing and RNA-binding activity, and dephosphorylation inhibited these activities[8] (Table I). Furthermore, p54 was found to be activated through oxidation by glutathione disulfide (GSSG), and inhibited by reduction with reduced glutathione (GSH). Moreover, antagonistic effects of phosphorylation and redox state were observed. Kinase pretreatment of p54 before GSSG resulted in the highest levels of activation, whereas other combinations led to less active or inactive enzyme preparations. These experiments suggested that both phosphorylation and SH-group oxidation determine the level of activity of p54, and that the temporal order of treatments is important.[8]

In the following sections we provide information about protocols for phosphorylation and redox experiments used to investigate the effects of phosphorylation and redox poise on the RNA-binding and endonucleolytic activity of p54.

In Vitro Phosphorylation of p54

To demonstrate that p54 is subject to phosphorylation *in vitro,* the purified protein is are incubated with [γ-^{32}P]ATP in the presence or absence of a protein kinase. Ten-microliter samples of purified p54 (10–20 ng) are added to kinase buffer [130 m*M* Tris-HCl (pH 8.0), 40 m*M* KCl, 4.5 m*M* $MgCl_2$, 2 m*M* $CaCl_2$, 3.5 m*M* 2-mercaptoethanol, 0.5 m*M* dithiothreitol (DTT), 0.4 m*M* EDTA, 0.08 m*M* ATP] containing 5 μCi of [γ-^{32}P]ATP and 2 μg (38 U/mg) of the catalytic subunit of bovine heart kinase (protein kinase A, PKA; Sigma, St. Louis, MO). Samples are incubated in a total volume of 50 μl at 30° for 30 min and phosphorylated proteins are then analyzed by electrophoresis on 10% (w/v) SDS–polyacrylamide gels and subsequent autoradiography. In additional experiments[8] it was shown that p54 is also subject to phosphorylation by a Ser/Thr-specific protein kinase prepared from mustard chloroplasts (protein tyrosine kinase, PTK)[16] and the general outline of the procedure could be easily adapted to test other protein kinases as well.

For use in RNA processing (Fig. 2A) and binding experiments (UV cross-linking; not shown), purified p54 is phosphorylated as described above, except in the presence of 0.15 m*M* unlabeled ATP, and omitting [γ-^{32}P]ATP. Large-scale preparations of phosphorylated p54 are prepared as follows: 300 μl of purified p54 (0.5 μg) is incubated in kinase buffer with 0.15 m*M* ATP and 60 μg (38 U/mg) of PKA in a total volume of 1 ml. Using Centricon-C10 columns (Amicon, Danvers, MA), samples are concentrated by centrifugation at 2800 *g* (5500 rpm; Sorvall SS34 rotor) for 6 hr at 4°. Dialysis buffer [30 m*M* Tris-HCl (pH 7), 100 m*M* NaCl, 5 m*M* 2-mercaptoethanol, 0.5 m*M* EDTA, and 20% (v/v) glycerol] is added to the resulting concentrate (80–150 μl) to a final volume of 300 μl and the preparation is then dialyzed overnight against 2 liters of the buffer. Aliquots are quick-frozen in liquid nitrogen and stored at −80° without significant loss of activity.

In Vitro Dephosphorylation of p54

The p54 protein (10 μl; 10–20 ng) is dephosphorylated in 1× CIAP buffer [20 m*M* HEPES–KOH (pH 8.5), 10 m*M* KCl, 0.1 m*M* $MgCl_2$, 0.01 m*M* $ZnCl_2$] containing 0.3 U of calf intestine alkaline phosphatase (CIAP; Roche, Rahway, NJ). The reaction is incubated in a total volume of 50 μl at 30° for 30 min and subsequently dialyzed against buffer containing 30 m*M* Tris-HCl (pH 7), 100 m*M* NaCl, 5 m*M* 2-mercaptoethanol, 0.5 m*M* EDTA, and 20% (v/v) glycerol. Larger amounts of dephosphorylated p54 are prepared by incubating 300 μl of purified protein (0.5 μg) in CIAP buffer with 10 U of immobilized CIAP (Sigma) at 30° for 30 min in a total volume of 1 ml with occasional shaking. After brief centrifugation, the supernatant is concentrated by centrifugation in a Centricon-C10 cell at 2800*g* (5500 rpm, Sorvall SS34 rotor) for 6 hr at 4°. The resulting concentrate (80–150 μl) is brought to a final volume of 300 μl by adding dialysis buffer and is then dialyzed against 2 liters of the buffer overnight. Aliquots are quick-frozen in liquid nitrogen and stored at −80°.

To obtain control samples, reactions with purified p54 are prepared and treated in the same way as outlined for phosphorylation and dephosphorylation reactions, but with neither ATP nor kinase (or CIAP) present.

[16] S. Baginsky, K. Tiller, and G. Link, *Plant Mol. Biol.* **34,** 181 (1997).

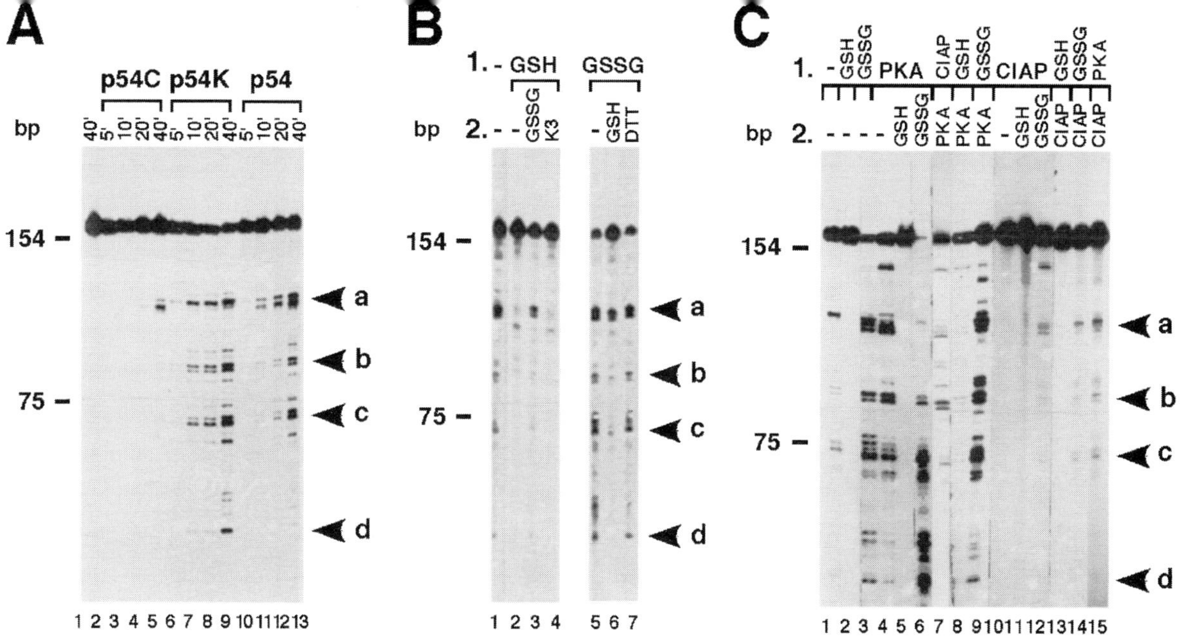

FIG. 2. Endoribonuclease p54 is regulated by phosphorylation and redox state. (A) Phosphorylation affects the *in vitro* processing activity of purified p54. ^{32}P-Labeled 3′ transcripts of the chloroplast *trnK* gene were generated by *in vitro* transcription using T7 RNA polymerase (Promega, Madison, WI). The *in vitro* transcripts were incubated with CIAP-treated p54 (p54C; lanes 2–5), PKA-treated p54 (p54K, lanes 6–9), or untreated p54 (lanes 10–13) for the times (min) indicated at the top. Lane 1, control lane with no protein. Glutathione specifically affects p54 processing activity. (B) Redox reversibility of p54 processing activity with *trnK* 3′ RNA. Lane 1, control lane with untreated p54. Lanes 2–4, p54 incubated with reduced glutathione (GSH) alone (lane 2) or subsequently treated with either oxidized glutathione (GSSG, lane 3) or menadione (K3, lane 4) before processing. Lanes 5–7, p54 treated with oxidized glutathione (GSSG) alone (lane 5) or subsequently treated with either GSH (lane 6) or DTT (lane 7). (C) Phosphorylation and redox state together control p54 processing activity. The purified p54 protein was pretreated as indicated in the first line at the top (row 1) and then as listed below (row 2). Thereafter, RNA processing reactions containing *trnK* 3′ RNA were started by using the pretreated protein. – No treatment of p54; PKA, phosphorylation by protein kinase A; CIAP, dephosphorylation by alkaline phosphatase; GSH, GSSG, treatments with the reduced and oxidized forms of glutathione (20 m*M*), respectively. Processing products are marked by lower-case letters.

In Vitro Redox Assays of p54

RNA-binding and processing assays are carried out in standard reaction mixtures containing p54 (see above). They are pretreated, however, with a 20 m*M* concentration of oxidant menadione (K3; Sigma), cystine (CysCys; Sigma), or oxidized glutathione (GSSG; Sigma), or with reductant 2-mercaptoethanol (EtSH; Sigma), dithiothreitol (DTT; Sigma), cysteine (CysSH; Sigma), or reduced glutathione (GSH; Sigma). The pretreatments are at room temperature for 10 min before the addition of labeled RNA. Stock solutions (100 m*M*) of the redox reagents are prepared in 50 m*M* Tris-HCl, pH 8.0. *Escherichia coli* thioredoxin (5 μM; BRL, Gaithersburg, MD) is reduced with a 5000 *M* excess of DTT or is oxidized with a 5000-fold excess of menadione before use.

To confirm that p54 processing activity is specifically modulated by glutathione, redox reversibility assays are carried out. Purified p54 is preincubated with 20 m*M* oxidized or reduced glutathione, menadione, or DTT for 5 min and is subsequently treated with equimolar amounts of the indicated redox reagent for an additional 5 min (Fig. 2B).

To test the extent to which phosphorylation and redox state act together in the control of p54 activity *in vitro,* processing experiments are carried out with p54 that has been pretreated in various combinations (Fig. 2C). In experiments involving initial treatment with the kinase or CIAP, followed by the redox-reactive reagent, the two steps are separated by reisolation of the protein as described above. In contrast, when p54 is first treated by redox reagents and then subjected to phosphorylation or dephosphorylation, the latter treatments can be carried out without prior reisolation of p54. In the case of phosphorylation, after preincubation with the respective redox reagent, a mixture is added containing 2 μl of 10× kinase buffer, 0.2 μl of 20 n*M* ATP, and 0.5 μg of PKA (20-μl final reaction volume; containing 1× kinase buffer, 0.2 m*M* ATP) and the sample is incubated at 30° for 30 min. Similarly, dephosphorylation after redox treatment is done by adding a mixture of 2 μl of 10× CIAP buffer and 20 units of CIAP in a final reaction volume of 20 μl, and incubation of the sample is continued at 30° for 30 min. To start the subsequent RNA-binding or processing reactions, radioactively labeled RNA is added and the samples are incubated at room temperature according to the protocols given above.

Acknowledgments

We thank Sacha Baginsky for the protein kinase from mustard chloroplasts that was used in some of the p54 phosphorylation experiments. This work was supported by the Deutsche Forschungsgemeinschaft and the Fonds der Chemischen Industrie, Germany.

Section V

Viral Ribonucleases

[35] E^{rns} Protein of Pestiviruses

By MARCEL M. HULST and ROB J. M. MOORMANN

Introduction

Together with flaviviruses and hepatitis C virus (HCV), an important human pathogen, the pestiviruses, are classified as a genus within the family Flaviviridae.[1] Viruses in this family are small, enveloped, positive-strand RNA viruses. All three currently known pestiviruses, classical swine fever virus (CSFV), bovine viral diarrhea virus (BVDV), and border disease virus (BDV), are important disease agents of livestock. Although these three pestiviruses are structurally, antigenically, and genetically closely related they infect different host species.[2] BVDV and BDV can infect ruminants and pigs. CSFV infections are restricted to pigs. Similar to the two other genera of the family of Flaviviridae, pestiviruses contain one large open reading frame that is translated as a single polyprotein.[3–6] Structural proteins and nonstructural (NS) proteins are formed after cleavage of the polyprotein by cellular and viral proteases.[7] Except for short stretches present in two nonstructural proteins, sequence homology at the amino acid level between the three genera of this family is not significant.[8] However, with respect to organization of the RNA genome and the function of the NS proteins there are many similarities between pestiviruses and HCV.[1] Moreover, the NS proteins, encoded in the 3′ half of the genome, are arranged in the same order and have similar functions. Striking differences in genome organization between pestiviruses and HCV are found in the 5′ region of the genome, where the structural genes are encoded. Compared with HCV, the pestivirus genome encodes two additional proteins, an N-terminal autoprotease, N^{pro}, and the envelope protein E^{rns}. The latter protein is a unique viral protein. Comparison of the amino acid sequences of E^{rns} of pestiviruses with amino acid sequences in databases identified two short stretches, both eight amino acids in length, that are homologous to the active site domains of ribonucleases of

[1] F. A. Murphy, C. M. Fauquet, D. H. L. Bishop, S. S. Ghabrial, A. W. Jarvis, G. P. Martelli, M. A. Mayo, and M. D. Summers, *Arch. Virol. Suppl.* **10,** 415 (1995).

[2] E. A. Carbrey, W. C. Stewart, J. L. Kresse, and M. L. Snijder, *J. Am. Vet. Med. Assoc.* **169,** 1217 (1976).

[3] P. Becher, A. D. Shannon, N. Tautz, and H.-J. Thiel, *Virology* **198,** 542 (1994).

[4] M. S. Collett, R. Larson, C. Gold, D. Strick, D. K. Anderson, and A. F. Purchio, *Virology* **165,** 191 (1988).

[5] G. Meyers, T. Rümenapf, and H.-J. Thiel, *Virology* **171,** 555 (1989).

[6] R. J. M. Moormann, P. A. M. Warmerdam, B. van der Meer, W. Schaaper, G. Wensvoort, and M. M. Hulst, *Virology* **177,** 184 (1990).

[7] T. Rümenapf, G. Unger, J. H. Strauss, and H.-J. Thiel, *J. Virol.* **67,** 3288 (1993).

[8] R. H. Miller and R. H. Purcell, *Proc. Natl. Acad. Sci. U.S.A.* **87,** 2057 (1990).

0076-6879/01 $35.00

the RNase T2 family.[9,10] Enzymatic tests of purified E^{rns} proved that these stretches are involved in ribonuclease activity.[9,10] In this chapter the specific properties of this RNase activity and its function in relation to the life cycle of pestiviruses are reviewed.

Amino Acid Sequence and Structure of E^{rns}

The E^{rns} gene encodes a polypeptide backbone of 227 amino acids with a molecular mass of 25.4 kDa (E^{rns} of CSFV strain C[11]). The three-dimensional structure of E^{rns} is formed in the endoplasmic reticulum (ER)–Golgi compartment by intramolecular disulfide linkages and intermolecular disulfide linkages between two monomers.[12] In addition, the protein backbone of E^{rns} is heavily glycosylated. E^{rns} of strain C contains six or seven N-linked glycosyl groups, which represent about 50% of the total mass. As analyzed by sodium dodecyl sulfate–polyacrylamide gel electrophoresis (SDS–PAGE), the complex N-glycosylated homodimers run as a protein of approximately 100 kDa.[7] Together with the two other glycoproteins, E1 and E2 (E2 homodimers and E1–E2 heterodimers), E^{rns} homodimers are associated with the envelope.[12] Animals infected with pestiviruses raise antibodies against E2 and E^{rns}. Although E^{rns} is less immunogenic than E2, like E2, E^{rns} is indispensable for viral attachment and entry into cells.[13] E^{rns} lacks a transmembrane-spanning domain, and association with the viral envelope is accomplished by an as-yet unknown mechanism. A substantial part of E^{rns} is not bound to virions and is secreted into the extracellular environment.[7] Figure 1A presents an alignment of the E^{rns} amino acid sequence of CSFV strain C and of BVDV strain NADL.[4,11] Few differences in E^{rns} amino acid composition are found between different CSFV strains. The same is also true, however, to a lesser extent, for ruminant pestivirus strains (BVDV and BDV). The N-terminal and C-terminal cleavage sites, the two RNase domains, and the cysteine residues (including residues adjacent to these cysteines) are well conserved in all three pestiviruses. Most sequence differences between E^{rns} of CSFV and BVDV are found in the C-terminal half of the protein. For example, in the most C-terminal part of CSFV E^{rns} (the last 40 amino acids) most of the positively charged residues are arginines. In contrast, in BVDV E^{rns} lysines are the predominant positively charged residues in this region. Studies have shown that E^{rns} of CSFV and BVDV can mediate initial binding of virus particles to cells by interaction with membrane-associated heparan sulfate.[14,15] For CSFV it was established that this positively charged region is important for interaction with heparan sulfate.[14]

[9] M. M. Hulst, G. Himes, E. Newbigin, and R. J. M. Moormann, *Virology* **200,** 558 (1994).

[10] R. Schneider, G. Unger, R. Stark, E. Schneider-Scherzer, and H.-J. Thiel, *Science* **261,** 1169 (1993).

[11] R. J. M. Moormann, H. G. P. van Gennip, G. K. W. Miedema, M. M. Hulst, and P. A. van Rijn, *J. Virol.* **70,** 763 (1996).

[12] H.-J. Thiel, R. Stark, E. Weiland, T. Rümenapf, and G. Meyers, *J. Virol.* **65,** 4705 (1991).

[13] M. M. Hulst and R. J. M. Moormann, *J. Gen. Virol.* **78,** 2779 (1997).

A

```
                                                                             *
CSFV-C        1 ENITQWNLSDNGTNGIQHAMYLRGVNRSLHGIWPGKICKGVPTHLATDVELKEIQGMMDA
BVDV-NADL     1 """"""""Q""""E"""R""FQ""""""""""""E"""T"""S"""""I"""T"H"""""
                       **            *                           *   *
CSFV-C       61 SEGTNYTCCKLQRHEWNKHGWCNWHNIDPWIQLMNRTQADLAEGPPVKECAVTCRYDKDA
BVDV-NADL    61 ""K""""""R""""""""""""""Y""E"""LV""""""N"T""Q"PR"""""""""RAS
                                 *                *
CSFV-C      121 DINVVTQARNRPTTLTGCKKGKNFSFAGTVIESPCNFNVSVEDTLYGDHECGSLLQDAAL
BVDV-NADL   121 "L"""""""DS""P""""""""""""""ILMRG""""EIAAS"V"FKE""RI"MF""TT"

CSFV-C      181 YLVDGMTNTIENARQGAARVTSWLGRQLRTAGKRLEGRSKTWFGAYA
BVDV-NADL   181 """""L""SL"G""""T"KL"T"""K""GIL""K""NK"""""""""
```

B

```
domain 1  T2     TIHGLWPD
          Rh     TLHGLWPD
          S2     TIHGLWPD
          CSFV   SLHGIWPG

domain 2  T2     EWNKHGTC
          Rh     EWSKHGTC
          S2     EYVKHGTC
          CSFV   EWNKHGWC
```

FIG. 1. (A) Alignment of the amino acid sequence of E^{rns} of CSFV strain C (Moormann *et al.*[11]) with that of E^{rns} of BVDV strain NADL (Collett *et al.*[4]). The two stretches responsible for RNase activity are underlined. Conserved cysteine residues are marked with an asterisk above the sequence. (B) Amino acid sequences of the domains responsible for RNase activity for members of the RNase T2 family. T2, RNase T2 of the fungus *Aspergillus oryzae* (Kawata *et al.*[18]); Rh, RNase Rh of the fungus *Rhizopus niveus* (Horiuchi *et al.*[17]); S2, the S_2-glycoproteins of the plant *N. alata* (McClure *et al.*[16]); CSFV, envelope protein E^{rns} of CSFV strain C (Hulst *et al.*[9]).

Figure 1B compares the two amino acid regions in E^{rns} (shown for CSFV strain C[9]), which are essential for RNase activity, with the two active site domains of several RNases of the RNase T2 family: The S_2-glycoproteins of the plant *Nicotiana alata,*[16] RNase Rh of the fungus *Rhizopus niveus,*[17] and RNase T2 of the fungus *Aspergillus oryzae.*[18] No significant homology between other parts of the amino acid sequence of E^{rns} and the complete sequences of these RNases are apparent.

[14] M. M. Hulst, H. G. P. van Gennip, and R. J. M. Moormann, *J. Virol.* **74,** 9553 (2000).

[15] M. Iqbal, H. F. Flick-Smith, and J. McCaulley, *J. Gen. Virol.* **81,** 451 (2000).

[16] B. A. McClure, V. Haring, P. R. Ebert, M. A. Anderson, R. J. Simpson, F. Sakiyama, and A. E. Clarke, *Nature (London)* **342,** 955 (1989).

[17] H. Horiuchi, K. Yanai, M. Takagi, K. Yano, E. Wakabayashi, A. Sanda, S. Mine, K. Ohgi, and M. Irie, *J. Biochem.* **103,** 408 (1988).

[18] Y. Kawata, F. Sakiyama, and H. Tamaoki, *Eur. J. Biochem.* **176,** 683 (1988).

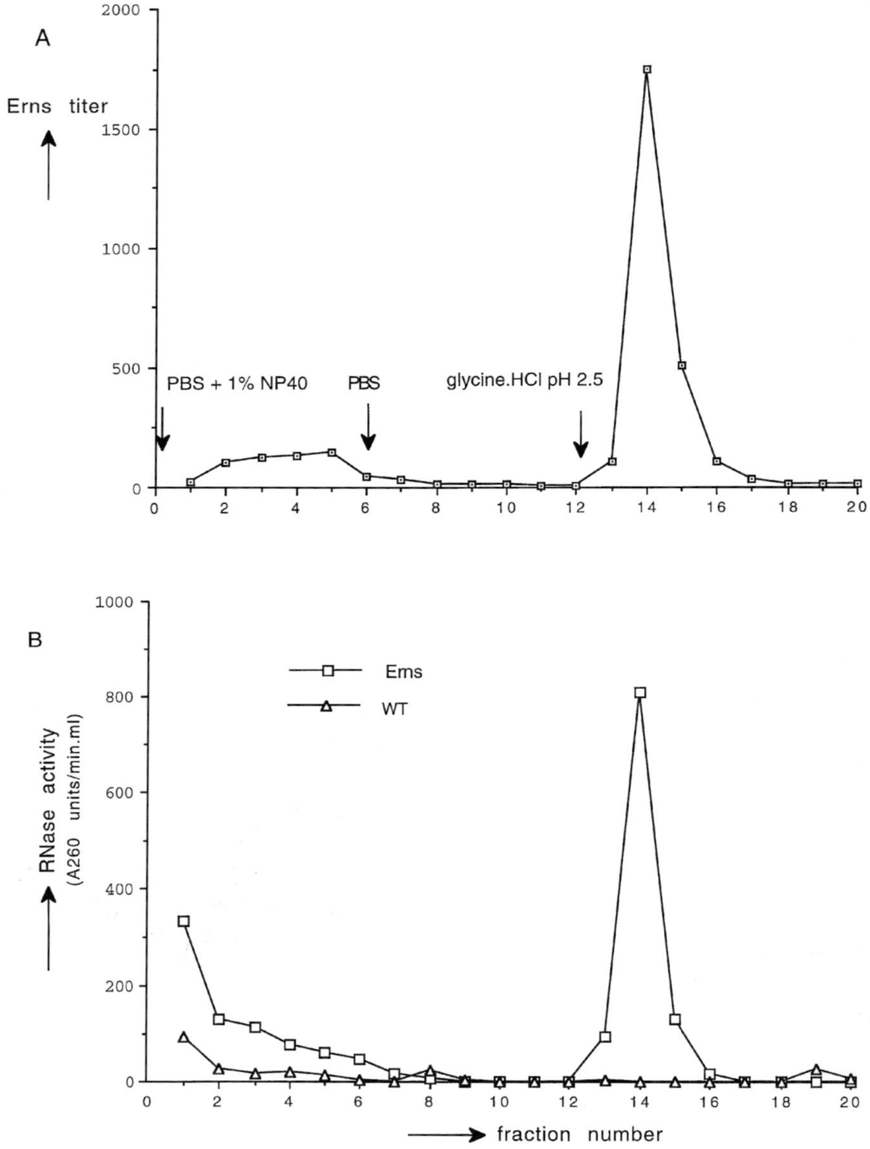

FIG. 2. Immunoaffinity chromatography of cytoplasmic extracts prepared from insect cells infected with a recombinant baculovirus that expresses E^{rns} of CSFV strain C (Erns) or with a wild-type baculovirus (WT) (Hulst *et al.*[9]). A monoclonal antibody directed against E^{rns} of strain C was coupled to CNBr-activated Sepharose 4B (Pharmacia). Half of the beads were mixed with the wild-type extract

Properties of RNase Activity of E^{rns}

The properties of the RNase activity of CSFV E^{rns} were determined with purified native E^{rns},[10,19] and E^{rns} purified from insect cells.[9,19] Almost 100% pure preparations of native and insect cell derived E^{rns} were prepared by immunoaffinity chromatography. In Fig. 2 the chromatography of recombinant E^{rns} of strain C from insect cells is shown.[9] RNase activity coeluted precisely with the E^{rns} protein. Because only limited amounts of native E^{rns} can be purified from pestivirus-infected cells some of these data were generated only with insect cell-derived E^{rns}. However, optimum conditions for RNA cleavage and substrate specificity were determined for both native and insect cell-derived preparations.[9,19] Only minor, insignificant, differences between these preparations were found.

E^{rns} exhibited a broad pH optimum ranging from pH 4.5 to 7.0.[9,19] At pH 4.5 and 7.0, enzyme activity is nearly 100%. At pH 3.5 and 8.0 enzyme activity was 25 and 40% of the maximum activity, respectively.[19] Also, a broad temperature optimum ranging from 35 to 60° for enzyme activity was observed.[9,19] Several potential RNase inhibitors were tested. RNasin (an RNase A, B, and C inhibitor), EDTA, and EGTA did not reduce enzyme activity.[9,19] The failure of the chelators EDTA and EGTA to inhibit activity indicated that enzyme activity is not dependent on binding divalent metal ions in the active site. However, submillimolar concentrations of Zn^{2+} and Mn^{2+} efficiently inhibited RNA cleavage, probably by disturbing the conformation of E^{rns}.[19] E^{rns} readily cleaves its own viral RNA,[9,19] and exhibits a clear preference for uridine-rich sequences.[19] Poly(rA), poly(rC), and poly(rG) substrates were not cleaved whereas cleavage of poly(U) is about 17 times faster than cleavage of yeast RNA.[19] Also, double-stranded RNA (dsRNA) was not cleaved. The specific activity of native E^{rns} and insect cell-derived E^{rns}, determined with yeast RNA, is about 500 A_{260} units min^{-1} mg^{-1} and comparable to the specific activity of other RNases of this family.[9,16,19] The fact that the specific activity of completely deglycosylated E^{rns} was reduced by no more than 30 to 40%, indicates that the *N*-glycans are dispensable for enzyme activity.[19]

---

and the other half were mixed with the E^{rns} extract. After incubation for 16 hr at 4° under light shaking, the beads were packed in 1-ml columns and sample eluates were collected. Both columns were washed with 6 ml of PBS containing 1% (v/v) NP-40 followed by a wash with 6 ml of PBS. Bound proteins were eluted with 0.1 *M* glycine hydrochloride, pH 2.5, and these fractions were immediately neutralized with 25 μl of 3 *M* Tris-HCl, pH 10.0, per milliliter. Fractions of 1 ml were collected during the entire washing and elution procedures. (A) Fractions collected from the E^{rns} column were analyzed for E^{rns} in a direct ELISA (Hulst *et al.*[9]). (B) Fractions from the E^{rns} and wild-type columns were assayed for RNase activity by using Torula yeast RNA (Sigma) as substrate. After cleavage at pH 4.5 and 37° acid-soluble nucleic acids were measured by UV spectrometry.

Mutations that Inactivate RNase Activity

For ribonucleases of the RNase T2 family it has been shown that the histidine residues in the two conserved domains are essential for RNase catalysis.[18,20] Mutational studies showed that this is also true for E^{rns}.[21,22] The following E^{rns} mutant proteins were tested for RNase activity: (1) substitution of histidine in the first domain by a lysine or a leucine,[21,22] (2) substitution of histidine in the second domain by lysine or leucine,[21,22] (3) substitution of both histidine residues by lysines or leucines (double mutants),[21,22] (4) deletion of histidine in the first or in the second domain,[22] and (5) a double mutant in which histidine in the first domain was replaced by a lysine and histidine in the second domain was replaced by a leucine.[22] All these mutated E^{rns} proteins lost their ability to cleave RNA. Except for the mutant in which the histidine in the first domain was deleted, all these mutated E^{rns} proteins still functioned properly as viral surface protein,[21,22] indicating that changes in these domains do not have a severe effect on protein folding.[21,22] Interactions between other amino acids of these domains or forming of disulfide bridges may have more impact on protein folding of E^{rns}.

The tertiary structure of E^{rns} homodimers is complex. E^{rns} is folded at the lumenal site of the ER and therefore formation of a correct tertiary structure including the active site is probably also dependent on the addition of N-linked glycosyl-groups and formation of intra- and intermolecular disulfide bridges. In RNase T2 the cysteine in the second (most C-terminal) domain (Cys-118) forms a disulfide bond with a cysteine located 10 positions downstream of the last amino acid of the first domain (Cys-68).[18] For E^{rns} no data are available yet concerning pairing of cysteines. If a disulfide bridge is involved in the active site of E^{rns}, formation of a disulfide bridge between Cys-38 and Cys-82 is most obvious. Cys-82 in the second domain is conserved for these RNases and residue 38 is the closest cysteine downstream of the first domain. Experiments with purified native and insect cell-derived E^{rns} showed that disconnection of the disulfide bridges by 2-mercaptoethanol and dithiothreitol (DTT), and disturbance of protein–protein interactions with low and high concentrations of salt, did not affect the RNase activity of E^{rns}.[19] This suggests that the conformation of the active site of E^{rns} was not disturbed by breakage of all intramolecular and intermolecular disulfide bonds and by dissociation of E^{rns} dimers. Strong interactions between amino acids of both domains and between

[19] J. M. Windisch, R. Schneider, R. Stark, E. Weiland, G. Meyers, and H.-J. Thiel, *J. Virol.* **70,** 352 (1996).

[20] J. Royo, C. Kunz, Y. Kowyama, M. Anderson, A. E. Clarke, and E. Newbigin, *Proc. Natl. Acad. Sci. U.S.A.* **91,** 6511 (1994).

[21] M. M. Hulst, F. E. Panoto, A. Hoekman, H. G. P. van Gennip, and R. J. M. Moormann, *J. Virol.* **72,** 151 (1998).

[22] G. Meyers, A. Saalmüller, and M. Büttner, *J. Virol.* **73,** 10224 (1999).

amino acids in other parts of E^{rns} are probably more important to maintain the correct tertiary structure of the active site.

Role of RNase Activity of E^{rns} in Virulence and Pathogenesis

Infections of animals with pestiviruses can run an acute or chronic course. Acute infections with pestiviruses induce a severe leukopenia resulting in a suppression of the host immune system. Infected animals become susceptible to infection with secondary pathogens, which often results in a fatal course of the disease. Early in infection, depletion of B lymphocytes in particular is probably the main cause of weakening the host immune defense.[23] It has been shown that many RNases may serve as cytotoxic agents. They can exhibit immunomodulatory and/or antitumor effects.[24] Significant amounts of free E^{rns} (not bound to virions) circulate in the blood of infected animals. Therefore, it was speculated that E^{rns}, and/or the RNase activity of E^{rns}, may be one of the factors responsible for induction of this immunosuppression.[9,10,22,25] This question has been addressed. *In vitro* experiments with E^{rns} produced in insect cells showed that E^{rns} induces apoptosis in lymphocytes of several species, while epithelial cells, although susceptible to pestivirus infection, were not affected.[25] Remarkably, apoptosis was also induced by E^{rns} in which the RNase activity was inactivated by replacement of the histidine in the second domain by a lysine.[26] In these experiments RNase-active and inactive E^{rns} were sequestered to the cell surface of lymphocytes, probably by interaction with heparan sulfate (HS), and could not be detected inside lymphocytes. Binding of E^{rns} to HS on the lymphocyte cell surface might facilitate interaction with a cell-specific signaling receptor that triggers apoptosis. An RNase-dependent cytotoxic action, however, may be masked by activation of this specific surface receptor or by the tight binding to heparan sulfate. This tight binding could also prevent efficient internalization. Further studies with recombinant E^{rns} proteins that have no or a low affinity for HS are needed to evaluate this.

In vivo studies showed that pigs inoculated with a recombinant virus in which the RNase activity of E^{rns} was inactivated (by deletion of the histidine in the second RNase domain) recovered from infection and survived.[22] This attenuation was correlated with the failure of this virus to induce depletion of B lymphocytes. It was postulated that the RNase activity of free E^{rns} circulating in the blood of infected pigs might be responsible for the decrease in B lymphocytes in infected

[23] M. Susa, M. Konig, A. Saalmüller, M. J. Reddehase, and H.-J. Thiel, *J. Virol.* **66,** 1171 (1992).

[24] G. D'Alessio, *Trends Cell Biol.* **3,** 106 (1993).

[25] C. J. Bruschke, M. M. Hulst, R. J. M. Moormann, P. A. van Rijn, and J. T. van Oirschot, *J. Virol.* **71,** 669 (1997).

[26] C. J. Bruschke, Pathogenesis and Vaccinology of BVDV Infections. Ph.D. Thesis. State University of Utrecht, Utrecht,The Netherlands, 1998.

pigs.[22] If so, before this enzymatic action, E^{rns} must be internalized, a process most likely mediated by a B-cell-specific determinant.

Besides the possible cytotoxic action of E^{rns} toward the host immune system there is also a possible role for E^{rns} RNase activity in regulation of RNA synthesis in infected cells. Recombinant viruses in which the RNase activity of E^{rns} was inactivated by replacement of the histidine in the first or in the second domain by a lysine were fully viable. However, compared with their RNase-active parent virus (a vaccine strain), which is not cytopathogenic in cell culture, these viruses were cytopathogenic because of induction of apoptosis.[21] This induction of apoptosis was independent of the level of processing of the cleavage site between the non-structural proteins NS2 and NS3, a hallmark of cytopathogenicity for ruminant pestiviruses.[27] There are two possible mechanisms by which the RNase activity of E^{rns} might regulate RNA synthesis in infected cells. First, the RNase activity of E^{rns} might downregulate synthesis of cellular RNA. Viral replication can benefit from the degradation of messenger RNA because sufficient production facilities and nutrients are now available for translation of viral proteins. Without degradation of messenger RNA, a shortage of nutrients may arise. This shortage may result in the observed induction of apoptosis triggered by host cell factors.[21] Second, E^{rns} RNase activity may shutoff viral RNA synthesis in a late stage of infection. Viral shutoff results in a limited production of virus particles per cell. This may explain the persistent character of pestivirus infection in cells and in the natural hosts. For both possible actions, free E^{rns} (not bound to virions) must be transported to the nucleus or the cytoplasm, either by sorting at a specific site in the ER–Golgi compartment or by the release and direct internalization at the apical membrane of the cell (recycling of E^{rns}).

For all of the possible mechanisms regarding the cytotoxic action of E^{rns} discussed above, interaction with specific host proteins, involved in signal transduction, internalization, or intracellular transport, is essential before E^{rns} can exhibit its cytotoxic action. However, in all these studies mutated proteins were used. The complex tertiary structure of E^{rns} and with this the ability of E^{rns} to interact properly with their host counterparts may be affected, even by minor differences introduced by these mutations. Therefore, as mentioned by Meyers *et al.*,[22] it will be difficult to obtain solid evidence of the role of the RNase activity of E^{rns} in the pestiviral life cycle.

RNase Activity of E^{rns} as Diagnostic Tool

Pestiviruses, and especially CSFV, are economically important diseases. Outbreaks of CSFV occur intermittently in Europe as well as in other parts of the world. An active screening program combined with eradication of infected pigs is

[27] G. Meyers, N. Tautz, E. J. Dubovi, and H.-J. Thiel, *Virology* **180,** 602 (1991).

used to control outbreaks of the disease. For this purpose screening tests must be able to detect all CSFV strains and must be able to discriminate between CSFV and the two other pestiviruses that infect pigs. Currently, the main serological tests used specifically detect antibodies directed against conserved epitopes (present on all CSFV strains) on envelope protein E2 of CSFV.[28] Such tests, mostly enzyme-linked immunosorbent assays (ELISAs), are relatively fast, sensitive, and easy to perform. Several (commercial) screening tests have been developed on the basis of E^{rns}.[29] In all these tests serum antibodies directed against E^{rns} are detected. Probably as a result of immunosuppression, early in infection, levels of E^{rns}- and E2-specific antibodies in the blood of pigs are low and difficult to detect. The fact that free E^{rns} circulates in the blood of infected pigs is advantageous. Detection of E^{rns} by a CSFV-specific, sensitive, and reliable RNase assay may detect infected animals earlier than serological tests do. So far, tests based on the detection of E^{rns} RNase activity have been used only for research purposes.[9,10,19,21,22] In these assays the cleavage of high molecular weight RNA substrates by E^{rns} is measured by the UV absorbance of acid-soluble nucleic acids. To develop a diagnostic test that specifically detects E^{rns} RNase activity in blood samples, specific conditions must be used. RNA hydrolyzes at low pH (4.5) and addition of RNasin and chelators (EDTA and EGTA) may reduce background RNase activity from host RNases present in blood. The use of poly(U) as substrate will not only improve sensitivity but will also reduce background activity. An acceptable level of background activity was obtained when E^{rns} RNase activity was measured in crude cell extracts,[22] indicating that direct measurement of E^{rns} RNase activity in blood samples may be possible. However, direct measurement of E^{rns} activity does not allow differentiation between pestiviruses. Enrichment of CSFV E^{rns} from blood samples by binding to solid-phase immobilized antibodies, specifically directed against CSFV E^{rns} (which do not cross-react with E^{rns} of BVDV and BDV), can not only make such a test specific for CSFV but may also further improve sensitivity and reduce background RNase activity.[21] Antibodies used in such a test must recognize conserved epitopes on CSFV E^{rns}.[29] Because the binding of some, but not all, antibodies can neutralize RNase activity,[19] it is preferable that RNA cleavage by E^{rns} be performed in an environment in which E^{rns} dissociates from the immobilized antibodies. A low pH of 4.5 or high salt concentrations may be suitable for this. The most difficult problem to overcome for such a test is probably a practical aspect. Large quantities of samples must be measured simultaneously in a relatively short period. The use of poly(U)-coupled dyes as substrate may make possible fast measurement of numerous blood samples in 96-well plates by a low-cost ELISA reader.

[28] G. Wensvoort, R. Bloemraad, and C. Terpstra, *Vet. Microbiol.* **17,** 129 (1989).

[29] A. J. de Smit, Classical Swine Fever Virus: Efficacy of Marker Vaccines and Laboratory Diagnosis. Ph.D. Thesis. State University of Utrecht, Utrecht,The Netherlands, 2000.

During outbreaks of CSFV, slaughter of infected and suspected herds and the imposing of quarantine restrictions result in large economic losses. Vaccination of pigs in affected areas with recombinant vaccines that allow discrimination between vaccinated and infected pigs (so-called marker vaccines) will reduce transmission and thereby economic losses. Several of these vaccines have been developed. For example, a subunit vaccine based on E2,[29,30] and a live recombinant virus vaccine that does not express E^{rns},[31] are capable of inducing a protective immune response in pigs. For this latter vaccine infectious virus, used for inoculation, is produced in a cell line that constitutively expresses E^{rns} (*in trans* complementation). In pigs infected with field virus, E^{rns} circulates in the blood whereas in pigs inoculated with these marker vaccines no E^{rns} is present in the body fluids. Thus, field virus-infected herds can be differentiated from vaccinated herds by using a sensitive test that is capable of detecting the RNase activity of E^{rns} in blood samples. Furthermore, in combination with such a test, the attenuated RNase-negative CSFV viruses may also be suitable as (live) marker vaccines.[21,22]

[30] M. M. Hulst, D. F. Westra, G. Wensvoort, and R. J. M. Moormann, *J. Virol.* **6,** 5435 (1993).
[31] M. N. Widjojoatmodjo, H. G. P. van Gennip, A. Bouma, P. A. van Rijn, and R. J. M. Moormann, *J. Virol.* **74,** 2973 (2000).

[36] Herpes Simplex Virus vhs Protein

By JAMES R. SMILEY, MABROUK M. ELGADI, and HOLLY A. SAFFRAN

Introduction

The virion host shutoff (vhs) protein encoded by herpes simplex virus (HSV) gene UL41 is responsible for the rapid shutoff of host protein synthesis that occurs during the earliest stages of HSV infection.[1,2] In this chapter we summarize our current understanding of the mechanism of action and regulation of vhs activity, and provide a detailed description of a simple and convenient *in vitro* assay for vhs-dependent ribonuclease activity. Because we have not cited all the articles that have contributed to our present understanding of vhs function, we refer the reader to the introductions of two more recent papers for more detailed background information.[3,4]

[1] G. S. Read and N. Frenkel, *J. Virol.* **46,** 498 (1983).
[2] A. D. Kwong, J. A. Kruper, and N. Frenkel, *J. Virol.* **62,** 912 (1988).
[3] M. M. Elgadi, C. E. Hayes, and J. R. Smiley, *J. Virol.* **73,** 7153 (1999).
[4] M. M. Elgadi and J. R. Smiley, *J. Virol.* **73,** 9222 (1999).

METHODS IN ENZYMOLOGY, VOL. 342
0076-6879/01 $35.00

vhs is a structural component of the HSV virion that is synthesized late in infection and packaged into the tegument of the mature virus particle (the space between the envelope and the nucleocapsid).[5,6] It is then delivered into the cytoplasm of newly infected cells after fusion of the virion envelope with the host plasma membrane, where it triggers host shutoff before the onset of *de novo* viral gene expression. vhs-induced shutoff is characterized by strong inhibition of host protein synthesis, disruption of preexisting polyribosomes, and accelerated turnover of host mRNAs.[7,8] Although the causal interrelationships between these three effects have yet to be completely defined, the simplest interpretation of the available data is that vhs degrades host mRNA, thereby causing polysome disruption and translational arrest (see below). However, it is worth noting that vhs-induced translational arrest can precede over mRNA degradation under certain conditions,[9] raising the possibility that vhs inhibits protein synthesis through more than one mechanism.

The vhs-dependent shutoff system exhibits little specificity, destabilizing most, if not all, cellular and viral mRNAs in the infected cell.[10,11] The rapid decline in host mRNA levels presumably helps viral mRNAs gain access to the cellular translational apparatus. In addition, the relatively short half-life of viral mRNAs contributes to the sharp transitions between the successive phases of viral protein synthesis, by tightly coupling changes in the rate of transcription of viral genes to altered mRNA levels. These effects likely enhance virus replication, and may account for the finding that vhs mutants display a 10-fold reduction in virus yield in tissue culture.[1] vhs also plays a critical role in HSV pathogenesis: vhs mutants are severely impaired for replication in the cornea and central nervous system of mice, and cannot efficiently establish or reactivate from latency.[12–14] vhs may additionally help the virus evade host defense mechanisms, by reducing the levels of cellular proteins that mediate antiviral responses. For example, vhs contributes to the resistance of HSV-infected cells to cytotoxic T lymphocytes,[15] and vhs mutants display enhanced virulence in mice lacking interferon receptors.[16] vhs homologs are found in all of the alpha (neurotropic) herpesviruses that have been

[5] C. A. Smibert, D. C. Johnson, and J. R. Smiley, *J. Gen. Virol.* **73,** 467 (1992).

[6] J. McLauchlan, C. Addison, M. C. Craigie, and F. J. Rixon, *Virology* **190,** 682 (1992).

[7] M. L. Fenwick and M. M. McMenamin, *J. Gen. Virol.* **65,** 1225 (1984).

[8] R. J. Sydiskis and B. Roizman, *Science* **153,** 76 (1966).

[9] N. Schek and S. L. Bachenheimer, *J. Virol.* **55,** 601 (1985).

[10] A. A. Oroskar and G. S. Read, *J. Virol.* **61,** 604 (1987).

[11] A. D. Kwong and N. Frenkel, *Proc. Natl. Acad. Sci. U.S.A.* **84,** 1926 (1987).

[12] L. I. Strelow and D. A. Leib, *J. Virol.* **69,** 6779 (1995).

[13] L. Strelow, T. Smith, and D. Leib, *Virology* **231,** 28 (1997).

[14] L. I. Strelow and D. A. Leib, *J. Virol.* **70,** 5665 (1996).

[15] M. A. Tigges, S. Leng, D. C. Johnson, and R. L. Burke, *J. Immunol.* **156,** 3901 (1996).

[16] D. A. Leib, T. E. Harrison, K. M. Laslo, M. A. Machalek, N. J. Moorman, and H. W. Virgin, *J. Exp. Med.* **189,** 663 (1999).

characterized to date, but are absent from beta and gamma herpesviruses (which establish latency in other cell types). vhs therefore likely plays a key role in the interaction between herpesviruses and postmitotic neurons. The vhs system provides a striking and readily dissected example of gene regulation at the level of mRNA stability in mammalian cells, and may therefore help illuminate the host mRNA turnover pathways that regulate cell growth, differentiation, and oncogenesis.

Mechanism of vhs Action

Several lines of evidence strongly suggest that vhs is either a ribonuclease, or a required subunit of a ribonuclease that also includes one or more cellular subunits: (1) vhs displays at least two regions of amino acid sequence similarity with the FEN-1 family of nucleases that are involved in DNA replication and repair in eukaryotes and archaebacteria (Fig. 1)[17]; (2) extracts of HSV-infected cells and partially purified virions contain a ribonuclease activity,[18–20] and this activity is eliminated when the UL41 gene is inactivated by mutation; (3) the ribonuclease activity present in extracts of partially purified virions is inhibited by anti-vhs antibodies[20]; and (4) vhs induces endoribonucleolytic cleavage of exogenous RNA substrates when it is produced as the only HSV protein in a rabbit reticulocyte lysate (RRL) *in vitro* translation system.[3,20] Although the foregoing data are highly suggestive, a definitive demonstration of whether vhs is itself a ribonuclease will require characterization of highly purified and biologically active protein. Previous attempts to characterize the activity of vhs purified from bacterial or baculovirus overexpression systems have been hindered by the insolubility of the protein thus produced. G. S. Read and colleagues have made substantial progress in overcoming this problem. These investigators found that vhs forms a specific complex with the newly recognized eukaryotic translation initiation factor eIF4H (P. Feng, D. N. Everly, and G. S. Read, personal communication, 2000). The soluble vhs–eIF4H complex that forms when these proteins are coexpressed in *Escherichia coli* has been partially purified and shown to display ribonuclease activity; in contrast, vhs–eIF4H complexes containing some mutant versions of vhs are devoid of activity (D. N. Everly, P. Feng, and G. S. Read, personal communication, 2000). Although it is not yet clear whether the eIF4H subunit is required for the ribonuclease activity of the vhs–eIF4H complex, these data provide the strongest evidence to date that the vhs protein is an integral part of the vhs-dependent ribonuclease.

Most of our current knowledge about the mechanism of action of the vhs-dependent ribonuclease has emerged from *in vitro* studies of complex extracts

[17] A. J. Doherty, L. C. Serpell, and C. P. Ponting, *Nucleic Acids Res.* **24,** 2488 (1996).
[18] C. R. Krikorian and G. S. Read, *J. Virol.* **65,** 112 (1991).
[19] C. M. Sorenson, P. A. Hart, and J. Ross, *Nucleic Acids Res.* **19,** 4459 (1991).
[20] B. D. Zelus, R. S. Stewart, and J. Ross, *J. Virol.* **70,** 2411 (1996).

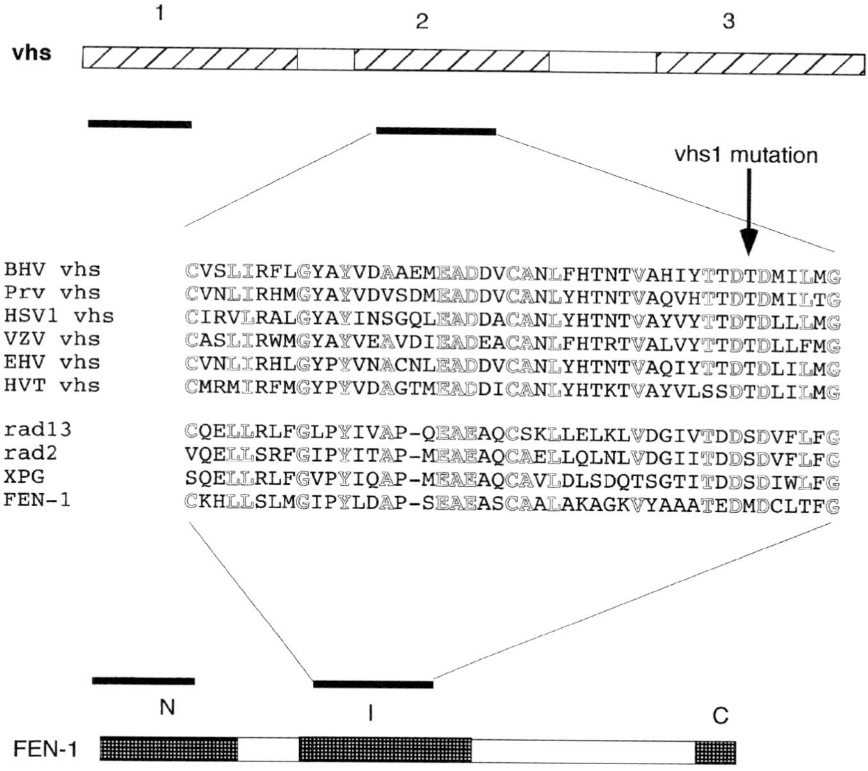

FIG. 1. vhs displays amino acid sequence similarity to the FEN-1 family of nucleases. The vhs homologs of alpha herpesviruses display three regions of strong sequence conservation (indicated as regions 1, 2, and 3), as do FEN-1 nucleases (regions N, I, and C). Portions of vhs conserved regions 1 and 2 show similarity to FEN-1 regions N and I, respectively (indicated by the thick lines). An alignment of the homologous segments of vhs region 2 and FEN-1 region I is presented. The position of the inactivating vhs 1 point mutation (Thr-214 → Ile) is indicated. BHV, PrV, HSV1, VZV, EHV, HVT: bovine herpesvirus 1, pseudorabies virus, herpes simplex virus type 1, varicella-zoster virus, equine herpesvirus type 1, and herpesvirus of turkeys, respectively. Rad13, Rad2, XPG, and FEN-1 are from *Saccharomyces cerevisiae, Schizosaccharomyces pombe, Homo sapiens*, and *Homo sapiens*, respectively. The diagram is not precisely to scale.

that contain a large number of other proteins. As noted above, vhs-dependent ribonuclease activity can be detected in extracts of HSV-infected mammalian cells, partially purified virions, and RRL containing translated vhs. In all these cases, activity is resistant to the RNase inhibitor RNasin and requires Mg^{2+}. Zelus and co-workers characterized the cleavage products of β-globin mRNA in some detail and concluded that the reaction proceeds through endoribonucleolytic cleavage events.[20] More recent work from our laboratory has confirmed this conclusion, by showing that precisely matching sets of 5′ and 3′ products are produced at a variety of cleavage sites.[3] The *in vitro* reaction displays little selectivity, in that a wide range of RNAs serve as substrates. Moreover, cleavage is not obviously sequence specific: of 37 cleavage sites examined at the nucleotide sequence level, 19 occur between purine residues; the others are GC (6), UG (5), AC (2), GU, CU, UU, UC, and CG (1 each).[4]

Although vhs displays little sequence specificity *in vitro* and targets most, if not all, cellular and viral mRNAs *in vivo,* other cytoplasmic transcripts such as rRNA, tRNAs, and 7SL RNAs are spared during HSV infection.[11,18,21] These observations raise the possibility that mRNAs are targeted for selective degradation *in vivo* through one or more features that distinguish these transcripts from other cytoplasmic RNAs. Zelus *et al.* suggested that the 3′ poly(A) tail might serve as a preferred site for vhs action.[20] However, we find that activity in the RRL *in vitro* system is not greatly affected by the presence of a 3′ poly(A) tail in the RNA substrate,[3] and Karr and Read have made the same observation in extracts of HSV-infected cells.[22] In addition, the 5′ cap is not required for the reaction in RRL or extracts of partially purified virions.[3,20] These observations apparently eliminate a role for two of the most obvious structural features that distinguish most mRNAs from other cellular transcripts. Moreover, activity partitions with the postribosomal supernatant in RRL and extracts of infected cells,[3,18,19] demonstrating that ribosomes are not required to recruit vhs activity to mRNAs. Notwithstanding the foregoing findings, work has uncovered two strong indications that the vhs-dependent nuclease may target specific functional or structural features of mRNA substrates. First, the initial sites of endoribonucleolytic cleavage are nonrandomly clustered over the 5′ quadrant of signal recognition particle α-subunit mRNA in the RRL system.[3] Consistent with this finding, vhs degrades the 5′ end of HSV thymidine kinase (tk) mRNA before the 3′ end is affected *in vivo*.[22] Second, we have shown that the internal ribosome entry sites (IRES elements) of two picornaviruses [encephalomyocarditis virus (EMCV and poliovirus)] act to target vhs-induced cleavage events to multiple sites within a narrow zone located just 3′ to the IRES, irrespective of sequence context or the location of the IRES in the RNA.[4] IRES elements are highly structured *cis*-acting sequences found in some cellular and many viral mRNAs that promote cap-independent translational initiation, by recruiting initiation factors required for loading of the 40S ribosomal subunit.[23] The two distinct modes of initial cleavage revealed by these studies (5′ proximal, and IRES directed) raise the possibility that vhs targets mRNAs by interacting with one or more components of the translational apparatus that are

[21] A. A. Oroskar and G. S. Read, *J. Virol.* **63,** 1897 (1989).
[22] B. M. Karr and G. S. Read, *Virology* **264,** 195 (1999).
[23] R. J. Jackson and A. Kaminski, *RNA* **1,** 985 (1995).

delivered to the RNA before loading of the 40S ribosomal subunit. Consistent with this idea, G. S. Read and colleagues have found that vhs interacts with the newly characterized translation initiation factor eIF4H[24,25] in the yeast two-hybrid system and in mammalian cells (P. Feng, D. N. Everly, and G. S. Read, personal communication, 2000). The role of eIF4H in translational initiation has yet to be precisely defined; however, it appears to act in collaboration with other eIF4 factors before loading of the 40S subunit.[24,25] Although the cap independence of the vhs reaction on non-IRES substrates in RRL might be taken to argue against this idea, translation in RRL is relatively cap independent, and so these data do not exclude the hypothesis.

Regulation of vhs Activity during Infection

vhs significantly destabilizes viral mRNAs in infected cells, and even targets it own mRNA for destruction in the RRL system.[3] These observations raise an interesting question: how do HSV mRNAs accumulate to high levels in infected cells in the face of vhs action? This question is especially pertinent at late times postinfection, when high levels of new vhs protein are made for incorporation into progeny virions.[5] M. Fenwick and colleagues[26,27] proposed that the solution lies in temporal control of vhs activity during infection. Specifically, Fenwick suggested that a newly synthesized viral protein partially dampens the activity of vhs delivered by the infecting virion, thereby allowing viral mRNAs to accumulate after host mRNAs have been degraded. We have shown that vhs specifically binds to the virion transcriptional activator VP16,[28] and provided genetic evidence that this interaction downregulates vhs activity.[29] VP16 is well known for its ability to activate transcription of the viral immediate-early genes, through its association with the host factors Oct1 and host cell factor (HCF).[30] We found that viral mRNAs are grossly destabilized during infection in the absence of VP16, leading to virtually complete translational arrest midway through the infection cycle.[29] This defect was corrected by transcriptionally incompetent forms of VP16 that retain the ability to bind vhs, and was eliminated by inactivating the vhs gene of the VP16 null mutant. Moreover, cells constitutively expressing VP16 were rendered resistant to virion-induced shutoff mediated by superinfecting HSV. Taken in combination, these results revealed a major and unanticipated posttranscriptional regulatory

[24] G. W. Rogers, Jr., N. J. Richter, and W. C. Merrick, *J. Biol. Chem.* **274,** 12236 (1999).
[25] N. J. Richter-Cook, T. E. Dever, J. O. Hensold, and W. C. Merrick, *J. Biol. Chem.* **273,** 7579 (1998).
[26] M. L. Fenwick and S. A. Owen, *J. Gen. Virol.* **69,** 2869 (1988).
[27] M. L. Fenwick and R. D. Everett, *J. Gen. Virol.* **71,** 411 (1990).
[28] C. A. Smibert, B. Popova, P. Xiao, J. P. Capone, and J. R. Smiley, *J. Virol.* **68,** 2339 (1994).
[29] Q. Lam, C. A. Smibert, K. E. Koop, C. Lavery, J. P. Capone, S. P. Weinheimer, and J. R. Smiley, *EMBO J.* **15,** 2575 (1996).
[30] C. C. Thompson and S. L. McKnight, *Trends Genet.* **8,** 232 (1992).

function of VP16, and provided insight into how HSV evades one of its own host shutoff mechanisms. However, it is not yet clear exactly how VP16 dampens vhs activity.

VP16 is a major component of the virion tegument, and it is present in at least 10-fold molar excess over vhs. Presumably, the vhs–VP16 complex present in the tegument of the infecting virion must be disrupted during the earliest stages of infection in order to allow shutoff to proceed. The complex likely then reforms later during the lytic cycle, dampening vhs activity and allowing viral mRNAs to accumulate after the host transcripts have been degraded. Little is known of the factors that regulate the programmed disassembly and reassembly of the vhs–VP16 complex. It is interesting to note that these viral proteins bind distinct cellular partners, and that VP16 cannot simultaneously bind vhs and host Oct1/HCF.[28] One interesting possibility is that host proteins serve to displace vhs from VP16 (and vice versa). Further work is required to test this hypothesis. Some evidence suggests that the protein kinase encoded by the HSV-1 UL13 gene may contribute to these processes: UL13 null mutants display a vhs-deficient phenotype,[31] and UL13 appears to play a role in disassembly of the tegument during the earliest stages of infection.[32]

Methods

As described above, vhs-dependent ribonuclease activity can be readily detected in extracts of infected cells, partially purified virions, or in rabbit reticulocyte lysates (RRLs) containing pretranslated vhs. We have used the latter system extensively to characterize the mode of vhs-induced RNA decay.[3,4] The assay consists of three basic steps. First, RRL is programmed with *in vitro* transcripts encoding vhs, and translation is allowed to proceed. Second, reporter RNA is added to the lysate, and samples are withdrawn at various time points for analysis. Third, the fate of the reporter RNA is monitored by gel electrophoresis, primer extension, or other methods. The assay is simple, rapid, convenient, and highly sensitive. It has been used to detect the ribonuclease activity of the vhs proteins encoded by herpes simplex viruses types 1 and 2, and pseudorabies virus. The *in vitro* transcription and translation steps are accomplished with commercially available kits, adding to the convenience and reproducibility of the assay.

In Vitro Translation of vhs

Biologically active vhs protein is synthesized *in vitro* in nuclease-treated RRLs, usually obtained from Promega (Madison, WI). A key feature of our assay system

[31] H. Overton, D. McMillan, L. Hope, and P. Wong-Kai-In, *Virology* **202,** 97 (1994).
[32] E. E. Morrison, Y. F. Wang, and D. M. Meredith, *J. Virol.* **72,** 7108 (1998).

is the use of an *in vitro* translation vector (pSPUTK) that is optimized for maximal translational efficiency in the RRL system.[33] The plasmid pSP6vhs[3] contains a 1.8-kb *Nco*I–*Eco*RI fragment that includes the HSV-1 vhs open reading frame from pCMV vhs[34] inserted into pSPUTK. The resulting construct contains the vhs open reading frame under the control of the SP6 RNA polymerase promoter, fused to a modified 5′ untranslated region (UTR) derived from *Xenopus laevis* β-globin mRNA, and contains a consensus Kozak translation initiation signal.

Ten micrograms of pSP6vhs template DNA is linearized with *Eco*RI, and then purified with QIAquick columns (Qiagen, Valencia, CA) according to the manufacturer procedure. The linearized DNA is eluted in 30 μl, so that the final concentration is 330 ng/μl. *In vitro* transcription is performed with 1 μg of the linear template in a 20-μl reaction including 0.5 m*M* cap primer $m^7G(5')ppp(5')G$ (Pharmacia, Piscataway, NJ), 40 U of RNAseOUT (GIBCO-BRL Gaithersburg, MD), 5 m*M* ATP, CTP, GTP, and UTP, and 40 U of SP6 RNA polymerase [GIBCO-BRL or MBI Fermentas (St. Leon-Rot, Germany)], in the buffer provided by the manufacturer of the RNA polymerase. The reaction is allowed to proceed for 1 hr at 37°, and the template is then degraded by incubating it for 20 min at 37° with 10 U of RNase-free DNase I (Boehringer Mannheim/Roche, Indianapolis, IN). The RNA product is extracted once with phenol–chloroform and once with chloroform. RNA is precipitated with 95% (v/v) ethanol, washed in 70% (v/v) and then 95% (v/v) ethanol, and resuspended in 12 μl of diethyl pyrocarbonate (DEPC)-treated H_2O.

In vitro translation is performed with an RRL system, according to the manufacturer protocol (Promega). Basically, 2μl of vhs RNA template (approximately 2 μg) is incubated in lysate supplemented with amino acids (minus methionine), 40 U of RNase inhibitor, and 2 μl of [^{35}S]methionine (1175 Ci/mmol; New England Nuclear, Boston, MA). Control reactions are generated as outlined, except that mRNA is omitted from the translation reaction. Reactions are allowed to proceed for 90 min at 30°, and then are quick frozen in liquid nitrogen and stored at −80°. vhs activity is stable for at least 1 month under these conditions. To confirm protein production, 4% of each translation reaction is resolved on a 12% (w/v) sodium dodecyl sulfate (SDS)–polyacrylamide gel. Dried gels are exposed to X-ray film (Fuji, Tokyo, Japan) overnight, and proteins are analyzed for intensity and correct mobility.

Preparation of RNA Substrates

RNA substrates are generated by *in vitro* transcription, using SP6 or T7 RNA polymerase. Depending on the application, unlabeled, internally labeled, or 5′-cap-labeled reporter RNAs are generated. Substrates bearing either a 5′-triphosphate

[33] D. Falcone and D. W. Andrews, *Mol. Cell. Biol.* **11,** 2656 (1991).

[34] F. E. Jones, C. A. Smibert, and J. R. Smiley, *J. Virol.* **69,** 4863 (1995).

terminus or m^7GpppG 5′ cap can be used, with equivalent results.[3] Substrate RNAs are typically ~2000 nucleotides in length. Plasmid DNA templates used for production of substrate RNAs are linearized at an appropriate restriction endonuclease cleavage site and purified with QIAquick columns (Qiagen). *In vitro* transcription of 1 μg of linearized template DNA is performed in 20 μl of the RNA polymerase buffer supplied by the manufacturer, supplemented with 0.25 m*M* ATP, CTP, GTP, and UTP, and 40 U of RNAseOUT (GIBCO-BRL). For internally labeled transcripts, the GTP concentration is reduced to 0.125 m*M* and supplemented with 1 μCi of [α-^{32}P]GTP (3000 Ci/mmol; New England Nuclear). Reactions are incubated at 37° for 45 min, and then transcription is terminated by adding a 1/10 volume of RNA load buffer [50% (v/v) glycerol, 1 m*M* EDTA, xylene cyanol (10 mg/ml), bromphenol blue (10 mg/ml)]. RNA is loaded directly onto a 1% (w/v) agarose gel cast in 1× TBE (90 m*M* Tris–borate, 2 m*M* EDTA) containing ethidium bromide (1 μg/ml). After electrophoresis, the intact RNA band is visualized with a UV transilluminator and a minimal gel slice containing the full-length RNA transcript is excised with an RNase-free scalpel. RNA from the gel slice is electroeluted in a 100-μl 7.5 *M* ammonium acetate trap in a six-well v-channel electroelutor (International Biotechnologies Inc., New Haven, CT) at 100 V for 30 min in 0.5× TBE, according to the manufacturer instructions. RNA is then recovered from the salt by ethanol precipitation. Samples are washed with 70% (v/v) ethanol, and then with 95% (v/v) ethanol, and resuspended in DEPC-treated H_2O.

5′ Cap-labeled reporter RNAs can be generated from uncapped unlabeled runoff transcripts, using vaccinia virus guanylyltransferase (GIBCO-BRL) in the presence of [α-^{32}P]GTP. Approximately 500 ng of RNA in 50 m*M* Tris-HCl (pH 7.9), 1.25 m*M* $MgCl_2$, 6 m*M* KCl, 2.5 m*M* dithiothreitol (DTT), bovine serum albumin (BSA, 0.1 mg/ml), and 0.1 m*M* *S*-adenosyl-L-methionine are combined with 1–3 U of guanylyltransferase and 50 μCi of GTP in a 30-μl reaction for 45 min at 37°. The samples are then extracted once with phenol–chloroform and once with chloroform, and RNA is recovered by ethanol precipitation.

Labeled RNA substrates are counted (without scintillant) in a Beckman (Fullerton, CA) LS6500 scintillation counter and resuspended to 5000 Cerenkov cpm/μl. Because of radiolysis, uniformly labeled RNA substrates are used for only up to 1 week after synthesis. Unlabeled RNA substrates are resuspended in a smaller volume (~20 μl) and can be stored for longer periods at −80°.

We find that the use of gel-purified full-length RNA substrates is absolutely key to obtaining consistent and reproducible results in the vhs assay.

vhs Assay

Reporter RNA substrates are added to RRLs containing pretranslated vhs protein, and the reactions are incubated at 30°. For each desired time point, we typically use 5 μl of lysate and 1 μl of substrate RNA (prepared as described above). Active vhs preparations can be diluted in RRL (lacking added amino acids and mRNA), or in "retic" buffer [1.6 m*M* Tris–acetate (pH 7.8), 80 m*M* potassium acetate, 2 m*M* magnesium acetate, 0.25 m*M* ATP, 0.1 m*M* DTT] before the assay. Control reactions, using RRLs incubated without vhs mRNA, allow monitoring of endogenous ribonuclease activity in the system. Aliquots (5 μl) are removed at various times and immediately added to 200 μl of Trizol (GIBCO-BRL) containing 10 μg of *E. coli* tRNA (Sigma, St. Louis, MD) as carrier. The lysate is allowed to mix with the Trizol for 5 min and then 40 μl of chloroform is added. Samples are vortexed for 2 min, and the phases are separated by centrifuging for 10 min at room temperature in an Eppendorf microcentrifuge. 2-Propanol (110 μl) is added to the aqueous layer, samples are incubated for 10 min at room temperature, and then centrifuged for 15 min at 4°. RNA pellets are washed with 70% (v/v) ethanol, and then with 95% (v/v) ethanol, and resuspended in 25 μl of formamide load buffer [1× MOPS buffer: 200 m*M* 3-*n*-morpholinopropanesulfonic acid (pH 7.0), 50 m*M* sodium acetate, 5 m*M* EDTA], 16.7% (v/v) formaldehyde, 50% (v/v) formamide].

Agarose Gel Electrophoresis and Northern Blotting

RNA samples recovered from the vhs assay as described above are denatured for 15 min at 55° and subjected to electrophoresis through a 1% (w/v) agarose gel containing 6% (v/v) formaldehyde and 1× MOPS. At least one lane is loaded with 1× RNA loading buffer containing xylene cyanol and bromphenol blue in order to monitor migration. Electrophoresis is carried out at approximately 5 V/cm until the dye has run ~7 cm. RNA is transferred to a nylon membrane (GeneScreen Plus; New England Nuclear) overnight in 10× SSC (1.5 *M* sodium chloride, 0.15 *M* sodium citrate). After UV cross-linking (UV Stratalinker 2400; Stratagene, La Jolla, CA), ^{32}P-labeled RNA fragments are detected by exposure to Kodak (Rochester, NY) Biomax MS film at −70°. Alternatively, unlabeled RNA fragments cross-linked to GeneScreen Plus membranes can be detected by hybridization. Briefly, membranes are prehybridized in Church buffer [250 m*M* sodium phosphate buffer (pH 7.2), 7% (w/v) SDS, 1% (w/v) BSA, 1 m*M* EDTA] at 65° for 2–6 hr. Fresh Church buffer containing specific probe is used for overnight hybridization at 65°. The membrane is then washed twice for 20 min in 2× SSC–0.1% (w/v) SDS at 68° and twice for 20 min in 0.1× SSC–0.1% (w/v) SDS and subjected to autoradiography.

Results

RNA substrates are usually stable in control RRLs incubated under the conditions described above (see, e.g., Fig. 2B), and rapidly sustain multiple

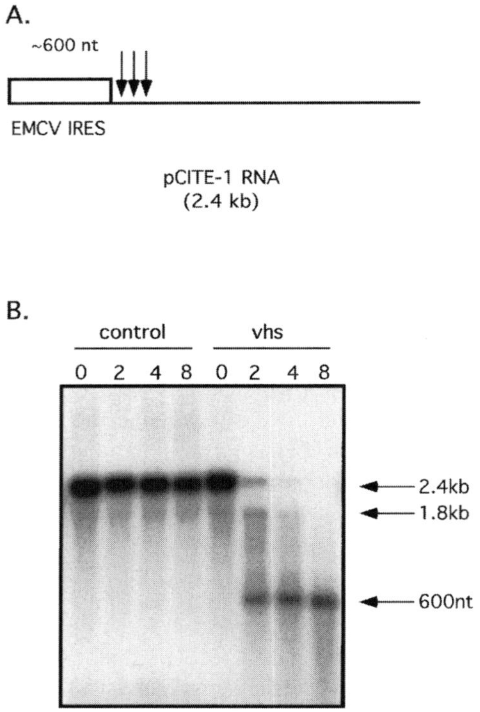

FIG. 2. vhs-induced cleavage of pCITE1 RNA. (A) Diagram of pCITE-1 RNA, indicating the position of the EMCV IRES and the locations of the initial preferred sites of vhs-induced cleavage. Cleavage immediately downstream of the IRES gives rise to 5′ and 3′ degradation intermediates of ~600 and 1800 nucleotides, respectively. (B) Internally labeled pCITE1 RNA was incubated in RRLs containing (vhs) and lacking (control) pretranslated vhs, and RNA samples extracted at the indicated times (minutes) were analyzed by agarose gel electrophoresis. Arrows indicate the mobility of the 3′ and 5′ degradation products. Note that the 600-nucleotide 5′ product bearing the IRES element is stable throughout the course of the reaction.

endoribonucleotytic cleavage events in reactions containing active vhs (Fig. 2B). With most RNA substrates, vhs activity produces a heterogeneous set of degradation intermediates, which are progressively reduced in size as the reaction proceeds.[3] The heterogeneous nature and marked instability of the products can render quantification of activity difficult, particularly during the earliest stages of the reaction (when only a small fraction of the substrate has been consumed), or in batches of RRLs that exhibit a high background of endogenous ribonuclease activity. However, this difficulty can be largely overcome by using an RNA substrate that bears the EMCV IRES element at its 5′ end (pCITE-1 RNA). As diagrammed in Fig. 2A, the EMCV IRES strongly targets vhs-induced cleavage events to a narrow zone located immediately 3′ to the IRES,[4] leading to the early production of discrete 5′ and 3′ degradation intermediates (~600 and 1800 nucleotides in length, respectively; Fig. 2B). The 3′ product is subject to further rounds of vhs-induced cleavage, but the 5′ fragment bearing the IRES is stable throughout the course of the reaction. Moreover, the endogenous ribonuclease activity present in RRLs does not give rise to a product with this electrophoretic mobility. Thus, the pCITE-1 substrate gives rise to a stable discrete 5′ product that is diagnostic of vhs activity. For these reasons, we recommend the use of the pCITE-1 RNA (or another transcript bearing a 5′ EMCV IRES) as the preferred substrate for detecting and quantifying vhs activity.

We have found that the level of background ribonuclease activity varies substantially between lots of commercially obtained RRLs. In addition, some lots of RRL generate high levels of vhs protein, but nevertheless display little vhs activity. It is possible that this latter variation stems from differences in the amount or activity of a required cellular cofactor. For these reasons, we recommend testing small samples of several lots of RRL, before deciding which lot to purchase in quantity.

Conclusion

The simple assay described in this chapter allows rapid detection of the vhs-dependent endoribonuclease, and characterization of its mode of action.

Acknowledgments

The National Cancer Institute of Canada, the Medical Research Council of Canada, and the Alberta Heritage Foundation for Medical Research supported the research in the authors' laboratory. We thank Dr. G. Sullivan Read for communicating unpublished data.

[37] Influenza Virus Endoribonuclease

By KLAUS KLUMPP, LISA HOOKER, and BALRAJ HANDA

Introduction

The influenza A virus, an enveloped, single-standed RNA virus of the Orthomyxoviridae family, contains an essential, cap-dependent endoribonuclease associated with the viral RNA polymerase complex. The genome of this virus consists of 8 single-stranded vRNA segments, which encode a total of 10 viral proteins.[1] The individual vRNA molecules are packaged into characteristically coiled structures through the interaction with multiple copies of the viral nucleoprotein (NP),

0076-6879/01 $35.00

which binds to the phosphodiester backbone of RNA molecules.[2–4] The viral RNA-dependent RNA polymerase (RdRp), a heterotrimer of the subunits PB1, PB2, and PA, binds to both ends of the NP-coated viral genomic RNA (vRNA), thereby forming a noncovalent, circular structure, termed the viral ribonucleoprotein (RNP).[5,6] The viral RdRp is responsible for transcription and replication of the viral genome, which occur in the nucleus of infected cells.

The viruses belonging to the Orthomyxoviridae and Bunyaviridae families are unique by using a virally encoded RNA endoribonuclease activity to generate short RNA primers for the initiation of transcription. This activity is associated with the viral RdRp proteins. The influenza virus heterotrimeric RdRp binds to capped mRNA or hnRNA molecules in the nucleus of infected cells and cleaves the capped host RNA molecules at a position between 9 and 15 nucleotides downstream of the cap structure.[7–10] The cleavage generates a 3′-OH group on the capped RNA fragment, which is then used by the RdRp as a primer for the initiation of transcription (Fig. 1). In this way, the RdRp provides its own viral mRNAs with cap structures, which are important for optimal efficiency of the downstream processes of RNA splicing, nuclear export, and translation. The cap-dependent endonuclease is a virus-specific activity with no known equivalent in eukaryotic cells. Therefore, it constitutes an interesting activity and possible target for drug development. Selective inhibitors of endonuclease and polymerase activities of the influenza virus RdRp have been reported.[11,12] Both types of inhibitors show antiviral activity in animal models.[13,14] For improved inhibitor

[1] R. A. Lamb and R. M. Krug, *in* "Virology" (B. N. Fields, D. M. Knipe, and P. M. Howley, eds.), p. 1353. Lippincott-Raven, Philadelphia, 1996.

[2] M. W. Pons, I. T. Schulze, G. K. Hirst, and R. Hauser, *Virology* **39,** 250 (1969).

[3] R. W. Compans, J. Content, and P. H. Duesberg, *J. Virol.* **10,** 795 (1972).

[4] F. Baudin, C. Bach, S. Cusack, and R. W. Ruigrok, *EMBO J.* **13,** 3158 (1994).

[5] M. T. Hsu, J. D. Parvin, S. Gupta, M. Krystal, and P. Palese, *Proc. Natl. Acad. Sci. U.S.A.* **84,** 8140 (1987).

[6] K. Klumpp, R. W. H. Ruigrok, and F. Baudin, *EMBO J.* **16,** 1248 (1997).

[7] S. J. Plotch, M. Bouloy, and R. M. Krug, *Proc. Natl. Acad. Sci. U.S.A.* **76,** 1618 (1979).

[8] S. J. Plotch, M. Bouloy, I. Ulmanen, and R. M. Krug, *Cell* **23,** 847 (1981).

[9] A. J. Caton and J. S. Robertson, *Nucleic Acids Res.* **8,** 2591 (1980).

[10] A. R. Beaton and R. M. Krug, *Nucleic Acids Res.* **9,** 4423 (1981).

[11] J. Tomassini, H. Selnick, M. E. Davies, M. E. Armstrong, J. Baldwin, M. Bourgeois, J. Hastings, D. Hazuda, J. Lewis, W. McClements, G. Ponticello, E. Radzilowski, G. Smith, A. Tebben, and A. Wolfe, *Antimicrob. Agents Chemother.* **38,** 2827 (1994).

[12] M. Tisdale, M. Ellis, K. Klumpp, S. Court, and M. Ford, *Antimicrob. Agents Chemother.* **39,** 2454 (1995).

[13] J. E. Tomassini, M. E. Davies, J. C. Hastings, R. Lingham, M. Mojena, S. L. Raghoobar, S. B. Singh, J. S. Tkacz, and M. A. Goetz, *Antimicrob. Agents Chemother.* **40,** 1189 (1996).

[14] M. Tisdale, G. Appleyard, J. Tuttle, D. Nelson, S. Nusinoff-Lehrmann, W. Al Nakib, J. Stables, D. Purifoy, K. Powell, and G. Darby, *Antiviral Chem. Chemother.* **4,** 281 (1993).

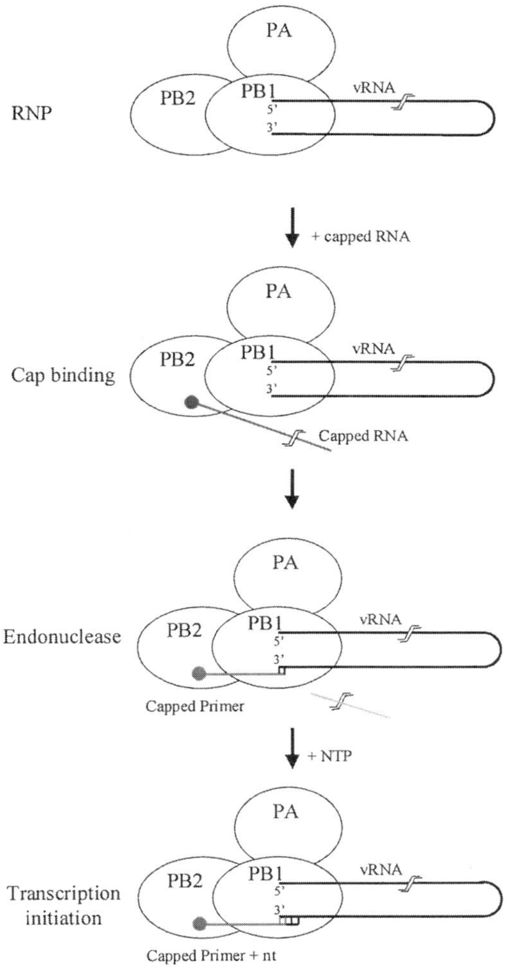

FIG. 1. Model of cap-dependent transcription initiation by influenza virus RNP. The polymerase subunit PB1 interacts with both ends of the genomic vRNA molecule. The RNA is bound by many copies of NP protein (not shown). The polymerase subunit PB2 binds to the cap structure of substrate RNA, which is then cleaved close to the 5′ end by the endonuclease function. The capped RNA product serves as a primer for the initiation of transcription.

design and assessment of drug selectivity it is important to characterize the endonuclease active site structure and identify the most closely related eukaryotic proteins.

The influenza virus endonuclease activity has been studied *in vitro* with RNP preparations from purified virus particles[8,15–18] or recombinant RdRp expressed in mammalian cells.[19–21] This chapter describes a robust procedure for the purification of influenza A and B virus RNPs, which contain influenza virus-specific endonuclease and RNA polymerase activities, and an optimization of incubation conditions under which to study the RNP-associated endoribonuclease activity *in vitro*.

Experimental Procedures

Buffers

Buffer A (lysis): 1% (w/v) Triton X-100, lysolecithin (1 mg/ml), 5 m*M* $MgCl_2$, 200 m*M* KCl, 10 m*M* dithiothreitol (DTT), 5% (v/v) glycerol, 40 m*M* Tris-HCl (pH 8), RNase inhibitor (40 U/ml)
Buffer B (capping): 50 m*M* HEPES (pH 8), 1.25 m*M* $MgCl_2$, 6 m*M* KCl, 10 m*M* DTT, RNase inhibitor (1 U/μl)
Buffer C (endonuclease reaction): 50 m*M* Tris-HCl (pH 8), 1 m*M* $MgCl_2$, 20 m*M* KCl, RNase inhibitor (1 U/μl), 0.3% (w/v) Triton X-100
Buffer D (RNP storage): 50% (v/v) glycerol, 50 m*M* NaCl, 20 m*M* Tris-HCl (pH 8), 5 m*M* 2-mercaptoethanol, 5 m*M* DTT

Choice of Substrate

The natural substrates for the influenza virus endoribonuclease are cellular mRNA and heterogeneous nuclear (hnRNA) molecules. The products of the endonuclease reaction are used as primers in viral mRNA synthesis. Therefore, the viral mRNAs contain 9–15 nucleotides of host-derived RNA sequence at their 5′ ends.[7,9,10,22] An analysis of viral mRNAs produced during influenza virus infection of mammalian cells showed no apparent sequence selectivity at the cap-proximal end of the host RNAs, but a certain preference for the sequence PyGCA↓N was suggested to exist at the cleavage site.[23] However, efficient cleavage after G and U residues has also been observed *in vitro* and *in vivo*. Consistent with the heterogeneity of target sequences *in vivo*, a number of artificial RNAs can serve as specific endonuclease substrates *in vitro*. α- and β-globin

[15] K. Kawakami, K. Mizumoto, and A. Ishihama, *Nucleic Acids Res.* **11,** 3637 (1983).
[16] A. Honda, J. Mukaigawa, A. Yokoiyama, A. Kato, S. Ueda, K. Nagata, M. Krystal, D. P. Nayak, and A. Ishihama, *J. Biochem.* **107,** 624 (1990).
[17] K. Klumpp, M. J. Ford, and R. W. Ruigrok, *J. Gen. Virol.* **79,** 1033 (1998).
[18] L. Doan, B. Handa, N. A. Roberts, and K. Klumpp, *Biochemistry* **38,** 5612 (1999).
[19] M. Hagen, T. D. Chung, J. A. Butcher, and M. Krystal, *J. Virol.* **68,** 1509 (1994).
[20] T. D. Chung, C. Cianci, M. Hagen, B. Terry, J. T. Matthews, M. Krystal, and R. J. Colonno, *Proc. Natl. Acad. Sci. U.S.A.* **91,** 2372 (1994).
[21] M. L. Li, B. C. Ramirez, and R. M. Krug, *EMBO J.* **17,** 5844 (1998).
[22] R. Dhar, R. M. Chanock, and C. J. Lai, *Cell* **21,** 495 (1980).
[23] R. A. Lamb and P. W. Choppin, *Annu. Rev. Biochem.* **52,** 467 (1983).

mRNA, reovirus mRNA, capped plant virus RNAs, capped poly(A) and poly(U), and a few other capped RNAs with random sequences have been described previously.[7,8,15,17,18,20,24–28] It has been shown that capped DNA molecules could also serve as substrates for the influenza virus endonuclease.[29]

The endonuclease substrate used in the experiments described below is chemically synthesized and enzymatically capped G20 RNA, 5′-m^{7}GpppGmGAAUAC UCAAGCUAUGCAUC-3′. The sequence of G20 is derived from SP6 polymerase transcripts of pGem7Zf(+) (Promega, Madison, WI), which have been described previously.[17,19] This RNA is cleaved by the influenza virus endonuclease at a single major site, nucleotide 11. The resulting capped oligonucleotide "G11" (m^{7}GpppGmGAAUACUCAAG) is a functional primer for the influenza virus polymerase.[17–20]

Preparation of Ribonucleoprotein

Most experiments have been done with influenza virus strain A/PR/8/34, which was grown in embryonated hens' eggs and obtained in purified form from Pasteur-Merieux (Marcy L'Etoile, France). The virus is stored in phosphate-buffered saline (PBS), 10% (v/v) glycerol, at −80° and at a total protein concentration of 10 mg/ml according to Bradford assay determination, using bovine serum albumin (BSA) as standard. The RNP purification procedure has also been used to purify active RNP from egg-grown influenza virus strain B/Lee/40, obtained from Charles River Laboratories (Preston, CT). The procedure of genomic RNP preparation from purified influenza virus is composed of two parts, virus lysis and glycerol gradient centrifugation, as outlined below. The following purification protocol for influenza virus RNP has been developed on the basis of previously published procedures.[8,30]

Procedure. Influenza virus is thawed at room temperature. The virus suspension is then incubated with 1× virus lysis buffer A at a final virus protein concentration of 4 mg/ml in a total volume of 2.5 ml for 2 hr at room temperature. The resulting lysate is loaded onto a 26-ml 30–60% (v/v) continuous glycerol gradient, buffered in 20 m*M* Tris-HCl (pH 8), 50 m*M* NaCl, 5 m*M* DTT, 5 m*M* 2-mercaptoethanol on top of a 2-ml 70% (v/v) glycerol cushion. The gradients are spun in a Kontron ultracentrifuge (Kontes, Vineland, NJ), SW28 rotor, 4°, at

[24] M. Bouloy, S. J. Plotch, and R. M. Krug, *Proc. Natl. Acad. Sci. U.S.A.* **75,** 4886 (1978).
[25] M. Bouloy, M. A. Morgan, A. J. Shatkin, and R. M. Krug, *J. Virol.* **32,** 895 (1979).
[26] M. Bouloy, S. J. Plotch, and R. M. Krug, *Proc. Natl. Acad. Sci. U.S.A.* **77,** 3952 (1980).
[27] M. Hagen, L. Tiley, T. D. Chung, and M. Krystal, *J. Gen. Virol.* **76,** 603 (1995).
[28] D. B. Olsen, F. Benseler, J. L. Cole, M. W. Stahlhut, R. E. Dempski, P. L. Darke, and L. C. Kuo, *J. Biol. Chem.* **271,** 7435 (1996).
[29] K. Klumpp, L. Doan, N. A. Roberts, and B. Handa, *J. Biol. Chem.* **275,** 6181 (2000).
[30] O. Rochovansky, *Virology* **73,** 327 (1976).

26,000 rpm for 16.5 hr. Fractions (2 ml) are collected from the bottom of the gradient. Fractions 7–9 contain the peak of functional RNP particles at an approximate glycerol concentration of 50% (v/v).

To prepare RNPs for storage and to remove copurifying divalent metal ions, the pooled RNP fractions are incubated for 15 min at room temperature with 7.5 m*M* EDTA, 1.5 m*M* phenanthroline and then dialyzed extensively against buffer D. After dialysis, the RNPs are stored at -20°. Under these conditions the RNPs retain endonuclease and polymerase activity for many months. The RNP preparation in buffer D can also be stored at -80° with comparable stability. The samples in buffer D (50-μl aliquots) are shock-frozen in liquid nitrogen. A high concentration of glycerol is important for stability under these conditions. Influenza polymerase activity, in contrast to endoribonuclease activity, appears to be sensitive to atmospheric oxygen. Therefore, the presence of a reducing agent in the storage buffer is essential, and repeated air exposure of stored RNP samples should be avoided.

Previously published protocols suggested an additional acid wash step to remove viral M1 matrix protein from the RNP particles.[31,32] M1 protein is assumed to form a polymeric protein shell between the membrane and the genomic RNP in influenza virus particles. The solubilization of aggregated M1 protein in the virus lysate by incubation at low pH was reported to significantly increase the purity of the final RNP preparation and the specific activity of polymerase in purified RNP.[32] Our modified procedure generates RNP peak fractions with only small amounts of M1 protein remaining even in the absence of an acid wash. The M1 protein is separated into two main areas on the glycerol gradient, soluble M1 protein at the top of the gradient (Fig. 2, fractions 10–13) and large precipitates in the 70% (v/v) glycerol pellet (Fig. 3)

We analyzed the effect of acid wash as an additional step in our RNP purification procedure. After virus lysis as described above, 1.75 *M* morpholineethanesulfonic acid (MES) was added to the lysate in 25-μl aliquots until it reached pH 5.5. At this pH the sample was incubated for 15 min at room temperature. The acidified lysate was then loaded onto a 26-ml 30–60% (v/v) continuous glycerol gradient, buffered in 50 m*M* MES–NaOH (pH 5.5), 50 m*M* NaCl, 5 m*M* DTT, 5 m*M* 2-mercaptoethanol on top of a 2-ml 70% (v/v) glycerol cushion. Centrifugation and fraction collection were as described above, except that the pH of the fractions was immediately readjusted to pH 7.5 by the addition of 120 μl of 1 *M* Tris-HCl, pH 9. The pH of the glycerol gradient fractions was controlled by using BDH (Poole, UK) triple pH indicator strips.

The acid wash had a significant effect on M1 solubilization. Figure 3 shows the large reduction in the amount of M1 in the pellet found at the bottom of the

[31] O. P. Zhirnov, *Virology* **176,** 274 (1990).

[32] J. E. Tomassini, *Methods Enzymol.* **275,** 90 (1996).

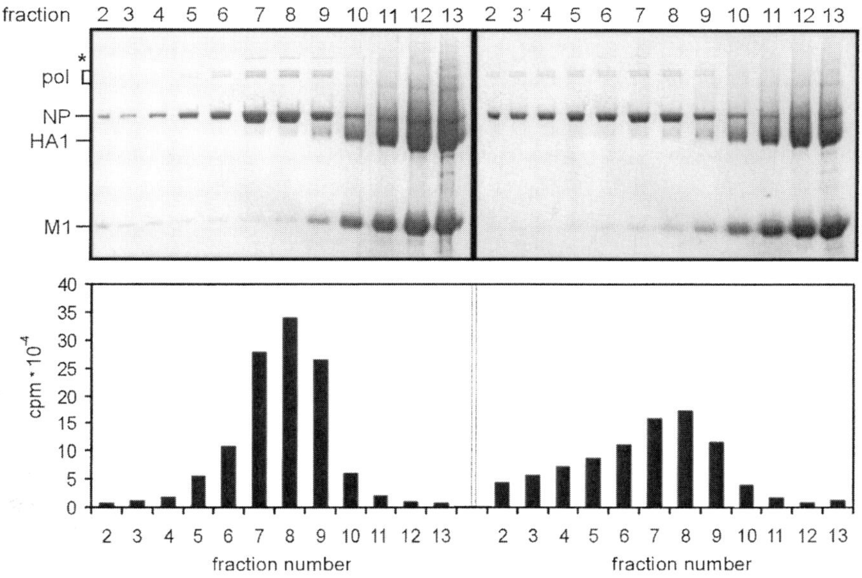

FIG. 2. Glycerol gradient purification and polymerase activity of RNP. The RNPs were prepared on a glycerol gradient at pH 8 without (left-hand side) or with acid wash (right-hand side). The gradient fractions were taken from the bottom (fraction 1) to the top (fraction 15) of the gradient. The proteins in gradient fractions 2–13 were separated on 10% (w/v) NuPage gels (Invitrogen) and stained with Coomassie blue. The migration positions of the three polymerase subunits (pol), NP, HA1, and M1 proteins, are indicated on the left. Four microliters of each fraction was used in a cap-dependent polymerase assay, measuring the amount of [^{3}H]UTP incorporated into RNA product.[18,32] Fractions 7–9 contain the RNP peak with regard to NP and polymerase protein concentration and polymerase activity under standard conditions. The acid wash procedure broadens the RNP peak, but also solubilizes large M1 protein aggregates present in fractions 2 and 3 after the standard procedure. The asterisk indicates an additional band with an apparent molecular mass of >100 kDa. Peptide analysis of this band suggests an undissociated dimer of NP.

glycerol gradient after acid wash. A significant amount of M1 protein can be found in the pellet and up to fraction 2 from the bottom of the gradient after virus lysis at pH 8, but not after acid wash (Figs. 2 and 3). However, the acid wash procedure did not result in any further reduction of the already low levels of M1 in the RNP peak fractions (Fig. 2). On the gradient at pH 5.5, the RNP particles appeared to be slightly less focused into peak fractions, but formed a broader distribution between fractions 2 and 10. In addition, the integrated level of polymerase activity from all gradient fractions after acid wash was lower than that obtained with the optimized procedure at pH 8 as described above.

During optimization of virus lysis time we observed a significant increase in RNP protein concentration in the peak fractions, when lysis was allowed to

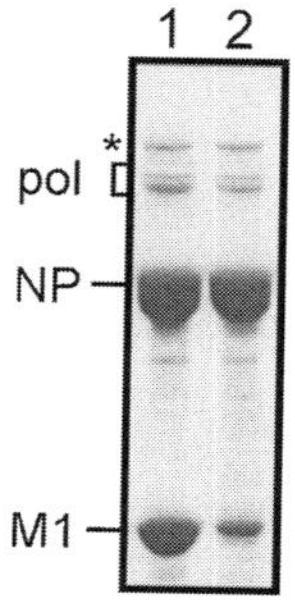

FIG. 3. Pellet fractions obtained after glycerol gradient centrifugation of influenza virus lysate without (lane 1) or with acid wash (lane 2). The proteins of the pellet were resuspended in PBS and protein loading buffer, separated on a 10% (w/v) NuPage gel, and stained with Coomassie blue. Significantly less M1 protein is found in the pellet after the acid wash procedure (lane 2).

proceed for up to 2 hr at room temperature (Fig. 4). The increase in total peak protein concentration correlated with a similar increase in RNA polymerase activity in these fractions. A correlation of polymerase activity was also seen with the relative amount of NP protein as determined from densitometric analysis of scanned, Coomassie-stained acrylamide gels. There was, however, no correlation between polymerase activity and the relative amounts of M1 protein present in any of the peak fractions. RNP samples 1–3 in Fig. 4 showed increasing polymerase activity, although the ratio of NP to M1 decreased. This observation is further exemplified by the glycerol gradient fractions 7 and 9 in Fig. 2, which show almost comparable polymerase activity, yet different amounts of cosedimenting M1 protein. Overall, we could not find evidence of an inhibitory effect of M1 protein on RNP activity in any of the gradient fractions.

Estimation of Ribonucleoprotein Concentration. At present the relationship between protein and RNA concentration and enzymatic activity in RNP preparations is not well understood. Previously, the RNA content was determined after phenol extraction and ethanol precipitation. The concentration of RNP particles was estimated by assuming a mean RNA molecule length of 1700 nucleotides (580 kDa) in the RNP preparation, containing a mixture of 8 genomic RNP species.[18] Alternatively, we have determined the relative protein concentration with Bradford reagent (Bio-Rad, Hercules, CA), using BSA as standard, or we calculated relative RNP concentration from direct measurement of UV light absorption at 260 nm. A conversion factor of 60 μg/ml for an RNP solution with an optical density (OD) of 1 was used to estimate RNP concentration by assuming the molecular mass of an averaged RNP to be 5600 kDa.[1,18] However, in our hands, these methods suggested significantly higher RNP concentrations than expected from active site titration in the endonuclease assay (see below and Fig. 6). This could be due to intrinsically low specific activity of the RNP preparations. It is also possible that there are significant differences in the specific activities of individual RNP species

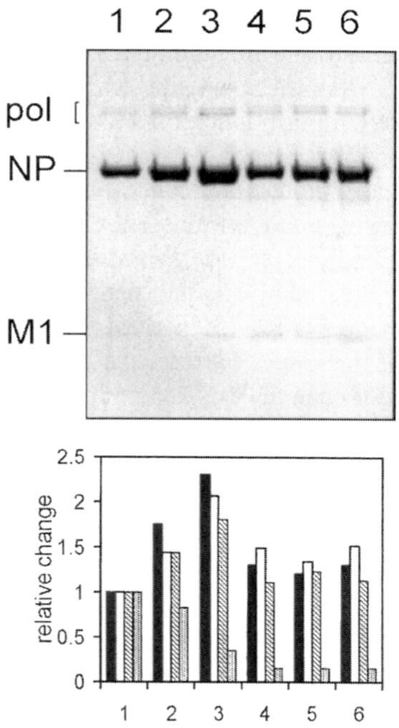

FIG. 4. Influence of lysis time on RNP yield. The 10% (w/v) NuPage gel shows the proteins of equivalent volumes of fractions 7 of six separate glycerol gradients. The gel was stained with Coomassie blue. The virus lysates were prepared without (lanes 1–3) or with acid wash (lanes 4–6) with a lysis time at room temperature of 15 min (lanes 1 and 4), 60 min (lanes 2 and 5), or 120 min (lanes 3 and 6). The corresponding graph shows the analysis of each fraction. Analyzed were polymerase activity (solid columns), NP concentration (open columns), total protein concentration (hatched columns) and ratio of NP to M1 protein (dotted columns). The yield from the preparation under standard conditions and 15-min lysis time (lane 1) was set as 1 and the relative changes are shown on the graph. Lysis time increased the yield of polymerase activity as well as protein concentration in the standard procedure without acid wash (lanes 1–3), but had no effect on yields after the acid wash procedure (lanes 4–6). Increased polymerase activity correlated best with NP and total protein concentration, but did not correlate with the ratio of NP versus M1 protein.

in the mixture of all genomic RNP segments. Generally, RNP preparations have been used in relative volumes for most biochemical experiments and active site titrations have been carried out to compare RNP preparations.

Substrate RNA Synthesis

Capped RNA oligonucleotides are synthesized in five steps.

1. The RNA backbone with a nucleotide sequence of choice is generated by solid-phase synthesis on controlled-pore glass (CPG) support.

2. The CPG-bound oligoribonucleotide is 5′-triphosphorylated with 2-chloro-4*H*-1,3,2-benzodioxaphosphorin-4-one and bis(tri-*n*-butylammonium) pyrophosphate followed by iodine oxidation.

3. The triphosphorylated oligoribonucleotide is cleaved from the solid support, deprotected, and desalted by gel filtration.

4. An N-7 methylated cap structure is added to the triphosphorylated RNA, using vaccinia virus capping enzyme guanylyltransferase (Life Technologies, Rockville, MD).

5. The capped RNA is purified by acrylamide gel electrophoresis.

Detailed protocols for these procedures have been published in this series[33] and are followed with minor changes. RNA synthesis is carried out on an Applied Biosystems (Foster City, CA) 392 DNA/RNA synthesizer with *tert*-butyl-dimethylsilyl (TBDMS)-protected monomers from Cambio (Cambridge, UK) with phenoxyacetyl (Pac), 4-isopropylphenoxyacetyl (i-Pr-Pac), and benzoyl acetyl (Bz/Ac) amino-protecting groups for adenine, guanine, and cytidine bases, respectively. The 5′-triphosphorylation of CPG-bound oligoribonucleotides is performed according to the procedures described in detail by the Brownlee and Olsen laboratories.[28,34] Bis(tri-*n*-butylammonium) pyrophosphate is prepared according to Ludwig and Eckstein.[35] The purity of the reagent and absence of phosphate are confirmed by ^{31}P NMR. As this reagent is unstable and must be synthesized and kept moisture free, it is generally freshly prepared and kept under argon at −20°. A 5′-N^7-methylated cap structure can be added to 5′-triphosphorylated RNA, using 10 units of vaccinia virus guanylyltransferase, 20 μM oligonucleotide, 1 mM *S*-adenosyl-methionine, and 1–5 μM GTP in buffer B in a total volume of 30 μl. The capping reaction is incubated for 3 hr at 37°. [α-^{32}P]GTP is included as a tracer to quantify the yield of capped RNA after gel purification on 16–20% (w/v) denaturing acrylamide : bisacrylamide (19 : 1) gels. We use 1 μM [α-^{32}P]GTP (3000 Ci/mmol) to obtain highly labeled cap structures and 10 μM [α-^{32}P]GTP (30 Ci/mmol) for low specific activity caps.

Use of Synthetic RNA for Study of Influenza Virus Endonuclease

Endonuclease Assay. In the standard assay, 0.1–0.5 nM cap-labeled RNA (specific activity, 10,000–20,000 cpm/fmol) is incubated with 1–5 nM RNP (according to RNP estimation using OD measurement at 260 nm) in buffer C in a total volume of 5 μl for 1–10 min at 30°. The reaction is stopped by addition of 5 μl of formamide loading buffer. The samples are incubated at 98° for 2 min and

[33] S. S. Carroll, F. Benseler, and D. B. Olsen, *Methods Enzymol.* **275,** 365 (1996).

[34] G. G. Brownlee, E. Fodor, D. C. Pritlove, K. G. Gould, and J. J. Dalluge, *Nucleic Acids Res.* **23,** 2641 (1995).

[35] J. Ludwig and F. Eckstein, *J. Org. Chem.* **54,** 631 (1989).

1- to 2-μl aliquots are then immediately loaded onto a preheated 38 × 50 cm, 20% (w/v) acrylamide gel of 0.4-mm thickness. The gel is run at 100 W for 2–3 hr, and then fixed for 15 min in 10% (v/v) ethanol, 10% (v/v) acetic acid, dried under vacuum at 80°, and exposed to a PhosphorImager cassette (Molecular Dynamics, Sunnyvale, CA). Band intensities are quantified with ImageQuant 5.0 software. To analyze coupled endonuclease–transcription initiation reactions, CTP is added to the endonuclease assay. The samples are treated identically to the endonuclease assay described above. To measure transcription initiation independently of endonuclease, chemically synthesized, capped G11 primer RNA can be used in the assay.[29] The apparent K_m values for CTP are 10 and 12 nM, respectively, under the enzyme excess conditions of this assay.

Endoribonuclease Activity of Influenza Virus Ribonucleoprotein. Figure 5 shows time courses for the endonuclease and coupled endonuclease–transcription initiation reactions catalyzed by influenza virus RNP-associated polymerase complex. The endonuclease of RNP generates an 11-mer capped RNA (G11). In the presence of CTP, the polymerase of RNP elongates the G11 primer by 1 nucleotide to generate a 12-mer RNA (G11+1nt). Under the reaction conditions, the substrate is fully depleted. Up to 80–90% of the radioactivity is converted into specific product bands G11 and G11+1nt, the residual 10–20% is spread into evenly distributed background degradation products of various lengths. The relative rates

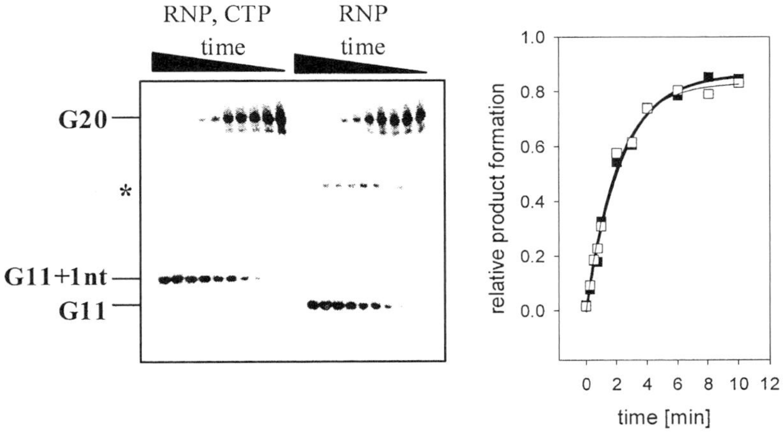

FIG. 5. Single turnover endonuclease and coupled endonuclease–transcription initiation reactions. RNPs (4 nM) were incubated with 0.32 nM ^{32}P-labeled G20 RNA for various times up to 10 min at 30°. One set of reactions contained 10 μM CTP as an additional substrate. *Left:* Analytical gel electrophoresis of the reaction products. The positions of RNA substrate (G20) and products of endonuclease (G11) and transcription initiation (G11+1nt) are indicated on the left. The asterisk indicates an additional, low-level endonuclease cleavage site at nucleotide 16 of G20. *Right:* The relative band intensities as determined from the gel could be fitted to a single exponential progress curve as shown. Endonuclease (RNP, G20 RNA), solid squares; transcription initiation (RNP, G20 RNA, CTP), open squares.

of the endonuclease and the coupled nuclease–initiation reactions are identical, indicating that the rate-limiting step under these conditions occurs before transcription initiation. The good fit to a single exponential equation is consistent with enzyme excess conditions in the assay. This can be confirmed by enzyme and RNA titrations. Figure 6b shows that with increasing concentrations of RNA the reaction kinetics can be shifted from single exponential toward biphasic curves. Burst kinetics have been observed previously for this reaction and suggest that product dissociation is rate limiting under substrate excess conditions.[28] The extent of the burst is directly dependent on the concentration of active sites. Figure 6a shows a doubling of burst size with double enzyme concentration in the assay. The estimated concentrations of endonuclease active sites in the experiment shown in Fig. 6 are 0.1 and 0.2 n*M*, corresponding to about 10% according to the amount of RNP estimated from OD measurement at 260 nm. At RNA concentrations below the concentration of active sites, the product formation kinetics can be fitted to single exponential curves. As expected, under enzyme excess conditions, at a constant concentration of RNA, the relative rates of substrate cleavage are not changed at different protein concentrations (Fig. 6c).

Optimization of Endonuclease Reaction Conditions

Divalent Metal Ions. The endonuclease reaction is dependent on the presence of divalent metal ions. The enzyme might employ a two-metal-ion mechanism of RNA cleavage, as has been proposed before on the basis of the observation of cooperative metal ion binding and synergistic activation of endonuclease activity by metal ion mixtures.[18] From a screen of mono-, di-, and trivalent metal ions, six divalent metal ions have been identified, which activate RNA cleavage in the standard assay conditions. Their relative activation potential is dependent on metal ion concentration, ionic strength, and pH. It has been reported that at 100 m*M* KCl and pH 8 the relative activities are $Mn^{2+} \geq Co^{2+} > Mg^{2+}$, Zn^{2+}, $Ni^{2+} > Fe^{2+}$.[18] Optimal activity with Mg^{2+} is achieved at about 10-fold higher metal ion concentrations as compared with the other five divalent metal ions (1 vs 0.1 m*M*). The activation of the endonuclease by Fe^{2+} is, however, difficult to assess in the presence of atmospheric oxygen, because of rapid oxidation of Fe^{2+} to Fe^{3+} and concurrent inhibition of the endonuclease reaction. Kinetic analysis of the endonuclease reaction at 20 m*M* KCl and pH 7 in the presence of various metal ions shows a high initial rate with Fe^{2+}, followed by protein inactivation (Fig. 7a). Cleavage in the presence of Mn^{2+} or Co^{2+} is faster than with Mg^{2+}, similar to the situation at pH 8. Zn^{2+} and Ni^{2+} show sigmoidal time courses, suggesting that a structural reorganization of the active site might be required for endonuclease activity to occur with these metal ions (Fig. 7a).

Monovalent Salt Concentration and Buffer Conditions. The velocity of the endonuclease reaction increases only slightly in the presence of monovalent salts up to 20 m*M* KCl, NaCl, or KCH_3CO_2. Inhibition of endonuclease starts to occur

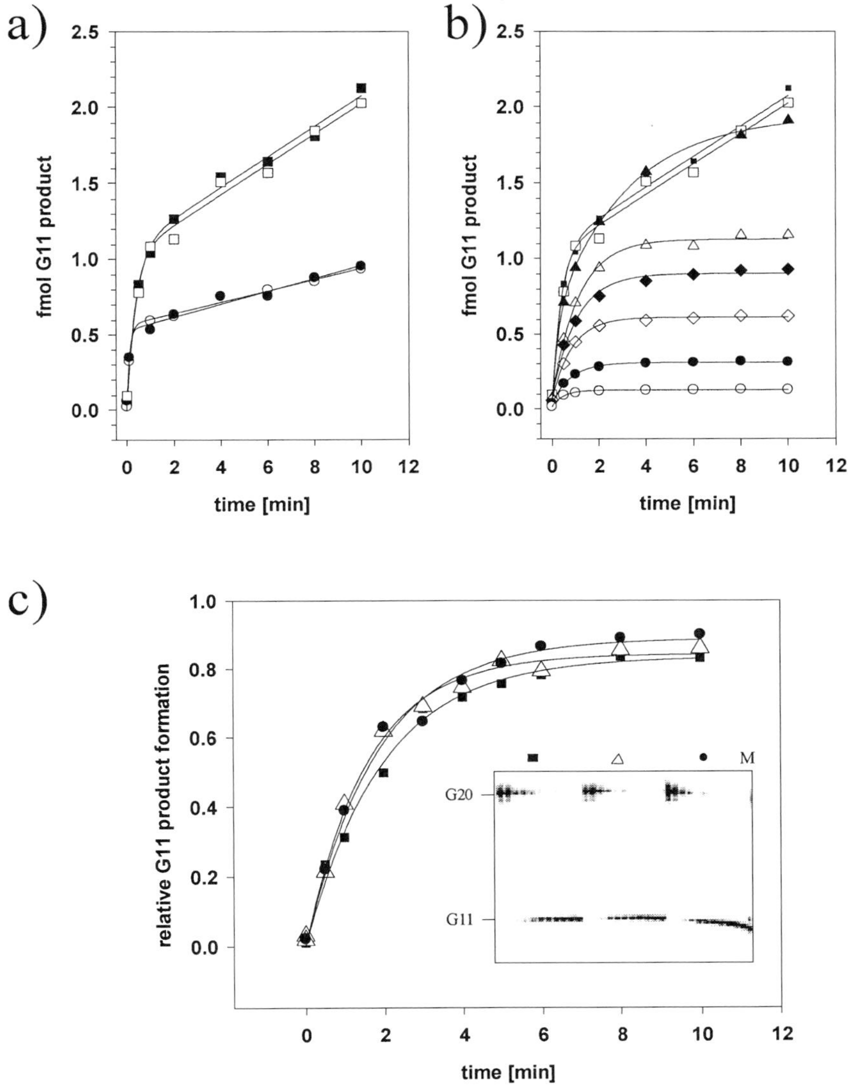

FIG. 6. Endonuclease active site titration. (a) RNPs (1 n*M*, circles; 2 n*M*, squares) were incubated at 30° with various amounts of G20 RNA: 2.7 n*M*, solid squares; 1.35 n*M*, open squares; 5 n*M*, solid circles; 1 n*M*, open circles. The amount of G11 product generated was calculated according to the analysis of band intensities on an analytical acrylamide gel. (b) RNPs (2 n*M*) were incubated at 30° with various amounts of G20 RNA: 2.7 n*M*, solid squares; 1.35 n*M*, open squares; 0.675 n*M*, solid triangles; 0.27 n*M*, open triangles; 0.2 n*M*, solid diamonds; 0.135 n*M*, open diamonds; 0.0675 n*M*, solid circles; 0.027 n*M*, open circles. The data were fitted to biphasic or single exponential progress curves. (c) G20 RNA (0.135 n*M*) was incubated at 30° with 0.5 n*M* (black squares). 1 n*M* (white triangles) or 2 n*M* (black circles) RNP. The relative band intensities of the G11 bands from the gel (*inset*) were fitted to single exponential progress curves. Lane M, capped G11 RNA size marker.

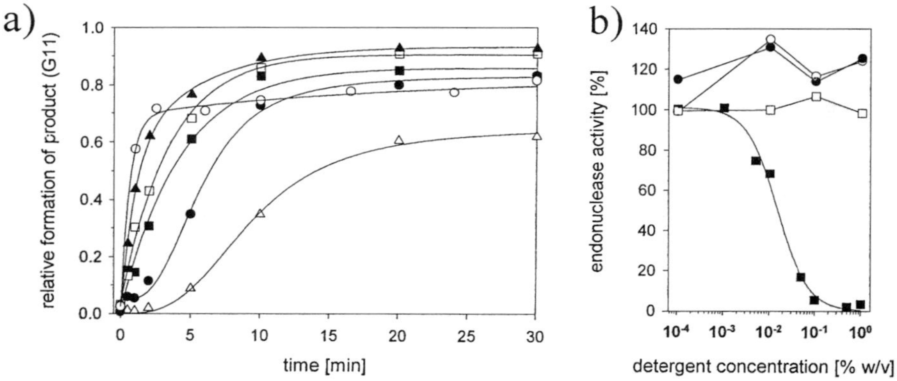

FIG. 7. Divalent metal ions and detergent in the endonuclease reaction. (a) G20 RNA (0.13 n*M*) was incubated at pH 7 and 30° with 1 n*M* RNP in the presence of freshly prepared 0.1 m*M* $FeCl_2$ (open circles), 0.1 m*M* $CoCl_2$ (solid triangles), 0.1 m*M* $MnCl_2$ (open squares), 1 m*M* $MgCl_2$ (solid squares), 0.1 m*M* $NiCl_2$ (solid circles), and 0.1 m*M* $ZnCl_2$ (open triangles). The endonuclease requires about 10-fold higher concentrations of $MgCl_2$ as compared with the other divalent metal ions to be optimally active. The data were fitted to biphasic (Fe^{2+}), double exponential (Co^{2+}), single exponential (Mn^{2+}, Mg^{2+}), and sigmoidal (Ni^{2+}, Zn^{2+}) progress curves. (b) G20 RNA (0.27 n*M*) was incubated for 3 min at pH 8, 30° with 1 n*M* RNP and the indicated concentrations of detergent. Endonuclease activity was calculated relative to the activity in the absence of detergent. Sodium deoxycholate (DOC), solid squares; CHAPS, open squares; Triton X-100, solid circles; NP-40, open circles.

at salt concentrations above 100 m*M*. We observed no significant difference in endonuclease activity when using Tris or HEPES to buffer the assay solutions at pH 7, 7.5, or 8. The reaction is fastest at high pH. Depending on the type of divalent metal ion, endonuclease activity can be reduced at high pH, presumably because of the formation of insoluble hydroxide salts. Half-maximal activity was reported at pH 6.8 for 1 m*M* $MgCl_2$ and at pH 6.2 for 0.1 m*M* $MnCl_2$.[18]

Detergent. RNP particles are prepared by detergent treatment of purified influenza virus. In a limited screen of detergent types, nonionic and zwitterionic compounds {3-[(3-cholamidopropyl)-dimethyl-ammonio]-1-propanesulfonate (CHAPS), Nonidet P-40 (NP-40), Tween 20, Triton X-100, octylglycoside} did not significantly activate or inhibit the endonuclease reactions at concentrations up to 1% (w/v). The ionic detergent sodium deoxycholate (DOC) inhibits the endonuclease reaction with an IC_{50} of 0.02% (w/v) (Fig. 7b). This is consistent with previous reports demonstrating the use of DOC to selectively dissociate the influenza polymerase subunits from the RNP particles.[6,36] For reasons of improved liquid handling, 0.3% (w/v) Triton X-100 is included in the standard endonuclease reaction buffer C.

[36] S. C. Inglis, A. R. Carroll, R. A. Lamb, and B. W. Mahy, *Virology* **74,** 489 (1976).

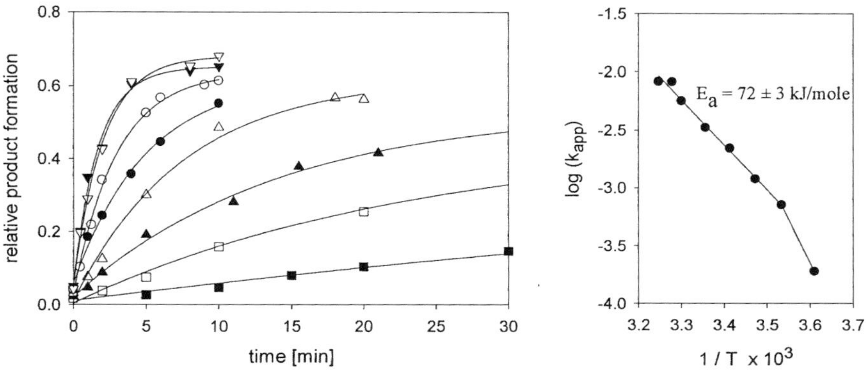

FIG. 8. Temperature dependence of the endonuclease reaction. *Left:* G20 RNA (0.2 n*M*) was incubated with 1 n*M* RNP. The incubation times were adjusted to the temperatures and ranged between 10 min (35–25°), 20 min (20°), 30 min (15°), and 120 min (10–4°). The data were fitted to single exponential progress curves, which are shown for times up to 30 min at left (4°, solid squares; 10°, open squares; 15°, solid upward triangles; 20°, open upward triangles; 25°, solid circles, 30°, open circles; 32°, solid downward triangles; 35°, open downward triangles). *Right:* The derived apparent rate constants were plotted according to the Arrhenius equation. The slope calculated from linear regression of the data corresponds to: $-E_a/RT$.

Temperature. The RNA cleavage rates by influenza virus endonuclease on RNP increase with temperature up to 35°. The enzyme is stable under the assay conditions for at least 1 hr at 30° and 32°, but significant inactivation is observed after 30 min of preincubation at 35°. More than 50% of activity is lost after 5 min of preincubation at 40° or 1 min at 45°. Interestingly, RNPs in storage buffer D are significantly more stable at elevated temperatures. In contrast to RNPs in reaction buffer C, the RNPs in storage buffer D lost no activity even after 1 hr of preincubation at 35° or 10 min at 40°, and more than 50% of activity was lost at 45° only after 10 min of preincubation.

The apparent RNA cleavage rates obtained from temperature scans can be plotted according to the Arrhenius equation (log k_{app} vs $1/T$.[37] From the slope of such graphs an apparent activation energy of $E_a = 72 \pm 3$ kJ/mol can be calculated for the RNA cleavage reaction (Fig. 8).

Effects of Other Reagents. Reduced RNA cleavage rates are observed at glycerol concentrations above 15% (v/v) with 20% activity remaining at 50% (v/v) glycerol. The cleavage rates are not affected by glycerol concentrations between 5 and 15% (v/v).

Spermidine is known to activate small phage RNA polymerases such as T7 and SP6 DNA-dependent RNA polymerases, possibly by increasing local concentrations of nucleic acid in the sample buffer.[38,39] Under the assay conditions,

[37] M. Dixon and E. C. Webb, "Enzymes." Longman Group, London, 1979.

spermidine had no effect on influenza virus endonuclease at concentrations up to 5 m*M*, but did inhibit the reaction at and above 10 m*M*. Similar results were obtained with other crowding reagents. Polyvinylpyrrolidone (PVP) had no effect on the reaction up to 36 mg/ml. Similarly, polyethylene glycol (PEG) 6000 had no effect up to 9% (w/v), but inhibited at 18% (w/v), the highest concentration tested. Dimethyl sulfoxide (DMSO) activated RNA cleavage by the endonuclease up to 2-fold at concentrations between 1 and 5% (v/v). No significant activation or inhibition of the reaction was observed at concentrations between 10 and 25% (v/v) DMSO.

Summary and Conclusions

The influenza virus polymerase complex contains two associated enzymatic activities, an endoribonuclease and a RNA-dependent RNA polymerase activity. Both activities have so far been observed only with the complete polymerase complex consisting of three subunits, PB1, PB2, and PA. This chapter describes a robust and optimized procedure for the purification of active influenza virus polymerase in complex with genomic RNA and the single-stranded RNA-binding protein nucleoprotein from influenza virus particles. It also explains the synthesis of capped RNA molecules as substrates of the influenza virus endonuclease. The enzymatic properties of influenza virus-derived endoribonuclease activity have been characterized with a model RNA substrate of 20-nucleotide length, termed G20 RNA. The rate of RNA cleavage under steady state conditions appears to be limited by product dissociation. Therefore conditions have been optimized to study the chemical step of RNA cleavage under single turnover conditions. The enzyme requires divalent metal ions for activity and can use Mn(II), Co(II), and Fe(II) efficiently at pH 7, Mg(II) with intermediate efficiency, and Ni(II) and Zn(II) with lower efficiency. The reaction progress curves show slow binding of Zn(II) and Ni(II) to the protein, suggesting a conformational change of the active site as a prerequisite for endonuclease activity in the presence of these two metal ions. Low concentrations of the detergent DOC inhibit the activity and also disrupt the trimeric polymerase complex, whereas other detergents do not have a significant effect on the activity.

Acknowledgments

We are grateful to Dr. Hanno Langen (F. Hoffmann-La Roche AG, Basel) for peptide analysis of RNP proteins, and to Dr. Noel A. Roberts for critical reading of the manuscript.

[38] O. C. Uhlenbeck, J. Carey, P. J. Romaniuk, P. Lowary, and D. Beckett, *J. Biomol. Struct. Dyn.* **1,** 539 (1983).

[39] J. F. Milligan, D. R. Groche, G. W. Witherell, and O. C. Uhlenbeck, *Nucleic Acids Res.* **15,** 8783 (1987).

[38] Bacteriophage T4 RegB Endoribonuclease

By MARC UZAN

Initial Findings

The development of bacteriophage T4 is largely determined by a precisely coordinated subversion of the host RNA polymerase to permit initiation from different classes of T4-specific promoters. The synthesis, in cascade, of regulatory elements that modify the transcription apparatus, ensures the temporal ordering of the developmental program. Early, middle, and late promoters can be distinguished on the basis of their characteristic sequences and the *trans*-acting factors controlling their activation. Control of phage development is also exercised at the posttranscriptional level, in particular when inhibition of the genetic activity is required. Extensive information on the T4 phage biology can be found in the monograph edited by Karam.[1]

In the early 1970s, Hall and colleagues isolated new bacteriophage T4 mutants that had acquired the ability to grow on medium supplemented with folic acid analogs. They hoped thus to identify new genes involved in global regulation of prereplicative (early and middle) gene expression. Several independently isolated T4 mutant phages were found to contain various large deletions in the region of genes *tk* and *denV* on the T4 chromosome. In each case, the deletion led to overproduction of several enzymes involved in nucleotide metabolism, including dihydrofolate reductase.[2] Evidence was obtained to indicate that the effect occured at the posttranscriptional level. Chace and Hall[2] suggested that the loss of at least one gene, called *regB* (for regulation), eliminated in all these deletions, could be responsible for the observed phenotype.

Years later, Uzan *et al.*[3] found that bacteriophage T4-infected *Escherichia coli* cells contain an endoribonuclease that cleaves some phage mRNAs in the middle of the sequence GGAG, producing 5′-OH RNA ends.[3] Because the first cuts were found in regions of the mRNA that interact with the initiating ribosome (the GGAG tetranucleotide is one of the most frequent Shine–Dalgarno sequences), we at once realized that, by destroying ribosome-binding sites, this nuclease would lead to functional inactivation of mRNAs, which we showed later (see below). It was shown that the nucleolytic cuts, detected as soon as 1.5 min after infection (at 30°), depend on the synthesis of one or more phage proteins, because addition of a potent protein synthesis inhibitor to the growth medium before infection prevents

[1] J. D. Karam, "Molecular Biology of Bacteriophage T4." ASM Press, Washington, D.C., 1994.
[2] K. V. Chace and D. H. Hall, *J. Virol.* **15,** 929 (1975).
[3] M. Uzan, R. Favre, and E. Brody, *Proc. Natl. Acad. Sci. U.S.A.* **85,** 8895 (1988).

0076-6879/01 $35.00

the processing.[3] Shortly afterward, Ruckman *et al.*[4] formally identified a T4 open reading frame (ORF) necessary for this nucleolytic activity among the ORFs lying within the overlapping region of the deletions described by Chace and Hall. We then showed that a point mutation in this ORF, leading to nuclease deficiency, allowed the phage to grow in the presence of pyrimethamine, the folate analog with which *regB* deletion mutants had been selected,[5] thus establishing this ORF as the *regB* gene postulated by Chace and Hall.

RegB Endoribonuclease: Sequence-Specific RNase Requiring Partner S1 Ribosomal Protein *in Vitro*

The T4 *regB* gene is transcribed from a typical early promoter, immediately upstream of the gene. It encodes a basic protein of 153 amino acids, rich in lysine and arginine, with a calculated p*I* of 8.8. It seems to be autoregulated because *regB* mRNA is efficiently cleaved by the RegB nuclease within its Shine–Dalgarno sequence (GGAG), and, to a lesser extent, at three places within its coding sequence.[4,6] As shown by polymerase chain reaction (PCR) analysis, the genome of four other T4-related phages (K3, RB70, T2, and T6) seems to be organized similarly in the *regB* region,[7] strongly suggesting that the *regB* gene also exists in these phages.

If protein synthesis is blocked before infection with T4, the only phage transcripts that are made are early transcripts. These are, of course, unprocessed by RegB because no T4 protein is made. However, if, before protein synthesis inhibition, the RegB nuclease is induced from a plasmid, specific cuts can be detected within the T4 transcripts.[8] Also, when expressed from plasmids, in uninfected cells, RegB is able to introduce specific cuts within its own mRNA or within other T4 transcripts synthesized from a second compatible plasmid.[9,10] Therefore, the RegB nuclease needs no other T4 protein for its activity.

In vitro, purified RegB protein cleaves artificial as well as natural RNAs specifically in the middle of the GGAG sequence, showing that the catalytic activity and the specificity are carried by the RegB polypeptide.[9] However, the first *in vitro* assays revealed that the cleavage efficiency was extremely low compared with the RegB activity *in vivo*. Of particular interest, Ruckman *et al.*[9] found that RegB activity can be stimulated about 100-fold by the *E. coli* 30S ribosomal subunit. The 50S subunit is without effect. These authors further showed that S1-depleted

[4] J. Ruckman, D. Parma, C. Tuerk, D. H. Hall, and L. Gold, *New Biol.* **1,** 54 (1989).
[5] B. Sanson and M. Uzan, *J. Mol. Biol.* **233,** 429 (1993).
[6] B. Sanson and M. Uzan, unpublished data (1994).
[7] F. Repoila, F. Tétart, J.-Y. Bouet, and H. M. Krisch, *EMBO J.* **13,** 4181 (1994).
[8] M. Uzan, unpublished data (1997).
[9] J. Ruckman, S. Ringquist, E. Brody, and L. Gold, *J. Biol. Chem.* **269,** 26655 (1994).
[10] B. Sanson, R.-M. Hu, E. Troitskaya, N. Mathy, and M. Uzan, *J. Mol. Biol.* **297,** 1063 (2000).

30S ribosomal subunits are inefficient but that purified ribosomal protein S1 is sufficient to achieve a level of stimulation similar to that obtained with the entire 30S subunits. S1 is an RNA-binding protein known to play an essential role during translation, "clamping" the ribosomes to mRNAs (extensive information about the S1 protein has been reviewed by Subramanian[11]) (see also below).

Specificity of RegB Endoribonuclease: Role of S1 Ribosomal Protein

Not all GGAG-carrying mRNAs are substrates for RegB. Figure 1 shows an alignment of 34 T4 mRNA sequences tested so far for their susceptibility to the RegB nuclease *in vivo*. This is not an exhaustive list. The upper part of Fig. 1 shows the sequences of the RNAs efficiently cleaved by RegB. It can be seen that in the vast majority of cases, cutting occurs in the middle of the tetranucleotide GGAG. In one case, a cut in the middle of a GGAU sequence was found. Most of the cleaved motifs are carried by early transcripts. All these sites are located in intergenic regions and many of them are Shine–Dalgarno sequences. The GGAG motifs found in coding sequences are, for the most part, poorly or not at all recognized (the cleaved coding sequences are not shown).[3,4,12]

Not all intergenic GGAG sequences are processed by RegB. A few early, most middle, and all late GGAG-carrying transcripts escape RegB processing during phage T4 development[10] (Fig. 1, bottom). Thus, although the RegB nuclease is made shortly after infection, most of the phage transcripts made after a few minutes of development escape its action, even when they contain the GGAG sequence. Sanson *et al.*[10] could detect RegB activity until at least 10 min after infection (at 30°), that is, toward the end of the middle period. This implies that GGAG-carrying early as well as middle transcript RNAs that escape RegB processing *in vivo* are not substrates for the enzyme, and this conclusion could apply to late mRNAs as well. However, because none of the GGAG-carrying late transcripts was found to be cleaved by RegB, it was possible that the nuclease was inactivated by a phage-encoded factor immediately before the burst of late transcription. The synthesis of RegB stops between minutes 3 and 4 of phage development,[6] so any mechanism able to inactivate RegB would lead to the complete disappearance of enzymatic activity. In fact, Sanson *et al.*[10] showed that the GGAG-carrying mRNAs that are unprocessed by RegB during infection are not substrates, regardless of the period of the T4 cycle when the transcripts are made. This conclusion is based on the observation that in a two-plasmid system, where in the same bacterial cell one plasmid provides the RegB nuclease and the other provides a T4 RNA to be tested (hence, in the absence of T4 infection), the same pattern of sensitivity or resistance

[11] A.-R. Subramanian, *Prog. Nucleic Acid Res. Mol. Biol.* **28,** 101 (1983).

[12] B. Sanson and M. Uzan, *FEMS Microbiol. Rev.* **17,** 141 (1995).

Processed T4 RNAs ↓ A+U

regB	pppAUCAGUUAAGAGGAGAAUAACAUGACUAUCAAUACAG	E	RBS
ORF *frd1*	CUUUUGAUUAAAGCUUGCUAAGGAGAAUAAAAUGAGAUUACAACGC	E	RBS
5' *cef*	UACUGAUACAGAAAACAACUUGGAGAAUAAAAUGGAAAAUUCGCT	E	RBS
motA	AACUACUGAUACAGAAUUACGGAGAUUAGAAAAUGUCUAAAGUAA	E	RBS
motB.+2	UAUAACACCAUAGCUACUGAGGAUAAUAAAAUGAAAAUUUAUCGU	E	RBS
ORF tRNA.4	pppAUCAACUACUGAGGAGAAUAAAAUGAAACGCUGUGAA	E	RBS
ORF*frd2*	AUUACCCUAUCAAAACUAAUGGAGAAAAGAAAAUGUUCGCACCUU	E	
43	GGGCUUCGGCCCCUUUAUUUGGAGUAUAAUAUAUCAAGAGCCUA	E	
39	CCUUUAGCUUUAUGAUUACCGGAGUAUAAUAUUCCCGAAACCAA	E	
motB	CCAGAGUAUAAUGGUCCCGUGGAGUAUAAAAUCUUUUUAACAAG	E	
5' *ndd*	GUGUGUAUUCAUGCGGCCUUGGAGUAGAAAAUAAUUUAGAGGAA	E	
ORF*45.2*	UAUUAGAGUAUAAUCUCUAUGGAGGAAAAACAUGGAAUAUUCAA	M	RBS
ORF *47.1*	UCGGCCCUUUAGCUUUAUACGGAGUUUGAUAUAAUGAUAUUUCU	M	RBS

Unprocessed or poorly processed T4 RNAs

denV	CUAUAGACCUAUCAACUACAGGAGAACACUAAAAUGACUCGUAUC	E	RBS
ORF*mrh.2*	AUAGAAUUAUCUUAUAGAGGAGAGUACUAUGUUAAAUCGUUGGA	E	RBS
ORF*mobD.5*	ACAACCAUCGGAUAAAGAGGAGAACAUCAUGAAAAUUGAAGCA	E	RBS
gene *1*	AUAAUCUUAAUUAAAUUUGAGGAGAAACACAUGAAACUAAUCUUUU	M	RBS
ORF *Y.-2*	UAAAUAGAUAUAAUUCAGAGGAGACAAUCAUGUCAGAUAAG	M	RBS
41	AAAAAAUUUAAAUUCUUUUAAGGAGUAAGUGUGGUAGAAAUUAUUU	M	RBS
denA	AGUUUUAAAAAAUACUUUUAAGGAGUGGGCCGCAAGGCCCAUUUUA	M	RBS
e	AAUACCUUCUAUAAAUACUUAGGAGGUAUUAUGAAUAUAUUUGAAAU	L	RBS
soc	AAUCAUGUAAUUUAAAUAAAGGAGAAUUACAUGGCUAGUACUCGCGG	L	RBS
t	UACGUAUAAAUAUCUUAAAAGGAGGGUCUAUGGCAGCACCUAGAA	L	RBS
ORF *5,1*	TTCCAATGAGCGCTGAUUUAGGAGAAUCCAUGGAAGGUUCUUCUA	L	RBS
4=50=65	pppAUCUAUUUAUAAGGAGAAUCCAAUGGCAUAUUCUGGA	L	RBS
5	AACGACCUUAUACGUAUAAUGGAGAAAUCUUAUGGAAAUGAUAAG	L	RBS
15	AAUAAUUAAAGCCCGGAUUAGGAGAAGUCAUGUUUGGUUAUUUU	L	RBS
20	pppCUAUAAUUCCCAUUUGGAGAAUACAAUGAAAUUUAAUGUAU	L	RBS
26	pppGAUAAAUAAAGGAGCUAAAUAUGUAUGAAUACAAAU	L	RBS
34	UUAUUAUUGAAAACACAAUAGGAGCCCGGGAGAAUGGCCGAGAU	L	RBS
35	GAGUUUUUGACCCUUCCACCGGAGCAUUAGUUGAUAGUAAGUCAUA	L	
38	AUAUCGUUCGUUUAGCAUAAGGAGAGGGGCUUCGGCCCUUCUAA	L	RBS
51	UAUUUAUAAAUAGAAUAAAAGGAGCAUCUAUGGCAAACAUUAUUCG	L	RBS
53	pppUAAUAUAUUUAUAAACAGGAGGGCCCAUGCUCUUUACAUUUUU	L	RBS

FIG. 1. Sequences of some processed and unprocessed T4 mRNAs *in vivo*. *Top:* RNA sequences efficiently cleaved by RegB. *Bottom:* Poorly processed or unprocessed RNA sequences. The arrow (*top*) indicates the site of cleavage; the GGAG motif as well as the cytosines are in boldface letters. The initiation codons are underlined. The (A+U)-rich region 3′ of the efficiently cleaved GGAG motifs is marked by a thick bar (*top*). ppp, The 5′ transcription initiation start of the transcripts; E, M, and L (right-hand side), sequences carried by early, middle, and late transcripts, respectively; RBS, ribosome-binding sites. In the sequences of *denA* and *35* mRNAs, arrows underline sequences that can base pair to form a stem–loop secondary structure, moving the Shine–Dalgarno sequence and the initiation codon closer together. [The results presented are from M. Uzan, R. Favre, and E. Brody, *Proc. Natl. Acad. Sci. U.S.A.* **85,** 8895 (1988); J. Ruckman, D. Parma, C. Tuerk, D. H. Hall, and L. Gold, *New Biol.* **1,** 54 (1989); B. Sanson and M. Uzan, *FEMS Microbiol. Rev.* **17,** 141 (1995); B. Sanson, R.-M. Hu, E. Troitskaya, N. Mathy, and M. Uzan, *J. Mol. Biol.* **297,** 1063 (2000); R.-M. Hu and M. Uzan, unpublished data (1998).]

to RegB is observed as is found during T4 infection. In particular, a late transcript is not processed in this system whereas an early transcript is. Moreover, *in vitro,* the naturally resistant GGAG sequences remain unprocessed or are only weakly processed, whereas the sensitive sequences are efficiently cleaved. Therefore, the information determining whether an RNA is cut by RegB is carried in *cis* by these transcripts.[10] The possibility that RegB is inactivated late in the phage cycle has not been ruled out, but this hypothesis is not necessary to account for the RegB resistance of late transcripts.

Looking for RegB Target Site

An important question raised by the preceding conclusion concerns the basis of the discrimination, among the GGAG-carrying RNAs, between substrates and nonsubstrates. Either a good substrate for RegB must contain additional sequence or structural elements, in addition to GGAG, or the GGAG motif is sufficient and the resistant molecules carry antideterminants. Sequence comparison of cleaved and uncleaved mRNAs (shown in Fig. 1) reveals that in addition to the strongly conserved GGAG sequence, the cleaved RNAs contain an (A+U)-rich region 3′ of –GGAG–, which is strikingly devoid of cytosine, whereas in the unprocessed transcripts cytosines are always found within this region and no nucleotide preference can be detected. Whether this sequence bias is related to RegB/S1 specificity is not yet known (see below).

In an attempt to better define the sequence/structure required for RegB cleavage, a SELEX (systematic evolution of ligands by exponential enrichment; see Tuerk and Gold[13]) run was carried out, on the basis of the selection of RNA molecules cleaved in the presence of RegB and the ribosomal protein S1.[14] About 40 sequences were selected and analyzed. They all contained the GGAG tetranucleotide (one exception was GGAC), with a preference for the sequence GGAGG (this bias is not found among the T4 natural mRNA sequences; see Fig. 1). No conserved sequence, besides the GGAG motif, was found among the different selected RNA molecules. In most cases, this tetranucleotide was found in the 5′ part of the randomized region and the region immediately 3′ of the GGAG motif was strikingly rich in A and C nucleotides. Thus, like in the natural T4 RNAs, the nucleotide composition of the region 3′ of –GGAG– is not random, suggesting that this region plays a role in RegB/S1 recognition. The difference in composition between the two types of molecules, natural and artificial, may mean that a structural element plays a role in the selection of sites.

Probing of RNA structures with chemical agents did not reveal any obvious differences between substrate and nonsubstrate T4 mRNAs. They all seem rather

[13] C. Tuerk and L. Gold, *Science* **249,** 505 (1990).

[14] V. K. Jayasena, D. Brown, T. Shtatland, and L. Gold, *Biochemistry* **35,** 2349 (1996).

unstructured in the vicinity of the GGAG sequence.[15] A possibility could be that not all the information necessary to define a good substrate is contained in *cis,* but just a part of it. This information could be sufficient, however, to attract a protein that would impose a structural constraint on the RNA. Only the structured RNAs would then be recognized by RegB. Thus, the system would require an RNA chaperone and it is tempting to suggest that such a role is played by the ribosomal protein S1. Arguments in support of this model are presented in the next section.

Role for S1 in Specificity of Reaction

Checking the above-described model demands that the structure of certain RegB substrates be resolved. Lebars *et al.*[16] derived several short RNAs (about 30 nucleotides long or less) from some of the sequences obtained in the SELEX procedure carried out by Jayasena *et al.*[14] A correlation between the kinetic constants obtained with these small RNAs, and their structure, as determined by nuclear magnetic resonance (NMR), establishes that when GGAG motifs are included in secondary structures, they are poorly cleaved in the absence of S1. For most of the molecules analyzed, the presence of S1 stimulates only weakly the cleavage reaction within these motifs. This result certainly accounts for part of the resistance of T4 RNAs to RegB. In group I intron RNAs, tRNAs and certain polycistronic mRNAs, several GGAG sequences are included in stable secondary structures, and some of them were shown to be refractory to RegB processing.[8]

However, many GGAG sequences, not included in RNA secondary structures, are resistant to RegB. For example, the leader regions of some GGAG-carrying early (*denV*) and late transcripts (gene *15*), fully resistant to RegB *in vivo,* cannot form stable secondary structures (see Fig. 1). Again, the analysis of a small SELEX-derived RNA turned out to be informative. Lebars *et al.*[16] found that a 30-mer RNA, in which the G in the 3′ position of the GGAG motif is base paired whereas the rest of the sequence is unpaired (in a loop), is efficiently and specifically cleaved by RegB in the absence of S1. Stimulation by S1 is only about 20%. A 16-mer RNA of sequence identical to the sequence surrounding GGAG in the 30-mer RNA, but completely unstructured (as checked by NMR), is poorly attacked by RegB, and the cleavages did not occur at the expected position but between the two G's of the GGAG motif, and at a few other positions in the molecule. The S1 ribosomal protein stimulates only slightly and equally each of these nonspecific cuts.[16] Thus, RegB alone is able to perform efficient and specific cleavages provided two conditions are met: (1) that most of the GGAG sequence is unpaired and (2) that this motif is partly constrained. This result suggests that, when S1 is required for an efficient

[15] R.-M. Hu and M. Uzan, unpublished data (1998).

[16] I. Lebars, R.-M. Hu, J.-Y. Lallemand, M. Uzan, and F. Bontems, *J. Biol. Chem.* **276,** 13264 (2001).

cleavage (with the T4 natural transcripts, e.g.), the role of this protein is to promote this constraint on the RNA, presumably by interacting with a target site located in the vicinity of the GGAG sequence. This site would seem to be missing in the 16-mer derived from the SELEX RNA molecule and, presumably, in all unprocessed T4 mRNAs, but would be present in most early transcripts. It is tempting to suggest that the (A+U)-rich sequence found 3′ of the well-cleaved GGAG motifs, in natural T4 mRNAs, is implicated in S1 recognition.

There is no indication that S1 copurifies with RegB.[9,15] Furthermore, a direct interaction between these two proteins could not be detected. For example, the S1 protein is not retained on a column of RegB-bound Ni-agarose, whether in the presence or the absence of *motA* RNA.[17] In a filter-binding assay, under conditions in which S1 binds RNA, RegB alone binds target RNA poorly, if at all.[14] These data strongly suggest that S1 helps the RegB nuclease via its binding to RNA. The fact that the amount of RNA retained on filters increases when both RegB and S1 proteins are added[14] may be interpreted in the same sense: S1 could convert the RNA into a form capable of being recognized by RegB.

The role of the S1 ribosomal protein is not limited to translation. S1 seems to be a multifunctional protein involved in several unrelated processes that all use the RNA-binding and, possibly, the single-stranded DNA-binding properties of the protein. As a component of the phage Qβ replicase, it is required for the specific recognition and binding of the enzyme onto Qβ RNA.[18,19] By repressing translation of its own mRNA, S1 autoregulates its synthesis.[20] This protein might participate in transcription termination/antitermination.[21] Also, as a poly(A)-binding protein, it might play a role in mRNA turnover.[22] S1 is found bound to the Redβ protein of phage λ, involved in phage recombination. Yet, the role of this association, if any, is unknown.[23] The *E. coli* S1 ribosomal protein contains six homologous RNA-binding domains, each of about 70 amino acids. The two N-terminal domains are involved in binding to the ribosome whereas the four C-terminal domains are devoted to mRNA interactions. Similar domains can be found in a large number of RNA-associated proteins from bacteria to humans, in particular in certain RNases (*E. coli* RNase E, polynucleotide phosphorylase, and RNase II). The structure of the S1 RNA-binding domain from *E. coli* polynucleotide phosphorylase was determined by NMR.[24] The S1 domain is structurally similar to a domain of the

[17] E. Troitskaya, R.-M. Hu, and M. Uzan, unpublished data (1997).

[18] F. Meyer, H. Weber, and C. Weissmann, *J. Mol. Biol.* **153,** 631 (1981).

[19] T. Blumenthal and G. G. Carmichael, *Annu. Rev. Biochem.* **48,** 525 (1979).

[20] J. Skouv, J. Schnier, M. D. Rasmussen, A. R. Subramanian, and S. Pedersen, *J. Biol. Chem.* **265,** 17044 (1990).

[21] J. Mogridge and J. Greenblatt, *J. Bacteriol.* **180,** 2248 (1998).

[22] M. P. Kalapos, H. Paulus, and N. Sarkar, *Biochimie* **79,** 493 (1997).

[23] T. V. Venkatesh and C. M. Radding, *J. Bacteriol.* **175,** 1844 (1993).

[24] M. Bycroft, J. P. Hubbard, M. Proctor, S. M. V. Freund, and A. G. Murzin, *Cell* **88,** 235 (1997).

E. coli major cold shock protein, CspA, whose action as an RNA chaperone is well documented.[25] In line with our model, the CspA protein was shown to facilitate the action of RNases A and T1.[25]

Does the S1 protein recognize a specific sequence or structure? Translation initiation and elongation of most, if not all, *E. coli* transcripts seem to require this ribosomal protein,[26] suggesting that S1 can interact with a large variety of sequences. The S1 protein is known to present a high affinity for polypyrimidine polymers[11] but, in addition, various sequences were found to interact specifically with S1 *in vitro*. U-rich sequences are targets for S1 in the *E. coli ssb* mRNA and in phage RNAs Qβ and fr.[27,28] S1 protects U-rich as well as $(CAA)_n$ sequences in some plant viral mRNAs.[29] Also, S1 associates with poly(A) tails 3′ of *E. coli* mRNAs.[22] Mogridge and Greenblatt[21] found that S1 interacts specifically and with high affinity with *boxA* RNA, a 12-nucleotide-long sequence involved in *E. coli* and phage λ transcription termination/antitermination. High-affinity ligands generated by SELEX against purified S1 protein contain pseudoknots with conserved sequences in the loop regions, one of which is a Shine–Dalgarno sequence.[30] These data strongly suggest that S1 protein accommodates more than one nucleic acid-binding site and evidence supporting this notion has been presented.[11,22,31] The number of S1 domains required in RNA recognition and binding is unknown, but the slight sequence differences among the four C-terminal domains might account for the observed diversity of interaction. The (A+U)-rich sequences found immediately downstream of the well-cleaved GGAG motifs (Fig. 1) have not been identified among the S1-binding sequences.

Role *in Vivo*

A classic way to address the question of the role of RegB during T4 infection is to examine the consequences of a mutation that inactivates the enzyme. The *regB* mutant phage used throughout the physiological studies contains a missense mutation that converts the arginine codon at position 52 (CGC) into a leucine codon (CUC). This change completely abolishes RegB activity as assessed by the total absence of processing of T4 early transcripts, *in vivo*. This mutation probably arose spontaneously (see Ruckman *et al.*[4]). Other missense, nonsense, or single-base-pair deletions in *regB* exist but were not used for this purpose.[4,6,32]

[25] W. Jiang, Y. Hou, and M. Inouye, *J. Biol. Chem.* **272,** 196 (1997).
[26] M. A. Sørensen, J. Fricke, and S. Pedersen, *J. Mol. Biol.* **280,** 561 (1998).
[27] S. Goelz and J. A. Steitz, *J. Biol. Chem.* **252,** 5177 (1977).
[28] I. V. Boni, D. M. Isaeva, M. L. Musychenko, and N. V. Tzareva, *Nucleic Acids Res.* **19,** 155 (1991).
[29] N. V. Tzareva, V. I. Makhno, and I. V. Boni, *FEBS Lett.* **337,** 189 (1994).
[30] S. Ringquist, T. Jones, E. E. Snyder, T. Gibson, I. Boni, and L. Gold, *Biochemistry* **34,** 3640 (1995).
[31] D. E. Draper and P. H. von Hippel, *J. Mol. Biol.* **122,** 321 (1978).
[32] B. Miroux and M. Uzan, unpublished data (1996).

Regulation of Translation of Many Prereplicative T4 Genes by RegB Nuclease

Thirty-two percent of *E. coli* genes have the GGAG motif as Shine–Dalgarno sequence (339 of 1055[33]) and this proportion applies to the T4 early genes as well. Removal by RegB of half of this strong Shine–Dalgarno sequence is expected to result in a dramatic decrease in the translatability of the messages. This is in fact the case.[5,10] As shown for the early *motA* transcript, inhibition of translation is rapid and coincides with the onset of processing. Hence, functional inactivation of an mRNA via an endonucleolytic attack within the ribosome-binding site is an efficient way to turn off gene expression irreversibly. Although a fraction of the intergenic GGAG motifs escapes RegB processing (see above), it is likely that many early transcripts are inactivated as a consequence of RegB cleavage within their Shine–Dalgarno sequences. During phage T4 development, the RegB-mediated translational repression mechanism is superimposed on a strong transcriptional inhibitory mechanism that shuts off early promoters.[5,34–36] The combination of the two regulatory circuits leads to a limitation of the synthesis of some early proteins strictly to the first minute of phage development.

In contrast, a few other proteins, including T4 DNA polymerase, are made in lesser amounts in the absence of RegB, suggesting that RegB processing upstream of the translation initiation region can improve mRNA translation, presumably as a consequence of a change in the RNA conformation.[10] Therefore, although the RegB nuclease downregulates the synthesis of many prereplicative proteins, it also activates the synthesis of a few others.

Facilitation of Degradation of Most Early mRNAs, But Not Middle or Late mRNAs, by RegB

RegB nuclease also participates in phage mRNA turnover. Sanson and Uzan[5] found that it destabilizes transcripts from the early *comCα* region 2- to 3-fold. In fact, the turnover of most early transcripts must be affected by RegB because bulk T4 early mRNA is 3- to 4-fold destabilized after wild-type infection as compared with *regB* mutant infection. In contrast, the average lifetime of middle and late mRNAs is unaffected by RegB, in agreement with the fact that most middle transcripts and all late mRNAs carrying the GGAG motif escape RegB processing[10] (see Specificity of RegB Endoribonuclease, above).

The mechanism by which RegB facilitates mRNA breakdown is still unclear. The fact that, in the absence of RegB, early transcripts are 4-fold stabler than

[33] K. E. Rudd and T. D. Schneider, *In* "A Short Course in Bacterial Genetics: A Laboratory Manual and Handbook for *Escherichia coli* and Related Bacteria" (J. H. Miller, ed.), pp. 17.19–17.45. Cold Spring Harbor Laboratory Press, Cold Spring Harbor, New York, 1992.

[34] C. Pène and M. Uzan, *Mol. Microbiol.* **35,** 1180 (2000).

[35] M. Uzan, E. Brody, and R. Favre, *Mol. Microbiol.* **4,** 1487 (1990).

[36] M. Uzan, J. Leautey, Y. d'Aubenton-Carafa, and E. Brody, *EMBO J.* **2,** 1207 (1983).

middle and late mRNAs is intriguing. A simple explanation may be that the sequences transcribed during the first 3 min after infection contain fewer accessible cleavage or entry sites for host degradation enzymes than those transcribed subsequently. The large size of polycistronic early mRNAs in the absence of RegB[6,37] likely provide some protection. This suggests that RegB accelerates mRNA decay by creating, within long polycistronic mRNAs, new 5′ ends that an RNase can utilize to initiate a cascade of cuts with 5′- to -3′ polarity. In support of this view, cleavages have been detected in (A+U)-rich regions six to eight nucleotides downstream of three efficiently processed GGAG motifs. These cuts do not occur in a *regB* mutant infection.[5] Despite the fact that the cleaved sequences are reminiscent of the sequences recognized by RNase E,[38] these secondary cleavages still occur after infection of various RNase E-defective host mutants.[8] Alternatively, the 3′ ends created by RegB processing might be targeted for polyadenylation, thus facilitating polynucleotide phosphorylase-mediated 3′-exonucleolytic degradation (see Coburn and Mackie[39] for a review on prokaryote mRNA degradation mechanisms). It is not yet known whether T4 mRNAs can be polyadenylated but, some time ago, Hurwitz *et al.*[40] found that after infection with the closely related bacteriophage T2, host poly(A) polymerase activity is inhibited.

RegB Action on Host RNAs

Escherichia coli translation is dramatically inhibited after T4 infection by a mechanism that still remains unclear. The discovery of a T4-induced ribonuclease activity immediately suggested that it might take part in this mechanism. Any specific effect of RegB on bacterial metabolism must be studied in the absence of T4 infection, which causes various severe perturbations to the cellular metabolism of the host. To avoid this the *regB* gene was placed under the control of the inducible *lac* promoter, in the low copy number plasmid pCL1920. On RegB induction from the resulting recombinant plasmid, pREGB (described in Sanson *et al.*[10]), the colony-forming capacity drops 10^5-fold.[6] This strong poisoning effect on *E. coli* cell growth suggested that at least some bacterial RNAs might be RegB targets. On induction of the *regB* gene from plasmid pREGB, *E. coli* K12 cells filament extensively, suggesting that at least one protein involved in septation is underproduced. Indeed, we found that the transcript of *ftsZ* (a gene involved in the early steps of septation) is efficiently cleaved by RegB within its Shine–Dalgarno sequence (GGAG).[15] *minE*-deficient mutants also filament and the Shine–Dalgarno

[37] T. Hsu and J. D. Karam, *J. Biol. Chem.* **265,** 5303 (1990).

[38] S. N. Cohen and K. J. McDowall, *Mol. Microbiol.* **23,** 1099 (1997).

[39] G. A. Coburn and G. A. Mackie, *Prog. Nucleic Acid Res. Mol. Biol.* **62,** 55 (1999).

[40] J. Hurwitz, J. J. Furth, M. Anders, P. J. Ortiz, and J. T. August, *Cold Spring Harb. Symp. Quant. Biol.* **26,** 91 (1961).

sequence of this gene is GGAG as well. However, the susceptibility of the *minE* transcript to RegB was not analyzed. To determine whether most bacterial mRNAs are affected by RegB, we measured the half-life of bulk host mRNA before and after RegB induction from the same plasmid and found that it is not changed.[15] This may mean that RegB targets are rare in *E. coli* cells, in agreement with the narrow specificity of the enzyme (see above).

RegB Purification and *in Vitro Assay*

RegB Purification

The first purification procedure described by Ruckman *et al.*[9] involves the use of a T4 phage specially engineered to overproduce the protein as a source. Briefly, after lysis of the cells by passages through a French press and clarification of the lysates by centrifugation, the proteins precipitated with 30 to 80% ammonium sulfate saturation are dissolved and dialyzed before fractionation through an anion-exchange chromatography (Q-Sepharose). The pool of fractions containing the RNase activity are then further fractionated through a hydrophobic column (phenyl-Sepharose) and subsequently through a cation-exchange chromatography (S-Sepharose). Ruckman *et al.*[9] noticed that the purified RegB protein is soluble in the presence of at least 0.1 *M* NaCl.

To easily purify enough protein for enzymological studies, the *regB* gene is cloned in plasmid pET-7, a pBR322 derivative.[41] To be able to purify the nuclease in one step, a sequence is inserted at the 5′ end of the gene, resulting in the addition of a tail of 12 amino acids containing 6 consecutive histidines at the N terminus of the protein.[10] In the resulting recombinant plasmid, pARNU2, the gene is transcribed from a phage T7 promoter that can be activated on infection of the cells with λCE6, a λ phage carrying the T7 RNA polymerase gene.[42] Five hours after infection of the strain JM101(pARNU2) with λCE6, cells are harvested and resuspended in a sodium phosphate buffer containing 0.1 m*M* antiprotease 4-(2-aminoethyl)benzenesulfonyl fluoride hydrochloride (AEBSF). The cells are broken with a French press and the lysates are clarified by centrifugation at 20,000*g* for 30 min at 4°. The supernatant is incubated with Ni-NTA resin (Qiagen, Valencia, CA) for 30 min (at 4°) with continuous gentle stirring. The resin is packed in a column and washed with a buffer containing 40 m*M* imidazole. RegB is eluted with 0.5 *M* imidazole. Purified RegB protein is stored at −20° in 20 m*M* Tris-HCl (pH 8.0), 200 m*M* NaCl, 0.2 m*M* dithiothreitol (DTT), and 50% (v/v) glycerol. Under these conditions, RegB remains active for months.[10]

[41] A. H. Rosenberg, B. N. Lade, S.-W. Chui, J. J. Dunn, and F. W. Studier, *Gene* **56,** 125 (1987).
[42] F. W. Studier and B. A. Moffatt, *J. Mol. Biol.* **189,** 113 (1986).

RegB in Vitro Assay

The first *in vitro* studies of RegB activity used the native enzyme with either a short, 10-mer artificial RNA or a 112-nucleotide T7 transcript, both containing a GGAG motif.[9] 5′-^{32}P-labeled substrates are incubated for various times with the enzyme, in the presence or the absence of purified S1 ribosomal protein, and the products are visualized by autoradiography after electrophoresis on thin polyacrylamide–urea gels. The reactions are carried out in a buffer containing 50 m*M* Tris-HCl (pH 8.4), 1 m*M* DTT, and 0.1 m*M* EDTA. An S1- to -RegB molar ratio of 2 was generally used throughout J. Ruckman's studies,[9] while Sanson *et al.*[10] set this value to 1. These values have been determined experimentally to produce the best specific cleavage efficiencies. In fact, if it is assumed that S1 stimulates the RegB activity via its binding to a specific site on the RNA (see Specificity of RegB Endoribonuclease, above), the S1 concentration should be set rather as a function of the RNA concentration and target site density. More information about the nature (and the size) of the S1 target sites must be obtained to be able to interpret correctly the effect of S1 on the different RNA substrates.

In addition to the two main results mentioned above (see RegB Endoribonuclease), namely, the RegB carries the catalytic site and it is stimulated by the S1 ribosomal protein, the *in vitro* assay has shown that RegB works without metal ions and is inhibited by Mg^{2+}. Mg^{2+} (1 m*M*) is already slightly inhibitory, and at 5 m*M* the enzyme is 55% inhibited.[15] It is not yet known whether Mg^{2+} acts directly on the enzyme or through its capacity to stabilize RNA structures.

Conclusions and Perspectives

In general, bacteriophages utilize host ribonucleases to process and degrade their mRNAs. Krisch and colleagues indeed showed that the host endoribonuclease, RNase E, plays a major role in bacteriophage T4 mRNA degradation throughout phage development.[43,44] So why was a phage-encoded endoribonuclease added to this concert of host RNases?

PCR analysis of the genome of several T4-related phages showed that the *regB* gene is probably maintained in some of these phages, although the sequences need to be determined to assess the degree of conservation among the different genes.[7] Hence, it is reasonable to assume that, in spite of the fact that it is not an essential gene under laboratory conditions, the *regB* gene confers a selective advantage on phage growth in nature. The available data support a role for RegB as a general posttranscriptional regulator of the prereplicative period. RegB modulates the translatability of many prereplicative genes. By destroying ribosome-binding sites, this nuclease inactivates mRNAs functionally, providing a novel mechanism

[43] E. A. Mudd, P. Prentki, D. Belin, and H. M. Krisch, *EMBO J.* **7,** 3601 (1988).

[44] E. A. Mudd, A. J. Carpoussis, and H. M. Krisch, *Genes Dev.* **4,** 873 (1990).

for turning off gene expression. By facilitating early mRNA breakdown, it also contributes to decrease the translation capacity of this pool of mRNAs. A plausible hypothesis is that RegB facilitates the transition between early and subsequent phases of T4 gene expression. When phage middle transcripts start to be synthesized, at 3 min postinfection (at 30°), they compete for the ribosomes with the abundant early messages, and posttranscriptional mechanisms that free the translation apparatus from the early mRNAs might well offer an advantage for phage growth. The observed positive effect of RegB on the synthesis of some proteins including the T4 DNA polymerase, a typical middle protein, suggests that RegB could also affect other middle transcripts positively, thus further facilitating the early/middle phase transition.

How RegB discriminates among the diverse GGAG-carrying RNAs is not yet understood. The analysis of the system is complicated, but it is also made particularly interesting by the fact that efficient processing of natural and some artificial RNAs *in vitro* requires a second protein, the S1 ribosomal protein. However, the stimulation factor by S1 can vary greatly according to the substrate used. In Specificity of RegB Endonuclease (above), we presented experiments showing that a high efficiency of cleavage can be obtained in the total absence of S1 protein. Our data are compatible with a model according to which S1, by interacting with a specific sequence not present in all GGAG-carrying RNAs, would commit these RNA molecules to RegB processing by changing their conformation. The S1 ribosomal protein would play the role of an RNA chaperone but for only a fraction of the RNA molecules.

To test this model, many questions will have to be answered. Does S1 interact only with the RNA, and with which sequence or structure? Which kinetic parameter of the RegB reaction does S1 influence? Which part of the S1 protein is involved in RNA interaction? How many S1 domains are implicated? Is it possible to detect conformational change in the RNA after binding of S1 to the RNA molecule? Given the widespread distribution of the S1 domain in evolution, we believe that understanding in detail how the S1 protein contributes to help RegB will shed light on the role that the S1 domain plays in other processes.

We do not know whether the ribosomal protein S1 is involved *in vivo* in RegB processing, and if it is, if it participates in the reaction as a free protein or in association with the ribosome. On the basis of the strong similarity between RegB target sites and translation initiation regions, Sanson and Uzan[5] speculated that the RegB nuclease binds to the 30S ribosomal subunit and utilizes the ability of this subunit to recognize Shine–Dalgarno-containing sequences (within true and false translation initiation regions), for increasing the concentration of tetranucleotide GGAG in the vicinity of RegB. This model implies that the nuclease should cut more efficiently GGAG-containing sequences to which the 30S ribosomal subunits bind preferentially; conversely, it should be less efficient (or not active) on sequences to which the 30S subunits bind poorly (or not at all). The finding, by Ruckman and colleagues that S1 significantly enhances RegB activity *in vitro* seemed to be in favor of this model. Moreover, J. Ruckman observed a strong positive correlation between the relative levels of RegB processing within a series of ribosome-binding sites containing the GGAG motif as Shine–Dalgarno sequence, and the efficiency of translation initiation (quoted in Ruckman *et al.*[9]). The isolation of *E. coli* mutants resistant to RegB (e.g., induced from a plasmid) might help to identify which host factor is involved in the RegB cleavage reaction and, hence, should help to answer the question of whether RegB processing is coupled to translation *in vivo*.

Acknowledgments

I thank Drs. J. Plumbridge, F. Bontems, S. Laalami, and R. d'Ari for discussions and help during the preparation of the manuscript.

Author Index

C

D

E

H

I

J

K

L

M

N

O

P

Q

R

S

T

U

V

W

X

Y

Z

Subject Index

A

B

C

L

M

N

R

S

T

V

X